Advances in Intelligent Systems and Computing

Volume 371

About this Series

The series “Advances in Intelligent Systems and Computing” contains publications on theory, applications, and design methods of Intelligent Systems and Intelligent Computing. Virtually all disciplines such as engineering, natural sciences, computer and information science, ICT, economics, business, e-commerce, environment, healthcare, life science are covered. The list of topics spans all the areas of modern intelligent systems and computing.

The publications within “Advances in Intelligent Systems and Computing” are primarily textbooks and proceedings of important conferences, symposia and congresses. They cover significant recent developments in the field, both of a foundational and applicable character. An important characteristic feature of the series is the short publication time and world-wide distribution. This permits a rapid and broad dissemination of research results.

More information about this series at http://www.springer.com/series/11156

Theodor Borangiu
Editor

Advances in Robot Design and Intelligent Control

Proceedings of the 24th International Conference on Robotics in Alpe-Adria-Danube Region (RAAD)

Springer

Editor
Theodor Borangiu
Faculty of Automatic Control and Computer
Science (Automatica)
University Politehnica of Bucharest
Bucharest
Romania

ISSN 2194-5357 ISSN 2194-5365 (electronic)
Advances in Intelligent Systems and Computing
ISBN 978-3-319-21289-0 ISBN 978-3-319-21290-6 (eBook)
DOI 10.1007/978-3-319-21290-6

Library of Congress Control Number: 2015943804

Springer Cham Heidelberg New York Dordrecht London

Printed on acid-free paper

Springer International Publishing AG Switzerland is part of Springer Science+Business Media (www.springer.com)

Preface

The 24th International Conference on Robotics in Alpe-Adria-Danube Region, RAAD 2015, was held in the conference centre of Grand Hotel Continental, Bucharest, Romania, on 27–29 May 2015. The conference brought together academic and industry researchers in robotics from the 11 countries affiliated to the Alpe-Adria-Danube space, Austria, Croatia, Czech Republic, Germany, Greece, Hungary, Italy, Romania, Serbia, Slovakia and Slovenia, and their worldwide partners in a collegial and stimulating environment.

According to its tradition, RAAD 2015 covered all important areas of research, development and innovation in robotics, including new trends such as bio-inspired and cognitive robots, visual servoing of robot motion, human–robot interaction and personal robots for ambient assisted living.

Papers were solicited in topics related to new theories, advanced design of robot mechanics and intelligent control architectures and development of robot applications, including but not limited to:

- Novel design and applications of robotic systems; micro- and nano-scale robots
- Dexterous grasping, handling and intelligent multi-finger hand control
- Intelligent cooperating and service robots; bio-inspired and swarm robotic systems
- Human–robot interfaces; natural-like interaction of humans with robots
- Advanced robot control; robot vision systems and visual servoing techniques; intelligent information technologies for cognitive robots
- Mobile, humanoid and walking robots
- Robot integration in holonic manufacturing; underwater and spatial robots
- Medical robots and bionic prostheses; robots for ambient assisted living
- Education in robotics; history of automation and robotics

Human activities in many sectors are nowadays supported or replaced by robots, which range from standard robots for industrial or service applications to autonomous robots for complex activities, such as underwater and space exploration. The great versatility and flexibility of nowadays robots allow them to be employed in different sectors, to perform a diversity of tasks.

A number of papers included in this volume report advances in robot control and integration in production and services. Solving constraint satisfaction problems allows for robust, safe planning of multiple robots in manufacturing. The sustainability of robotized processes is analysed by monitoring energy consumption at operation level and consequently reconfiguring robot speed and acceleration or conducting robot allocation scenarios in an efficient way. In order to adapt themselves to the environment and characteristics of material flows, robot systems are often equipped with vision systems.

Vision-guided robot motion using look-and-move and visual servoing methods provide best performances in the generation of accurate, task-oriented motion patterns. Integrating visual quality control services in manufacturing environments allows product traceability. In the context of agent-based manufacturing, some papers approach the problem of planning cooperative activities in robot teams.

The problem of dual- or multiple-arm robot cooperation for handling objects in service tasks is approached from the point of view of path planning, kinematics and movement synchronization. Two solutions can be recognized: hybrid position–force control in dual-arm tasks and I/O communication with master–slave synchronization.

There are also analysed new methods of using robots in interaction with humans (natural interaction) to provide assistance services. Using depth sensors, the robots are able to detect the human operator and to avoid collisions. Collision avoidance uses depth sensors which monitor the activity outside and inside the multi-robot system workspace using skeleton tracking (e.g. with the Kinect sensor), allowing the robot to detect collisions and stop the motion at the right time.

Papers in the conference address the development of software interfaces for natural-like interaction of humans with personal robots. This type of interaction is considered for communication (models of hand gestures are established that allow many natural gestures to be interpreted by the personal robot), emulation of human skills and routine tasks (extracting reusable task knowledge from visual observation of human gestures, learning dexterous operations from human demonstration). These papers describe spatial and temporal modelling of communicative and manipulative gestures; hand gestures analysis and recognition based on multiple-image processing; reusable task knowledge extraction from visual observation of human performance and action reproduction (human tasks emulator); and visual servoing for motion tuning.

The conference topics address theoretical principles and methods, implementing solutions and tools for visual servo control of robot manipulators in grasping tasks. Guidance vision is presented as an advanced motion control method, which provides flexibility to robots integrated in manufacturing cells with unstructured environment and in line quality inspection, e.g. in the glass industry.

The received papers have been grouped into nine parts:

Part I—Robot Integration in Industrial Applications
Part II—Grasping Analysis, Dexterous Grippers and Component Design
Part III—Advanced Robot Motion Control

Part IV—Robot Vision and Sensory Control
Part V—Human-Robot Interaction and Collaboration
Part VI—Modelling and Design of Novel Mechanisms and Robotic Structures
Part VII—Robots in Medicine and Rehabilitation
Part VIII—Tracking Systems and Unmanned Aerial Vehicles
Part IX—Autonomous Task Learning, Motion Planning and Scheduling

All these aspects are treated in the special issue "Advances in Robot Design and Intelligent Control—Proceedings of the 24th International Conference on Robotics in Alpe-Adria-Danube Region" of the Springer series *Advances in Intelligent Systems and Computing*, which we hope you will find useful reading.

Bucharest
May 2015

Theodor Borangiu

Contents

Part I Robot Integration in Industrial Applications

On Applying CSP for Coordination of a Multi-robot Holonic Manufacturing Execution System 3
Doru Panescu and Gabriela Varvara

Establishing Optimal Energy Working Parameters for a Robotized Manufacturing Cell 13
Silviu Raileanu, Theodor Borangiu and Florin Anton

ROS Based Safety Concept for Collaborative Robots in Industrial Applications 27
Stephan Kallweit, Robert Walenta and Michael Gottschalk

Experimental Platform for Performance Tests of Compliant Robotic Devices 37
Jaroslav Hricko

Cooperative Assembly Using Two Industrial Robots 47
Mădălin-Petru Sbanca and Gheorghe-Leonte Mogan

The Beginning of the Automation 59
Cesare Rossi

Part II Grasping Analysis, Dexterous Grippers and Component Design

Gripping Analysis of an Underactuated Finger 71
Francesco Penta, Cesare Rossi and Sergio Savino

Dynamic Behaviour of an Underactuated Finger 79
Vincenzo Niola, Cesare Rossi, Sergio Savino and Pavel Potapov

Swivel Walker with Electromotor Module and Designation of Stability 89
Mikuláš Hajduk and Jozef Varga

Resistance Feedback of a Shape Memory Alloy Wire 97
Daniela Maffiodo and Terenziano Raparelli

Study of 3-Jaw Gripper Architectures 105
Giuseppe Quaglia and Luca Girolamo Butera

Flexible Actuator for Biomorphic Applications: Performances and Energy Consumption Evaluation 115
Andrea Manuello Bertetto, Carlo Ferraresi, Luigi Antonio Besalduch, Roberto Ricciu and Andrea Cadeddu

Part III Advanced Robot Motion Control

Optimizing Trajectory Points for High Speed Robot Assembly Operations 127
Florin Anton, Silvia Anton, Silviu Raileanu and Theodor Borangiu

Cost Function-Free Optimization in Inverse Kinematics of Open Kinematic Chains 137
József K. Tar, László Nádai, Imre Felde and Imre J. Rudas

Control and Coordination System Supported by Biologically Inspired Method for 3D Space "Proof of Concept" 147
Ján Zelenka and Tomáš Kasanický

Preliminary Ideas on the Odometry of an Omni-directional Mobile Robot 157
Ioan Doroftei, Ionel Conduraru and Vasile Horga

Study of Controlled Motion of Exoskeleton Moving from Sitting to Standing Position 165
Sergey Jatsun, Sergei Savin, Andrey Yatsun and Andrei Malchikov

Robotic System Equipped with Catapult 173
Sergey Jatsun, Oksana Loktionova, Lyudmila Vorochaeva
and Alexander Vorochaev

Part IV Robot Vision and Sensory Control

**A Control Predictive Framework for Image-Based Visual
Servoing Applications** 185
Corneliu Lazar and Adrian Burlacu

**Motion Leap Compared to Data Gloves in Human
Hand Tracking** 195
Constantin Cătălin Moldovan and Ionel Staretu

Considerations for Robot Vision in Glass Production 203
Anton Ružič

**Rotation Angle Determination of a Rectangular Object
Using an Infrared Sensorial System** 213
Tony Stănescu, Diana Savu and Valer Dolga

Part V Human-Robot Interaction and Collaboration

**Individuals with Autism: Analysis of the First Interaction
with Nao Robot Based on Their Proprioceptive
and Kinematic Profiles** 225
Pauline Chevalier, Brice Isableu, Jean-Claude Martin
and Adriana Tapus

**Human Robot Collaboration for Folding Fabrics Based
on Force/RGB-D Feedback** 235
Panagiotis N. Koustoumpardis, Konstantinos I. Chatzilygeroudis,
Aris I. Synodinos and Nikos A. Aspragathos

**Case Studies for Education in Robotics: From Serious
Games to "Technology to Teach Technology" Platforms** 245
Monica Drăgoicea and Theodor Borangiu

**Designing a Multimodal Human-Robot Interaction Interface
for an Industrial Robot** 255
Bogdan Mocan, Mircea Fulea and Stelian Brad

Part VI Modelling and Design of Novel Mechanisms and Robotic Structures

Modelling of the Hexapod Mobile Robot Leg Using Matlab SimMechanics 267
Sorin Mănoiu Olaru and Mircea Niţulescu

Static Analysis of Rotary Positioning Modules for Technological Head of the Robot 277
Jan Semjon, Marek Vagas and Vladimir Balaz

A New Hyper-Redundant Arm and Control System 287
Viorel Stoian, Ionel Cristian Vladu and Ileana Vladu

Dynamic Model of a Discrete Planar Tentacle Robot 297
Mihaela Florescu and Mircea Ivanescu

How to Use 3D Printing for Feasibility Check of Mechanism Design 307
Marco Ceccarelli, Giuseppe Carbone, Daniele Cafolla and Mingfeng Wang

Gibbs-Appell Equations of Motion for a Three Link Robot with MATLAB 317
Dan B. Marghitu and Dorian Cojocaru

Part VII Robots in Medicine and Rehabilitation

Kinematic Behaviour of a Novel Medical Parallel Robot for Needle Placement 329
Bogdan Gherman, Doina Pîslă, Gabriel Kacso and Nicolae Plitea

Optimal Planning of Needle Insertion for Robotic-Assisted Prostate Biopsy 339
Doina Pîslă, Bogdan Gherman, Florin Gîrbacia, Călin Vaida, Silviu Butnariu, Teodora Gîrbacia and Nicolae Plitea

Towards Robot-Assisted Rehabilitation of Upper Limb Dysfunction 347
Irina Voiculescu, Stephen Cameron, Manfred Zabarauskas and Piotr Kozlowski

Simulation and Control of a Robotic Device for Cardio-Circulatory Rehabilitation 357
Carlo Ferraresi, Daniela Maffiodo and Hamidreza Hajimirzaalian

Part VIII Tracking Systems and Unmanned Aerial Vehicles

Aiming Procedure for the Tracking System 369
Karol Dobrovodský and Pavel Andris

Trajectory Generation with Way-Point Constraints for UAV Systems 379
Florin Stoican and Dan Popescu

Kinematics-Based Localization of a Skid-Steer Vehicle 387
Rocco Galati, Ivan Giannoccaro, Arcangelo Messina and Giulio Reina

Dynamic Task Planning of Aerial Robotic Platforms for Ground Sensor Data Collection and Processing 397
Grigore Stamatescu, Dan Popescu and Cristian Mateescu

Improving Communication Efficiency of Hybrid UAV-WSN Systems 407
Maximilian Nicolae, Dan Popescu, Radu Dobrescu and Cristian Mateescu

Top Viewing Human Tracking in Shopping Centres 417
Petros G. Vasileiou, Nefeli Lamprinou, Emmanouil Z. Psarakis, Giannis Tzimas and Nikos Achilleopoulos

Part IX Autonomous Task Learning, Motion Planning and Scheduling

Motion Planning and Scheduling with Stochastic Demands 429
Elias K. Xidias and Philip N. Azariadis

Autonomous Learning of Internal Dynamic Models for Reaching Tasks 439
Tadej Petrič, Aleš Ude and Auke J. Ijspeert

Adaptation of Motor Primitives to the Environment Through Learning and Statistical Generalization 449
Miha Deniša, Aleš Ude and Andrej Gams

Extended Kalman Filter (EKF)-Based Local SLAM in Dynamic Environments: A Framework 459
Horaţiu George Todoran and Markus Bader

Motion Analysis of a Robotic Wheelchair 471
Ionuţ Geonea and Nicolae Dumitru

On Building Remotely Operated Underwater Robot-Explorer with Bi-manual Poly-articular System 481
Aleksandar Rodić, Ilija Stevanović, Miloš D. Jovanović and Đorđe Urukalo

Author Index 491

Part I
Robot Integration in Industrial Applications

On Applying CSP for Coordination of a Multi-robot Holonic Manufacturing Execution System

Doru Panescu and Gabriela Varvara

Abstract The use of distributed constraint satisfaction problem for the coordination of holonic schemes that include multi-robot systems is investigated here. The formalism needed to apply the constraint satisfaction approach for manufacturing problems is discussed and exemplified in a case study. An implementing solution is also introduced. The proposed method determines safe operation even in cases with conflicting goals, without using any centralized component.

Keywords Holonic manufacturing execution system · Distributed constraint satisfaction problem · Coordination · Planning · Multi-robot systems

1 Introduction

Holonic manufacturing systems (HMSs) are among the promising approaches for solving the requirements on adaptability, agility, scalability and optimality imposed to companies at present [1, 2]. HMSs can conduct to a suitable trade-off between centralized and distributed control. The holonic method is in a great deal based on multiagent systems (MASs). Even so, the mechanisms of MASs must be adapted in order to be adequate for holonic schemes [3] and such an adaptation is considered here. We refer to the shop-floor level of a company, meaning the focus is on holonic manufacturing execution systems (HMESs) [2, 4]. In such a case, holons' coordination must be related with planning [1, 2], which becomes a distributed one. This aspect is even more important when an HMES contains a multi-robot system, due to certain additional issues, such as robots' collision avoidance and synchronization.

D. Panescu (✉) · G. Varvara
Department of Automatic Control and Applied Informatics,
"Gheorghe Asachi" Technical University of Iasi, Iasi, Romania
e-mail: dorup@ac.tuiasi.ro

G. Varvara
e-mail: gvarvara@ac.tuiasi.ro

T. Borangiu (ed.), *Advances in Robot Design and Intelligent Control*,
Advances in Intelligent Systems and Computing 371,
DOI 10.1007/978-3-319-21290-6_1

The new investigated possibility regards the use of constraint satisfaction problem (CSP), specially its distributed version (DisCSP) [5]. The established research goals are presented in the next section. Then, the proposed method is described using a representative example. The paper ends with some implementation details, a few conclusions and ideas for future work.

2 Related Work and Research Goals

There are different treatments for associating planning/coordination and DisCSP [6–8]. They concern coupling DisCSP with a classical planning technique to get individual plans of agents, or a method that is based on adapting the well-known A* algorithm (this would be difficult to apply in HMESs due to the problem of finding proper cost functions), or a mechanism that applies distributed planning and CSP through graph merging. Between the raised issues, one has to consider: finding a suitable formalization both for the planning part and the CSP one, studying the degree of coupling between agents and obtaining appropriate implementation schemes. Another important point is to observe the influence of how information is gathered and shared; this specifically applies for HMESs, as holons are specialized entities and possess certain types of knowledge [9]. Taking into account all these concerns we established the following research aspects, which will be tackled in the rest of this paper:

- To study the possibility of coupling a well established multiagent planning approach—the one based on the *Belief Desire Intention* (BDI) agent's architecture [10]—with DisCSP, into the framework of a holonic scheme;
- To find a solution suitable for HMESs with respect to the way information is shared/communicated between different types of holons;
- To compare a previously devised holonic solution that was based on a centralized component with the new approach that eliminates this additional entity.

3 A Difficult Case for HMES: A Scenario with Two Related Goals

Starting from our previous research [2, 11], we consider a holonic scheme where holarchies must be decided by product/order holons according to the manufacturing goals they have to solve. Our holonic scheme, named HAPBA, besides other things differs from PROSA regarding the product holons, as they can be involved in establishing goals for resource holons [2]. Each holonic agent is endowed with a library of plans, as being of a BDI type [10]. Thus, a received goal is mapped to a corresponding plan or plan pattern, after the BDI reasoning process. A goal means a certain product to be obtained, and to find the related plan signifies to settle the required sequence of actions, together with the needed actors. An increased flexibility is obtained when partially ordered plans are used [12]. It means the sequence of actions composing a plan is not entirely settled, but the order between certain actions is not established. This happens when the order of actions does not matter, meaning they can be carried out in any order or even in parallel. Such a form of plans is beneficial for a distributed treatment, too. As an example, two such plans are presented in Fig. 1a; it is supposed that they are selected by two product holons, according to the goals they have to fulfil. The first plan (for product holon PH_1) is composed of four actions. Action a_{11} must be executed first; when it is finished actions a_{21} and a_{31} can be started and carried out in any order; after their end, action a_{41} must be run and thus the plan is fulfilled. The plan for the second product holon (PH_2) contains two actions—a_{12} and a_{22}—which can be executed in any order. As already told, the two holons establish these plans by combining planning with the BDI mechanism, as explained in [2, 10, 13]. In fact, planning is not ended until appropriate actors/resources are found for all actions included in plans.

These plans regard some experiments performed with a manufacturing system containing two industrial robots (see Fig. 1b) [11]. In the tested scenarios, the resource holons are the two robots composing a multi-robot system, storage devices, a conveyor, a computer vision system and a machine tool (these last two devices do not appear in Fig. 1b).

Let us consider the difficult case, namely when holons PH_1 and PH_2 are part of the same manufacturing environment and they try to materialize their plans in the same time by involving the same resource holons. This refers to the case when the two robots in Fig. 1b have to fill in two pallets, working together in a common area. To be able to make a comparison, we consider the same scenario as in [2], namely the existence of four resource holons ($RH_1 \div RH_4$). The capability of each resource holon concerns the actions it can achieve, the correspondence being as follows: $RH_1 \rightarrow \{a_{11}, a_{41}, a_{12}\}$ (it means holon RH_1—a robot resource holon can perform the actions a_{11}, a_{41} and a_{12}, which in our case refer to part placement on pallet, see

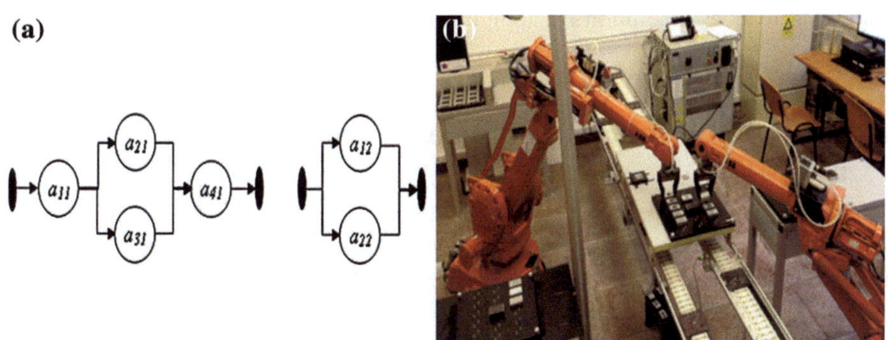

Fig. 1 **a** Plans of two product holons. **b** The experimental manufacturing environment

Fig. 1b); $RH_2 \rightarrow \{a_{11}\}$ (the second robot holon); $RH_3 \rightarrow \{a_{21}, a_{22}\}$ (the computer vision resource holon, achieving part recognition); $RH_4 \rightarrow \{a_{31}\}$ (the machine tool resource holon, performing part processing operations). All resources have a unitary capacity, i.e. they are not able to simultaneously carry out two actions (these can be done only in sequence). As presented in [2], when holons PH_1 and PH_2 concurrently try to establish holarchies for solving their goals according to the above presented plans and by applying the Contract Net Protocol (CNP) as negotiation scheme [2, 13], the HMES can fail. Namely, though the existing resources are enough to fulfil the goal for one of the two product holons, it can happen that both product holons end their negotiation with the resource holons with a negative result. For example, this may occur when RH_1 makes a bid for action a_{11} as needed by PH_1 and RH_3 makes a bid for action a_{22} as requested by PH_2.

Thus, PH_1 fails in assigning resources for its plan, because there is no resource holon to carry out its action a_{21} (RH_3 is committed to PH_2), and PH_2 fails because it receives no bid for action a_{12} (RH_1 is committed to PH_1). As stated in [2], one possibility to avoid such a wrong operation of an HMES is to add a centralized component. This can be a staff holon [2, 9, 13], which is supposed to provide the needed synchronization so that the two product holons should not block one another.

The use of a staff holon can be criticized for more reasons. It supposes the introduction of a centralized part, which must possess knowledge on the plans of holons that are working for establishing their holarchies. Then, the staff holon must know the capabilities of resource holons so that it should detect potential conflicts. This can happen when the plans of two/more managers within the CNP framework address the same resources; in such a case, the staff holon decides to block the access of one or more managers to contractors until the first manager settles its holarchy. It is clear that this mechanism is safe, but it can cause a bottleneck regarding the information to be acquired and handled by the staff holon. Then, it must be also taken into account that the privacy of holons is somehow violated: product holons must transmit information on their plans towards the staff holon, while resource holons must communicate their capabilities to the same staff holon. So, we propose an alternative for obtaining a safe operation that excludes the staff holon and reaches a solution by applying DisCSP.

4 A DisCSP Solution for Obtaining Coordination in HMES

CSP is a method for solving problems when these are expressed as finding the appropriate assignment for a set of variables [5]. The domain for each variable is a priori known and represented by a finite set. A problem is specified by a set of constraints. The problem is solved when an assignment of all variables is found so that it complies with all constraints. In DisCSP, a distributed approach is taken with more agents that handle distinct variables. Restrictions can be classified in different categories: implicit constraints, those being given when the problem is described, and implied constraints; these last ones regard prohibited combinations of variable values which are discovered during the search process. Such a new revealed constraint is called *nogood* [5, 7]. Moreover, there are intra-agent constraints that concern the variable or variables of a single agent and inter-agent constraints that imply variables from more agents.

A significant point for applying CSP is to transpose the problem according to the specific, imposed formalism [7]. In our case, one must express the HMES planning/coordination problem according to the CSP pattern. Because holonic approach is a distributed one, the DisCSP should be naturally used. For the considered holonic system, as stated above, the problem is to obtain coordination for the two product holons; so, the holonic agents for PH_1 and PH_2 will handle distinct variables. Our solution is that the value for such a variable should be the instantiated plan of the corresponding holon, namely a set of pairs: action—actor, according to the relation:

$$P_i = ((a_1, act_1), \ldots (a_n, act_n)) \tag{1}$$

In (1) n represents the number of actions of plan P_i. By applying (1) for the plans of holons PH_1 and PH_2, one obtains (see Fig. 1):

$$P_{PH_1} = ((a_{11}, act_{11})), (a_{21}, act_{21}), (a_{31}, act_{31}), (a_{41}, act_{41})) \tag{2}$$

$$P_{PH_2} = ((a_{12}, act_{12})), (a_{22}, act_{22})) \tag{3}$$

If holons PH_1 and PH_2 have received the bids from the resource holons, then they can establish the domains for their variables. For the above described scenario these are:

$$D_{PH_1} = \{((a_{11}, RH_1)(a_{21}, RH_3), (a_{31}, RH_4), (a_{41}, RH_1)); ((a_{11}, RH_2), (a_{21}, RH_3), (a_{31}, RH_4), (a_{41}, RH_1))\} \tag{4}$$

$$D_{PH_2} = \{((a_{12}, RH_1), (a_{22}, RH_3))\} \tag{5}$$

The constraints to be considered are as follows. As presented in Fig. 1a, an optimal plan for PH_1 would be the one when actions a_{21} and a_{31} are carried out in the same time (they are overlapping). Thus an implicit constraint for the variable which is P_{PH_1} should be (see 2):

$$act_{21} \neq act_{31} \tag{6}$$

One can observe that the above constraint is an intra-agent one, as it regards only the holon PH_1; namely, according to the way a variable is defined (see 1), it has as value a set of pairs, and between the elements of these pairs certain restrictions can appear. The constraint imposes for the two actors to be different, so that their activity can overlap. The similar corresponding constraint for PH_2 is:

$$act_{12} \neq act_{22} \tag{7}$$

The need of coordination between the product holons can be expressed through an inter-agent constraint. Namely, the same actor is not allowed to be involved in plans of both product holons; in this way any possible blockage is avoided. Thus, the following constrained is got:

$$act_{i1} \neq act_{j2}; \quad i \in \{1, 2, 3, 4\}; \quad j \in \{1, 2\} \tag{8}$$

Based on this set up, the two product holons can apply the DisCSP algorithm to find a solution for their coordination. Taking into account that in the manufacturing environment distinct goals may have different priorities a priori known, the *asynchronous backtracking algorithm* (ABT) can be used [5]. This means the priorities of holonic agents are fixed, as imposed by the importance of the goals they manage.

Nevertheless, ABT must be slightly modified when applied in an HMES; this can be explained using the case study that was previously introduced. Figure 2 presents the interaction diagram for holons PH_1 and PH_2 when they apply ABT (the problem is the one expressed by relations 2 ÷ 8). The common notation for CSP is used, and the *Ok* and *Nogood* messages exchanged between agents are presented. PH_1 is considered as the higher priority holonic agent. Thus, it chooses one of the two possibilities for its variable assignment; let it be the first value from D_{PH_1} (see 4), which was labelled p_1 in Fig. 2. PH_1 sends a first *Ok* message according to the value decided for its variable. Due to the constraint (8), PH_2 cannot find an assignment for its variable and so it replies with a *Nogood* message. Consequently, PH_1 changes the assignment of its variable and sends a new *Ok* message (p_2 represents the second element of D_{PH_1}). Again, PH_2 is not able to find a value satisfying constraint (8), and so a new *Nogood* message is issued towards PH_1. After that, if ABT is strictly applied, then the two holonic agents will report that the problem has no solution. Instead of this, we propose to modify the ending condition for ABT. Namely, when one or more agents with higher priority can find

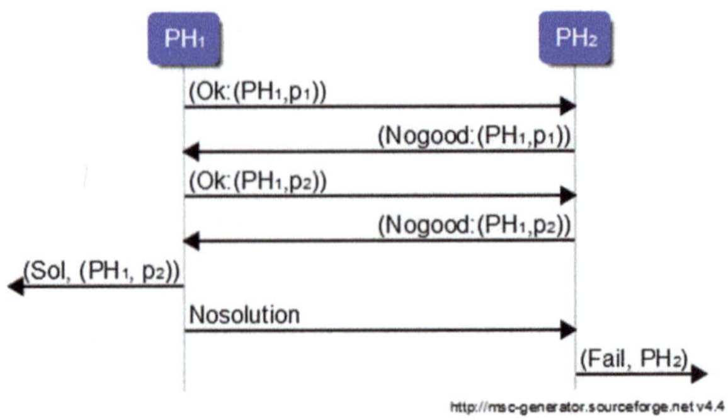

Fig. 2 The interaction diagram for product holons

assignments that do not violate any intra-agent constraint and any inter-agent constraint between them, then they report a partial solution. Only the lower priority agents that are not able to find assignments for their variables consistent with the constraints determined by them and higher priority agents must report a fail. This condition expresses a case more general than the one in Fig. 2, where there are only two holonic agents; i.e., from a group of holonic agents, it can happen that a subgroup (those having the higher priorities) succeeds to accomplish their goals, while the others fail. This generalization is obvious; the only complication consists in correctly expressing constraints on the number of available resources.

The final part of the interaction diagram must be understood as PH_1 informs the entity from which it received the goal that it could establish a holarchy to fulfil it (this corresponds to using RH_1, RH_2, RH_3 and RH_4), while PH_2 issues a message about its failure in solving the goal. After the goal of PH_1 is achieved and resource holons are released, if PH_2 further keeps its goal, it will be able to fulfil it. An important point is that the proposed DisCSP based mechanism rends impossible the case when both product holons fail, and this without using a staff holon.

5 An Implementing Possibility

In order to verify the above presented method, a distributed constraint solver was used, namely DisChoco 2 [14]. It consists of an open source Java library designed to offer DisCSP solutions based on several algorithms, including ABT and its variants. The main advantages of using this distributed constraint reasoning software platform, as mentioned in [15], refer to an easy access to DisCSP solutions and benchmarks with default parameters ready to use by industry practitioners on one hand, and the ability to work with an open source framework in order to develop new constraint programming techniques, benefiting from the associated testing facilities on the other hand. DisChoco 2, as MAS platform, can be used for simulation purposes on a single machine or for implementing real distributed frameworks, when its independence from the communication system is the main gain. Its library of generators for satisfaction/optimization problems provides a benchmark for testing new developed algorithms.

For the considered case study, a simulation experiment based on a network with six ABT agents was performed. The agent's solution is obtained through a centralized Choco solver. After that, an ABT Master object is needed; this will control the global search in order to obtain a solution based on the message exchange between agents. In this implementation, each agent refers to a single action and manipulates (as variable) its associated resource index or indexes, as illustrated by the following code lines:

```
agents[0]=(ABTAgent)disCSP.makeAgent("A1","");// a11
variables[0]=agents[0].makeInternalVar(new Integer[]{1,2}
);//RH1, RH2
```

In order to impose the restrictions corresponding to relations (6) and (7), external constraints involving several agents are defined, as illustrated by the next source lines:

```
agents[1].neqY(agents[1].makeExternalVar(variables[2]));
//act21!=act31
agents[4].neqY(agents[4].makeExternalVar(variables[5]));
//act12!=act22
```

Associated to the agents' network definition, an ABT master object is instantiated and transmitted to a DisCSP simulation solver:

```
ABTMaster master=(ABTMaster)Protocols.getMaster(Protocols. ABT);
DisProblem disCSP=new DisProblem("Planificator", master);
DisCSPSolver solver = new DisSolverSimulator(disCSP);
solver.solve();
```

It will monitor the global execution, sending messages to launch/stop the search and collaborating for distributed algorithm progress. Execution could be associated with recording several metrics, useful to assess the search complexity, through the maximum number of non-concurrent constraint checks, cumulated CPU time, total number of exchanged messages and their maximum/total size, and each agent's individual effort measured by CPU time and specific costs, as depicted below:

```
Solution with class simple.dcsp.abtfamily.abt.ABTMaster
  a11       a21       a31       a41       a12       a22
(1.1=1/1)(2.1=3/1)(3.1=4/1)(4.1=1/1)(5.1=1/1)(6.1=3/1)
    =act11    =act21    =act31    =act41    =act12    =act22
Statistcs{msg=28.0, termination=12.0, MaxMsgsSize=9.0, cumulat-
edCpuTime=0.936006, ncccs=0.0, nbInferedNogood =0.0, nbAssign-
ments=6.0, AgentOrdering=12.0, CCs=0.0, MsgsSize=54.0,
CPU(s)=0.156001,OK=6.0, NCCPUTime(ms)=0.0, time(s)=128.991041}
```

6 Conclusion and Future Work

According to the proposed research goals, we can conclude that it is possible to obtain a planning and coordination scheme for HMESs including multi-robot systems, which must handle goals creating conflicts by involving an adapted form of DisCSP. This approach brings benefits in what concerns the way different holons use their own knowledge without the need to centralize the information towards a staff holon.

The briefly sketched implementation scheme shows the viability of the proposed solution. Nevertheless, it is to further analyse which is the most efficient communication mechanism, knowing that DisCSP can conduct to a great number of messages between holons for a long search. As future research, we intend to use the possibilities offered by the considered open-source software platform to compare different DisCSP algorithms (there is an entire class of ABT versions) and thus to find the one being the most suitable for HMESs. Furthermore, simulations are to be transposed in our experimental manufacturing environment, where agents will be deployed to the physical distributed network.

References

1. Borangiu, Th., Răileanu, S., Trentesaux, D., Berger, T.: Semi-heterarchical agile control architecture with intelligent product-driven scheduling. IMS'10 **10**, 108–113 (2010)
2. Panescu, D., Pascal, C.: On a holonic adaptive plan-based architecture: planning scheme and holons' life periods. Int. J. Adv. Manuf. Tech. **63**(5–8), 753–769 (2012)
3. Panescu, D., Varvara, G., Pascal, C., Sutu, M.: On the design and implementation of the resource holons in a PROSA based architecture. In: Proceedings of the IEEE 13th International Conference on Intelligent Engineering System, pp. 101–106. Barbados (2009)
4. Valckenaers, P., Van Brussel, H.: Holonic manufacturing execution systems. CIRP Ann. Manuf. Technol. **54**, 427–432 (2005)
5. Yeoh, W., Yokoo, M.: Distributed problem solving. AI Mag **33**(3), 53–65 (2012)
6. Brafman, R., Domshlak, C.: From one to many: planning for loosely coupled multi-agent systems. In: Proceedings of ICAPS, pp. 28–35 (2008)
7. Panescu, D., Pascal, C.: A constraint satisfaction approach for planning of multi-robot systems. In: Proceedings of ICSTCC, October 17–19, pp. 157–162. Sinaia, Romania (2014)
8. Durkota, K., Komenda, A.: Deterministic multiagent planning techniques: experimental comparison. In: Proceedings of DMAP-ICAPS'13 (2013)
9. Van Brussel, H., Wyns, J., Valckenaers, P., Bongaerts, L., Peeters, P.: Reference architecture for holonic manufacturing systems: PROSA. Comput. Ind. **37**, 255–274 (1998)
10. Pascal, C., Panescu, D.: HAPBA—A BDI agent based solution for holonic manufacturing execution systems. In: Borangiu, T., Thomas, A., Trentesaux, D. (eds.) Service Orientation in Holonic and Multi Agent Manufacturing and Robotics, pp. 57–70. Springer, Berlin (2013)
11. Panescu, D., Pascal, C.: HAPBA—a Holonic Adaptive Plan-Based Architecture. In: Borangiu, Th., Thomas, A., Trentesaux, D. (eds.): Service Orientation in Holonic and Multi-Agent Manufacturing Control, pp. 61–74. Springer, Berlin (2012)
12. Ghallab, M., Nau, D., Traverso, P.: Automated Planning: Theory and Practice. Morgan Kaufmann, Los Altos (2004)

13. Panescu, D., Pascal, C.: On the staff holon operation in a holonic manufacturing system architecture. In: Proceedings of ICSTCC, October 12–14, pp. 427–432. Sinaia, Romania (2012)
14. DisChoco 2.0 platform homepage. http://dischoco.sourceforge.net/
15. Wahbi, M.: Algorithms and ordering heuristics for distributed constraint satisfaction problems. Ph.D. Thesis, Universite Mohammed V-Agdal, Rabat (2012)

Establishing Optimal Energy Working Parameters for a Robotized Manufacturing Cell

Silviu Raileanu, Theodor Borangiu and Florin Anton

Abstract The paper proposes a method for measuring energy consumption of single phase equipment. Although it can be used on different domains the solution targets the shop floor level of robotized manufacturing systems. The paper focuses thus on the energy consumption of industrial robots for different operation types. This solution is extended in the second part of the paper for a set of workstations cooperating in order to minimize energy consumption at global level for a batch production.

Keywords Industrial robot · Energy consumption · Sustainable manufacturing · HMES

1 Introduction

The increased usage of automated manufacturing equipment in industry led to an equivalent increase in energy consumption in this sector, especially due to the complexity of the material transforming processes [1]. Thus, the need to conserve, monitor [2, 3] and optimize energy [4] rises as a direct request for coping with ascending energy costs but also because using and developing green solutions is encouraged by governments, European Union [5] and other bodies around the globe.

S. Raileanu (✉) · T. Borangiu · F. Anton
Department of Automation and Applied Informatics, University Politehnica of Bucharest, Bucharest, Romania
e-mail: silviu.raileanu@cimr.pub.ro

T. Borangiu
e-mail: theodor.borangiu@cimr.pub.ro

F. Anton
e-mail: florin.anton@cimr.pub.ro

T. Borangiu (ed.), *Advances in Robot Design and Intelligent Control*,
Advances in Intelligent Systems and Computing 371,
DOI 10.1007/978-3-319-21290-6_2

In literature there are several solutions reported for the optimization of energy consumption: *analytical solutions* [6, 7] which, based on the physical characteristics of processing resources, offer a mathematical expression which links the input parameters to the energy consumption, and *empirical solutions* [8] which make use of measuring equipment to determine the energy consumption. The disadvantage of the first solution is that it relies on a model which in the majority of cases is an approximation of the real system being usually developed for long time operations. In our present research we are interested in counting energy consumption for compact periods of time for which the model does not apply well, and in realizing a knowledge base which will be used to take resource allocation decisions. Thus, we have chosen to implement an energy optimization solution based on smart meters [9].

There are equipment for continuous measurement of energy consumption (smart meters, watt meters) but since manufacturing is operation-based these equipment have to be integrated in a smart way into the shop floor in order to compute the energy consumed for each operation; this requires to integrate smart meters and physical processes with control applications that govern process execution at shop floor level. The aim of this paper is to analyse in real time the energy consumption for equipment which is frequently used in automated manufacturing cells—industrial robots. The analysis of energy consumption has been studied in literature for both household [10] (characterized by connectivity) and industrial applications [11] (characterized by precision) but the novelty of our approach is that it integrates the two modes into a single, easy to procure/configure system which is used to analyse standard robot operations both in assembly and continuous path following applications. Thus we analyse the movement of single joints, the movement of the controlled point (end tip point) and two manufacturing specific operations: **pick and place** and **continuous path movement**.

The remainder of the manuscript is organized as follows: section two presents the structure of the energy measurement and operation execution system; section three presents the scenarios used to analyse energy consumption for a single industrial robot, the collected data and selection of optimal operating parameters based on this data; section four extends the energy optimization solution to a set of industrial robots working together in a job-shop manufacturing layout, and the last section presents the conclusions and perspectives of future research.

2 Energy Measurement and Operation Execution Monitoring

The framework used to analyse and optimize energy consumption contains three elements as described in Fig. 1: the *monitored resource*, the *smart energy meter* with network connection and the *application* used to synchronize the data received from the previous two parts.

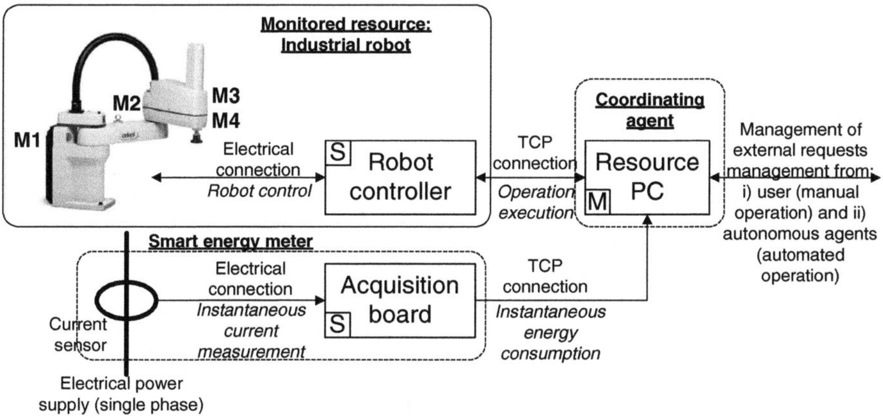

Fig. 1 The energy measurement system

2.1 Monitored Resource

The **monitored resource** consists of an Adept Cobra S600 [12] industrial robot (ROB) executing handling operations. From an energy consumption point of view its characteristics are: 4 degrees of freedom realized through 4 alternative current motors (M1–M4 in Fig. 1, J1–J4 in Fig. 2 right) supplied through a common power amplifier. It runs a TCP server consisting of a single loop (LOOP_ROB) able to receive handling commands identifying specific operations: *DRIVE*—actuating a single robot joint; *MOVE*—joint interpolated motion; *MOVES*—straight line motion; *PP* (Fig. 2, left)—pick and place: a combination of joint interpolated and straight line motions which moves an object between two points (`pick`, `place`); *CPF*—continuous path following (Fig. 2 right), and after motion execution it responds with a confirmation.

LOOP_ROB:

1. Wait for a TCP connection.
2. Extract operation and parameters (speed and acceleration/deceleration).
3. Start timer for measurement during operation execution.
4. Execute operation based on the code and parameters extracted at 2.
5. Confirm execution by sending an acknowledgement containing the execution time, the distance covered by the tool and by each joint.

2.2 Smart Energy Meter

The energy consumed by a manufacturing resource (a robot) is computed by integrating the instantaneous power (current x voltage) over time. In this case we will compute the apparent power as the root mean square of the current I_{RMS} (Eqs. 1, 2)

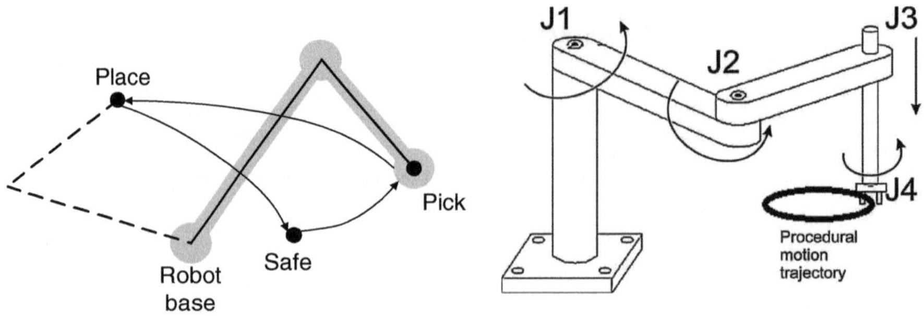

Fig. 2 PP (*left*) and procedural motion (*right*) configurations

multiplied by the root mean square of voltage U_{RMS} (which has a fixed value of 230 V) [13]. Thus the equipment used to measure energy is composed of a non-invasive current transformer connected to the analogue input of an acquisition board which does all the energy calculations and offers this information as a service over Ethernet.

$$I_{RMS} = \sqrt{\frac{\sum_{all\ samples} I^2}{number\ of\ samples}} \quad (1)$$

The implementation of the **smart energy meter** (SEM), see Fig. 1, consists of a non-invasive current sensor (accuracy 1 %) [14] connected to an Arduino Ethernet board [15] running the LOOP_SEM service which reads the instantaneous current value I, then computes the I_{RMS} using Eq. 1 while storing the time at which the value was calculated. With these values the energy consumed is computed as follows:

$$W = I_{RMS} * U_{RMS} * dt = I_{RMS} * 230 * (t - t') \quad (2)$$

where t is the current time and t' is the time when the previous current measurement was done.

LOOP_SEM:

1. Compute I_{RMS} from a continuous set of samples (Eq. 1)
2. Compute the consumed energy (Eq. 2)
3. Wait for a TCP connection and based on the received command provide the requested information (instantaneous current, energy, time)

2.3 Coordinating Agent

A coordinating agent is used to synchronize the continuous energy consumption with operation execution on products in order to compute the energy consumed by a robot during an operation. It acts as a master over the SEM and ROB executing in parallel the *energy monitoring function* (LOOP1) and the *robot operation execution* (LOOP2). Thus, LOOP2 uses information (energy consumption data) computed by LOOP1.

LOOP1:

1. Continuously poll energy consumption data

LOOP2:

1. Manually or automatically (depending on the operating mode) set the operating parameters and the operation type that has to be executed by the robot.
2. Request operation start and start execution log.
3. Wait acknowledgement of operation execution.
4. Stop log and store data: operation type, speed, accelerations, samples (used to analyse energy patterns: maximum intensity, profile, a.o.).
5. Update operation execution model.

3 Energy Consumption Analysis for Different Types of Robot Motions

3.1 Description of the Testing Scenarios

Industrial robots are integrated in manufacturing cells to automate two types of operations: *material handling* (through pick_and_place (PP) operations [16]) and *material processing* (through path following/procedural motions). The target manufacturing cell consists of resources executing both assembly operations realized through PP operations and material processing operations realized through continuous path following (CPF) operations.

The parameters for energy measurement in the case of a PP operation and of a procedural motion (CPF) are as follows:

(i) Total distance covered between the three points (starting point—`safe`, picking point—pick, destination—`place` and `back`) is 1820 mm and was realized using different speed limits and accelerations while the total distance in the continuous path following is ~500 mm,
(ii) All experiments have been done using the same handled objects characterized by a small weight (<100 g) which is negligible compared to the maximum payload of the robot (5 kg).

3.2 Measurements and Analysis for a Single Joint

For the first scenario the energies consumed by each joint were measured as described in Table 1.

As expected, the main contribution in energy consumption is brought by J1 (see Table 1).

3.3 Measurements and Analysis for Simple Tool Movements

The second scenario consists of a comparison between a joint interpolated motion (MOVE) and a straight line motion in Cartesian space (MOVES), both executed between the pick and place points (Fig. 2, left).

As can be seen from Fig. 3, a straight line movement consumes more energy and in our particular case (the PP sequence in Fig. 2) this energy increase is influenced by the location of the point of interest, the robot being forced to heavily accelerate and decelerate at the middle of the movement to maintain a straight line. Also, this is the reason why the MOVES command with 90 and 100 % is not possible due to high inertia.

3.4 Measurements and Analysis for the Pick and Place Procedure

The third scenario consists of a comparison for the PP operation (the basic operation used in assembly) executed using different speed and acceleration limitations.

Table 1 Energy consumed by each robot joint

Joint	Energy consumed (Ws)	Distance covered
J1	278	80°
J2	36	80°
J3	30	80 mm
J4	0 (negligible)	180°

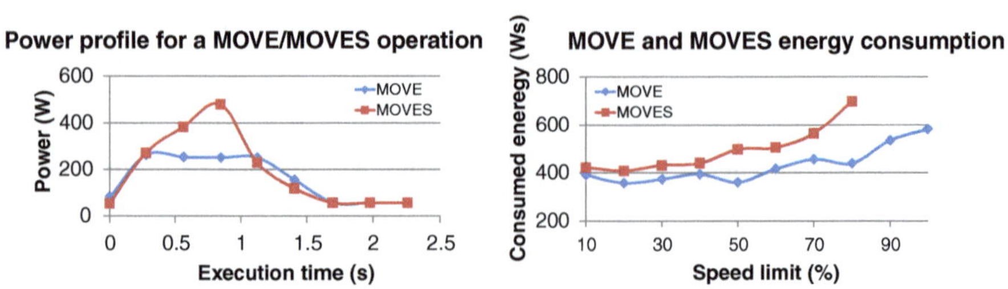

Fig. 3 Power difference between a MOVE and a MOVES robot motion instruction

Its basic power profile is depicted in Fig. 4 from where it can be observed that the energy consumption rises three times according to the *approach pick*, *approach place*, *move safe* operations.

Using the same interest points (`pick`, `place`, `safe`) the PP operation was executed considering different speed (10–100 %) and acceleration (20, 50 and 100 %) limitations. The energy consumption (Fig. 5) and the operating times (Fig. 6) are described below.

The conclusions which can be drawn from this scenario are: (1) increasing acceleration increases consumption; (2) increasing the operation speed limit has two effects: (i) an increase of the acceleration duration which causes an increase of energy and (ii) a decrease of operation time which decreases energy (P*dt).

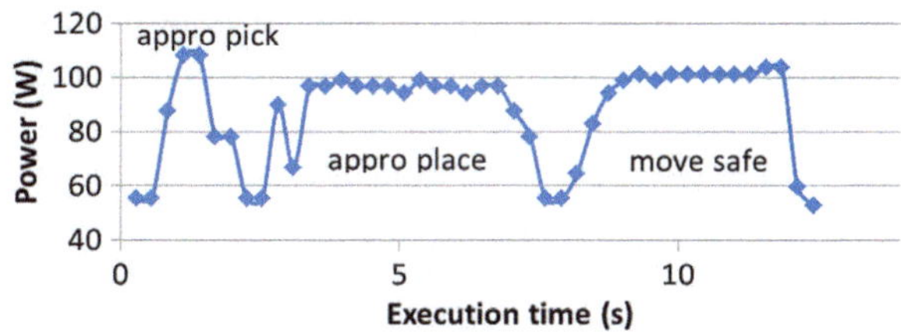

Fig. 4 PP energy consumption profile

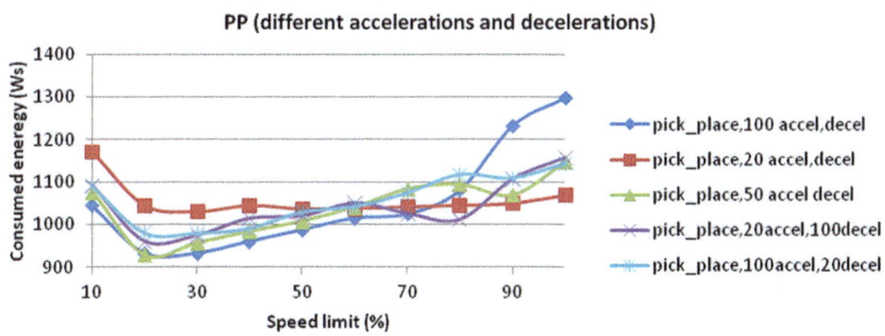

Fig. 5 PP energy consumption for different accelerations

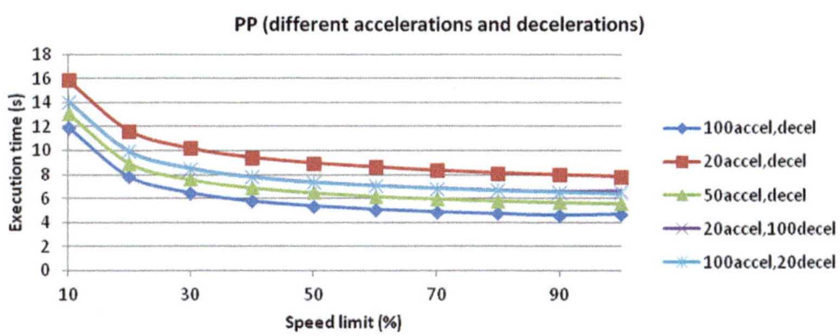

Fig. 6 Execution time for a PP operation for different speeds and accelerations

By combining these two effects, and due to the fact that acceleration determines a non-linear increase of energy it results that less energy is consumed for an intermediary speed, which is experimentally confirmed; (3) energy consumption varies if the same experiment is executed several times. However, the standard deviation for all experiments does not exceed 5 %.

3.5 Measurements and Analysis for Continuous Path Movement

The fourth scenario consists in monitoring the process of continuous path generation and following (Fig. 2 right). The data obtained from this experiment allows drawing the power profile (Fig. 7), the energy consumption and execution time diagrams for a variable speed limit (10–100 %), see Fig. 8.

The conclusion which can be drawn from this scenario is that the additional energy consumed for continuous path following has small variations (Fig. 7) especially due to the fact that the accelerations are eliminated once the cruise speed is attained. The power profile is influenced by the path's shape and by the robot structure. In our particular case (Fig. 2 right) the robot is following a circular path starting from near the base and extending towards the exterior. This continuous path energy consumption (Fig. 7) generates two maximum power consumption points

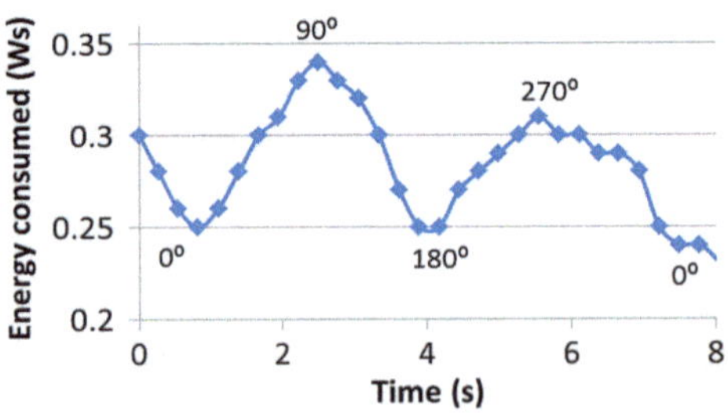

Fig. 7 Power profile for a procedural motion operation consisting of a circular path

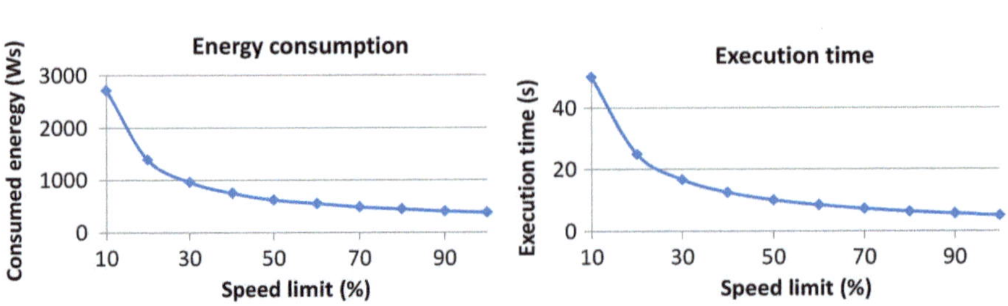

Fig. 8 Procedural motion energy consumption and execution time

corresponding to 90° and 270° of the circle and one minimum located at 180°, where the robot is extended. The energy is the product between the time and the average power. The average power does not differ significantly for different speed limitations during a continuous path movement but the time decreases, which generates an energy reduction.

4 Evaluating the Energy Consumption for a Multi-robot Manufacturing System

The energy measurement principle and operation execution system proposed in Sect. 2 were developed in order to be replicated and integrated over a set of robotic resources working together in a flexible manufacturing cell, where they are disposed as described in the shop floor layout depicted in Fig. 9 to allow cost effective job-shop production sequences. Energy measurement and operation execution monitoring are done at workstation level (a SEM_*i* device is used to monitor energy consumption at workstation P*i*, $1 \le i \le 4$). Each workstation P*i* contains one robot (P1, P2) or two robots and a machine tool (P3, P4). These resources can execute a set of operations characterized by different parameters (speed, acceleration, execution time, timeliness and energy consumption).

During initial, centralized production planning, job scheduling and resource allocation, the operation type (defined by the *product agent*) is mapped to a PP or

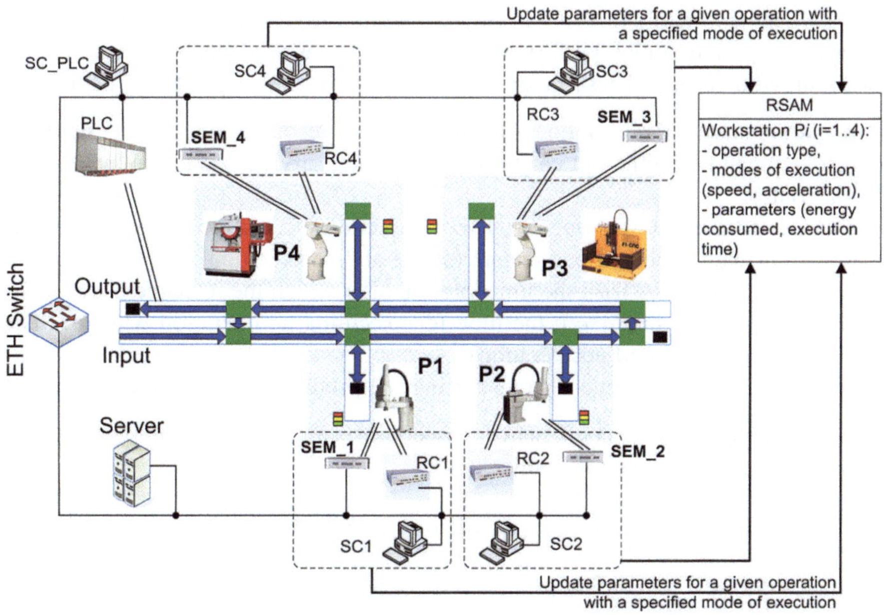

Fig. 9 Robotized manufacturing system layout

CPF robot motion task along with its speed and acceleration constraints which are set in a range of values leading to reasonable execution time and consumed energy. Thus, the energy measurement and operation execution system are used to provide continuously updated data for a Resource Service Access Model (RSAM) which describes the particularities of all operations that can be executed in the manufacturing cell and their execution mode [host robot, operation type, modes of execution (speed, acceleration, execution time, timeliness and energy consumption)] (Fig. 9). Based on this model the optimization of robot scheduling can be done with different objective functions such as: consumed energy, makespan, total production cost, a.o. [17, 18].

The manufacturing cell features a high degree of flexibility meaning that some operations can be executed by different robot workstations, but with different energy footprints due to the fact that resources are different (SCARA and vertical articulated robots, 3- and 4-axis CNC milling machines). This flexibility justifies operation scheduling and resource allocation processes, which in this case are launched at the beginning of production or whenever a resource changes its state or when the conditions met initially at the moment of scheduling differ from the current conditions at execution time. This difference is detected using a final interaction protocol between the product to be executed and the resource executing the operation [19].

The parameters collected from the previous operations (resource, operation, execution time, acceleration limit, speed limit, execution time stamp) are stored and used as input data for the next resource scheduling. Thus, resource scheduling is done by computing a preference coefficient that describes the degree of preference for executing a given operation on a given resource (robot). The coefficients are a sum of weighted factors (K1–K3 in Eq. 3), resulted from experiments; their choice is justified by the initial performances and their awards/penalties accumulated during previous usage. These factors, which are offered by the RSAM include:

- *resource status* ([*rs*], online or offline), and resource operations status (a resource can be online but cannot offer a certain service due to the lack of raw materials);
- *energy consumption for operation* [*ec*]: the energy consumed for executing an operation with specified parameters (speed and acceleration);
- *resource availability* [*ra*]: the time interval after which the resource is available;
- *service execution time* [*set*]: the time for executing an operation with specified parameters (speed and acceleration);
- *accessibility* [*tt*]: the time of transporting the product to the selected resource.

Hence, the preference coefficient for operation i ($\mathrm{Val\_op}_i$) is computed as follows, if the resource status is "online":

$$\mathrm{Val\_op}_i = \mathrm{K1} * ec + \mathrm{K2} * ra + \mathrm{K3} * set + \mathrm{K4} * tt \tag{3}$$

The result of the scheduling process is an ordered series of operations along with the chosen resource for each operation and optimal operating parameters consisting of robot processing speed and acceleration.

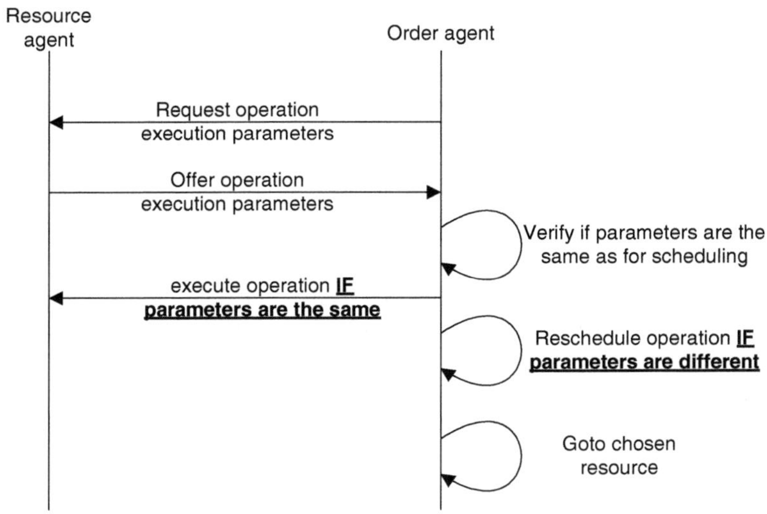

Fig. 10 Detailed description of the final interaction protocol (*right* side of Fig. 1)

The sequence diagram detailing the final interaction between an *order agent* and a *resource agent* (which will integrate the functionality of energy measurement) is presented in Fig. 10. The outcome of the final dialog can be either the execution of the operation or its rescheduling on another robot if the current conditions (resource status and operation execution time, energy consumption and quality) differ significantly from the initial ones.

A different energy consumption pattern, besides the invalidation of the last schedule, may signal a faulty motor prompting thus preventive resource maintenance [20].

5 Conclusion

This paper proposes a cost-effective solution for monitoring the energy consumption of an industrial single phase robot system. The energy is determined by measuring the instantaneous current consumed for sets of standard operations. The paper focuses on a four degree of freedom industrial robot executing four types of operations: single joint movement, movement in linear joint and Cartesian space, pick and place motion sequence and continuous path movement.

As seen from these measurement experiments, the consumed energy increases at higher operating speeds but the robot acceleration is the parameter that most significantly influences the consumed energy. This is why it is advisable even for handling operations to define and execute continuous motion paths in order to limit trajectory discontinuities which involve accelerations, avoiding thus motor overloads, overheating and consequently reducing the frequency of maintenance.

As expected, moving the robot at maximum or minimum speed does not optimize energy consumption, the solution lying somewhere in between these extreme values and depending heavily on the robot structure, the payload and the serviced locations. In the case of handling operations the optimal energy consumption is obtained in the range of speed between 20 and 30 % of the maximal robot speed value (Fig. 5).

The computation of the optimal motion operating parameters is difficult to be done analytically (due to the difficult access to motors to estimate their real characteristics); this is why an experimental solution which measures in real time energy consumption and adjusts the cruise speed and acceleration is a better one. By implementing this solution an improved monitoring of the motion process (allowing also the computation of the final energy foot print) was obtained.

Future research will address the following directions: (i) testing the accuracy of the SEM equipment and improving its performance if necessary and (ii) collecting energy data over longer periods of robot motion execution to estimate energy consumption variations in time and update consequently the resource's performances which will be used in product scheduling (resource allocation).

Acknowledgments This work was partially supported by the Sectoral Operational Programme Human Resources Development (SOP HRD), financed from the European Social Fund and the Romanian Government under the contract number POSDRU/159/1.5/S/137390/of the University Politehnica of Bucharest.

References

1. Saidur, R.: A review on electrical motors energy use and energy savings. Renew. Sustain. Energy Rev. **14**, 877–898 (2010)
2. Prabhu, V., Jeon, H.W., Taisch, M.: Simulation Modelling of Energy Dynamics in Discrete Manufacturing Systems, Service Orientation in Holonic and Multi-Agent Manufac-turing and Robotics, Studies in Computational Intelligence, vol. 472, pp. 293–311 (2013)
3. Diouri, M.M., Dolz, M.F., Gluck, O., Lefevre, L., Alonso, P., Catalan, S., Mayo, R., Quintana-Orti, E.: Assessing power monitoring approaches for energy and power analysis of computers. Sustain. Comput. Inf. Syst. **4**(2), 68–82 (2014)
4. Bi, Z.M., Wang, L.: Optimization of machining processes from the perspective of energy consumption: a case study. J. Manuf. Syst. **31**, 420–428 (2012)
5. www.ec.europa.eu/programmes/horizon2020/en/h2020-section/secure-clean-and-efficient-energy, consulted in Feb 2015
6. Pellicciari, M., Berselli, G., Leali, F., Vergnano, A.: A method for reducing the energy consumption of pick-and-place industrial robots. Mechatronics **23**, 326–334 (2013)
7. Meike, D., Ribickis, L.: Industrial robot path optimization approach with asynchronous FLY-BY in joint space. In: Proceedings of IEEE International Symposium on Industrial Electronics, pp. 911–915 (2011)
8. Chemnitz, M., Schreck, G., Kruger, J.: Analyzing energy consumption of industrial robots. In: Proceeding of Emerging Technologies and Factory Automation (ETFA), pp. 1–4 (2011). Print ISBN: 978-1-4577-0017-0
9. Depuru, S., Wang, L., Devabhaktuni, V.: Smart meters for power grid: challenges, is-sues, advantages and status. Renew. Sustain. Energy Rev. **15**(6), 2736–2742 (2011)

10. www.openenergy.com, consulted in Feb 2015
11. Paryanto, P.M.B., Kohl, J., Merhof, J., Spreng, S., Franke, J.: Energy consumption and dynamic behavior analysis of a six-axis industrial robot in an assembly system. In: 5th CATS 2014—CIRP Conference on Assembly Technologies and Systems, vol. 23, pp. 131–136 (2014)
12. http://www.adept.com/products/robots/scara/cobra-s600/general, consulted in Feb 2015
13. Akagi, H., Watanabe, E.H., Aredes, M.: Instantaneous Power Theory and Applications to Power Conditioning. Wiley, New York (2007)
14. www.echun-elc.com/En/ProductView.asp?ID=442, Split Core Current Transformer ECS1030-L72, consulted in Feb 2015
15. http://arduino.cc/en/Main/ArduinoBoardEthernet, consulted in Feb 2015
16. Borangiu, Th.: Advanced Robot Motion Control, pp. 1–450. Romanian Academy Press and AGIR Press, Bucharest (2003). ISBN 9732709681, 973-8130-99-9
17. Prabhu, V.V.: Performance of real-time distributed arrival time control in heterarchical manufacturing systems. IEEE Trans. **32**, 323–331 (2003)
18. Pipattanasomporn, M., Feroze, H., Rahman, S.: Multi-agent systems in a distributed smart grid: design and implementation. In: Proceedings of IEEE PES'09 Power Systems Conference and Exposition, pp. 1–8 (2009)
19. Truong, V.T.D., Sato, Y., Inoguchi, Y.: Performance evaluation of a green scheduling algorithm for energy savings in cloud computing. In: IEEE International Symposium on Parallel and Distributed Processing, Workshop and Ph.D. Forum (IPDPSW), pp. 1–8 (2010)
20. Prabhu, V.V.: Services for competitive and sustainable manufacturing in the smart grid. In: Borangiu, T., Trentesaux, D., Thomas, A. (eds.): Springer book series "Studies in Computational Intelligence", Service Orientation in Holonic and Multi-Agent Manufacturing Control, vol. 402, pp. 227–240 (2012)

ROS Based Safety Concept for Collaborative Robots in Industrial Applications

Stephan Kallweit, Robert Walenta and Michael Gottschalk

Abstract The production and assembly of customized products increases the demand for flexible automation systems. One approach is to remove the safety fences that separate human and industrial robot to combine their skills. This collaboration possesses a certain risk for the human co-worker, leading to numerous safety concepts to protect him. The human needs to be monitored and tracked by a safety system using different sensors. The proposed system consists of a RGBD camera for surveillance of the common working area, an array of optical distance sensors to compensate shadowing effects of the RGBD camera and a laser range finder to detect the co-worker when approaching the work cell. The software for collision detection, path planning, robot control and predicting the behaviour of the co-worker is based on the Robot Operating System (ROS). A first prototype of the work cell shows that with advanced algorithms from the field of mobile robotics a very flexible safety concept can be realized: the robot not simply stops its movement when detecting a collision, but plans and executes an alternative path around the obstacle.

Keywords Collaborative robot · Human-Robot interaction · Safety concept · Workspace monitoring · Path planning · RGBD camera · ROS

S. Kallweit (✉) · R. Walenta
Institute for Mobile Autonomous Systems and Cognitive Robotics (MASCOR), University of Applied Sciences Aachen, Aachen, Germany
e-mail: kallweit@fh-aachen.de

R. Walenta
e-mail: robert.walenta@alumni.fh-aachen.de

M. Gottschalk
ID Ingenieure & Dienstleistungen GmbH, Euskirchen, Germany
e-mail: gottschalk.m@id-engineerin.com

T. Borangiu (ed.), *Advances in Robot Design and Intelligent Control*,
Advances in Intelligent Systems and Computing 371,
DOI 10.1007/978-3-319-21290-6_3

1 Introduction

Modern production of consumer goods will develop from mass production of identical items to customized products in the near future [1], a so called “mass customization”. The flexibility of a production line can be increased by using a combination of a flexible human worker and an industrial robot [2]. The robots task is the disposing of all repeatable jobs, which are too time consuming, too heavy for the human co-worker or need a relatively high positional accuracy. Several aspects need to be considered for a collaborative workspace, as shown in Fig. 1. The human co-worker is the flexible and fault tolerant partner, guiding the robot to his next mission. In our study a typical “Pick-and-Place” task is analysed. Although this is the basic scenario for a variety of assembling tasks, the drawback is the threat generated by a fast moving robot [3].

Various safety concepts exist in the industrial environment, which are even commercially available like camera based monitoring systems [4]. The system detects an intruder in the working area of the robot and stops the running procedure. This is not beneficial for a real collaborative approach: the process would be most of the time switched off. Collaboration between the human co-worker and the industrial robot is only possible if the robot perceives the dynamically changing environment and reacts to the new situation accordingly [5]. In the field of mobile robotics, this scenario became already a reality: the Google self-driving car is licensed to autonomously drive in several states in the US [6]. Another example from the field of logistics are Kiva Systems, recently acquired from Amazon [7], which deploy autonomous mobile robots for commissioning tasks. As a consequence, advanced algorithms from mobile robotics can be applied to standard industrial tasks, like e.g. assembly or pick-and-place jobs.

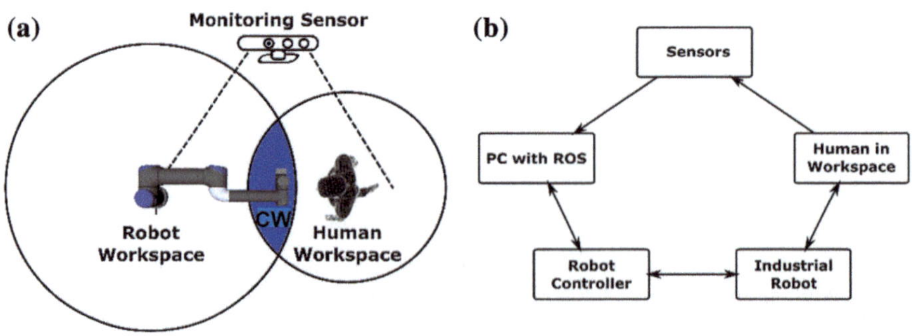

Fig. 1 **a** Collaborative workspace (CW) between human and robot monitored by multiple sensors; **b** interaction diagram showing the concept used for safe human robot collaboration

2 Safety Requirements

The operation of a collaborative robot system is strictly regulated. There are different regulations for each country: e.g. in Germany a license is required according to a combination of DIN-EN-ISO 10218 [8] and DIN-EN-12100 [9]. Here the fundamental risks are described which arise from the operation of a robot system without protection fences. Procedures are explained to discover and eliminate dangerous threats in order to protect the human co-worker. To operate a robot in a collaborative mode, the system needs to fulfil the following criteria:

- The robot needs to be realized in a light-weight design.
- An independent controller monitors the robot motion.
- Sensors are monitoring the position of the human co-worker.
- Safety classification must be done according to DIN EN ISO 13849 [10].

Figure 2 shows different "Hazard Zones" (D1–D3) which are generated considering the mentioned DIN standards, the speed of the robot, the latency time of the sensors and the distance between co-worker and robot base point. Zone D1 is the zone with highest hazard. The co-worker cannot leave quickly enough D1, if the robot makes an unpredicted movement. The only safe state for this situation is to stop completely the motion. Zone D2 is the collaborative zone. Here the interaction with the co-worker is allowed and all the sensor devices are monitoring continuously the human co-worker and the robot. If a threat is detected in D2, the system goes into a safety stop.

2.1 General Principles for the Design

An industrial robot is controlled by a user program which defines a specific task. Each part of this program, which initiates an action of the robot itself, has to be monitored by a safety function. This function needs to be classified as category 3 with "performance level d", which means:

- Failures cannot lead to a loss of safety functions.
- Failures have to be recognized at an early state.

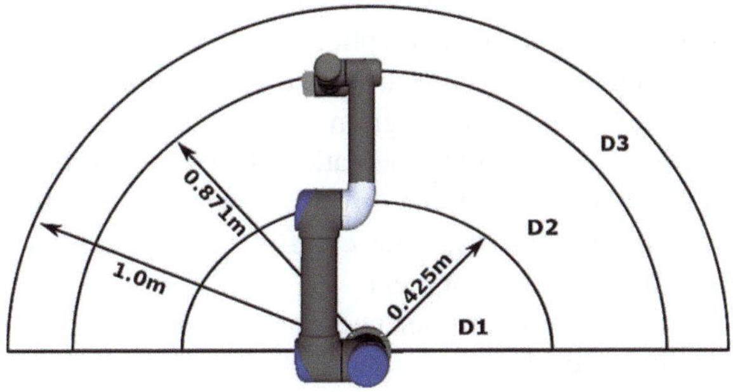

Fig. 2 Different "Hazard Zones" considering sensor latency and distance

- Every safety function leads to a safe state.
- The safe state is kept until the threat is relieved.

"Performance level d" indicates a probability of failures occurring when the system is used. This probability is below 10^{-6} for a critical failure/hour in the case of level d [10]. Safety functions are used for risk reduction. For a collaborative system the control of the robot motion is of great importance. This includes the ability to stop, to control the speed and to limit the free working area of the robot. The system needs an emergency stop and a safety stop. An emergency stop shuts the system down and cuts it from the energy source. A safety stop is only stopping the motion and waiting until a safe operation is guaranteed. The TCP velocity needs to be continuously controlled. It must be ensured that a maximum speed of 250 mm/s is not exceeded, when the co-worker is entering zone D3. The software must implement a limitation of the robot working range. When the robot reaches this limit, the robot has to stop. This simple method is an instrument for reducing the interaction area with the human co-worker and defines the area where the robot is allowed to move.

2.2 *Collaboration Scenarios*

Different collaboration tasks need different setups like described in [11]. The most interesting scenario for collaboration is a direct, joint interaction between human co-worker and robot. The other possible collaboration setups include e.g. forbidden areas, where a robot is integrated in a production line and some parts need to be loaded and unloaded to the production cell or heavy robot tools handled, where the human co-worker is just shortly exposed to the robot working area.

3 System Architecture

Several tasks can be associated to a system for robot collaboration. These tasks can be formulated quite similarly to the tasks a mobile robot has to solve; it is thus natural to use similar strategies. The main tasks for mobile robots are [12]: (a) Locomotion; (b) Perception; (c) Localization, and (d) Navigation.

Locomotion is realized by the robot standard hardware controller. When using an external control program which generates control commands for the joint positions, an external kinematic solver is needed. We use the Open Source library KDL [13] to solve the inverse kinematic iteratively for an UR5 light-weight industrial robot. The robot controller has a software interface to ROS [14] which simplifies the development.

The system controller needs to acquire enough data to generate a model of the current state. This perception of the environment is often done with 3D vision systems; in our setup we use a simple RGBD camera, several proximity sensors and a laser range finder. These sensors are supported by ROS or simple interfaces are developed with the rosserial package [14].

Another module is needed for collision detection, which is part of the localization task. Here the Open Source library FCL [15] is used, to allow a probabilistic collision check of static and dynamic objects in 3D environments. For collision avoidance a path planning algorithm is necessary for navigation. A common library for path planning within ROS is OMPL [16], which uses a local and a global planner to generate possible 3D paths of the kinematic structure and to check the closer surrounding of the robot. The path planning was simulated online in a virtual environment using MoveIt [17]. Here all the necessary geometrical constraints are provided as URDF [18] data sets. Figure 3 shows the main components of the generated system.

3.1 Robot Cell Setup

A laser range finder, mounted in front of the robot cell observes the plane orthogonal to the base coordinate plane of the robot and detects the human co-worker when entering D3. The RGBD camera is placed ~2 m above the working area for collision detection, i.e., detecting objects in all three Zones (D1–D3). Using depth images from the RGBD camera it is possible to monitor nearly the robot's complete working area.

Virtual fences limit the possible working area of the robot from three directions: in front of the robot cell an interaction with the robot is possible (D3 and D2); this area is partially monitored by the RGBD camera and some additional proximity

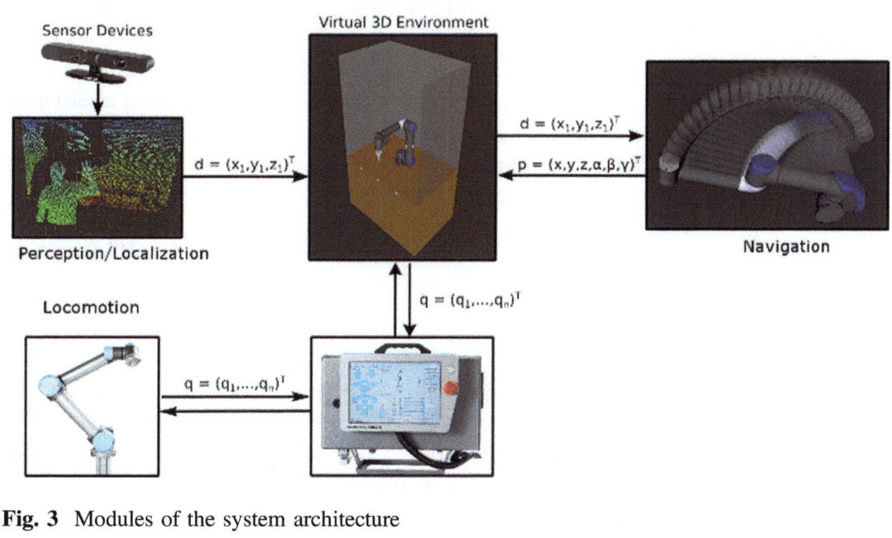

Fig. 3 Modules of the system architecture

sensors with a range of ~800 mm (Sharp GP2 Y0A21). The sensor signals S1–S3 are used for the safety state machines.

The robot itself is equipped with a sensor belt in order to detect objects near the kinematic chain. This belt consists of four of the already mentioned low cost proximity sensors (S4–S7), which have a lower latency time than the RGBD camera. Depending on the detected collision threat, the movement of the robot stops.

4 Safety Controller

The task of the safety controller is to perceive the working area and to monitor the robot and human co-worker. The implementation of object detection, collision avoidance, path planning and of the state machines was done in ROS using MoveIt. Due to the ROS communication engine, it is possible to exchange data for the UR5 robot via a simple TCP/IP connection. The robot hardware controller is expecting new joint positions, speed and acceleration values with a data rate of 125 Hz as a ROS client. The new data is FIFO transferred and directly sent to the servo motors.

As a test case, we implemented a simple Pick and Place task with several PTP and LIN movements. It is worth mentioning that MoveIt doesn't support LIN movements, so the target points for a linear movement need to be calculated in advance and sent to the robot controller. A simple state machine is controlling the Pick and Place process. The transitions can only be triggered if non-blocking calls are used. That prevents from using the `move()` and `execute()` functions of the MoveGroup. One possibility to use the state machine is using MoveGroupAction, which starts the ActionServer and publishes an event when the action is triggered.

The safety functions are implemented as state machines. As soon as the human co-worker is detected in D3, the system reduces the speed to 250 mm/s. If the co-worker enters D2 and he is closer than 200 mm to the robots upper arm, the system is stopped. It is as well stopped, when entering D1. If a colliding movement of the robot in D2 or D3 is detected, the system stops and calculates an alternative path around the colliding object. If successful, the planned path is executed. Figure 4 depicts the safety state machines where the signals S1–S7 are used for additional intruder detection not covered by the RGBD and laser range finder data.

5 Results

A successful path planning with colliding objects is generated using the mentioned setup. When the human co-worker stretches his arm into the collaboration area D3-D2, the arm is detected as an obstacle but as the bounding box of the arm is not interfering with the planned trajectory, a re-planning is not necessary. The robot moves with reduced speed of 250 mm/s to the target position. If the arm is detected

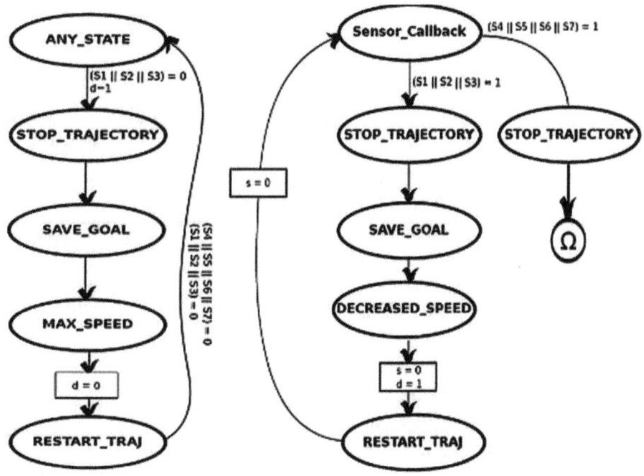

Fig. 4 Safety functions as state machines

as a colliding object, the system is stopped for a short time for re-planning the path around the obstacle and—if successful—moves the robot using the new trajectory.

Other collaboration scenarios are as well possible, e.g. a pick-and-place application, where the robot should pick the object from the co-workers hand. Here the hand of the co-worker is not classified as a collision object, because he is in a special collaboration zone. If the hand would be defined as a collision object, the robot wouldn't be able to reach the target position. This collaboration zone has only a certain limited height ~50 mm, so that it is possible to interact with the robot. If the dimension of this zone is increased too much, it will be dangerous for the co-worker.

In order to get a first indication of the latency time of only the collision detection system, the time between a trigger from the laser range finder and the complete stop of the robot was measured using internal time stamps from ROS; proximity sensors were ignored for this test. These time stamps are indicators which should be interpreted with care, due to the limited real time possibilities of ROS. This data is a first estimate of a possible threat for the co-worker. Figure 5 shows the setup of the measurement: the robot is repeatedly moving from starting position A to target position D on a circular trajectory. The co-worker enters the zone D3 via B1 or B2, disturbing the trajectory. The time for a complete stop of the robot when detecting the collision object is measured. The laser range finder doesn't stop the robot; only the safety controller based on the collision detection without the proximity sensors is used.

Three different speeds for the robot were used: 10, 50 and 100 %. The mean values $\bar{t}$ and the standard deviations s_d were calculated from 30 measurements (Table 1).

Obviously the latency time depends on the point where the co-worker enters the collaboration zone but does not depend on the robot's speed. The latency time between the two investigated points differs by ~0.2 s. During the test sequence the robot always stopped before the co-workers hand, so that the measured difference in latency time for B1 and B2 is due to different braking ramps of the robot hardware controller.

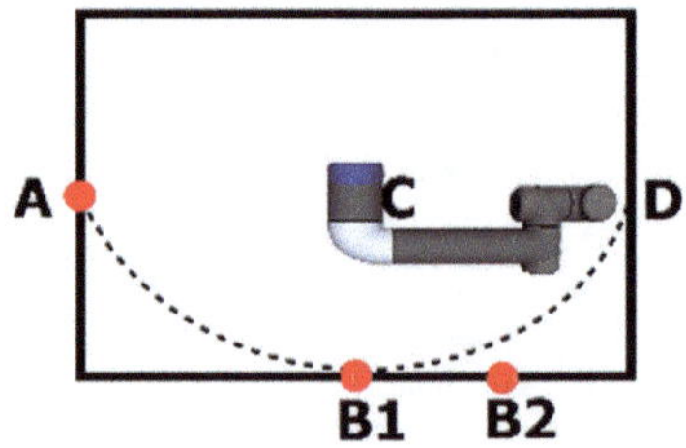

Fig. 5 Setup for measuring the latency time of the collision detection module

Table 1 Average latency times for different intrusion points and robot speeds

Speed (%)	B1: $\bar{t}$ in s	B2: $\bar{t}$ in s	B1: s_d in s	B2: s_d in s
100	0.7	0.96	0.2	0.17
50	0.63	0.9	0.19	0.1
10	0.74	0.97	0.16	0.1

6 Conclusion and Future Aspects

The successful setup of a collaborative robot system using the ROS framework shows that combining algorithms from mobile robotics with industrial tasks is suitable for future automation purposes. By using open source libraries a very flexible low cost collaborative system was developed. The used sensors are common, low cost devices which are not suitable for industrial safety applications, but the capabilities of combining 3D sensors with some additional proximity devices are considered.

The system is still not fast enough to detect quick movements of human co-workers due to the latency time of the RGBD camera and the high computational load. Here ROS is beneficial again: ROS scales quite well on multicore machines due to the fine-grained structure. Future multicore CPU power can be easily used.

Next steps in the research will be: the use of faster and more accurate RGBD cameras, e.g. some TOF cameras with higher frame rates and higher resolution. A decentralized approach with a cluster of powerful embedded systems will increase the overall performance. The use of a different kinematic solver can as well improve the global performance: KDL solves iteratively, so the algorithm spends more time in some kinematic positions compared to an analytic solver.

References

1. The Economist: The third industrial revolution, vol. 12, no. 16, http://www.economist.com/node/21553017 (2012)
2. Bascetta, L., Ferretti, G., Rocco, P., Ardo, H., Bruyninckx, H., Demeester, E., Di Lello, E.: Towards safe human-robot interaction in robotic cells: an approach based on visual tracking

and intention estimation. In: IEEE/RSJ International Conference on Intelligent Robots and Systems (IROS), pp. 2971–2978 (25–30 Sept 2011)
3. Bortot, D., Born, M., Bengler, K.: Directly or on detours? How should industrial robots approximate humans? In: Proceedings of 8th ACM/IEEE International Conference on Human-Robot Interaction (HRI), pp. 89–90 (3–7 Mar 2013)
4. Pilz, M.: SafetyEye, Internet http://www.hft-stuttgart.de/Studienbereiche/Vermessung/Bachelor-Informationslogistik/Aktuell/Veranstaltungen/inflogtag2014/SafetyEYE_HFT-Stuttgart_09-04-14_genehmigt.pdf (2014)
5. Balan, L., Bone, G.M.: Real-time 3D collision avoidance method for safe human and robot coexistence, intelligent robots and systems. In: IEEE/RSJ International Conference, pp. 276–282 (2006)
6. CBS Detroit: New Law Allows Driverless Cars on Michigan Roads. 28 Dec 2013
7. Kucera, D.: Amazon Acquires Kiva Systems in Second-Biggest Takeover. Available at: http://bloom.bg/Gzo6GU (2012)
8. DIN EN ISO 10218-1: Safety requirements for industrial robots, (Industrieroboter Sicherheitsanforderungen) (2012)
9. DIN EN ISO 12100: Safety requirements for machines—common design guiding principles (Sicherheit von Maschinen - Allgemeine Gestaltungsleitsätze) (2011)
10. DIN EN ISO 13849: Safety requirements for machines—safety related parts of control units (Sicherheit von Maschinen—Sicherheitsbezogene Teile von Steuerungen) (2010)
11. Maschinen Markt: Neue Norm soll Sicherheit zwischen Mensch und Roboter gewährleisten, http://www.maschinenmarkt.vogel.de, Internet (2014)
12. Siegwart, R., Nourbakhsh, I.: Introduction to Autonomous Mobile Robots. MIT Press, Cambridge (2004)
13. Orocos Kinematics and Dynamics, http://www.orocos.org/kdl, Internet (2014)
14. Quigley, M., Conley, K., Gerkey, B., Faust, J., Foote, T., Leibs, J., Wheeler, R., Ng, A.: ROS: an open-source robot operating system. In: ICRA Workshop on Open Source Software (2009)
15. Pan, J., Chitta, S., Manocha, D.: FCL: a general purpose library for collision and proximity queries. In: IEEE International Conference on robotics and automation (ICRA), 2012, pp. 3859–3866 (2012)
16. Sucan, I., Moll, M., Kavraki, E.: The open motion planning library. IEEE Robot. Autom. Mag. **19**(4), 72–82 (2012)
17. Chitta, S., Sucan, I., Cousins, S.: MoveIt!, IEEE Robot. Automat. Mag. **19**(1):18–19 (2012)
18. Kunze, L., Roehm, T., Beetz, M.: Towards semantic robot description languages. In: IEEE International Conference on Robotics and Automation (ICRA), 2011, pp. 5589–5595, (2011)

Experimental Platform for Performance Tests of Compliant Robotic Devices

Jaroslav Hricko

Abstract The design process of mechatronic devices is usually performed with Model-Based Design (MBD) methods. This approach enables obtaining properties and functional behaviour of designed devices at lower costs of prototyping. Building prototypes is usually the last operation before the real manufacture of products and includes several experimental tests for parameter verification. As prototype building is the most expensive part of the design process, the Hardware-in-the-loop (HIL) and Software-in-the-loop (SIL) simulation methods are strongly recommended to decrease final production cost of designed devices. This paper deals with designing a specific test bed for performance quality evaluation of compliant robotic devices. The proposed equipment enables parameter verification and evaluation of differences between mathematical models of flexible parts and their real characteristics. Thus, the structures and parameters of the compliant devices of interest can be optimally designed. The paper also discusses the utilization mode of the proposed testing and performance evaluation equipment for a designed micro-gripper.

Keywords Mechatronic device · Compliant mechanism · Mathematical model · Stiffness · Robot gripper

1 Introduction

Currently the design process of mechatronic devices is based on knowledge integration of from main technical fields: mechanics, electronics and informatics. Therefore, design methods include the mathematical modelling and simulation of the whole device and the main interactions between functional components of designed device. Well-known design methods of mechatronic devices are: V-model [1]

J. Hricko (✉)
Institute of Informatics, Slovak Academy of Sciences, Banská Bystrica, Slovakia
e-mail: hricko@savbb.sk

T. Borangiu (ed.), *Advances in Robot Design and Intelligent Control*,
Advances in Intelligent Systems and Computing 371,
DOI 10.1007/978-3-319-21290-6_4

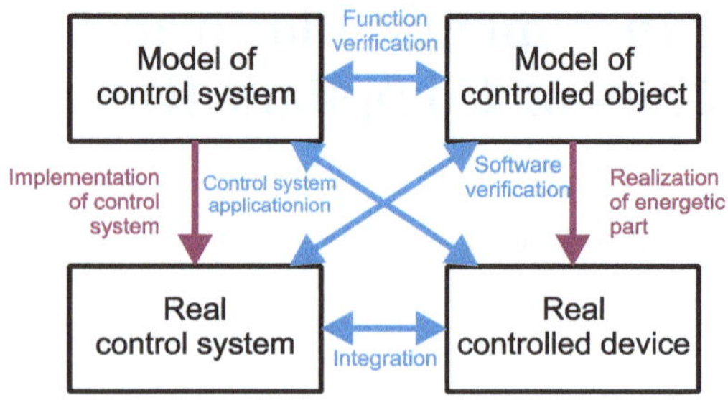

Fig. 1 Combinations of models and real objects

and Model-Based-Design [2, 3]. These methods include verification of the desired functional performances of prototypes, which is executed before physical realization of the device. Before performing this verification procedure, a careful analysis of the mathematical model is required.

Methods used for experimentation are based on combinations of physical models/prototypes and mathematical models in order to express functional characteristics in a most reliable way. Such approaches enable knowing the performance characteristics and expected parameters of a final product. There are two principal approaches [4]:

- *Software-in-the-Loop* (SIL) integrates the models of system's components in one common simulation background. Real components and simulation models could work together in real-time or off-line processes. This SIL simulation concept can be used in the design phase as a useful tool for the analysis of system dynamics including properties of drivers. It is useful for tests and simulation of control algorithms.
- *Hardware-in-the-Loop* (HIL) integrates physical components and models of other parts into one common simulation background and enables real time simulation. It is used mainly in cases when real functional performances will be verified.

Mutual interaction of both methods and models are shown in Fig. 1.

2 Experimental Platform for Prototype Testing

The most adequate performance characteristics of a device can be obtained in hybrid approach (theoretical model— real characteristics). Considering the specific group of compliant robotic devices this approach is really needed; such compliant devices include mechanically flexible parts (joints/arms) as important building elements. Flexural characteristics of these elements usually differ from theirs mathematical models and the cumulated error in performance can reach up to 5–10 % [5, 6].

The proposed experimental test bed allows carrying out functional tests and measurements on compliant devices such as grippers, force/displacement amplifiers, etc. There are some functional requirements to be met:

- Performing experimental tests with micro positioning devices (measurements of small displacements should be enabled);
- Tests of dynamic properties;
- Experiments with real size devices;
- Force measurements.

The purpose of this platform is performing preliminary tests in laboratory conditions; for this reason the following design criteria have been considered:

- Interaction with higher control systems (e.g., computer with MATLAB);
- Integration of position/force feedback (sensors);
- Using standard industrial components;
- Switching between global/modular approach;
- Keeping low cost of the platform and physical samples.

Considering the previous requirements, the platform includes drivers/actuators, sensors for displacement feedback and a control unit supporting connectivity with higher control systems, eventually in multitasking regime for multiple actuators.

2.1 Principal Components

The test platform is developed mainly for assisting the design of compliant robotic devices as for instance grippers and force/displacement amplifiers. For this reason the main active components are linear actuators. Although experiments have been made on this platform with 1-D.o.F. linear actuator, in principle the platform can be modified to accommodate up to 3-D.o.F. motion systems. According to Howell [7] and Smith [8], most of devices are currently built in planar configurations and usually work in one or two D.o.F.

The main component of the proposed platform is the linear actuator as the principal driving device of the motions of compliant—functional elements. In our case, because experiments will be made in the dimensional scale up to 10:1, the imposed characteristics of a linear actuator should assure the numerical values: min. stroke 5 mm, applied force 5 N and higher, positioning accuracy—at least micrometres, and highest possible working frequency.

Two physical principles were compared when choosing linear actuators: piezoelectric and electromagnetic. An actuator working on piezoelectric principle exhibits main parameters [9] with the values: stroke 80 mm, maximum motion speed 15 mm/s, resolution <1 nm, maximum force 7 N. An actuator working on electromagnetic principle [10] exhibits the following values for its parameters: stroke 10 mm, maximal motion speed 400 mm/s, resolution given an optical sensor

30 μm, positioning accuracy ±90 μm, maximum force 5.5 N. Considering the costs of both devices the electromagnetic drive was finally chosen.

The intended experimental tasks require measurements for position feedback of the actuator and for displacement measurements of other experimental devices as well. The optical encoder integrated in the linear actuator is used for open loop control. The output displacements of the tested devices can be calculated and verified from their mathematical-flexural models. Other possible sources of position measurements/feedback are contactless sensors. Some types of sensors have specific requirements: for example ultrasonic and capacitive sensors need relatively big surfaces for measuring, inductive sensors require metallic surfaces/parts, etc.

In order to assure maximal generality in position sensing, the platform uses a webcam device. Sensing relatively small areas (covered by a mechanism's end element) is done by processing the images of the end element's displacement. For example, when scanning a 10 mm × 10 mm area with a relatively cheap webcam of resolution 640 × 480 pixels, one pixel represents approximately 20 μm. On the other hand, it is more difficult to apply this method for fast moving objects; therefore, visual measurements are currently used only for evaluating deflection.

2.2 *Control System (HW, SW)*

The linear actuator and its controller are originally designed for industrial applications. In the proposed experimental platform, the connection of the motor controller with a higher control system uses digital inputs and outputs and supervised by a PLC. The number of digital inputs and outputs was increased when the motor controller didn't support direct computer control.

In order to expand the computer's digital and analogue port capability, data acquisition cards were used; they are fully supported by MATLAB and its real-time toolboxes [11]. Another method of interconnection of a computer with an industrial controller is using a dedicated microprocessor. Between the microprocessor's TTL logic and the 24 V motor controller's logic a galvanic isolation must be inserted as convertor. The block scheme of this solution is shown in Fig. 2b.

The Arduino microcontroller was used for the test platform [12], with galvanic isolation by optical couplers. The platform includes six digital computer outputs (inputs for the controller) and four digital inputs. Currently the galvanic isolation module is realized on an extension board which allows adding or removing further electronic modules.

A Graphical User Interface (GUI) for real-time control of linear actuators was developed in the MATLAB environment. MATLAB supports Arduino boards by running an on board server program which responds to commands arriving via serial port, executes the commands, and, if needed, returns a result [13]. The MATLAB function `timer` was adjusted at the working frequency of 20 Hz and used for real-time control. The GUI application supports two modes for controlling linear actuators: manual and automatic. In manual mode, the operator must know all

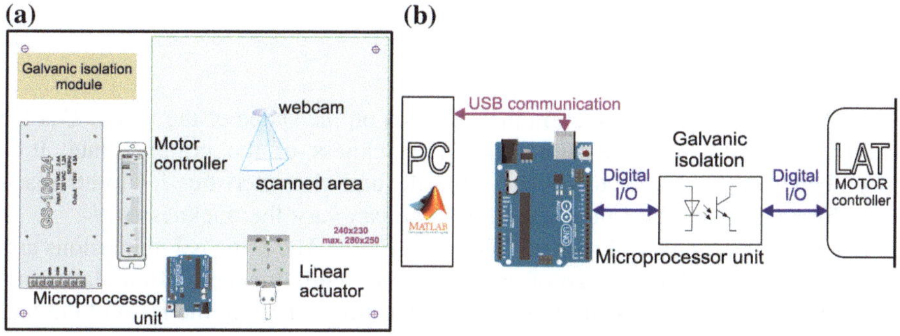

Fig. 2 **a** Placement of components on the test platform. **b** Block scheme of interconnection between industrial motor controller and computer

control instructions to have direct access to the parallel I/O connector. The actual states of the controller inputs and outputs are graphically individualized by the colour of the button and simultaneously drawn in a monitor window. The automatic mode enables setting desired values, only the current values being displayed.

3 Using the Experimental Platform for Micro-gripper Design. A Case Study

The possibilities of using the experimental platform are documented in an example of micro-gripper design [14–16]. The dimensions proposed for the gripper's body are: width 5.0 mm, height 6.2 mm, distance of fingers 0.2 mm, thickness of flexure joint 0.05 mm and thickness of the flexure plate 0.5 mm. The elastic material is aluminium alloy.

In principle, for manufacturing of such complex form there are two technologies that can be used: precise machining or 3D printing. Standard 3D printers support only specific types of materials like PLA or ABS, but these materials exhibit small flexibility. Polyamide seems to be a suitable material that can be processed by 3D printers. Naturally, if another material is used, different values of Young's elasticity modulus must be taken into account when calculating the flexural characteristics.

For experimentation, the physical sample of the micro-gripper was built at a scale of 20:1. As the Young's modulus of material for this model sample differs from polyamide, compliance/stiffness analysis is needed. It includes the calculation of compliance/stiffness of the elastic structure and building flexural model of the gripper. This type of gripper consists of relatively rigid parts interconnected by elastic joints having the form of circular double-side notches. Flexural characteristics of the whole gripper depend on the stiffness/compliance of its particular elastic connections. The coefficients of the compliance matrix for such joints can be expressed by the relation [5]:

$$C_{yFy} = \frac{12}{Ew} I_2 \tag{1}$$

where I_2 is the integral of function that depends on the shape of the notch, E is the Young's elasticity modulus and w is the thickness of the gripper plate. It is remembered that, in order to get the same flexural characteristics for joints made from different materials/Young's modules one can vary the thickness w.

The form of the physical model is shown in Fig. 3. The principal dimensions are: width of gripper—80 mm, high of gripper—154.2 mm, thickness of the flat elastic material—3 mm, thickness of all elastic joints manufactured in the form of circular notches—1 mm.

The kinematic relations between input motion of actuator and displacement of fingers, together with forces were analysed for the proposed elastic mechanisms of the gripper. The micro-gripper consists of two symmetric mechanisms as shown in Fig. 3b. The gripper mechanism transforms the displacement of the actuator u_{yOA} into motions of the finger tips. For the preliminary calculation, the simplified relation between input motion of the linear actuator and output movement of the gripper finger u_{xOG} (see Fig. 3b) is:

$$u_{yOA} = \frac{u_{xOG}}{\sqrt{(L_1 + L_5)^2 + (L_2 + L_4 - L_6)^2}} (L_2 + L_4) \tag{2}$$

The precise compliance analysis includes the calculation of deflections in desired points, and was done in COMSOL Multiphysics.

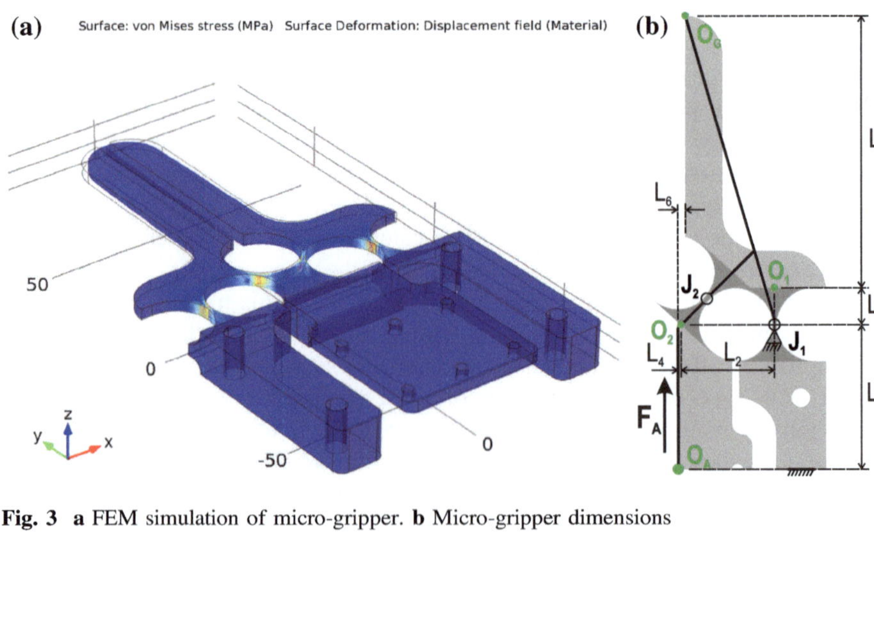

Fig. 3 **a** FEM simulation of micro-gripper. **b** Micro-gripper dimensions

The principal condition for flexure's design is that input force/energy exerted by the actuator should deflect the mechanisms and produce the desired force between fingers, expressed by the relation:

$$F_A = 2 \cdot \left(u_{yOA} K_G + T_{GA} F_G\right) \tag{3}$$

where F_A is the actuating force, K_G is the gripper stiffness, T_{GA} is the transformation matrix between points O_G and O_A and F_G is the desired force between gripper fingers.

The micro-gripper's symmetry enables the calculations only of its half part, the results being directly applied to the right half part [17].

For preliminary experiments the finger force was considered zero. In this case the actuating force overcomes only the stiffness of the two symmetric mechanisms.

Considering the linear flexural characteristics, the actuating force F_A, the stiffness K_G and the input displacement of the actuator u_{yOA} are related by:

$$F_A = K_G u_{yOA} \tag{4}$$

Then, calculating the stiffness for the whole gripper mechanism will be done as:

$$K_G = 2\left(T_{2A}\left(T_{12}\left(T_{01} C_{J1} T_{01}^T\right) T_{12}^T + C_{J2}\right) T_{2A}^T\right)^{-1}, \tag{5}$$

where T_{ij} are the transformation matrixes between i and j system coordinates defined in points O_i, and C_{Ji} are the compliance matrixes of particular joints.

4 Conclusion

An experimental platform for performance tests of compliant robotic devices has been described (see Fig. 4a). The purpose of this test bed is to verify the performance characteristics of various compliant devices frequently used in robotics (e.g. mechanisms, grippers and mechanical amplifiers). The testing allows applying HIL and SIL approached as suitable tools in designing complex mechatronic devices.

In order to perform the experiments, interconnection between actuators and control system was possible in MATLAB environment, which made possible working with standard industrial actuators. At present, the actual realization of the platform allows only relative low working frequencies; future research will permit frequency increase.

An example of the platform's application in micro-gripper development is included in the paper. The platform enables performing tests and verification of control algorithms (actuator control, position sensing and image processing). Finally, some theoretical considerations are made for the optimized design of compliant robotic devices.

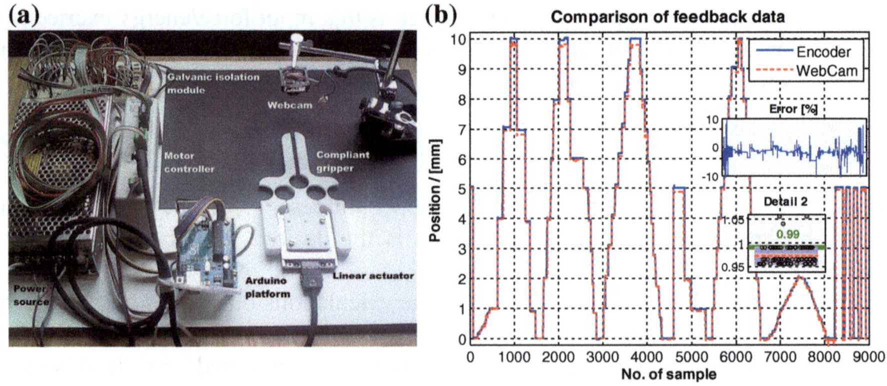

Fig. 4 **a** Experimental platform. **b** Comparison of data from encoder position feedback and webcam (*detail 2* shows distribution of measured data from webcam for 0.99 mm position)

Both described approaches were used as position feedback and comparison of their results is shown in Fig. 4b. The modularity of the designed experimental platform enables the integration of other types of sensors, e.g. force transducers.

Acknowledgments This paper presents the research work supported by the national scientific grant agency VEGA under project No.: 2/0048/13 "(Micro) Electro-mechanisms for robotics and extremely work spaces (environments)" and by the company "SMC Priemyselná Automatizácia s. r.o."

References

1. VDI Richtlinie 2206 Entwicklungsmethodik für mechatronische Systeme. Verein Deutscher Ingenieure, Düsseldorf, Beuth Verlag, GmbH, Berlin, June (2014)
2. Basten, T., Hamberg, R., Reckers, F., Verriet, J.: Model-Based Design of Adaptive Embedded Systems. Embedded Systems, Vol. 22, XIII, 306 p. ISBN 978-1-4614-4821-1 (2013)
3. Barbieri, G., et al.: A model-based design methodology for the development of mechatronic systems. Mechatronics **24**(7), 833–843. ISSN 0957-4158 (2014)
4. Boršč, M., Hricko, J., Thurský, B.: Using of the HIL simulation for mechatronics systems design and development. In: State-of-the-Art in Mechatronics, Vol. II, pp. 237–266. Simulation Research Press, Alpen aan den Rijn (2010). ISBN 978-90-807898-4-5
5. Lobontiu, N., et. al.: Corner-filleted flexure hinges. J. Mech. Des. **123**(3), 346–352 (2000). ISSN 1050-0472
6. Yong, Y.K., Lu, T.-F.: Kinetostatic modeling of 3-RRR compliant micro-motion stages with flexure hinges. Mech. Mach.Theory, **44**(6), 1156–1175 (2009). ISSN 0094-114X
7. Howell, L.L.: Compliant Mechanisms, 459 p. Wiley-IEEE (2001). ISBN 047138478X
8. Smith, T.S.: Flexures: Elements of Elastic Mechanisms, 448 p. Gordon and Breach Science Publishers, Philadelphia (2000). ISBN 90-5699-261-9
9. PiezoMotor Uppsala AB, Piezo LEGS® Linear 6N. http://www.piezomotor.com/app/content/uploads/150010_LL10.pdf

10. SMC Corporation, Card Motor—Series LAT3, id.: CAT.ES100-96B. http://ca01.smcworld.com/catalog/New-products-en/pdf/es100-96-lat3.pdf
11. http://humusoft.cz/produkty/datacq/. Cited 12 Jan 2014
12. http://www.arduino.cc/. Cited 15 Jan 2014
13. http://www.mathworks.com/hardware-support/arduino-matlab.html. Cited 20 Jan 2014
14. Celler, P., Hricko, J.: Design of simple micro gripper. Technical report BP, FM TnUAD v Trenčíne, Trenčín, 75 s (in Slovak) (2009)
15. Hricko, J., Havlík, S.: Small-size robotic tools with force sensing. In: Modern Machinery Science Journal: 20th International Workshop on Robotics in Alpe-Adria-Danube Region, pp. 73–77, special edition (2011). ISSN 1805-0646
16. Hricko, J.: Modelling compliant mechanisms-comparison of models in MATLAB/SimMechanics vs. FEM. In: RAAD 2012: 21st International Workshop on Robotics in Alpe-Adria-Danube Region, pp. 57–62. Edizioni Scientifiche e Artistiche, Naples (2012). ISBN 978-88-95430-45-4
17. Havlík, Š.: Contribution to solving adaptivity of robotic systems in interaction with environment. SAV DrSc thesis, Banská Bystrica (in Slovak) (1993)

Cooperative Assembly Using Two Industrial Robots

Mădălin-Petru Sbanca and Gheorghe-Leonte Mogan

Abstract The paper presents some design aspects concerning robot cooperation methods in order to accomplish dexterous tasks using two industrial robots with different controllers. The goal of this research is to present assembly strategies of two objects using an external master-slave controller for the robots actions. The method described in the paper presents a new solution for automatic robot collaboration and interaction with user interface.

Keywords Industrial robot · Robot cooperation · Force control · SCADA system · Automatic control system

1 Introduction

Compared with humans, robots have the disadvantage of lacking decision in critical situations. During the reported research, we studied the robots' behaviour by assimilating them with human's behaviour. Automatic robotic arms can assembly complex planar parts using torque-controlled DLR light-weight on-board cameras (eye-in-hand configuration). One of the objects is considered in a fix position and the other one is manipulated by the robot [1].

To realize an algorithm for assembly operations with one or more robots, the first step consisted in analysing human's behaviour. According to [2], a solution is hybrid robot control. The next step is to implement the decision procedure in the robot controller [3], which involves both the robot and the operator. Using a virtual environment, the human's reaction and skills can be learned [4–6]. In some cases delay problems occur [7] as a reaction of the robot positioning oscillations [8].

M.-P. Sbanca (✉) · G.-L. Mogan (✉)
Transilvania University of Brasov, 29, B-dul Eroilor No 29, Brasov, Romania
e-mail: madalinsbanca@gmail.com

G.-L. Mogan
e-mail: mogan@unitbv.ro

T. Borangiu (ed.), *Advances in Robot Design and Intelligent Control*,
Advances in Intelligent Systems and Computing 371,
DOI 10.1007/978-3-319-21290-6_5

2 Assembly Methods

The most efficient method of studying the assembly process is the "peg in the hole" problem. First, the shape of the peg will be considered and then the hole's chamfers system [9]. The orientation of the peg is also very important to establish the approach method [10, 11]. Researches were made with different feedback types, like infrared sensors [12] or vision cameras [1, 13, 14]. Furthermore, as feedback of the closed control loop, sometimes the impedance of the robot motors can be used [15–17].

Considering the necessary equipment for realizing the assembling process including robotic arms with 3–6 DOF, there were proposed such robotic structures [16–18] based on 3 DOF, with different implementations. The closest approach to this paper was the use of complex peri-robotics structures [16, 19] which are able to realize complex assembly task. Our aim is to use less equipment still obtaining accurate results and an optimized cycle time. Two possibilities will be considered: the first one is to keep the peg in a perpendicular position to the hole's plane and to implement a different searching trajectory in the robot; the second option is to lean the peg [11, 13] and follow the force feedback impulse.

3 System Description

The goal of this study was to realize a collaborative robot system. To achieve this, an assembly process using one robot was extended to multi robot usage. The first step to achieve the goal is to control the process using a master controller external with respect to the existing robots controllers. The second step is to use the force feedback signals from a sensor. Thus, the possibility to connect two different robots to a master controller with the aim of realizing collaborative tasks was investigated.

3.1 Hardware Presentation. The System's Architecture

The system developed (Fig. 1) to satisfy the collaborative robot behaviour uses some of the equipment existing in the Robotics Lab. of the Product Design and Robotics Department of Transylvania University of Brasov. The main system components are two ABB industrial robots, IRB 1600-145 and IRB 2400L both having their own controller. This means the robots are totally independent to each other. They are used in our study like slaves components. The system includes a force sensor JR3 type 100M40A; this is a wrist sensor able to measure forces in the x, y and z directions as well as the corresponding moments with a frequency of 10 kHz. There is also a PC-based master controller, connected via Ethernet to the robot controllers.

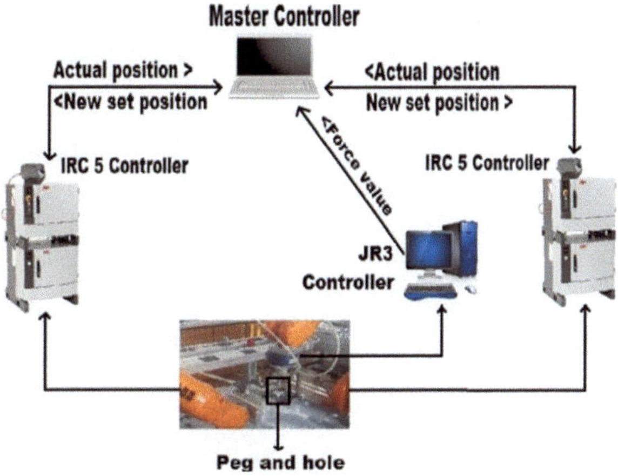

Fig. 1 Hardware architecture of the system

3.2 *Software Description—Automatic System Control*

The force control uses a special PCI connection allowing execution of an application in Visual Basic acquiring online the force values; it allows calibrating the force cell and if is necessary adding new offsets. These values are then transferred to the platform using UDP protocols at a frequency of 20 Hz. RAPID instructions are used to make the connection between robot programs and the C# master application. The main software part is the platform realized in C# software, which interconnects all the tools used in the study (Fig. 2). The platform has also a user interface to make the parameterization of the global control process, and is acting like a hierarchical automatic control system. The inputs of the system are the parameter specified by the user. The feedback is composed by the actual positions of the robots and the force cell values for all three directions. The output of the system is represented by the new positions of both robots, generated by the master application according with the implemented control laws. Generally, the control system can be implemented with different methods [15, 19]. The most popular hybrid control systems are the ones with continuous action, linear, of P, PI, and PID type, but there might be used nonlinear control laws too, such as bi-positional and tri-positional.

4 Case Study

The most important goal of this study is to realize a master controller for different robot types making the collaboration between them more natural in most of the situations. This study presents a different approach of robots control provided by the new developed platform. For the case study the peg in hole operation is analysed, with the tools presented in Figs. 3 and 4. Both of them have a small chamfer,

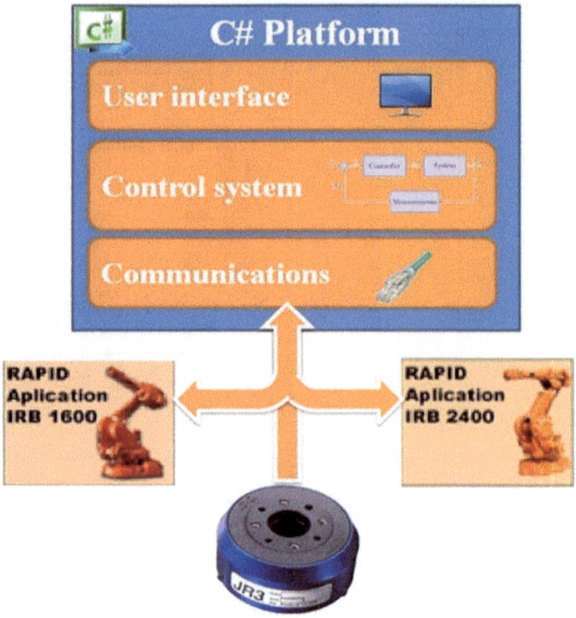

Fig. 2 Software structure

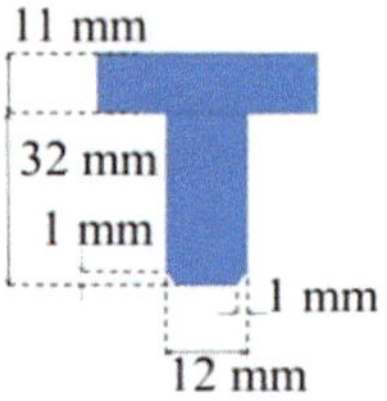

Fig. 3 Peg size

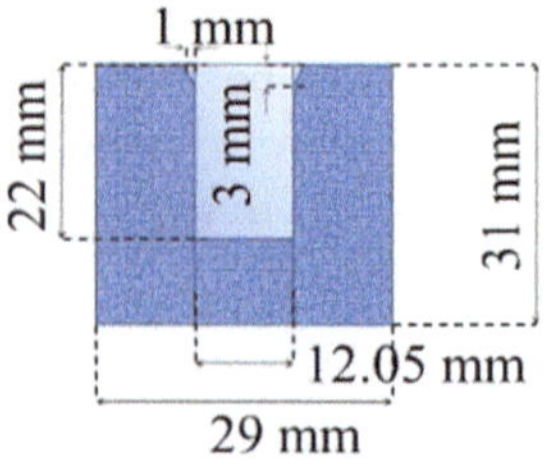

Fig. 4 Hole size

because of the tolerance between the peg size and the hole size which is only 0.05 mm, and we don't want to lean the peg to find the hole position.

This case study used a better version of the platform in comparison with the previous personal research [20]. Here are implemented two parallel communications with each robot, one is used for the actual positions and new set positions and

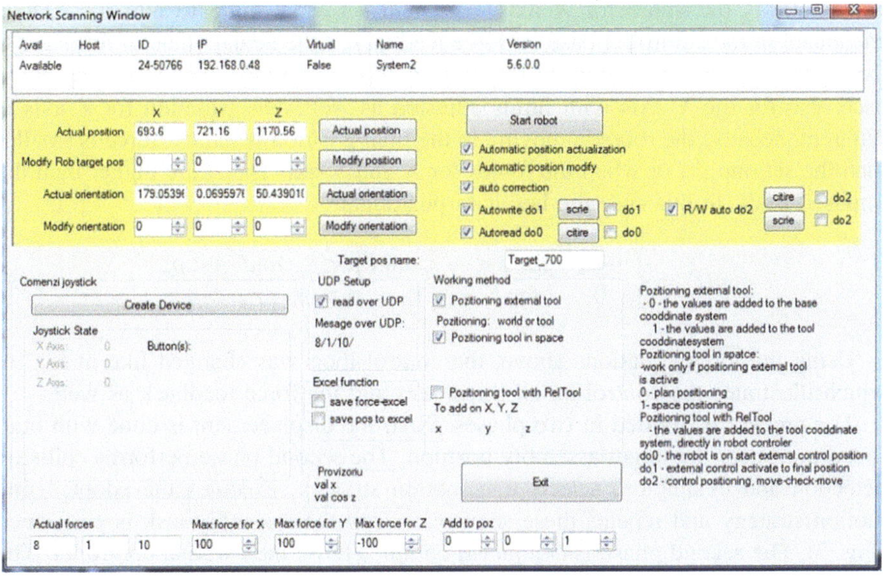

Fig. 5 User interface

the second one is used to synchronize the updates of the positions with the force feedback.

The user interface is presented in Fig. 5. The user is thus able to make the parameterization of the process. He can also choose the positioning method and the feedback signal. It is possible to select manual positioning or manual update of actual robot positions, synchronize the signals or manually manipulating the robot using a peripheral controller (teach box).

The most important component of this application is the automatic collaborative control system of the two robots. Considering f_r as force feedback, s, $\dot{s}$ and $\ddot{s}$ the actual position, speed and acceleration, s_c, $\dot{s}_c$ and $\ddot{s}_c$ the set position, speed and acceleration and M, B and K the mass, damping and stiffness matrix the process control can be formulated like in the following equation:

$$f_r = M(\ddot{s} - \ddot{s}_c) + B(\dot{s} - \dot{s}_c) + K(s - s_c) \tag{1}$$

Considering the present equipment (see chapter "ROS Based Safety Concept for Collaborative Robots in Industrial Applications"), we selected for this study the tri-positional and bi-positional control laws. These are applied for the X and Y axes as well for the Z axis like in the Eq. 2:

$$u(t) = \begin{cases} -u_m & for\ \varepsilon < -\varepsilon_p \\ 0 & for\ -\varepsilon_p < \varepsilon < +\varepsilon_p \\ +u_m & for\ \varepsilon > +\varepsilon_p \end{cases} \tag{2}$$

where u(t) is the value for X axis, $-u_m$ and $+u_m$ the value to add for X axis calculated in the controller base on force feedback, ε the actual value of force for X axis and $\mp\varepsilon_p$ are limits adjusted by the user in the interface. The same Eq. (1) is used also for the Y axis with limits adjusted as well. The equation for Z axis is different, because the robot is moving to the final position when the force is smaller than the set one (ε) or when the forces for X and Y axis (μ, ρ) are bigger than the imposed limit. In this case the law is bi-positional.

$$u(t) = \begin{cases} +u_m & \text{for } \varepsilon < -\varepsilon_p \text{ and } \mu < \mu_p \text{ and } \rho < \rho_p \\ 0 & \text{for } \varepsilon > -\varepsilon_p \text{ or } \mu > \mu_p \text{ or } \rho > \rho_p \end{cases} \quad (3)$$

Using the three equations above, the control loop was changed like in Fig. 6, which illustrates the control on all three axes and the force feedback as well.

The process is divided in two phases. First, the displacement is done with high speed until near the target assembly position. The second phase performs collision detection and evaluation, selects fine motion strategy, executes the adopted fine motion strategy and repeats these sequences until the assembly task is performed (Fig. 7). The second phase is conducted completely by the external controller. The

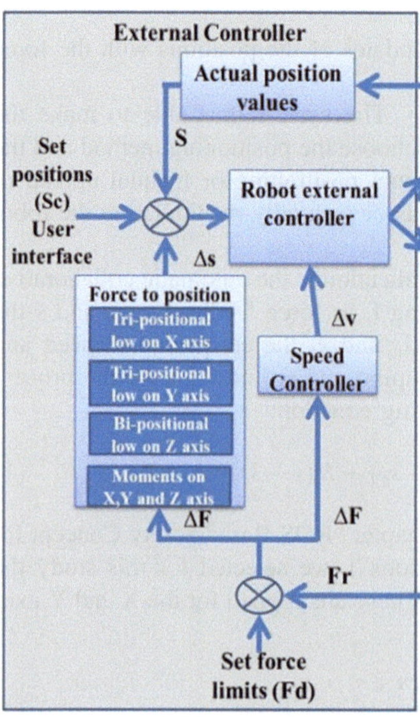

Fig. 6 Configuration of the automatic control system

Fig. 7 Process phases

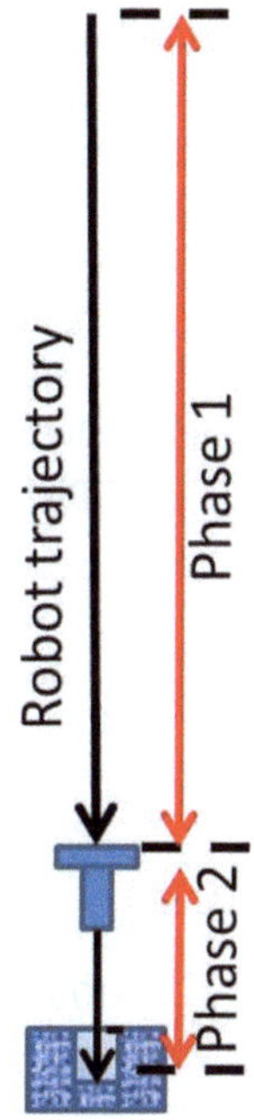

sample rate for the force sensor and for the positions refresh is 20 Hz. The master sends to both robots new positions reading the synchronization signals from the robots. In this way robot position can be precisely controlled. The robot starts to move and the master reads the force value and waits for positions reaching or force limit signal occurrence. Thus, robot collision is avoided. The process cycle time depends on the set force limits at the user interface, on the robot accuracy and parts' location in the gripper.

The robot trajectory is also influenced by parameter scaling in the C# platform. For the first test the position increment was set to 1 mm. The measured values of the force and positions of second robot were saved with a frequency of 10 Hz. Analysing the results shown in Fig. 8, big oscillations in robot positions can be observed. These are the results of force oscillations with values between ±100 N, having the switching limits set to 2.5 N in the user interface.

After the position type was modified by the external controller in terms of converting the parameters to float type, the test was repeated. In order to observe the difference caused, various force limits were imposed. Both the limits were adjusted considering the materials, and the robots torque limits to avoid crashing the robots or the parts. The force cell scale is between −400 and +400 N. For the second test the limit was set at 2.5 N and for the last test 12.5 N. Furthermore, Fig. 9 presents the second test with minimum increment value to 0.1 mm. Improvements on the oscillations range around ±25 N in comparison with ±100 N from the first test can be observed. Also the time has been optimized for the second test. For the cycle time we have to consider the deepness of the hole which is 22 mm with the peg hold until contact between the peg and the hole on Z axis occurs.

Considering the results, for the last test the limits were set to 12.5 N, in order to obtain smaller oscillation on the robots positions. This action has no influence on used part materials. Without oscillations the movement along the Z axis will be much faster. Only small automatic adjustments could be observed in the position. The complete cycle time is around 10 s (Fig. 10).

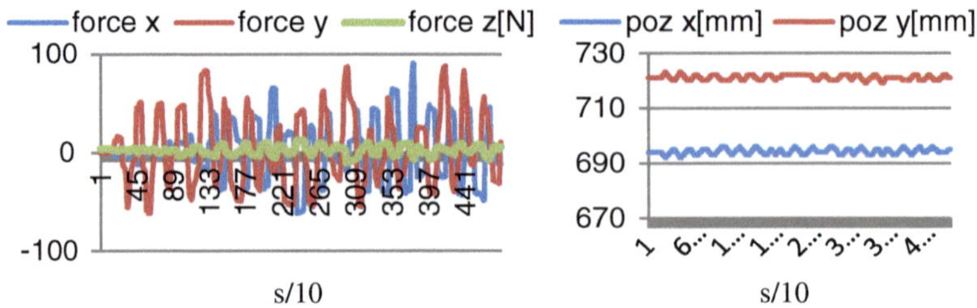

Fig. 8 Measured forces and positions

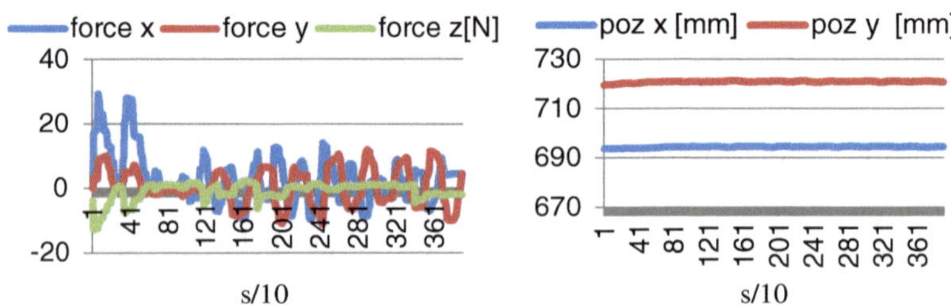

Fig. 9 Measured forces and positions after optimization

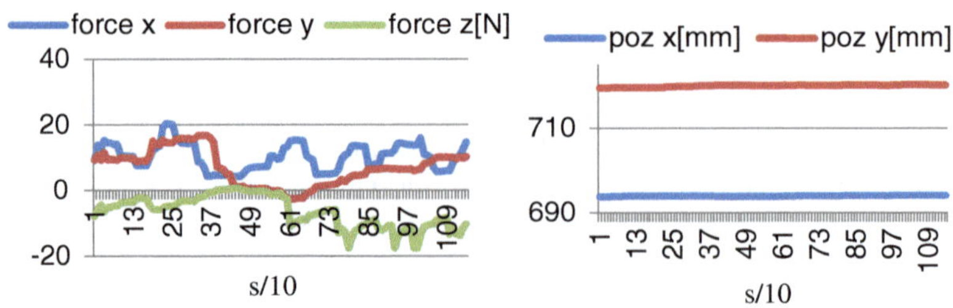

Fig. 10 Measured values of forces and positions for last test

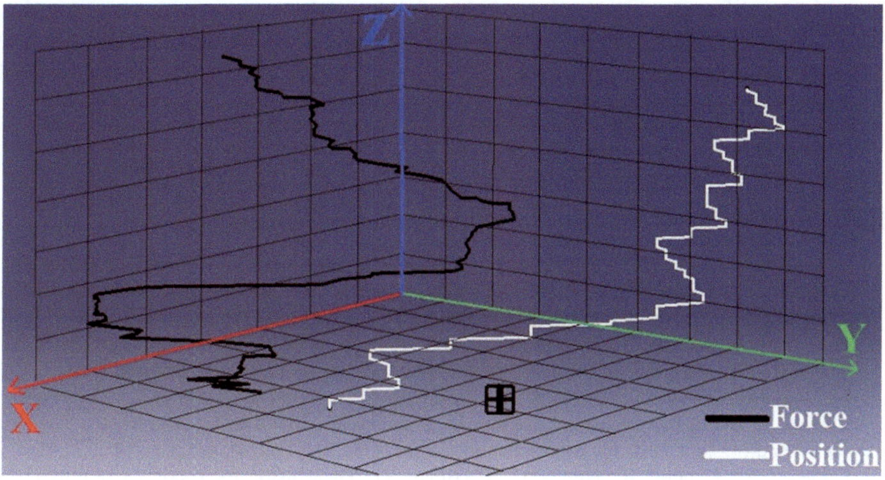

Fig. 11 Positions dependency on force feedback

Data from the force sensor and position of the second robot were already presented. The position of the peg holder can be observed only on the X and Y axes, because the goal was to observe the improvements of the automatic control system during algorithm modifications. The force values on all three directions were transformed into three dimensional points and then a trajectory was generated. This trajectory was compared with the robot trajectory generated by the external control system. Figure 11 presents the directions of the trajectories which are similar. This confirms a strong dependency relation between these two parameters in the external master controller.

5 Conclusions and Future Work

This study opens a new direction in the control of assembly processes. The novelty of the reported research is the cooperation between two robots in order to accomplish assembly tasks, reducing the cycle time.

Another advantage brought by this structure is flexibility because of the interactive, multi-modal user interface. There is no need to fix the robot base or to cooperate with the human during the process; this is because the conceived platform is automatically connecting to the robot. Using different feedback sensors, the robots can be used to accomplish dexterous tasks that only human persons can do at present.

The future research work consists in the study of more complex assembly processes and in extending the platform to allow connectivity to other feedback sensors and utilization of a greater number of collaborative robot manipulators, possibly of different types (e.g., produced by different manufacturers). Another future research line could investigate the possibility to implement and analyse different control system laws like PID or fuzzy logic, and integrate the flow of computing data into hierarchical multitasking dynamic robot motion control.

References

1. Giordano, P., Stemmer, A., Arbter, K., Albu-Schaffer, A.: Robotic assembly of complex planar parts: an experimental evaluation. IROS IEEE, pp. 3775–3782 (2008)
2. Horiguchi, Y., Kurono, K., Nakanishi, H., Sawaragi, T., Nagatani, T., Noda, A., Tanaka, K.: Ecological interface design for teaching assembly operations to industrial robot. IFAC IEEE, pp. 442–447 (2010)
3. Lauren, S., Buch, P., Sorensen, L., Kraft, D., Petersen, H., Elekilde, L., Schultz, U.: Towards error handling in a DSL for robot assembly tasks. DSLRob (2014). arXiv:cs/1411.7148
4. Cortesao, R., Koeppe, R., Nunes, U., Hirzinger, G.: Data fusion for robotic assembly tasks based on human skills. Robotics, IEEE, pp. 941–952 (2004)
5. Kavasaki, H., Furukawa, T., Ueki, S., Mouri, T.: Virtual robot teaching in the assembly work environment for multi-fingered robots. WAC'08, World IEEE, pp. 1–7 (2008)
6. Hirche, S., Stanczyk, B., Buss, M.: Haptic tele-assembly over the internet. EuroHaptics, Technische Universität München (2004)
7. Imaida, T., Yokokohji, Y., Doi, T., Oda, M., Yoshikawa, T.: Ground-space bilateral teleoperation of ETS-VII robot arm by direct bilateral coupling under 7-s time delay condition. IEEE Trans. Robot. Autom. **20**, 499–511 (2004)
8. Dong, S., Naghdy, F., Chen, Y.: Six d.o.f haptic rendered simulation of the peg-in-hole assembly. International Conference on Manufacturing Excellence (2003)
9. Callegari, M., Suardi, A.: On the force-controlled assembly operations of a new parallel kinematics manipulator. Proceedings of Mediterranean Conference on Control and Automation, pp. 18–20 (2003)
10. Debus, T., Dupont, P., Howe, R.: Contact state estimation using multiple model estimation and Hidden Markov models. Int. J. Robot. Res. **23**, 399–413 (2004)
11. Chatpar, S., Branicky, M.: Search strategies for peg-in-hole assemblies with position uncertainty. IEEE/RSJ International Conference on Intelligent Robots and Systems, Vol. 3, pp. 1465–1470 (2001)
12. Paulos, E.: Informed peg-in-hole insertion using optical sensors. SPIE 2059 (1993)
13. Wang, J., Zhang, H., Fuhlbrigge, A.: Force control technologies for new robotic applications. TePRA IEEE, pp. 143–149 (2008)
14. Yoshimi, B., Allen, P.: Active, uncalibrated visual servoing. IEEE Int. Conf. Robot. Autom. **1**, 156–161 (1994)
15. Bruzzone, L., Molfino, R., Zoppi, M., Zurlo, G.: The PRIDE prototype: control layout of a parallel robot for assembly tasks. IASTED International Conference on Modelling, Identification and Control (2003)
16. Broenink, J., Tiernego, M.: Peg-in-hole assembly using impedance control with a 6-DOF robot. Simulation in Industry, Proceedings of 8th European Simulation Symposium (1996)
17. Bruzzone, L., Molfino, R., Zoppi, M.: Modelling and control of peg-in-hole assembly performed by a translational robot. IASTED International Conference on Modelling, Identification and Control (2002)

18. Asad, M., Raza, A., Farooq, U., Ishaq, M., Ahmad, B.: Design and development of microcontroller based autonomous peg placement robot. IEEE ICCSIT, pp. 486–490 (2010)
19. Lopes, A., Almeida, F.: A force–impedance controlled industrial robot using an active robotic auxiliary device. Robot. Comput. Integr. Manuf. **24**, 299–309 (2008)
20. Sbanca, M., Mogan, G.: Cooperation of two industrial robots for manipulating rigid objects using SCADA systems. Appl. Mech. Mater. **555**, 312–319 (2014)

The Beginning of the Automation

A Brief Review on the Automatic Devices in the Hellenistic Age

Cesare Rossi

Abstract Some examples of the most interesting automatic devices of the Hellenistic Age are presented. These devices can be considered the first examples of automatic mechanical systems, thus representing the precursors of the automation; they clearly show how the concept of automation was present in the minds of ancient scientists and engineers since the II century B.C. or even earlier. The devices also clearly show the surprising modernity of the knowledge of the ancient scientists and engineers, both conceiving the mechanical designs and the building of them. Moreover, the presented device spans a rather wide range of fields of applications. The automatic systems that were chosen are presented by grouping them by inventor. The latter are: Archimedes, Heron of Alexandria, Ctesibius, Dionysius of Alexandria and Philon of Byzantium.

Keywords Ancient automatic devices · Automata · History of science

1 Introduction

The idea of building automatic devices is probably as old as the human thought. The oldest clues about these ideas are probably found in the Greek mythology. For instance, in the Iliad it is told that the god Hephaestus (Iliad, XVIII, vv. 519–525) built some "automata" that helped him in his smith works. Another legend tells that king Minos used a bronze mechanical giant, Talos, forged by Hephaestus, to patrol the isle of Crete. Moreover, the word "automatic" comes from the ancient Greek αυτοματος, an adjective meaning literally self-moving or self-acting. So, also in ancient times, the term and the concept had practically the same meaning we use today.

C. Rossi (✉)
Department of Industrial Engineering, University of Naples "Federico II", Naples, Italy
e-mail: cesare.rossi@unina.it

T. Borangiu (ed.), *Advances in Robot Design and Intelligent Control*,
Advances in Intelligent Systems and Computing 371,
DOI 10.1007/978-3-319-21290-6_6

As far as we can recognize of the past, automatic devices were first designed by the scientists/engineers of the Hellenistic Age; this is to say in the Mediterranean Area and in a period of time that strictly ranges between 323 B.C. and 31 B.C. but, from a cultural point of view ranges between the III century B.C. and the II century A.D. In this period of time, the human knowledge had a very considerable "jump" in almost any field of the culture and put the basis of the society in which we live today. Concerning the scientific and technical knowledge in particular, most of the inventions developed in this period of time were beaten not earlier than in the XVIII century. So, it seems interesting to propose a brief review on some examples of these early automatic devices.

2 Heron of Alexandria

Heron of Alexandria was probably the best known designer of automatic devices in ancient times. Although very few is known about his life, he is well-known for a wide number of his inventions in different fields; probably the widest fields of his inventions is the automation.

The main treatises by Heron about automatics (Pneumatica and Automata) were translated during the Italian Renaissance by Baldi [1] (1553–1617), and by Aleotti [2] (1546–1636); both describe a wide number of automatic devices.

The most famous automatic device by Heron is probably the mechanism to open and close the doors of a temple shown in Fig. 1 with a virtual reconstruction.

A fire was lit on the brazier F; so, the hot air heated the water in the pressure tank S. The pressure in this tank pushed the water in the mobile water container C through a U-shaped pipe. The mobile water container was connected to the temple doors by means of ropes or chains wrapped in coils on the door hinges. As the water container was filled with water, because of its weight the ropes were unrolled and

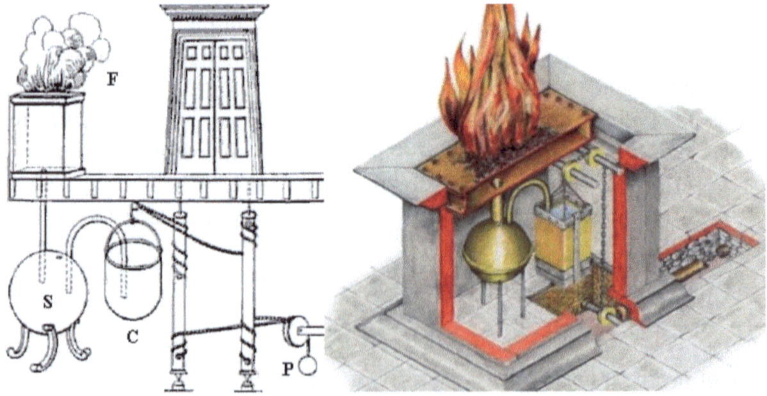

Fig. 1 Heron's mechanism for temple doors and virtual reconstruction, adapted from [17]

the doors were opened. When the fire was extinguished, the steam in the pressure tank condensed, hence the pressure in it decreased and the water was sucked up from the water container. As soon as the weight of the latter decreased enough, the counterweight P acted on the door hinges in the same way, but closing the doors.

Another interesting example of Heron's automata is reported in Fig. 2 where a reconstruction is shown made by Aleotti [2] in his "Gli artificiosi et curiosi moti spirituali di Hero Alexandinus" (the artificial and strange pneumatic motions By Heron of Alexandria), Ferrara 1589, of a famous Heron's automaton. This automaton was made up by two main characters: Heracles and a dragon. In few words: the dragon hisses, Hercules beats it with a club and the dragon spits on Hercules. The working principles can be deduced by Aleotti as it follows and is shown in Fig. 2. A water flow from S fills a tank H through a funnel T. When the tank H is filled the air flows through a small pipe M that is linked to the dragon's mouth and this one hisses. A rocker C can rotate on a pin O; one of its arms is linked to a cone B and to a rope E, while the other arm is linked to a water container Z; the latter, if empty weighs less than cone B. As the water level in the tank goes up, the water fills the mobile water container Z through the U-shaped pipe X. When the mobile container Z is heavier than the cone B, the rocker C rotates clockwise and rope E moves Hercules' arms through a simple T-shaped mechanism, not represented. In this way the club is lifted up. At the same time through pipe Y tank A, pipe Q, and the cone R are filled.

Since the mobile water container is conical, when it reaches the bottom it turns upside down and the water in it is evacuated. Now cone B is heavier than container Z and the rocker rotates counter-clockwise. The rope R is tightened and the club beats the dragon's head. At the same time, cone B gets inside cone R that is full of water and so the pressure in the pipe Q rises. This pipe is linked to the dragon's mouth and so the latter spits a water jet onto Hercules.

To Heron was also ascribed the water distributor that represents the ancestor of the modern drink vending machine.

But Heron's most surprising automatic device is probably the programmable self-moving robot. The latter was described by Baldi [1]. These automata were conceived as "actors" in a puppet theatre; on the theatre proscenium all the

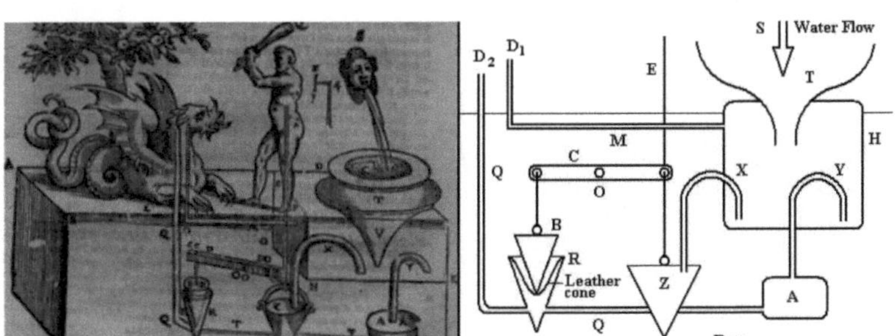

Fig. 2 Aleotti's reconstruction of Heron's automaton and working principle, from [2, 10]

performance was played by those automata that were, hence, programmable. In Fig. 3 drawings from Baldi's work are reported, showing the working principle and the counterweight motor that moved those automata. In the figure on the left, one can observe the counterweight that is located in a tank filled with millet or mustard seeds in order to regulate the counterweight motion, hence the automaton speed. The system that changes direction uses the programmability of motion concept: Fig. 3 on the right shows the axle of the driving wheels divided in two axle shafts that are independent one from the other; on each one of the shafts a rope is rolled. If the rope is rolled on one of the axle shafts in a different way from the other, when the counterweight goes down pulling the rope, one of the two driving wheels will rotate in different way from the other one. The programming of the motion can be also obtained by putting some knobs on the axle shaft like shown in Fig. 3; by these knobs it is possible to modify the rolling of the rope, in order to obtain different laws of motion for each wheel [3–5].

The counterweigh motors were widely used in ancient age for a number of applications such the theatre curtain lifter, the awnings of the Coliseum and most probably the internal motors of the siege towers [5].

A siege tower or "helepolis", as many ancient authors suggest [6–8], had internal motors so that it can be considered as a large size of the Heron's self-propelled automaton. These towers, in fact, very probably were powered by counterweigh motors. The scheme of a counterweigh motor for an helepolis is shown in Fig. 4; it essentially consists in a counterweigh that pulls a rope rolled on the wheel axle by means of a five-pulley block and tackle with 5 pulleys (Pentaspaston, described by Vitruvius [9] in I century B.C.).

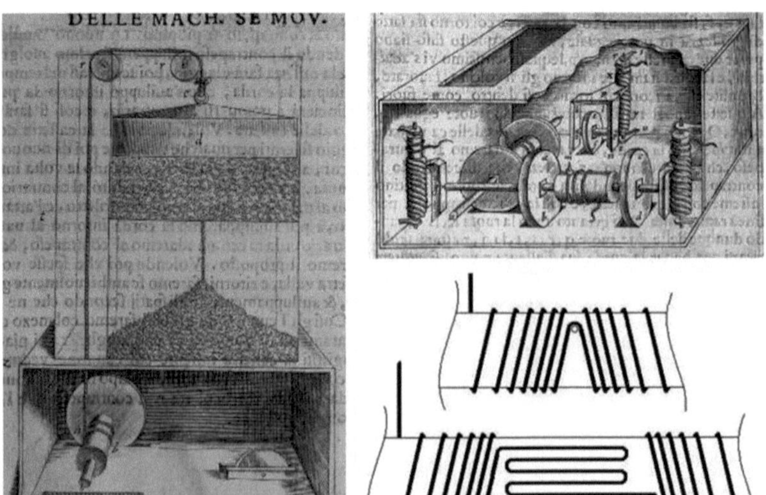

Fig. 3 Heron's programmable robot and working principle of the motion programming, adapted from [1, 10]

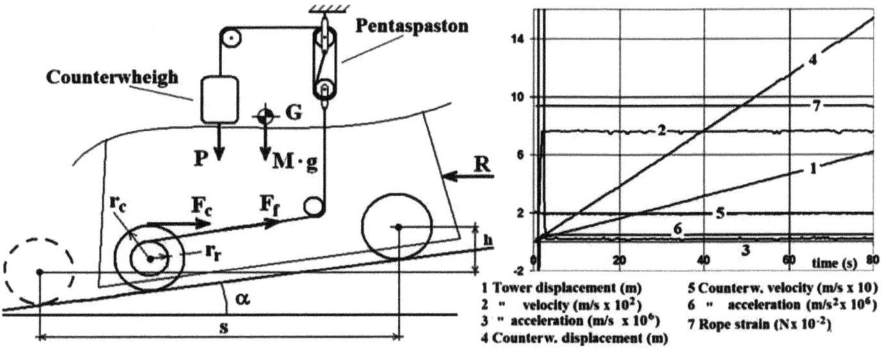

Fig. 4 Scheme of a counterweigh motor for a helepolis (*right*) and simulation results at constant counterweight velocity (*left*), adapted from [5]

A model of the tower was made by means of WM 2D™ code that permitted a dynamical study of the device; this was made in order to evaluate the reliability of the device itself. Since this investigation was already exposed in [5], in the following just the main results are reported. The data of the considered helepolis are:

- Mass of the helepolis = 40,000 kg;
- Mass of the counterweigh = 1000 kg;
- Radius of the wheels of the helepolis: $r_c = 1.5$ m;
- Radius of the axle shaft: $r_r = 0.8$ m;
- Slope $\alpha = 2°$.

From the data above it was computed that such a counterweigh, having a stroke of 20 m, could move the helepolis for 7.5 m. If the counterweigh was a water tank, it could have been unfilled once at the bottom of the stroke, then lifted again and finally re-filled by a pump or by a chain of buckles.

3 Ctesibius

Ctesibius (285–222 B.C.) was the director of the Library of Alexandria and is credited of a large number of inventions, several of them being automatic devices. Among these, one of the most interesting can be considered the water clock, shown in Fig. 5; in the figure a virtual reconstruction and technical drawings showing the working principle are reported. To understand the reason why an automatic device was required for a water clock, we must remember that the length of a roman hour was not constant since it was defined as 1/12 of the time between sunrise and sunset during the day and 1/12 of the time between sunset and sunrise during the night. Thus, the time duration of one hour was different from day to night (except at the equinoxes) and from a given day to another one. The water clock designed by Ctesibius solved this problem [10, 11]. A bottom tank is filled by a constant water

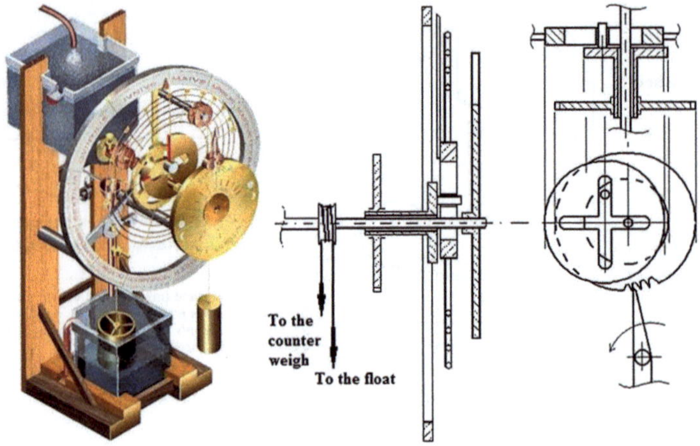

Fig. 5 Ctesibius' automatic water clock, adapted from [10]

flow from a top tank that is permanently maintained full. A yarn, wrapped in coil around the pointer axle, is connected to a floater and to a counterweight. The bottom tank is drained daily and the cycle starts again.

The dial is fitted on an off-centre shaft on which a gear having 365 teeth is fitted; once a day the float passes through a certain position and moves a rod that pushes one tooth; hence the dial completes a revolution in one year.

4 Dionysius of Alexandria and Philon of Byzantium

The invention of the repeating catapult is attributed to Dionysius of Alexandria (III century B.C.), and was described by Philon of Byzantium [10, 12–16]. It can be considered as a concentration of the most advanced mechanical kinematic and automatic systems of the time, many of which being still widely used. According to the author's recent reconstruction, based on previous works and from Philon's description, the device is really automatic. A pictorial reconstruction of it is shown in Fig. 6.

The device essentially consisted from a container holding within it a number of arrows, a cylinder feeding device and movement chain. According to Philon, the arrows were located in a vertical feeder F and were transferred one at a time into the firing groove by means of a rotating cylinder C, activated alternatively by a guided cam, in turn activated by a slide. One of the longer interior teeth T of the chain pulls the slide S which in turn pulls the cord, loading the coils of the motor. At the same time, an attached cam caused a 180° rotation in the direction of the cylinder, drawing an arrow from the loader and placing it in the channel in front of the rope. When the slide reached the rear of the weapon, the cog released it, while another opened the release mechanisms. An instant later, upon completion of sprocket

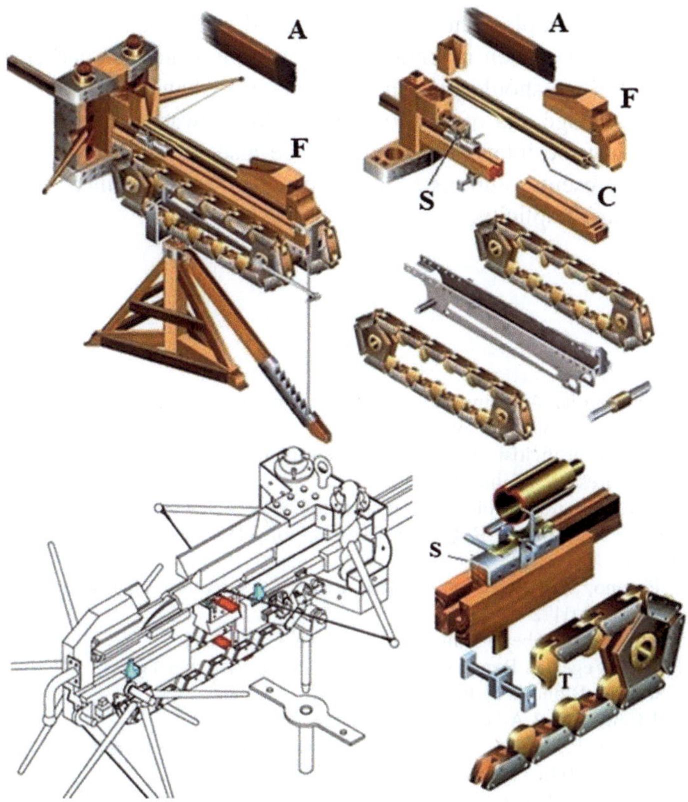

Fig. 6 Pictorial reconstruction and views of the automatic catapult, adapted from [10, 15]

rotation, the same cog coupled with the slide from underneath, pulling in the opposite direction. Near the top of the weapon, the second device closed the hook after it had retrieved the cord, while the feeder cylinder picked up another arrow from the feeder. A half rotation in the sprocket and the cycle was repeated. It must be observed that our reconstruction, based on our translation of the text by Philon, is really automatic; this because, differently from previous reconstructions, a simple rotation of the crank was sufficient to move the cylinder, the slide, the slide hooking mechanism and the trigger mechanism.

5 Conclusions

Some examples of automatic devices were presented, all belonging to the Hellenistic Age. This historical period, in fact, represents a period of time during which the human thought and knowledge had a deep improvement especially concerning technology and engineering (see e.g. [18–29]).

The presented examples show how advanced were our predecessors about 2000 years ago. Most of their inventions, in fact, were beaten not earlier than in the XVIII century. So, we should give them a tribute for having "placed the foundation" of our knowledge and of our comfortable life nowadays.

Moreover, it can be interesting (and may be also instructive) to observe how some devices represent solutions of problems that were obtained with a design showing a brilliant simplicity.

References

1. Baldi, B.: Di Herone Alessandrino de gli avtomati ouero machine se moventi libri due, Tradotti dal greco da B. Baldi, Venezia (1602)
2. Aleotti, G.B.: Gli artificiosi et curiosi moti spirituali di Erone. Bologna. The Archimedes Project, Digital Research Library (1589)
3. Sharkey, N.: The programmable robot of ancient Greece—New Scientist #2611, 4 July (2007)
4. Webb, B.: The first mobile Robot? Proceedings of TIMR 99 Towards intelligent Mobile Robots (1999)
5. Rossi, C., Pagano, S.: A study on possible motors for siege towers. J. Mech. Des. **133**, 1–8 (2011). ISSN 1050-0472
6. Gaio Giulio Cesare: De bello Gallico, liber II, XXX et XXXI
7. Publius Flavius Vegezius Renatus: Epitoma Rei Militaris, Liber IV, cap. XVII
8. Diodorus Siculus: Bibliotheca historica, liber XX
9. Vitruvius (I Cent. B.C–I A.D.): De Architectura, liber X
10. Rossi, C., Russo, F., Russo, F.: Ancient Engineers' inventions, precursors of the present. Springer, Berlin (2009). ISBN 978-90-481-2252-3
11. Russo, F., Rossi, C., Ceccarelli, M., Russo, F.: Devices for distance and time measurement at the time of the Roman Empire. HMM International Symposium on History of Machines and Mechanisms, Tainan, Taiwan, 10–14 Nov, pp. 101–114 (2008)
12. Shramm, E.: Die antiken Geshützen der Saalburg (reprint). Saalburg Museum, Bad Homburg (1980)
13. Marsden, E.W.: Greek and Roman Artillery Historical Development. Oxford University Press, Oxford, II (1969)
14. Soedel, V., Foley, V.: Ancient catapults. In: Scientific American, pp. 101–114 (1979)
15. Russo, F., Rossi, C., Russo, F.: Automatic weapons of the Roman Empire. EUCOMES'08 2nd European Conference on Mechanism Science, 17–20 Sept, Cassino, Italy (2009)
16. Rossi, C., Russo, F.: A reconstruction of the Greek-Roman repeating catapult. Mech. Mach. Theory **45**(1), 36–45, (2010). ISSN 0094-114X
17. Russo, F.: Il miracolo di Erone (The miracle by Heron, in Italian). Archeo, n. 201208, Aug, pp. 96–101, ISSN 1120-4559 (2012)
18. Russo, L.: La rivoluzione dimenticata. Il pensiero scientifico greco e la scienza moderna (reprint), Feltrinelli, Milano (2003). ISBN 88-07-81644
19. Penta, F., Rossi, C., Savino, S.: Mechanical behavior of the imperial carroballista. Mech. Mach. Theory **80**, 142–150 (2014). ISSN 0094-114X
20. Rossi, C.: Some examples of the hellenistic surprising knowledge: its possible origin from the east and its influence on later Arab and European engineers. Rivista Storica dell'Antichità **XLIV**, 61–84 (2014)
21. Scott, P.B.: La Rivoluzione Robotica. IAR Franco Muzzio Editore (1987). ISBN 88-7021-290-4

22. Chondros, T.G.: Archimedes life works and machines. Mech. Mach. Theory **45**(11), 1766–1775 (2010)
23. Dimarogonas, A.D.: Machine Design: A CAD Approach. Wiley, New York (Introduction—The machine: a historical design, 4–19) (2001)
24. Rossi, C., Unich, A.: A study on possible Archimede's Cannon. Rivista Storica dell'Antichità, Vol. XLIII (2013). ISSN 0300-340X
25. Chondros, T.G.: The development of machine design as a science from classical times to modern era. HMM 2008 International Symposium on History of Machines and Mechanisms, 11–14 Nov, Tainan, Taiwan. Springer, NL (2008). ISBN 987-1-4020-9484-2
26. Rossi, C.: On designs by Ancient Engineers. J. Mech. Des. **135**, 1–2 (2013). doi:10.1115/1.4024362, ISSN 1050-0472
27. Rossi, C.: Ancient throwing machines: a method to compute their performances. Mech. Mach. Theory **51**, 1–13 (2012). ISSN 0094-114X
28. Schmidt, W., Nix, L., Schöne, H., Heiberg, J.L. (eds.): Heron of Alexandria: "Perì Automatopoietikès" Heronis opera quae supersunt omnia, Leipzig (1899)
29. Mays, L.: Ancient Water Technologies. Springer, Berlin (2010). doi:10.1007/978-90-481-8632-7, ISBN 978-90-481-8631-0, e-ISBN 978-90-481-8632-7

Part II
Grasping Analysis, Dexterous Grippers and Component Design

Part II

[illegible]

Gripping Analysis of an Underactuated Finger

Francesco Penta, Cesare Rossi and Sergio Savino

Abstract The paper presents a theoretical study about the gripping of an underactuated tendon driver finger. The studies presented in the paper have been performed to analyse the capability of the proposed device to apply forces on the grasped object and to analyse the behaviour of the contact actions during the grasping. The analysis of two models is reported and the results of simulations are shown. The obtained results show a good efficiency of the designed finger.

Keywords Gripping analysis · Underactuated mechanism · Robotic finger

1 Introduction

A large number of existing grasping algorithms have been proposed in literature and many studies have been made to optimize the design of the fingers of a gripping member [1–3].

Simulation models with finite element and analysis systems were developed to evaluate the complex contact types of the human hand grasp [4]; kinematics and dynamics models were studied to calculate the force applied by fingers on points of contact in order to have stable grasp at reduced grasping complexity [5].

This contribution proposes a grasping analysis of an underactuated finger which is fitted (with four similar ones) in a mechanical underactuated hand that is patent pending. The latter is mainly conceived as hand prosthesis and can also be used as a

F. Penta · C. Rossi · S. Savino (✉)
Department of Industrial Engineering, University of Naples "Federico II",
Via Claudio 21, 80125 Naples, Italy
e-mail: sergio.savino@fastwebnet.it

F. Penta
e-mail: francesco.penta@unina.it

C. Rossi
e-mail: cesare.rossi@unina.it

T. Borangiu (ed.), *Advances in Robot Design and Intelligent Control*,
Advances in Intelligent Systems and Computing 371,
DOI 10.1007/978-3-319-21290-6_7

robot grasping device. A description of the hand was given in [6–12] and essentially is based on an adaptive scheme.

The main aim of the study is to propose a technique based on the Lagrangian approach in an energetic study, to analyse the finger gripping and to evaluate the suitability of the finger as a part of the whole hand, and to optimize its design.

2 Finger Grasping by Multibody Code

The grasping was analysed by a multibody code. Figure 1 depicts a model made up by multibody software: the model essentially consists of three rigid bodies representing the phalanxes, linked each to the other and to a fixed body by hinges that include some damping. The tendons are modelled by two pulley systems that are connected one to a linear actuator and the other to a spring. The elongated (up) and the fully rotated (below) finger are also shown in Fig. 1. To each of the phalanxes a constant force is applied, the direction of which is always orthogonal to the phalanxes' longitudinal axe. The force can act at the starting of the movement or when any phalanx reaches a given rotation; this simulates the grasping of an object. Figure 1 also shows the plot of the rotation of each phalanx and the force exerted by the actuator.

In Figs. 2 and 3 the forces exerted by the actuator are plotted versus the first phalanx rotation. In Fig. 2 the traction and contrast forces are equal and were applied from the start of the movement.

In Fig. 3 the forces to each phalanx are applied when the rotation of the proximal phalanx reaches 0.3 radians for F1, 0.6 radians for F2 and 1.4 radians for F3 (Fig. 1)

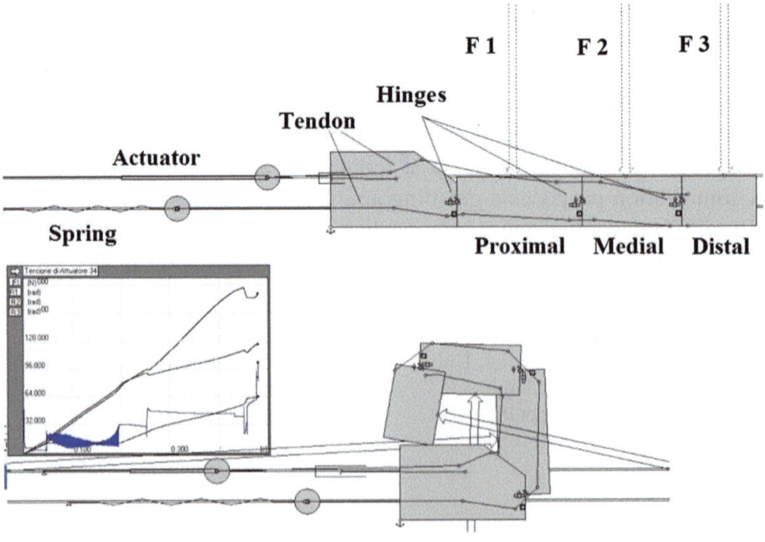

Fig. 1 Finger model by multibody software, with applied forces on the phalanges

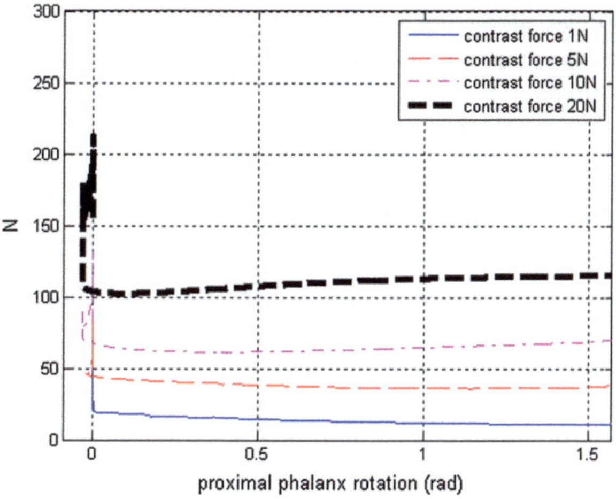

Fig. 2 Traction force with constant contrast force in the middle of phalanxes

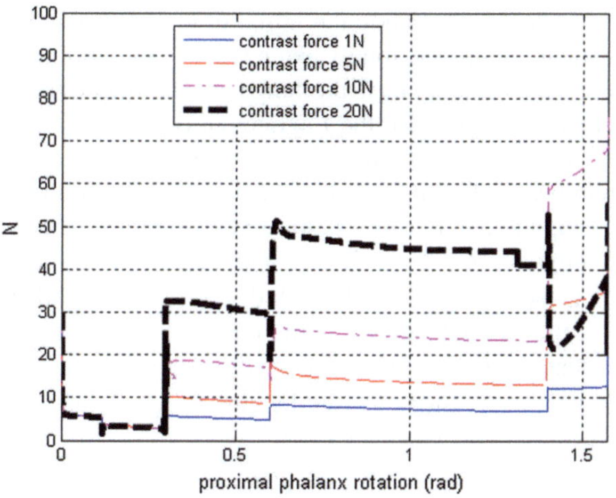

Fig. 3 Traction force when contrast forces are applied in sequence in the middle of phalanges

3 Mechanical Model of Finger Grasping

The finger studied in the present paper is a plane system having only three degrees of freedom. It is composed of four rigid links connected together by means of three friction-free hinges. Of these four links, one is fixed and acts as metacarpal phalanx; the remaining three are movable and function as proximal, medial and distal phalanxes.

The finger is equipped with a flexural and an extensor tendon, both having an end fixed to the distal phalanx. The extensor tendon is also elastically constrained to the metacarpal phalanx by a spring of stiffness equal to k, while to the free end of the flexural tendon an actuator applies the traction force causing the finger deformation. All the phalanxes, furthermore, are provided with a couple of cylindrical guides along which the tendons can freely slide when the finger deforms. We assume that both tendons are inextensible, free of thickness and have zero bending stiffness.

3.1 Mechanical Model

As Lagrangian coordinates of the finger we choose the absolute rotation φ_1 of its proximal phalanx, the relative rotation φ_2 between the proximal and the medial phalanx and finally the relative rotation φ_3 between the medial and the distal phalanx (see Fig. 4b). Hence, a finger configuration is univocally defined by the vector column

$$\boldsymbol{\varphi} = [\varphi_1, \varphi_2, \varphi_3]^T. \tag{1}$$

In the present study the initial or reference configuration of the finger is the one with all the fingers totally extended (see Fig. 4a), that is with all the rotations $\varphi_i = 0$. For sake of simplicity, the hinges of the finger are numbered consistently with the notation adopted for the Lagrangian coordinates: i will denote the hinge where the rotation φ_i occurs. Moreover, a movable phalanx is also identified by the same number of the hinge that is on its right side in the reference configuration.

Figure 5a, b show respectively enlarged views of the initial and deformed configurations of the flexure and extensor tendons close to the hinge i.

Denoting respectively with $\overline{d_i}$ and $\overline{s_i}$ the fixed distances from the centre of hinge i to the points P_i and Q_i in the initial configuration, and with d_i and s_i and the same distances in the deformed configuration, then from the Carnot theorem the following expressions for the lengths $\overline{P_iQ_i} = z_i$ and $\overline{R_iS_i} = 1_i$ are obtained:

$$z_i = \sqrt{d_i^2 + s_i^2 - 2 \cdot d_i s_i \cos(\Phi_i - \varphi_i)} \tag{2}$$

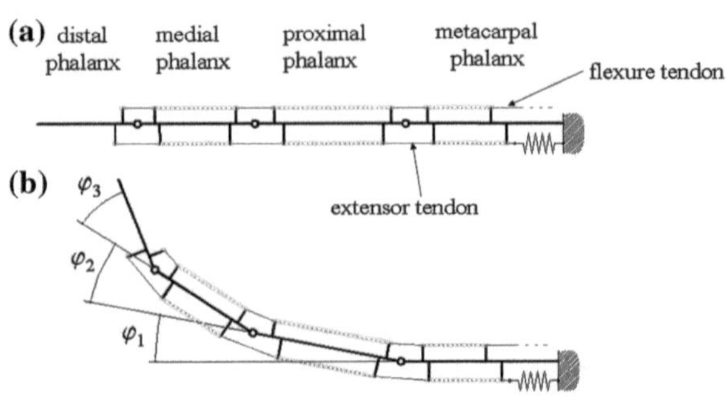

Fig. 4 Reference or initial configuration (**a**) and deformed configuration (**b**) of the finger

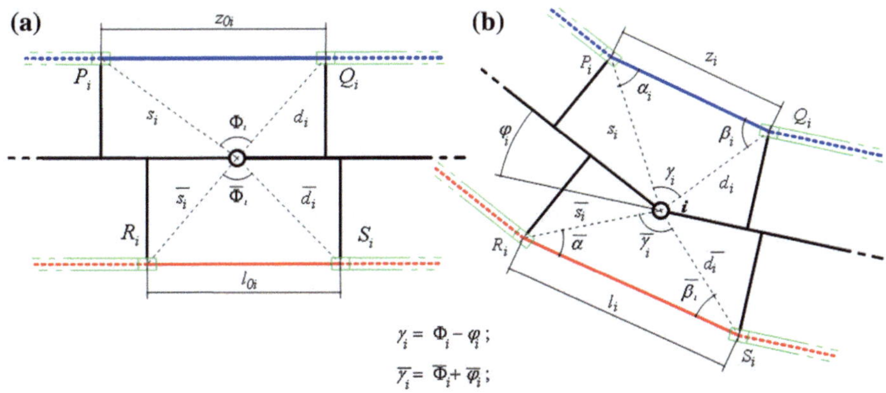

Fig. 5 Initial (**a**) and deformed (**b**) configurations of the flexure and extensor tendons

$$l_i = \sqrt{\bar{d}_i^2 + \bar{s}_i^2 - 2 \cdot \bar{d}_i \bar{s}_i \cos(\bar{\Phi}_i + \varphi_i)}, \tag{3}$$

In the equation above, the angles Φ_i and $\overline{\Phi_i}$ are defined in Fig. 5a.

When the finger reaches the configuration $\boldsymbol{\varphi}$ starting from its reference configuration, the flexure tendon free paths $\overline{P_iQ_i}$ experience the shortenings

$$\Delta z_i = z_{0i} - z_i, \quad (i = 1, 2, 3) \tag{4}$$

with z_{0i} the initial value of z_i (Fig. 5). As a consequence, the free end of the flexure tendon performs the displacement [considering Eq. (4)]:

$$u^{(f)} = \sum_{i=1}^{3} \Delta z_i^{(f)}. \tag{5}$$

Similarly, the free paths $\overline{R_iS_i}$ of the extensor tendon experience the length changes

$$\Delta l_i = l_i - l_{0i}, \; (i = 1, 2, 3) \tag{6}$$

with l_{0i} initial value of l_i, and the corresponding elongation of the spring constraining this tendon is:

$$\Delta l = \sum_{i=1}^{3} \Delta l_i. \tag{7}$$

We have analysed the finger behaviour during the grasping of a cylindrical object having the axis normally oriented to the finger plane. Under the assumption of negligibility of inertial effects and of frictionless contact between fingers and the lateral surface of the object, the equilibrium finger configurations $\boldsymbol{\varphi}$ have to satisfy both the stationary condition of the total potential energy Π,

$$d\prod = d\left(\frac{1}{2}k\Delta l^2 - \lambda \cdot u\right) = \sum_{i=1}^{3}\left(k\Delta l\frac{dl_i}{d\varphi_i} + \lambda \cdot \frac{dz_i}{d\varphi_i}\right) \cdot d\varphi_i, \quad \forall d\varphi_i \tag{8}$$

and the following additional constraint conditions for those phalanxes that are in contact with the object:

$$\eta_k - R = 0 \quad \text{with} \quad k \in C, \tag{9}$$

where η_k is the distance of the phalanx k from the cylinder axis, R the cylinder radius and C the set of k index values denoting phalanxes in contact with the object. This equilibrium problem is equivalent to searching for the stationary condition of the function:

$$\overline{\prod}(\boldsymbol{\varphi}) = \prod(\boldsymbol{\varphi}) - \sum_{k \in C} \rho_k(\eta_k - R), \tag{10}$$

where ρ_k are unknown Lagrange multipliers having the meaning of contact reactions exerted by the object on the phalanx k, see Fig. 6.

The finger equilibrium equations that are obtained by equating to zero partial derivative of the function $\overline{\prod}$ and the constraint equations (10) are non-linear in the rotations φ_i and should be solved numerically by a *small step incremental approach*: starting from the finger reference configuration, the actuator load λ is gradually increased by small steps $\Delta\lambda$. In each step, checks for contact detection are carried out in order to update the set C of phalanxes in contact; then equilibrium equations and constraint conditions are solved by the Newton-Raphson method assuming as first guess for the solution the equilibrium configuration attained at the previous load step.

Some numerical results regarding the finger behaviour during the free flexion and the grasping operation are synthetically presented in this section. These numerical simulations have been carried out with an increment $\Delta\lambda$ of the actuator load equal to 10^{-2} N. This value has been determined by trial and error and has

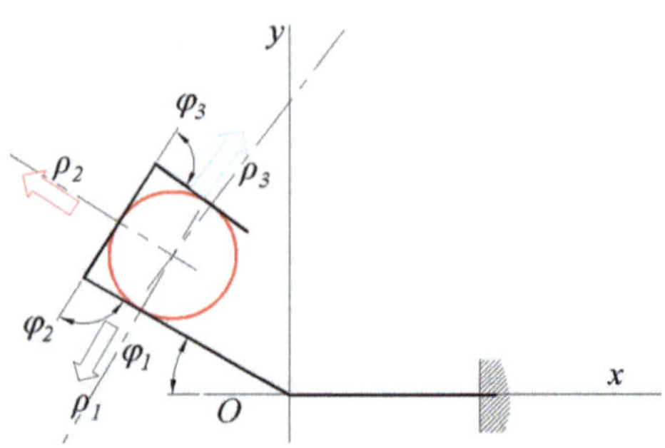

Fig. 6 Final deformed configuration of the finger

proven to be suitable to generate very small contact penetrations in each calculation step and to allow the Newton-Raphson algorithm to converge in very little iterations.

Figure 7a, b show respectively the diagrams of the absolute rotations α_i of the phalanxes and of the flexural tendon end displacement as function of the actuator force λ. Inspection of these figures indicates that during both the grasping operation of the object (black line) and the finger free flexion (coloured lines) S-shaped configurations are not assumed and that snapping and buckling phenomena do not occur.

An example of the capability of the proposed device to apply forces on the object through contacts is confirmed by the results in Fig. 8 where the predicted values of the contact forces ρ_i are diagrammed as function of the actuator force. When all phalanxes are in contact with the object, contact actions can be very easily controlled since a linear relationship is established among forces ρ_i and the flexural tendon traction λ.

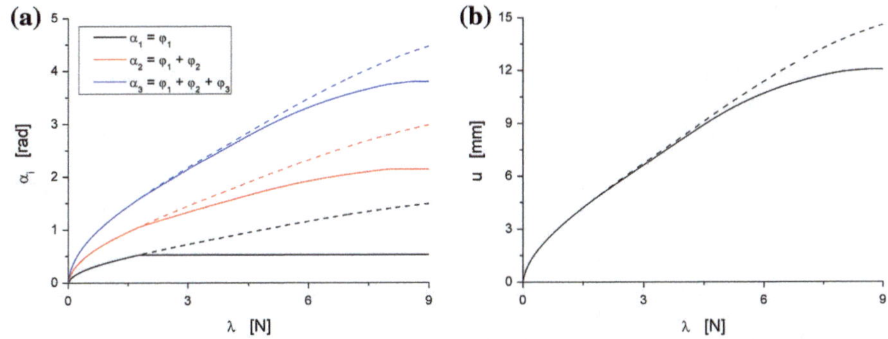

Fig. 7 **a** Absolute rotations of the phalanxes versus finger actuator force; **b** Displacement of the free end of the flexural tendon as function of finger actuator force

Fig. 8 Contact reactions versus finger actuator force

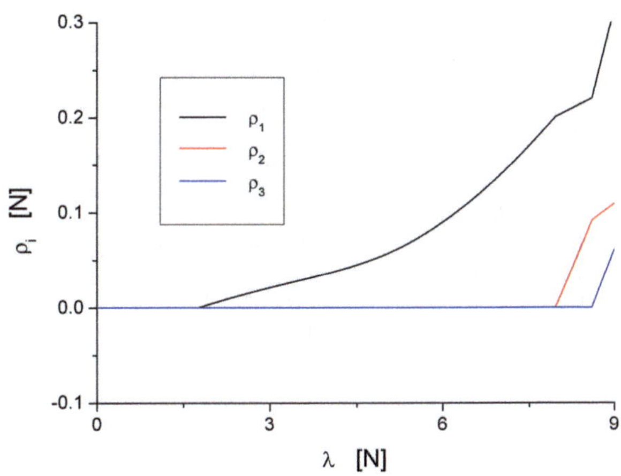

4 Conclusion

Models were developed to carry on the grasping analysis of an underactuated finger. The presented analyses show the suitability of such a finger to be used in hand prosthesis or in robot grasping end-effector. This is essentially because of the following aspects:

1. The finger shows a good self-adaptability to the object shape
2. The tendon system permits suitable grasping forces in the contact between the phalanxes' inner surface and the object's surface.
3. The forces above are obtained with relatively small traction forces of the tendon.

Further developments are in progress in order to optimize the tendons positions and the extension spring stiffness.

References

1. Ceccarelli, M.: Fundamentals of Mechanics of Robotic Manipulation. Kluwer Academic Publisher, Dordrecht (2004)
2. Zheng, Y., Qian, W.H.: Limiting and minimizing the contact forces in multifingered grasping. Mech. Mach. Theory **41**, 1247–1257 (2006)
3. Lanni, C., Ceccarelli, M.: An optimum design algorithm for mechanisms in two-finger grippers. In: Proceedings of the 13th WSEAS International Conference on SYSTEMS, pp. 63–70, Stevens Point, Wisconsin, USA (2009)
4. Ciocarlie, M., Miller, A., Allen, P.: Grasp analysis using deformable fingers. In: Proceedings of IEEE/RSJ International Conference on Intelligent Robots and Systems (IROS 2005), pp. 4122–4128, 2–6 Aug 2005
5. Leela Kumari, C., Sahu, D.: Force closure analysis of robotic hand using matlab. Int. J. Adv. Res. Comput. Sci. Softw. Eng. **2**(10), 71–76 (2012)
6. Rossi, C., Savino, S., Niola, V., Troncone, S.: An underactuated mechanical hand: a first prototype. In: Proceedings of the 23rd International Conference on Robotics in Alpe-Adria-Danube Region RAAD'14, 3–5 Sept 2014, Smolenice, Slovakia. ISBN 978-80-227-4219-1/IEEE Catalogue no. 34043 (2014)
7. Penta, F., Rossi, C., Savino, S.: An underactuated finger for a robotic hand. Int. J. Mech. Control **15**(2), 63–68. ISSN: 1590-8844 (2014)
8. Rossi, C., Savino, S.: An underactuated multi-finger grasping device. Int. J. Adv. Robot. Syst. doi:10.5772/57419 (2014)
9. Carbone, G. (ed.): Grasping in Robotics. Springer, Dordrecht (2013)
10. Rossi, C., Savino S., Niola V., Troncone, S.: A study of a robotic hand with tendon driven fingers. Robotica Int. J. 1–15 (2014)
11. Rossi C., Savino, S.: An underactuated multi-finger grasping device. Int. J. Adv. Robot. Syst. **11**, 20 (2014). Available on-line at http://www.intechopen
12. Carbone, G., Rossi C., Savino, S.: Performance comparison between Federica hand and LARM hand, submitted to InTech-IJARS

Dynamic Behaviour of an Underactuated Finger

Vincenzo Niola, Cesare Rossi, Sergio Savino and Pavel Potapov

Abstract The paper presents a theoretical study about the dynamic behaviour of an underactuated finger. By modelling the equation of dynamics, two kinds of fingers were studied: the first with equal phalanges, and the second with three different phalanges whose parameters are similar to those of a human finger. The results obtained for this last type of configuration are presented in the paper. The studies reported in the paper have been realized to study the dynamic behaviour of the underactuated finger; the main scope was to investigate the geometrical parameters which allow obtaining a more suitable closing sequence of the phalanges in order to better grasp objects.

Keywords Dynamics · Underactuated mechanism · Robotic finger

1 Introduction

Many mechanical systems used for grasping devices have been conceived and studied in the last decades. For grasping devices that use fingers, if the system's dimensions are comparable to those of a human hand, it is rather difficult to operate each phalanx of the finger by a (micro) motor. Underactuated finger mechanism is beneficial in the anthropomorphic applications in which it reduces the finger size,

V. Niola · C. Rossi (✉) · S. Savino
Department of Industrial Engineering, University of Naples "Federico II",
Via Claudio 21, 80125 Naples, Italy
e-mail: cesare.rossi@unina.it

V. Niola
e-mail: vincenzo.niola@unina.it

S. Savino
e-mail: sergio.savino@fastwebnet.it

P. Potapov
Moscow State Automobile and Road Technical University, Moscow, Russia

T. Borangiu (ed.), *Advances in Robot Design and Intelligent Control*,
Advances in Intelligent Systems and Computing 371,
DOI 10.1007/978-3-319-21290-6_8

weight and power consumption. For this reason, many studies were carried out on the underactuated systems development, and in particular systems moved with tendons.

Most of the previous research on underactuated mechanism is limited to grasping capacity, [1–3]. Newer studies propose mechanisms of the underactuated hand to perform both grasping and pinching operation, so that the finger can also pinch with the fingertip to hold an object, [4].

The present study is a part of investigations of an underactuated mechanical hand (patent pending) that can be used as human prosthesis or as a grasping device in general [5–8]. In order to optimize the hand performances, in this investigation a method to compute the dynamic behaviour of a single finger is presented. A computing example is also reported.

2 Dynamic Analysis

Considering the structure of the finger, the dynamic study of the mechanism will be referred to the convention of "Denavit and Hartenberg" for the arrangement of the frame systems associated with each of the phalanxes.

Figure 1 shows the adopted arrangement of the frames; the subscripts "M", "p", "m", "d" identify the four sets of frames, respectively integral to the metacarpal, the proximal phalanx, the medial phalanx and the distal phalanx.

By means of the Denavit and Hartenberg convention it is possible to identify the homogeneous transformation matrices that describe the relationships between the different parts of the mechanical structure that are in relative motion, [9].

In particular for two phalanges $i-1$ and i, the matrix $[^{i-1}A_i]$ defines the transformation between two frames $i-1$ and i integral with the phalanges [10]. In this way the relationships between the distal phalanx and the metacarpal is described by means of the following matrix:

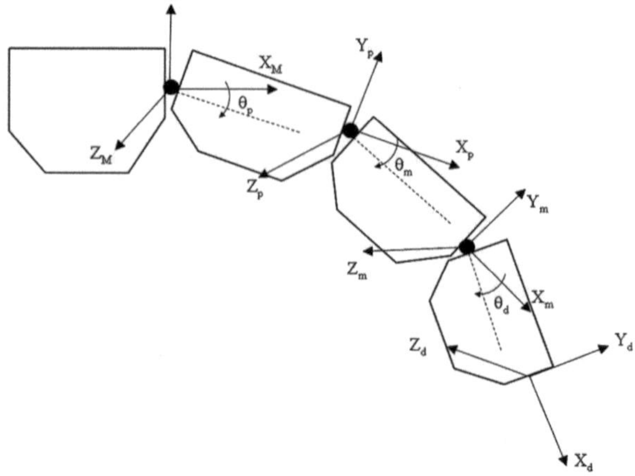

Fig. 1 "Denavit and Hartenberg" representation of the finger model

$$\left[{}^{M}A_{d}\right] = \left[{}^{M}A_{p}\right] \cdot \left[{}^{p}A_{m}\right] \cdot \left[{}^{m}A_{d}\right] \tag{1}$$

3 Dynamic Equilibrium of the Phalanx

Figure 2 shows the generic phalanx; in order to analyse its dynamic equilibrium, one must consider the following actions [11, 12]:

$[\Phi_i^{(e)}]$ actions of external forces applied to the phalanx i including the forces of inertia and the gravitational force;

$[\Phi_i]$ actions that the phalanx i exerts on the phalanx $i-1$, including reaction forces

$[\Phi_{i+1}]$ actions that the phalanx $i+1$ exerts on the phalanx i;

The dynamic equilibrium is expressed by the following relationship:

$$-[\Phi_i] + \left[\Phi_i^{(e)}\right] + [\Phi_{i+1}] = [0] \Leftrightarrow [\Phi_i] = \left[\Phi_i^{(e)}\right] + [\Phi_{i+1}] \tag{2}$$

In particular the action $[\Phi_i^{(e)}]$ can be expressed as:

$$\left[\Phi_i^{(e)}\right] = -\left[{}^{0}W_i\right][Y_i] + [Y_i]\left[{}^{0}W_i\right]^T + \left[W_g\right][Y_i] - [Y_i]\left[W_g\right]^T + [\Phi^*] \tag{3}$$

In the previous relation:

- $[{}^{0}W_i]$ is the absolute acceleration matrix of the phalanx i with respect to an inertial frame (0), projected on the frame that is connected to the phalanx i;
- $[Y_i]$ is an inertia matrix (pseudo-tensor of inertia) that takes into account the mass distribution of the phalanx with respect to the frame system i, integral to the phalanx;
- $[W_g]$ is the matrix of the gravity acceleration;
- $[\Phi^*]$ is the matrix that takes into account other external forces.

Equation (2) can be written in the form:

$$[\Phi_i] = \left[W_g\right][Y_i] - [Y_i]\left[W_g\right]^T - \left[{}^{0}W_i\right][Y_i] + [Y_i]\left[{}^{0}W_i\right]^T + [\Phi_{i+1}] + [\Phi^*] \tag{4}$$

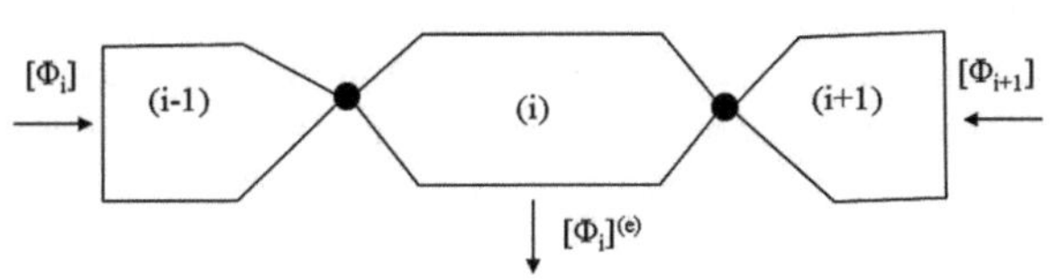

Fig. 2 Simplified diagram of the finger

By means of (4) it is possible to determine the generic action exchanged between two contiguous phalanges.

To determine the reactions at the joints, the actions between the last link and the one immediately preceding it will be calculated first, and then we will proceed back to the base. For this reason, the actions exchanged at a generic joint *i* shall be equal to the sum of the actions exchanged in the joints that follow it, according to the Eq. (5):

$$[\Phi_i]_0 = \sum_{k=i}^{n} \left([W_g]_0 [Y_k]_0 - [Y_k]_0 [W_g]_0^T - [{}^0W_k]_0 [Y_k]_0 + [Y_k]_0 [{}^0W_k]_0^T + [\Phi_k^*]_0 \right) \quad (5)$$

where the subscripts "*i*" identifies the generic link, the subscript "0" identifies an inertial system, while n is the total number of links in the mechanical structure.

In the case of the finger, denoting by the subscripts "d", "m" and "p" the matrix of actions relating respectively to the distal phalanges, medial and proximal, and denoting by the subscript "M" the inertial metacarpal system, it is possible to write the following system of equations:

$$\begin{cases} [\Phi_d]_M = \left[\Phi_d^{(e)}\right]_M \\ [\Phi_m]_M = \left[\Phi_m^{(e)}\right]_M + [\Phi_d]_M \\ [\Phi_p]_M = \left[\Phi_p^{(e)}\right]_M + [\Phi_m]_M \end{cases} \quad (6)$$

In the system (6) the terms in the first member of each equation, $[\Phi_i]$ with i = {d, m, p}, represent the matrix of actions that each phalanx exerts on the previous.

The terms $[\Phi_i^{(e)}]$ with i = {d, m, p}, represent the actions of external forces applied to the phalanx (i), and it can be expressed as:

$$\left[\Phi_i^{(e)}\right]_M = -[{}^MW_i]_M [Y_i]_M + [Y_i]_M [{}^MW_i]_M^T + [W_g]_M [Y_i]_M - [Y_i]_M [W_g]_M^T + [\Phi_i^*]_M \quad (7)$$

The generic matrix $[\Phi_i^*]_M$ contains the other actions which act on the phalanx *i* in the inertial system integral with the metacarpal (M); this matrix depends on the analogous matrix, $[\Phi_i^*]_{i-1}$, evaluated in the frame that is integral with the phalanx $i - 1$ by means of the relation:

$$[\Phi_i^*]_M = [{}^MA_{i-1}] \cdot [\Phi_i^*]_{i-1} \cdot [{}^MA_{i-1}]^T \quad (8)$$

The matrix $[\Phi_i^*]_{i-1}$ can be evaluated considering all the actions operating on the phalanx *i*. In particular, in our model, the actions that produce moments with respect to axes X and Y of the frame are neglected, and the main considered actions are:

- Action of the traction tendon;
- Action of the antagonist tendon;

- Action generated by forces applied on the face of the phalanx from the side of the palm;
- Action generated by the presence of mechanical stops for the motions of the phalanx.

4 Direct Dynamics

In a serial kinematic chain, the principle of virtual works states that the virtual work at each joint *i* must be zero for any virtual variation of the variable of the joint.

The virtual work for the generic joint can be evaluated by means of the following relation:

$$L_i = \left(\varphi_i + [\Phi_i] \otimes [L]_i\right) \cdot \delta\theta_i \qquad (9)$$

where:

- L_i = virtual work at the joint (i)
- $\delta\theta_i$ = virtual variation of the variable of joint θ_i
- $[\Phi_i]$ = matrix of actions;
- φ_i = action of the actuator *i* of the joint *i*;
- $[L]_i$ = matrix that describes the coupling of the joint *i*;
- $\otimes$ = the operator that indicates the pseudo scalar product.

So, to solve the direct dynamic problem, it is possible to consider the following relation:

$$\varphi_i + [\Phi_i] \otimes [L]_i = 0 \qquad (10)$$

In the case of the finger that is actuated by means of a traction tendon, there are not actuators for the three joints and the relation (13) becomes:

$$[\Phi_i] \otimes [L]_i = 0 \qquad (11)$$

In the case of the finger for each joint the matrix of the coupling has the form:

$$[L]_i = \begin{bmatrix} 0 & 1 & 0 & 0 \\ 1 & 0 & 0 & 0 \\ 0 & 0 & 0 & 0 \\ 0 & 0 & 0 & 0 \end{bmatrix}$$

So, in the case of direct dynamic of the finger, the relation (14) represents a system of three equations:

$$\begin{cases} [\Phi_d]_M(2,1)\cdot = 0 \\ [\Phi_m]_M(2,1)\cdot = 0 \\ [\Phi_p]_M(2,1)\cdot = 0 \end{cases} \tag{12}$$

By solving the system (12) with unknowns the angular accelerations of the three phalanges, it is possible to obtain a system of three differential equations that describe the motion of the finger, if all its geometric and inertial parameters, as well as the forces acting on it, are assigned.

$$\begin{cases} \ddot{\theta}_d(\dot{\theta}_d, \dot{\theta}_m, \dot{\theta}_p, \theta_d, \theta_m, \theta_p) = 0 \\ \ddot{\theta}_m(\dot{\theta}_d, \dot{\theta}_m, \dot{\theta}_p, \theta_d, \theta_m, \theta_p) = 0 \\ \ddot{\theta}_p(\dot{\theta}_d, \dot{\theta}_m, \dot{\theta}_p, \theta_d, \theta_m, \theta_p) = 0 \end{cases} \tag{13}$$

The system (13) represents a solution of the direct dynamic problem of the finger.

5 Example: Finger Having Geometrical Dimensions of the Human One

The dimensional parameters of the phalanges are crucial in the dynamic behaviour of the finger. In the following, an analysis of a model of finger having three different phalanges, whose parameters are similar to those of a human finger is presented (Table 1).

The elastic constant, the damping constant and the preload of the spring of the antagonist tendon were estimated with a simulation under the influence only of gravity. In these conditions, the values able to keep the finger extended were considered sufficient, and they were adopted for the successive simulations. These values were kept constant in following simulations. In particular an elastic constant of 31 N/m, a damping of 0.004 N/(rad/s) and a preload of 0.31 N were adopted.

5.1 Force Necessary for Closing the Finger

In Fig. 3a the phalanges behaviour is shown, with a traction force of 7.5 N.

Table 1 Parameters of the model

Phalanx	Mass (kg)	Length (m)—distance between the hinges
Proximal	0.01	0.045
Medial	0.0055	0.030
Distal	0.0030	0.025

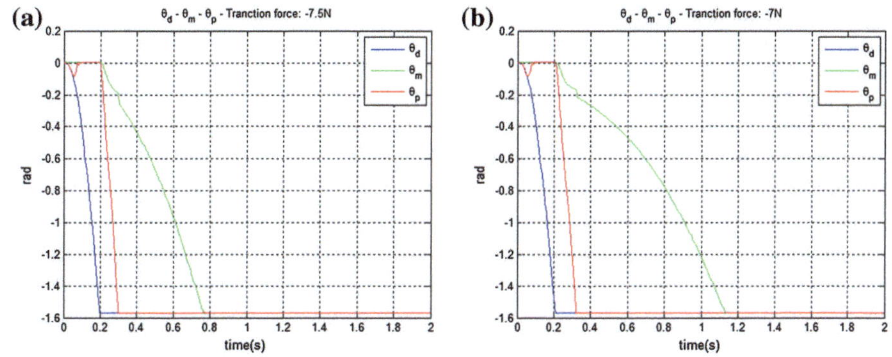

Fig. 3 Phalanges behaviour with traction force of: **a** 7.5 N; **b** 7 N

It can be observed that also for this model the closing sequence is d-p-m, while with the same traction force of the model with equal phalanges in this model the closing sequence occurs before.

In Fig. 3b the traction force is of 7 N, and this value is sufficient to complete the closing of the finger.

5.2 Closing Sequence of the Finger

If the sequence of closing of the finger is d-p-m, like shown in the Fig. 3c, the gripping of the object can be inadequate.

By means of the dynamic model, it was possible to study how to obtain a different sequence, in particular the possibility of obtaining a closing sequence of the phalanges p-m-d (proximal-medial-distal) was analysed.

To modify the dynamic sequence of closing the phalanges finger, it is necessary to modify the torques that act on the phalanges. This was made by keeping constant the traction force and by varying the geometrical position of the start and end of the guide of the traction tendon. In this way the torques acting on each phalanx can be varied with the same traction force and the closing sequence varies.

With a sequence of closing of the phalanges p-m-d, the torque on proximal and medial phalanges decreases when the distal phalanx exceeds a certain closing angle, so the proximal and the medial phalanges return to their initial configuration until the distal phalanx reaches its limit.

To prevent the occurrence of instability, the limit of closure of the distal phalanx was set. Figure 4 shows the dynamic behaviour of the finger when the distances along the Y-axis of the guides of all the phalanxes are decreased to 0.0015 m and with a limit of closure of distal phalanx of −45°.

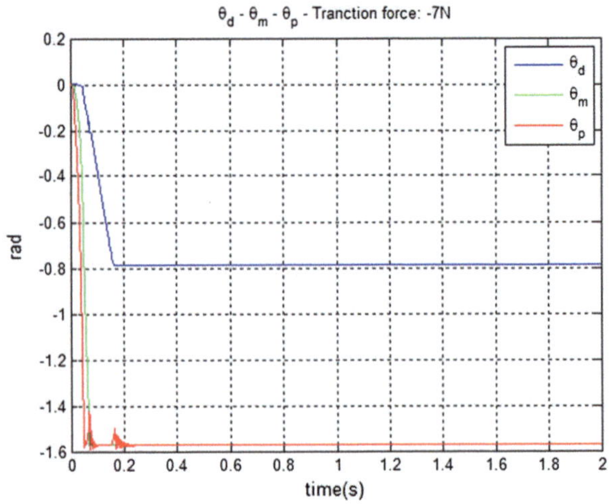

Fig. 4 Simulation with equal distances of 0.0015 m from the hinges along the Y-axis of the start and the end of the guide of each phalanx, and with a limit of the distal phalanx of -45°

6 Conclusion

A method to compute the dynamic behaviour of a single finger was presented; the method is based on the use of the matrixes of actions.

A computing example was also presented. The latter shows both the force necessary for closing the finger and a more suitable finger closing sequence to grasp objects. Both these aspects are very important to optimize the finger itself as far as the whole system parameters are concerned. In particular, the position of the tendon guides has a significant effect on the dynamic behaviour of the finger. The results also show that if only the distance from the hinges along the Y axis, of the start and end of the guide of the tendon are changed, the closing sequence of the finger changes. This information on the dynamic behaviour of the finger is very important to get a proper closure of the finger in the gripping operations.

These aspects, naturally, cannot be obtained by a mere kinematic investigation.

Further investigations are in progress in order to optimize the parameters above.

References

1. Laliberte, T., Gosselin, C.M.: Underactuation in space robotic hands. In: Proceeding of the 6th International Symposium on Artificial Intelligence and Robotics and Automation in Space, Canadian Space Agency, pp. 1–8. St-Hubert, Quebec, Canada, 18–22 June 2001
2. Norsinnira, Z.A. Yamaura, H.: Underactuated anthropomorphic finger mechanism for grasping and pinching with optimized parameter. J. Comput. Sci. **6**(8), 928–933. ISSN 1549-3636 (2010)
3. Ceccarelli, M., Tavolieri, C.: Design considerations for underactuated grasp with a one-d.o.f. Anthropomorphic finger mechanism. In: Proceedings of the 2006 IEEE/RSJ International Conference on Intelligent Robots and Systems. Beijing, China, 9–15 Oct 2006

4. Zollo, L., Rocecella, S., Tucci, R., Siciliano, B., Guglielmelli, E.: Biomechatronic design and control of an anthropomorphic artificial hand for prosthetic and robotic applications. IEEE/ASME Trans. Mechatron. **12**, 418–429 (2007)
5. Rossi, C., Savino, S., Niola, V., Troncone, S.: A study of a robotic hand with tendon driven fingers. Robotica, Article in Press. doi:10.1017/S0263574714001179, ISSN: 02635747 (2014)
6. Rossi, C., Savino, S.: An underactuated multi-finger grasping device. Int. J. Adv. Robot. Syst. **11**(20). doi:10.5772/57419 (2014)
7. Niola, V., Rossi, C. Savino, S.: A new mechanical hand: theoretical studies and first prototyping. Int. Rev. Mech. Eng. **8**(5), 835–844. ISSN: 19708734 (2014)
8. Niola, V., Rossi, C, Savino, S. Troncone, S.: An underactuated mechanical hand: a first prototype. In: Proceedings. of 23th International Workshop on Robotics in Alpe-Adria-Danube Region. Smolenice, Slovakia, 3–5 Sept 2014
9. Brown C.Y., Asada, H.H.: Inter-finger coordination and postural synergies in robot hands via mechanical implementation of principal components analysis. In: Proceedings of the 2007 IEEE/RSJ International Conference on Intelligent Robots and Systems. San Diego (CA, USA), Oct 29–Nov 2 2007
10. Carbone, G. (ed.): Grasping in Robotics. Springer, Berlin (2013)
11. Fu, K.S., Gonzalez, R.C., Lee, C.S.G. (eds.): Robotica. McGraw-Hill. ISBN 88-386-0617-X (1989)
12. Legnani, G (ed.): Robotica Industriale. CEA. ISBN: 9788808086310 (2003)

Swivel Walker with Electromotor Module and Designation of Stability

Mikuláš Hajduk and Jozef Varga

Abstract The aim of this paper is to propose a swivel walker design with electromotor module supporting the movement of handicapped people and determining the limit positions of the equipment. In the first part the paper describes the functional principle and design issues of the equipment. Initially, the design of the equipment was full mechanical, such devices being still used. The movement with mechanical swivel walker is physically demanding and for this reason we proposed a swivel walker design with electromotor modules to facilitate the movement of handicapped people. In its second part, the paper describes the stability principle of two legged equipment and determines the height CoG position limit to ensure swivel walker stability.

Keywords Swivel walker · Centre of gravity (CoG) · Electromotor module · Stability

1 Introduction

The first attempt to build up mechanical swivel walker was an experiment for children with amputated lower limbs. Initial trials of this equipment were made at the Californian University in 1963. This model consists of a base platform, pylon and a couple of rocking plates (foots). The experiment was not so successful and did not bring satisfactory progress. Child expended too much physical energy and the movement was very slow. Similar systems with improved mechanical structure were developed and are used still now; however, for people with higher degree of

M. Hajduk · J. Varga (✉)
Faculty of Mechanical Engineering, Department of Robotics,
Technical University of Košice, Košice, Slovakia
e-mail: jozef.varga.2@tuke.sk

M. Hajduk
e-mail: mikulas.hajduk@tuke.sk

T. Borangiu (ed.), *Advances in Robot Design and Intelligent Control*,
Advances in Intelligent Systems and Computing 371,
DOI 10.1007/978-3-319-21290-6_9

disabling moving by help of this walker is very hard. This was a first impetus to reach an easier movement using this walker. Therefore we suggested using electromotor modules to support the movements of disabled persons.

2 Functional Principle of Swivel Walker

Figure 1 shows the first functional principle of the walker. The person has to move his weighting centre (CoG) to one rocking plate and then the second rocking plate together with the base platform are turned about the vertical axis of the first rocking plate employing only inertial force. Every rocking plate is able to rotate about the pylon's vertical axis. Rotation is stopped after the limit position of the rocking plate is reached (limit stop) in each direction, and the rocking plate is returned back to the initial position by the force of the spring [1, 2].

3 Swivel Walker with Electromotor Module

The analysis of mechanical swivel walkers identified a number of deficiencies which led to the proposal of using electromotor modules which simplify the motion support for disabled people. The equipment shown in Fig. 2 consists of a base platform, rocking plates, bushing (combined axial and radial bearings) and return springs [3].

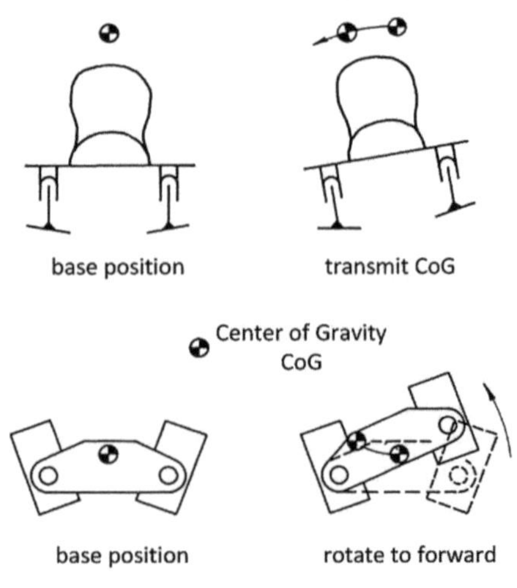

Fig. 1 First functional principle of the walker

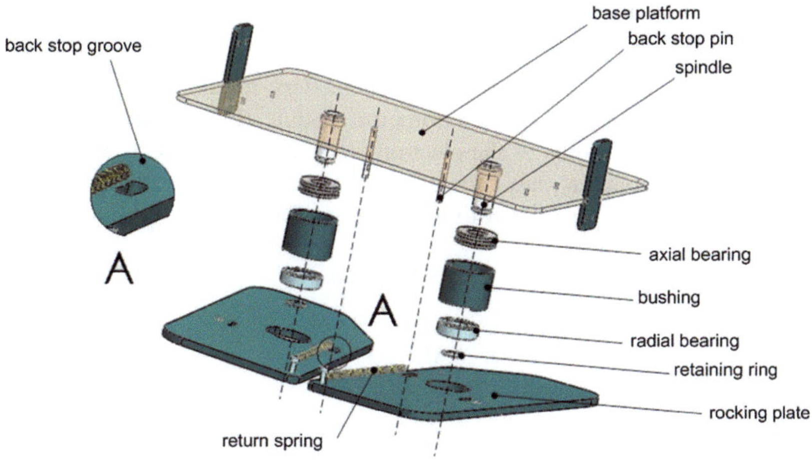

Fig. 2 Structure of the swivel walker

3.1 Design of the Electromotor Module

The design concept of the electromotor module is shown in Fig. 3. The module consists of a 24 V DC motor, a worm gear box, a motor frame, a tilting structure, a draw spring, a swivel eye joint and a stopper. The spring is drawing the stopper to the initial position. At every tilting of the structure, the stopper is returned back to the initial position and is thus prepared for the next touch with the base at a defined angle (the entire stopper's surface is in contact with the base). The module is mounted with two screws on the base platform. The electromotor module is built as right and left type of worm gear box.

Fig. 3 Design of the electromotor module

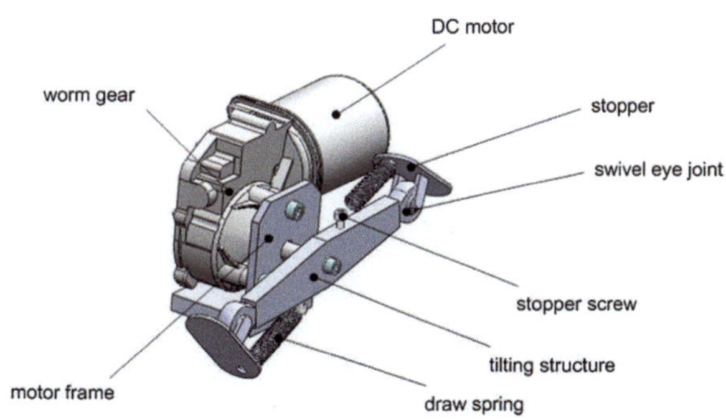

3.2 Main Parameters of Swivel Walker

The specification of the module's main parameters results from designing the swivel walker [4] equipped by electromotor module (Fig. 4). The angular displacement (α) and the radius of the electromotor module's displacement about the axis of the rocking plate r_M are known.

The step length was calculated by goniometric formulas as follows:

$$x = r_M - (r_M \cdot \cos\alpha) \tag{1}$$

$$k_M = \sqrt{(k'_M)^2 + x^2} \tag{2}$$

3.3 Tilting Simulation

In order to assure stability it is important to set up a certain value for the structure's tilting and a height level of the subject's CoG. For this reason, computer simulation in SolidWorks was done with the main concern in the path of the tilting structure.

The input parameters for simulation were the equipment's design data and the motor speed of 10 rpm. The simulation, shown in the figures below, was split into three stages.

In stage 0 shown in Fig. 5 left, the equipment is in the initial position when the tilting structure is in horizontal position at an angle γ_0. The central point of the tilting structure (arm) is at the height level z_0 above the base.

In stage 1 shown in Fig. 5 right, the stopper of the tilting structure is in contact with the base. There is a known distance between the axis of the stopper and the axis of the central point of the tilting structure in horizontal position $k'_M/2k_M'/2$,

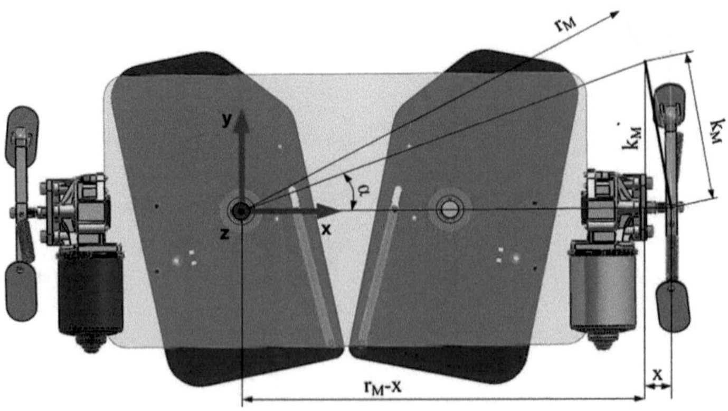

Fig. 4 Specifying the main parameters of the swivel walker from design considerations

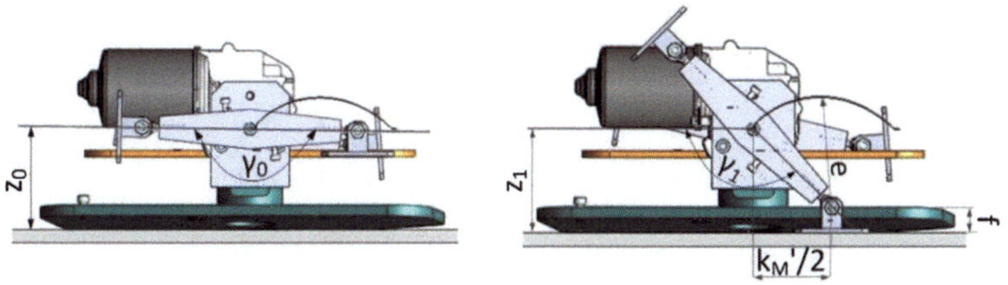

Fig. 5 Zero stage and first stage of the tilting simulation

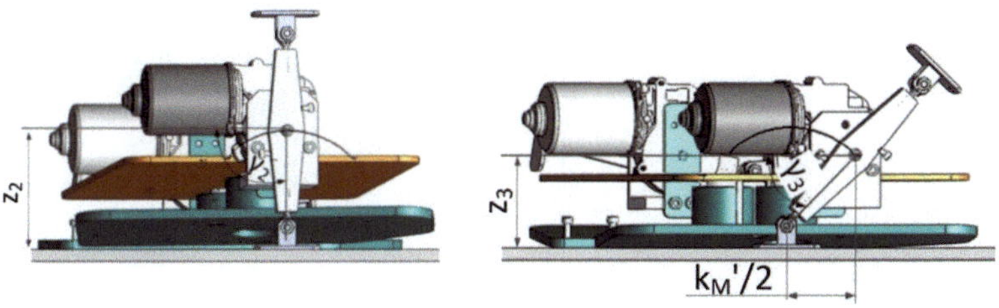

Fig. 6 Second and third stage of tilting simulation

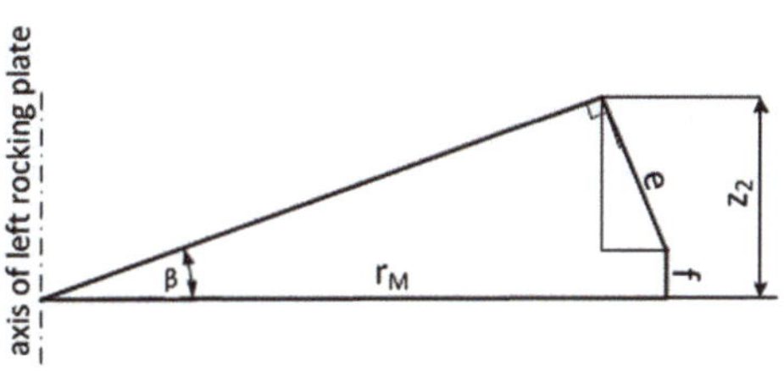

Fig. 7 Length of the tilting structure

height level of stopper f and vertical height level of center of tilting structure above base $z_1 = z_0$.

In stage 2 shown in Fig. 6, the tilting structure is in vertical position and the left rocking plate is in contact on its entire surface with the base. The centre point of the tilting structure according to the curved path is at height level z_2 above the base. In this stage the equipment is in the middle of the working cycle.

The length of the tilting structure, shown in Fig. 7, is calculated from the central point to the stopper axis where the equipment is tilted with the angle β.

Calculating the tilting structure e:

$$z_2 = r_M \times tg\beta \tag{3}$$

$$e = \frac{z_2 - f}{\cos\beta} \tag{4}$$

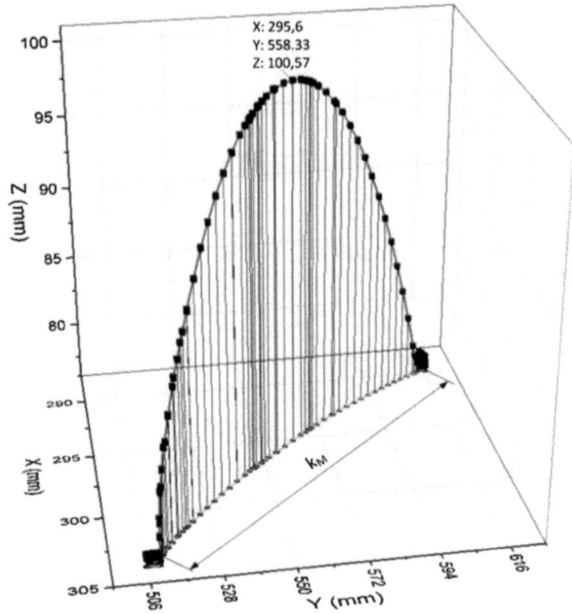

Fig. 8 Path chart of the tilting structure

In stage 3 shown in Fig. 6, the equipment is shown at the end of the tilting cycle and both rocking plates are touching with their inner edges with the base.

3.4 Graph of Tilting

The output of the simulation, shown in Fig. 8, is the path diagram of the tilting structure followed by the 3D movement of the centre point for a time t of 3 s.

4 Determination of Balanced Position

The balanced position is a relative status of the object body caused by force balancing. For this status there must be fulfilled the condition that all forces and torques applied to the object are balanced. Due to gravitation, the object must be stable in all conditions; even after tilting of the object, the CoG is still at the same height level and the axis of gravity intersects the support area [5, 6].

For bipedal structures there is a so called *double support position* [7]; the balance condition from design is such that the intersection point of the gravity axis with the support base is always near the center of the foot supporting area, see Fig. 9.

An important parameter for stability is the distance between the edge of the rocking plate and centre of the equipment y_{max}, which must be kept constant

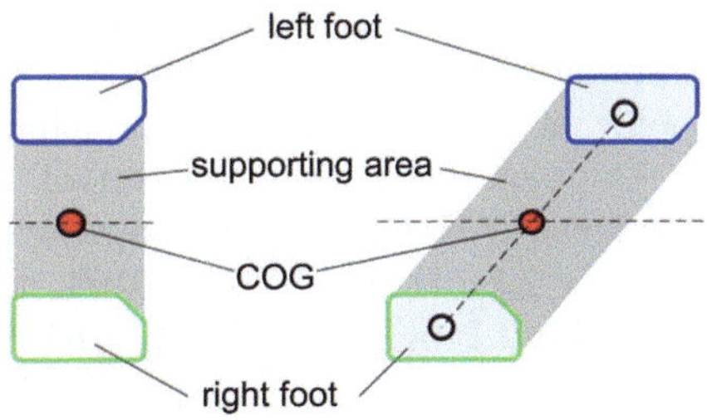

Fig. 9 Double support position

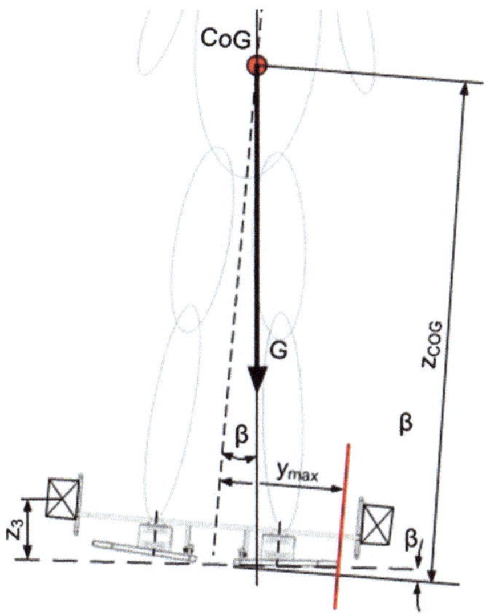

Fig. 10 Important parameters for stability

(Fig. 10). The gravity axis must not exit the rocking plate area because the system would become unstable [8].

In this way it is possible to set the maximum CoG height and consequently approximate the height of the subject for safe equipment stability:

$$z_{cog} = \frac{y_{\max}}{\tan \beta} \tag{5}$$

where y_{max} is the stability border and β is the angle of the tilted swivel walker.

5 Conclusion

This paper discusses the swivel walker design with electromotor module for assisting the motion of handicapped people. The simulation of the tilting process, split into 3 stages, was performed with SolidWorks software [9]. The results of simulation were implemented in 3D graph—the path chart of tilting structure.

Finally the CoG limit height of the subject was determined, and the condition ensuring stability of the system was established. The swivel walker with electromotor modules is designed in principal for people with high degree of disability.

This reported research is part of the project "Aplikovaný výskum systémov inteligentnej manipulácie priemyselných robotov s neorientovanými 3D objektmi (activity 2,1), ITMS: 26220220164, supported by the Research and Development Operational Program funded by ERDF.

References

1. Butler, P.B., Farmer, I.R., Poiner, R., Patrick, J.H.: Use of the Orlau swivel walker for the severely handicapped patient. Physiotherapy **68**(10), 324–326 (1982)
2. May, C.S., Broadhurst, M.J., Major, R.E.: Comparison of rocking edge spacing for two common designs of swivel walkers. Prosthet. Orthot. Int. **28**(1), 75–80 (2004)
3. Stallard, J., Lomas, B., Woollam, P., Farmer, I.R., Jones, N., Poiner, R., Miller, K.: New technical advances in swivel walkers. Prosthet. Orthot. Int. **27**(2), 132–138 (2003)
4. Swivel walker: http://www.prowalk.de/kinder-mobil-machen/swivel-walker.html/
5. Čirip, M., Hajduk, M., Čirip, M., Hajduk, M.: Proposal for stabilization and mathematical models of walking humanoid robots, Ph.D. thesis, University of Košice (2010)
6. Knoflíček, R., Haltmar, M.: Construction of mobile services robots (1–5), MM-Industrial spectrum, no. 2–6 (2000) (in Slovak)
7. Hajduk, M., Baláž, V., Sukop, M.: Sorting workstation with colour sensors. In: Proceedings of SAMI'05—3rd Slovakian-Hungarian Joint Symposium on Applied Machine Intelligence, pp. 353–357 Herľany, Slovakia, 21–22 Jan 2005. ISBN 963-7154-35-3 (2005)
8. Zubrzycki, J., Świć, A., Taranenko, V.: Mathematical model of the hole drilling process and typical automated process for designing hole drilling operations. Appl. Mech. Mater. **282**, 221–229 (2013)
9. Baláž, V., Sukop, M.: Multi-agent systems. In: Automation: Problems, Ideas, Decisions: Material International Educational-Technical 14–17 Sept, Sevastopol, SevNTU, 124–127. ISBN HMC23-52-10 (2005) (in Slovak)

Resistance Feedback of a Shape Memory Alloy Wire

Daniela Maffiodo and Terenziano Raparelli

Abstract The identification of a model able to relate the deformation of Shape Memory Alloy (SMA) wire to a state variable easily measured is widely studied in recent years. The electric resistance (ER) of a SMA wire varies depending on its martensite and austenite fraction, so the determination of the biunivocal relationship between the wire shortening and its electric resistance allows "previewing" the position of the SMA wire. In this way the resistance of the SMA wire can be used as feedback variable in a control system. A test bench was first used to determine the above mentioned relationship, and then this relation is used in a fuzzy control system to impose a desired position to a SMA wire and to check the response of the device. This type of control doesn't need a position sensor, which is a certain advantage in terms of cost, overall dimensions and weight for possible applications. The performances of this position control with resistance feedback are good and adequate for many applications; in particular it will be suitable for all applications which do not dispose of space to place a position sensor—like in aerospace applications or micro manufacturing.

Keywords Shape memory alloy wire · Resistance feedback · Fuzzy control

1 Introduction

Many studies have been carried out by various researchers with the aim of identifying a model to relate the deformation of Shape Memory Alloy (SMA) wire to a state variable of the same wire easily measured. The purpose of these researches is

D. Maffiodo (✉) · T. Raparelli
Department of Mechanical and Aerospace Engineering, Politecnico di Torino, Turin, Italy
e-mail: daniela.maffiodo@polito.it

T. Raparelli
e-mail: terenziano.raparelli@polito.it

T. Borangiu (ed.), *Advances in Robot Design and Intelligent Control*,
Advances in Intelligent Systems and Computing 371,
DOI 10.1007/978-3-319-21290-6_10

easy to understand: exactly knowing the deformation of the wire, see the position of one end of a straight wire when the other end is fixed, with an indirect measurement of an internal quantity of the wire, makes unnecessary the use of a position sensor to perform the control. In addition, a mathematical model effectively describing the complex behaviour of a SMA wire is very difficult to find; this is why researchers often prefer to consider the wire as a "black box", taking into account only the inputs and related outputs to design the control.

Cho et al. [1] studied the behaviour of a NiTiCu SMA wire, obtaining interesting results with an experimental set-up with strain gauge. Ma and Song [2, 3] developed an electrical resistance feedback control system for position regulation of a spring-biased NiTi SMA wire actuator.

Song et al. [4] designed a control for SMA actuator wires using as a feedback signal the value of the voltage drop across the wire, the relationship between voltage and position being modelled with neural network.

These are recent interesting solutions, for example the use of the inductance of the wire as the state variable [5]; but using resistance as the state variable is widely used [6, 7]; also the author worked on a position control with PWM modulation and resistance feedback signal [8]. Another work of Song et al. [9] considers the resistance feedback with an original neural network to relate position and resistance.

In the present research we decided to work on a resistance feedback using a fuzzy logic for the control. The fuzzy logic is particularly suitable in case of phenomena like the behaviour of SMA devices, in which there are uncertainties or no detailed information about the constitutive model, with strong time dependence and nonlinearities [10].

Ultimately the use of a resistance feedback control with fuzzy logic in SMA wires actuating devices will lead to reduce overall dimensions and weight and definitively save money. This kind of approach seems to be more and more required in robotic applications, such as in servo-systems of aerospace vehicles and aircraft [11] or precision manufacturing and assembly [12].

2 Resistance Feedback Control with Fuzzy Logic

The selected SMA wire (Nitinol HT wire, diameter 250 μm, length 200 mm) shows the contracted shape at temperatures beyond 70 °C. To obtain the return to the other crystalline form it is necessary to cool it and to apply a bias tension on the wire axis direction (at least 35 MPa). The heating is obtained by Joule effect, the cooling is on calm air. Having one wire end fixed, it is possible to consider the wire shortening, caused by heating, as the upwards of the free end of the SMA wire. The words "position control" will refer to the position of the free end with respect to the fixed one. The wire must be firmly constrained under mechanical strain and heated by electric current. A position sensor is used during the characterization phase to measure the reached position of the free wire end.

Figure 1 shows a sketch of the whole control test bench. The wire layout is simple and efficient: the SMA wire (1) is vertically arranged and connected at its ends to an insulated electric wire. On the upper side this wire is connected to the structure (2), on the lower side there is a suspended mass (3) of about 1 kg. The cursor of an inductive position sensor (4) is rigidly connected to the mass, sliding inside an external cylinder, fixed to the structure (LVDT Shaevitz E200). The arrow beside the wire indicates the heating/shortening direction, caused by Joule effect with a 12 V power supply (5).

A computer with acquisition board executes the position control of the SMA wire with Matlab/Simulink® programming. The amplification device (6) amplifies the low power signal from DAQ and acquires the electric current magnitude flowing into the SMA wire. In particular, the DAQ card makes the acquisition of both the signal of the potential drop between the SMA wire ends and of the potential drop at a known resistance in series with the SMA wire, and it transmits the command signal necessary to control the SMA wire. Moreover it acquires the position sensor signal where required. The resistance has a value much less than the resistance of the SMA wire in order to measure the current.

A previously developed fuzzy control [13] had a similar hardware but using the LVDT output as the feedback signal. The originality of the present study is on the comparison with a "previewed" position, instead of a measured position coming from a sensor, on the comparison block. This is possible because it is well known that the electric resistance of a SMA wire varies depending on its martensite and austenite fraction. This property can be used to create a biunivocal relationship between the wire end position (length) and its electric resistance; in this way it will be possible to "preview" the real position of the SMA wire. It is possible to

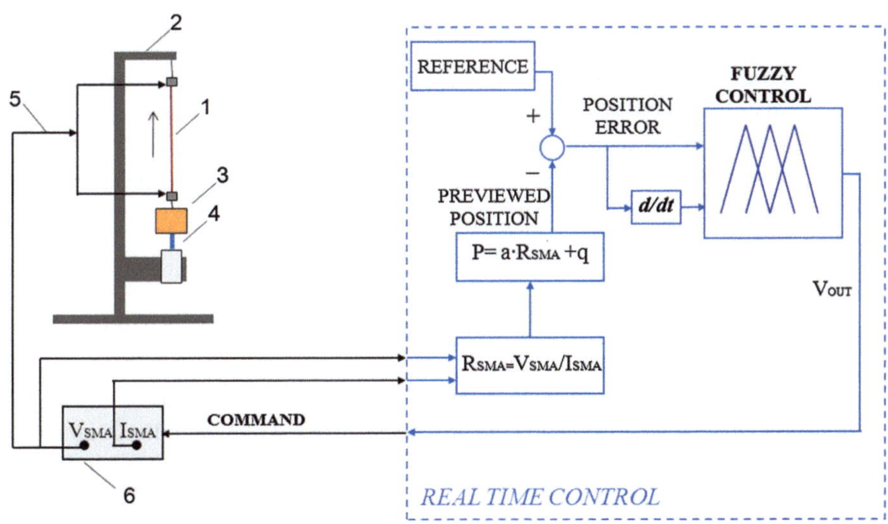

Fig. 1 Sketch of the test bench and its resistance feedback with fuzzy logic control

calculate the reached position by measuring the electric resistance thanks to a little resistance placed in series with the SMA wire and causing a little voltage drop. The first Ohm law allows calculating the current passing both through the resistance and the wire. The voltage drop V_{SMA} and the current I_{SMA} of the wire can be used to calculate the desired resistance: $R_{SMA} = V_{SMA}/I_{SMA}$

With the experimental tests above described it is possible to evaluate the relationship between the reached position y (SMA wire length) and its electric resistance R. This relationship was linearized, thus the equation $y = a \cdot R + q$ allows calculating the previewed position.

This control doesn't need a position sensor and this is a certain advantage in terms of money, overall dimensions and weight when using the device in applications. A possible drawback is the precision of this control that might be probably less accurate than that of a position feedback control.

3 Open Loop Tests

Experimental open loop tests were carried out in order to study the relationship between electric resistance and displacement of the Ni–Ti wire. These tests were necessary because the electric resistance variation is sensitive to many factors, such as heat treatment, exact alloy composition and manufacturing.

Tests under stress generated by different constant loads were performed. They showed that an increase of the load applied to the wire corresponds to a decrease of the amplitude of the hysteresis cycle; actually higher mechanical stresses cause a decrease of the R-austenite produced and an increase of the direct transformation from austenite to martensite. Moreover increasing the applied constant load causes an increase of the electric resistance corresponding to a fixed value of strain. The applied bias load was then increased until the hysteresis on the relationship between position and electric resistance nearly disappears. Figure 2 shows the relationship between position and electric resistance with a bias load of about 10 N. Data refers to 15 activation cycles with a sinusoidal input having the amplitude between 0.1 and 0.9 V and 1/60 Hz frequency. Figure 2 shows a very tight hysteresis cycle and a quite repeatable behaviour.

4 Relationship Between Position and Electric Resistance

In order to preview the SMA wire reached position it was decided to model the relationship between position y and electric resistance R with a simple straight line. Figure 3 shows a real curve (blue) and a linearized curve (red) with a = −12 and q = 43.

The model represents well the actuator stroke. As it is possible to see in the figure the wire maximum stroke is about 9 mm, but the linear behaviour is slightly

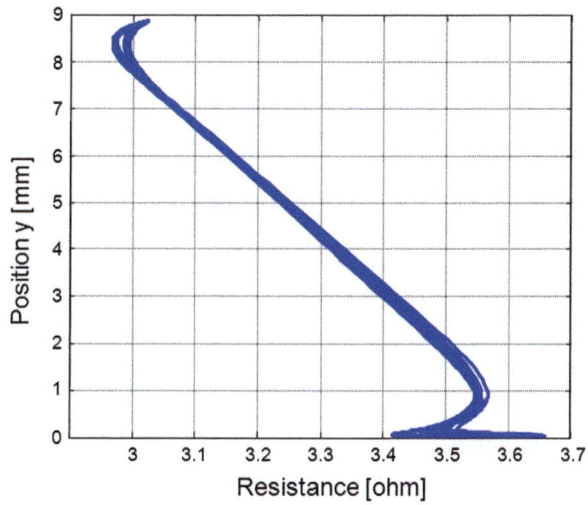

Fig. 2 Relationship between position and electric resistance (constant bias load of about 10 N)

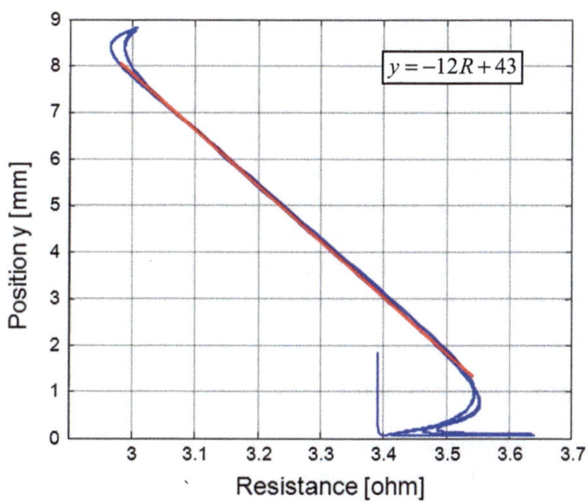

Fig. 3 Representative cycle for the linearization of the electric resistance/position relationship in case of 10 N constant load

reduced: the stroke that can be used during the closed loop resistance feedback is about 6 mm, from 1.5 to 7.5 mm.

The control was implemented in Matlab Simulink and is shown in Fig. 4. The error is generated by the comparison between the desired position and the previewed position, the real position being used as check information.

This kind of control is not better than a "normal" position control in terms of positioning precision, but it has the great advantage of not having the need to use a displacement sensor, which brings a number of advantages: less weight, overall dimensions and costs reduced.

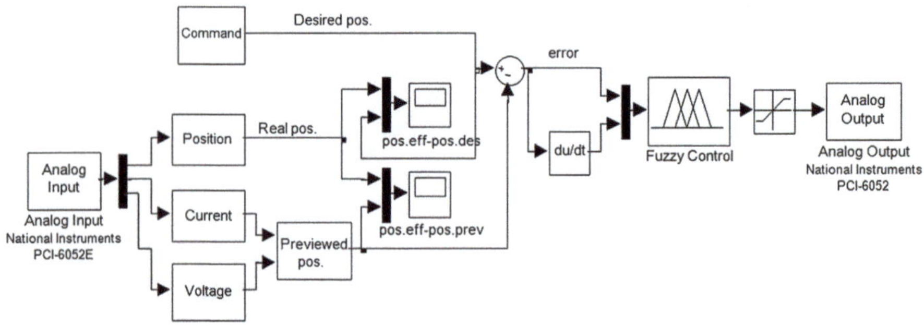

Fig. 4 Block Diagram of the resistance feedback control

5 Results

Various experimental tests have been carried out to validate the control system, in which the position sensor LVDT was used to compare the different results and not for the feedback.

Figure 5 shows one example of the test with square waveform. Here the desired position (dotted line) varies from 2 to 7 mm with frequency equal to 1/30 Hz; it is possible to compare the desired position with the previewed position and the reached position.

The first cycle is affected by errors due to the starting conditions and is not relevant; only when the desired position becomes greater than the previewed position the electric tension increases and the controller starts working effectively. As it can be seen from Fig. 5, there are little overshoots in the rise side (0.2 mm maximum), but the controller works well (with errors lower than 0.1 mm), and

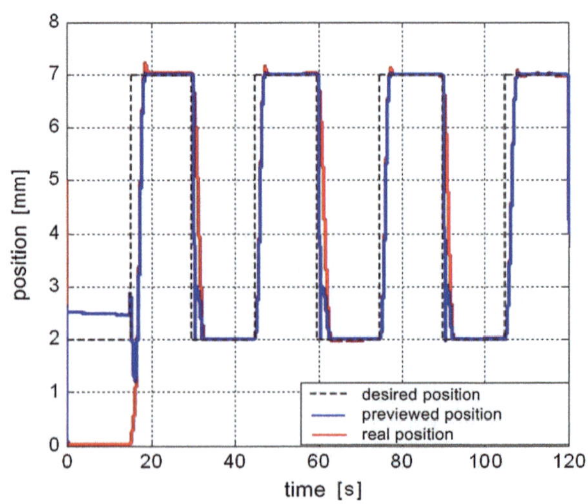

Fig. 5 Square wave with frequency 1/30 Hz for resistance feedback with fuzzy control, at constant load of 10 N; comparison between desired position (*dotted line*), previewed position (*blue*) and reached position (*red*)

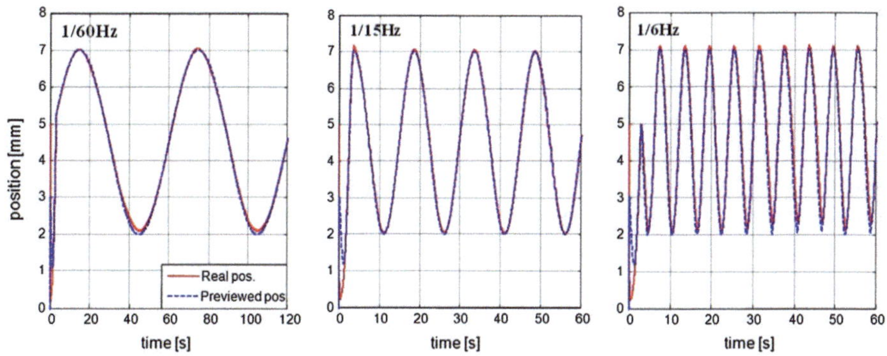

Fig. 6 Examples of tests with sine wave with 5 mm amplitude (from 2 to 7 mm of the wire stroke) with different frequencies (1/60, 1/15 and 1/6 Hz respectively) and constant load of 10 N

heating and cooling times are similar to the results obtained with the position feedback [13].

Figure 6 shows examples of tests with sine wave of 5 mm amplitude (from 2 to 7 mm of the wire stroke) at different frequencies (1/60, 1/15 and 1/6 Hz respectively). There is no unsteadiness, regardless the sine wave frequency.

Figure 7 shows the results for a test in which the reference position is a step function with single step height equal to 1 mm and step time of 20 s. There are five rising steps, corresponding to a heating phase, and five fall steps, corresponding to a cooling phase. The real position during rising is nearly the same as the desired position; whereas during the falling there are differences, due to the modelling of the hysteresis cycle.

Fig. 7 Example of test with step wave at constant load of 10 N

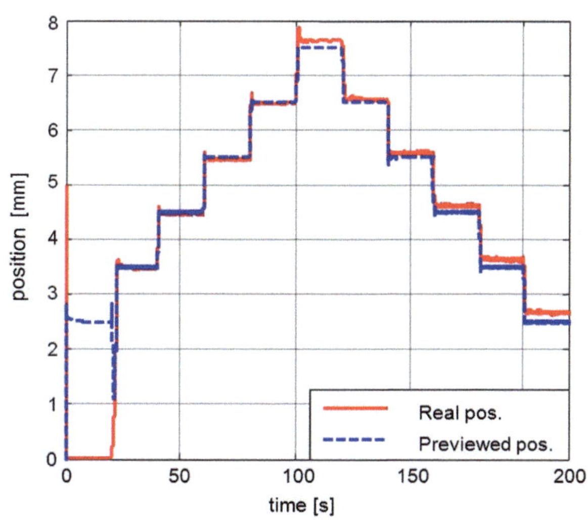

6 Conclusions

On the whole, the overall performances of the position control with resistance feedback are good and adequate for many applications; in particular it is suitable for all applications in which there is no physical space to place a position sensor. The possibility of controlling the position without a specific position sensor represents a serious advantage, causing a positive outcome in terms of simplicity of the device and cheapness.

References

1. Cho, H., Yamamoto, T., Takeda, Y., Suzuki, A., Sakuma, T.: Exploitation of shape memory alloy actuator using resistance feedback control and its development. Prog. Nat. Sci. Mater. Int. **20**, 97–103 (2010)
2. Ma, N., Song, G.: Control of shape memory alloy actuators using pulse-width pulse-frequency (PWPF) modulation. J. Intell. Mater. Syst. Struct. **14**, 15–22 (2003)
3. Ma, N., Song, G.: Control of shape memory alloy actuators using pulse width modulation. Smart Mater. Struct. **12**, 712–719 (2003)
4. Song, G., Chaudhry, V., Batur, C.: A neural network inverse model for a shape memory alloy wire actuator. J. Intell. Mater. Syst. Struct. **14**, 371–377 (2003)
5. Kim, H., Han, Y., Lee, D., Ha, J., Cho, K.: Sensorless displacement estimation of a shape memory alloy coil spring actuator using inductance. Smart Mater. Struct. **22**(2), 391–398 (2013)
6. Selvarani Ruth, D.J., Sunjai Nakshatharan, S., Dhanalakshmi, K.: Differential resistance feedback control of a self-sensing shape memory alloy actuated system. ISA Trans. **53**(2), 289–297 (2014)
7. Wang, T., Shi, Z., Liu, D., Ma, C., Zhang, Z.: An accurately controlled antagonistic shape memory alloy actuator with self-sensing. Sensors **12**(6), 7682–7700 (2012). doi:10.3390/s120607682
8. Raparelli, T., Zobel, P.B., Durante, F.: SMA wire position control with electrical resistance feedback. In: Proceedings of 3rd World Conference on Structural Control, Como, Italy, vol. 2, pp. 391–8 (2002)
9. Ma, N., Song, G., Lee, H.-J.: Position control of shape memory alloy actuators with internal electrical resistance feedback using neural networks. Smart Mater. Struct. **13**, 777–783 (2004)
10. Nguyen, B.K.: Modelling and control of shape memory alloy actuators by using Preisach model, genetic algorithm and fuzzy logic. Mechatronics **18**(3), 141–152 (2008)
11. Shi, Z., Wang, T., Liu, D., Ma, C., Yuan, X.: A fuzzy PID-controlled SMA actuator for a two-DOF joint. Chin. J. Aeronaut. **27**(2), 453–460 (2014). ISSN 1000-9361
12. Bassiuny, A.M., Ismail, M.A., Bakr, E.M.: Fuzzy incremental controller for SMA smart actuator with application to micro-positioning. J. Control Eng. Technol. **4**(1), 58–65 (2014)
13. Maffiodo, D., Raparelli, T., Belforte, G.: Fuzzy Logic Position Control of a Shape Memory Alloy Wire. Ventil 136–145, April, ISSN: 1318-7279 (2010/2)

Study of 3-Jaw Gripper Architectures

Giuseppe Quaglia and Luca Girolamo Butera

Abstract The growing use of robots in automated industrial processes involves the development of the solutions adopted for grasping objects. These devices or grippers are used for handling components and products that can have different shapes, materials and dimensions. Current demand for more cost-effective and flexible industrial processes requires smart grippers able to perform a variety of functions. In this paper we present a comparison of different architecture solutions for an industrial gripper with specific features. Using a 3-Jaw gripper model, we consider two different ways to grasp the object. In particular we introduce all requirements of the gripper, the process design for reaching the target of the project using a modular approach and the possible architectures of gripper.

Keywords Industrial manipulation · 3-jaw gripper · Design process

1 Introduction

The aim of this paper is to present a study that concerns the definition of architecture of a 3-Jaw industrial gripper to be used in a specific automatic process. The gripper's design requirements are described and the analysis and the comparison of different architectures that could be used are reported.

This paper proposes novelties concerning the methodology used to compare different architectures that can achieve the grasping task. A number of viable kinematic transmission systems for the synchronization of finger motion are also presented. An overview of the possible solutions is provided, specifying the design parameters and the pro and cons of the solutions.

G. Quaglia (✉) · L.G. Butera
DIMEAS, Politecnico di Torino, Turin, Italy
e-mail: giuseppe.quaglia@polito.it

L.G. Butera
e-mail: luca.butera@polito.it

T. Borangiu (ed.), *Advances in Robot Design and Intelligent Control*,
Advances in Intelligent Systems and Computing 371,
DOI 10.1007/978-3-319-21290-6_11

The literature on classification of mechanical grippers presents many solutions for manipulating objects that have a “disk” shape. For example there are radial impactive grippers, three-finger grippers with centring sliding fingers or with centring pivoted fingers [1, 2]. Robot hand architectures have also been developed, for example DLR II Hand [3], Barret Hand [4] but it is still rare to find these devices in an industrial process.

The present paper is focused on the design process, the modular approach and the comparison of possible solutions. These tools are proposed to satisfy the specific requirements: adaptability; centring accuracy; limited work space and weight; low-weight objects.

2 Design Requirements

This part of the paper presents the main design requirements for the gripper. Different aspects that concern the gripper and the working environment are considered.

- **Task and object description**: The gripper has to handle ring seals (Fig. 1) with defined physical and geometrical characteristics. It has to be guaranteed a secure handling of these objects during the execution of an automatic production cycle. Table 1 reports the main physical and geometric characteristics of three different work pieces, with maximum, intermediate and minimum outer radius.
- **Adaptability**: This requirement can be considered the main requirement. The modern production processes need devices with a high adaptability in order to satisfy different production necessities. Therefore a gripper that does not need a jaw change-over when the work piece changes, provides an added value to the whole production process: stops of the production machines are avoided and specific and expensive equipment for each work piece are not requested. This adaptability is mainly related to the r_{omax}/r_{omin} ratio that in this case is 2.5.

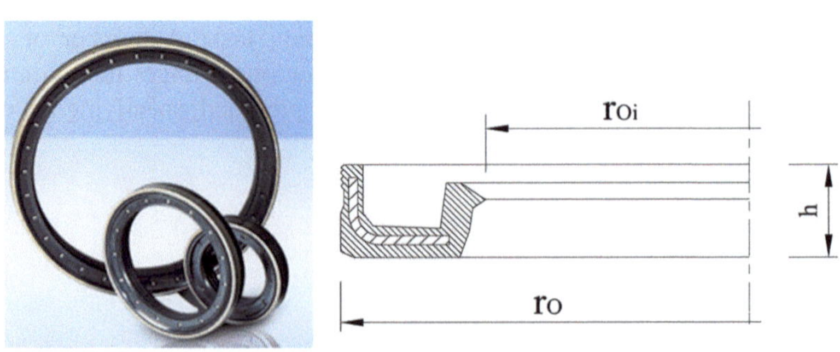

Fig. 1 Sample of grasped object

Table 1 Main characteristics of three different work pieces

Characteristic			
Geometry	Axial symmetry	Axial symmetry	Axial symmetry
Outer Radius, r_O (mm)	50	90	125
Inner Radius, r_{Oi} (mm)	38	80	109
Height, h (mm)	9	14.5	20
Weight (g)	60	140	440
Grasping force (N)	5	15	40

Other requirements concern the reliability: a precise, stable and secure grasping and the footprint and weight of the gripper.

3 Design of an Impactive Gripper

This section presents a comparison of the chosen gripper design solutions which meet the project requirements.

The choice of planar mechanisms for the gripper was influenced by the volume available with limited height (Fig. 2), the shape of the objects, the need to apply forces in radial direction and the limited surface to apply these grasping forces. Some mechanical architectures of gripper are here analysed and compared, although the analysis is not exhaustive.

3.1 Functions and Modularity

During the production cycle it is necessary to know the exact position of a grasped object. This requirement can be achieved with self-centring mechanisms with synchronization devices. Furthermore, grasping stability depends on the solution used to apply and distribute the grasping force (impactive force). Finally, there are some phases of the production cycle where the grasped objects have to be released

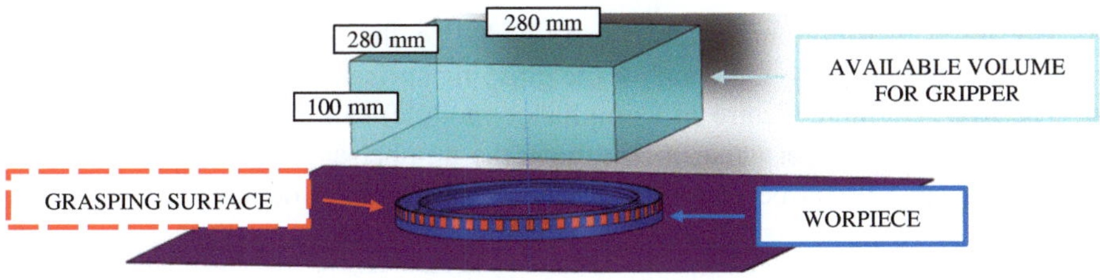

Fig. 2 Work environment representation

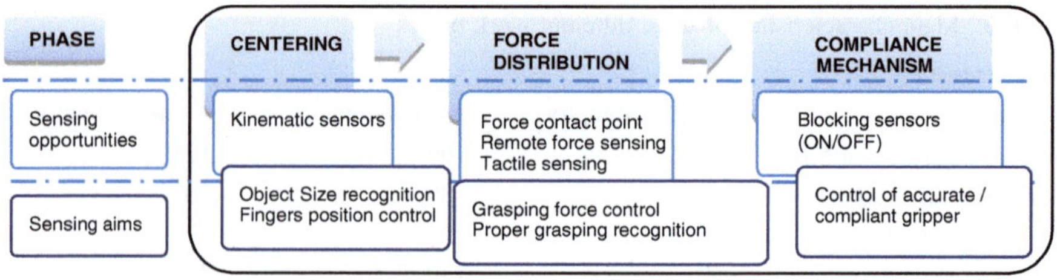

Fig. 3 Grasping process

to other working stations. This operation can be carried out by an RCC (Remote Center of Compliance) device that allows correcting alignment errors during peg-in-hole applications.

These three tasks may correspond to the three phases of the grasping process. In order to use a modular approach, it is possible to consider a relative subsystem for each phase that can be used in the gripper. Using this modular approach, the gripper can present one or more subsystems that may have different versions. In particular, for each module it is possible to install sensors that can provide relative information, useful to satisfy specific sensing aims (Fig. 3).

3.2 Gripper Subsystems

The gripper was divided into four subsystems in order to develop the design solution analysis:

- Finger architectures subsystem;
- Transmission and synchronization group;
- Force distribution device;
- Compliance module.

For each subsystem the most important design parameters are identified and some selection criteria are proposed. In this paper, due to the limited space available, only the first two subsystems that are responsible for the centring phase are presented in detail.

For the other two, that will be presented in future works, only a few comments:

- The force distribution device can be added to the gripper jaws, in order to increase the grasping force and/or the grasping surface, after the centring phase;
- The RCC (Remote Centre of Compliance module) can be inserted between the gripper and the robot wrist; it may have the number and type of dumped d.o.f. required, and it can be activated by a specific actuation device.

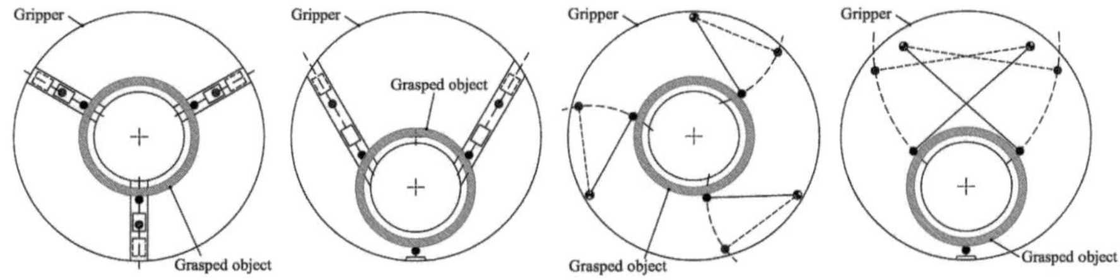

Fig. 4 Kinematic architectures for radial impactive grippers

3.2.1 Finger Architectures Subsystem

The architectures analysed are presented in Fig. 4: they are all capable of grasping on three points [5], by means of two or three movable fingers, using revolute joints or prismatic pairs.

The *3-mobile finger prismatic pair architecture* (Fig. 5), widely used, allows good centring of the cylindrical object (O_o work piece centre, O_G gripper centre), when it is actuated using a synchronized transmission system ($x_{1F} = x_{2F} = x_{3F} = r_O$). Another good property is due to the jaw approaching direction (described by the vector dS), that is always normal to the surface. Objects of different size (minimum, maximum radius r_{OMIN}, r_{OMAX}) receive grasping force on the edges of concentric triangles.

The *2 mobile finger prismatic pair architecture* (Fig. 6) has the advantage of reducing the number of mobile fingers but has some limitations:

- the grasped object's centre lies on a line and the position depends on r_O;
- the use of this architecture requires that the robot control system recognizes the object size and adapts the required center trajectory to this information;

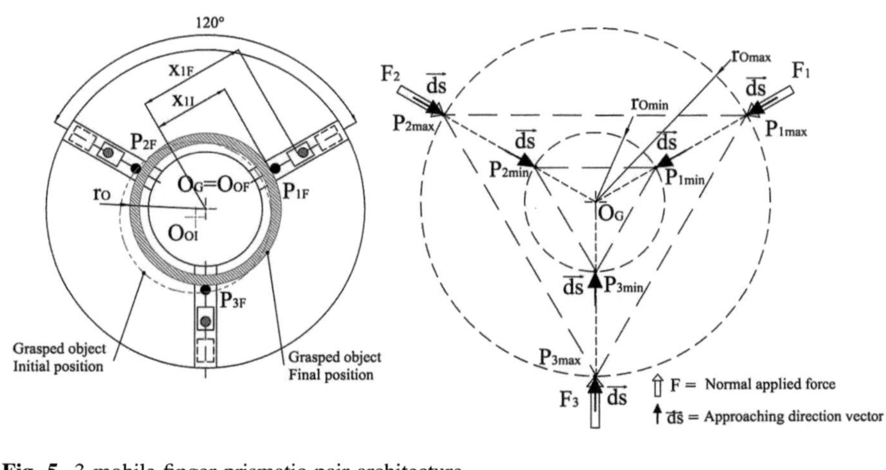

Fig. 5 3-mobile finger prismatic pair architecture

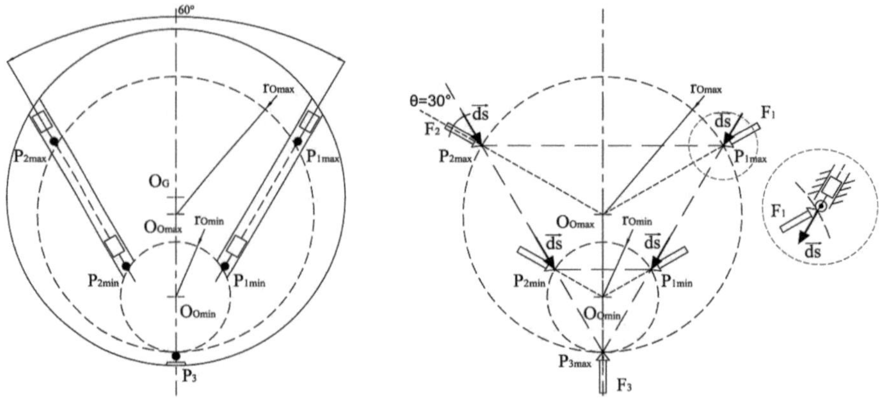

Fig. 6 2-mobile finger prismatic pair architecture

– the jaw approaching direction form an angle $\vartheta = 30°$ with respect to the normal direction at the contact point.

The *3-mobile finger revolute joint architecture* (Fig. 7) makes it possible to obtain a compact and cost-effective solution, operating on a wide range of object sizes, replacing prismatic pairs with revolute joints. The most important design requirement is the range of object size expressed by the ratio r_{OMAX}/r_{OMIN}, that in our case is 2.5. The triangle that describes the force's application points is rotating according to the object size but always has the same centre. The effect of choosing the finger length r_P, and the position of the finger revolute joint r_F, is presented in Fig. 8 left and right respectively.

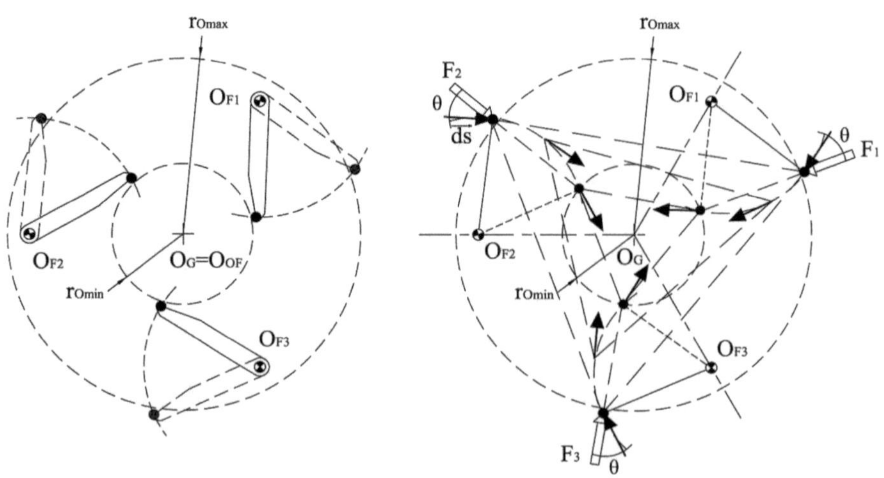

Fig. 7 3-mobile finger revolute joint architecture

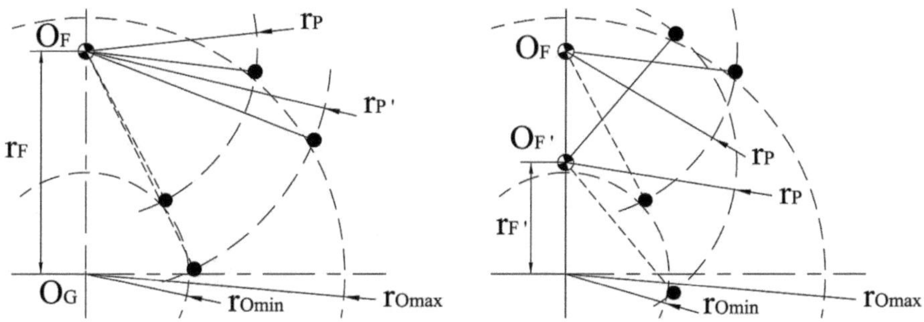

Fig. 8 Effect of finger length r_p and position of the finger revolute joint r_F

A design criterion for these parameters could be to minimize the approaching angle ϑ for all work piece sizes getting the same values ($\vartheta_{rmin} = \vartheta_{rmax}$) for minimum and maximum object radius.

The requirements lead to the graphic construction of Fig. 9, also showing that reducing ϑ leads to higher r_P values and so to greater overall dimensions of the gripper.

The *2-mobile finger revolute joint architecture* (Fig. 10) is a valid alternative to replacing the prismatic pair of Fig. 6. If the force application on the edges of the equilateral triangle is considered a good solution, the contact points must lie on an ideal contact point line: this condition can be approximated using the graphic constructions of Fig. 10 right.

3.2.2 Transmission and Synchronization Group

Figure 11 shows an example of solutions suitable when the motion is derived from a single rotary motor, illustrating the main systems for transmitting a rotary motion to three jaws when revolute pairs are used.

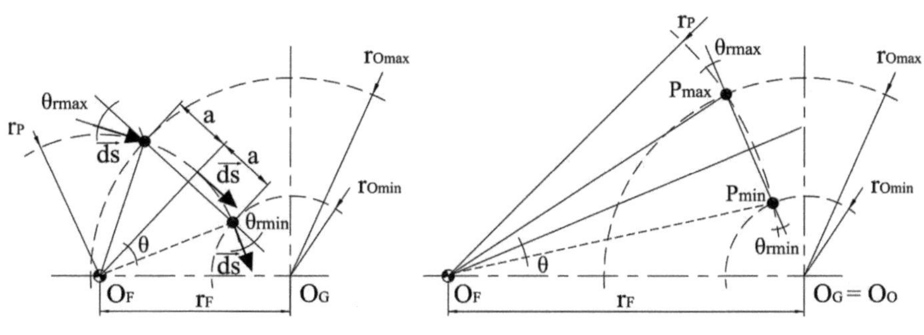

Fig. 9 Effect of approaching angle ϑ

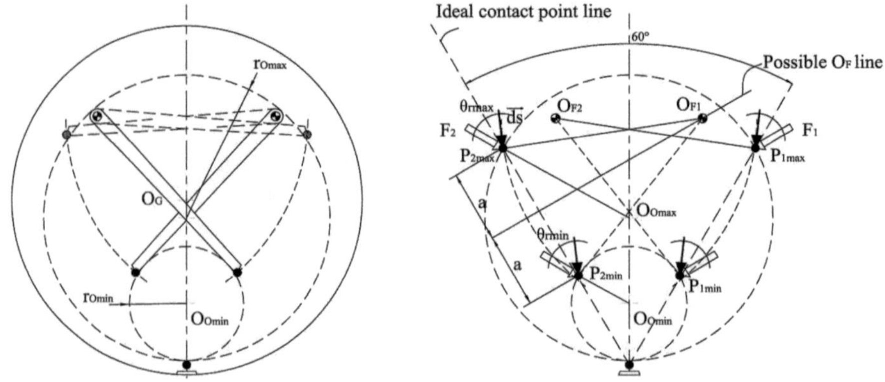

Fig. 10 2-mobile finger revolute joint architecture

The use of gear transmission (left), economic and compact, allows easy synchronization. It requires definition of the design parameters r_{TA}, r_{TF}, r_{TC}.

Systems that use articulated transmission (middle) are effective but have different equilibrium configurations that depend on the size of the grasped object. Furthermore a greater number of revolute pairs are used. In order to get the required range of motion ψ, the design parameters are δ, r_{M1}, r_{M2}, r_{M3}.

Cam systems (right) represent an effective alternative to gear transmission systems but they may present higher manufacturing costs. The design parameters are δ, r_B and the cam profile according to the geometrical transmission ratio desired.

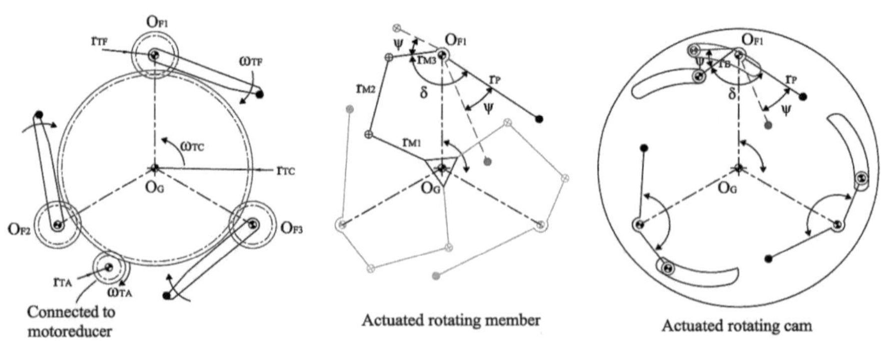

Fig. 11 Transmission and synchronization group

4 Conclusions

This work makes a comparative study of architectures of a 3-jaws gripper to meet two main requirements: adaptability to a wide variety of object sizes and the centring accuracy.

A modular approach was used. Two separate phases during work piece grasping were defined: the first is the *centring phase*, with limited force; the second is the *gripping phase*, where the gripping force increases and spreads over large surfaces in order to ensure accurate and safe handling. The proposed modular approach makes it possible to design devices with different performances by adding, changing and/or modifying these modules. For example the terminal parts of the jaws can be made rigid, or some of them can be activated in order to get the distribution of gripping force.

For reasons of brevity only the development of an RCC module was mentioned, that can be remotely activated. A detailed description of kinematic and dynamic synthesis, a control system and the experimental results will be presented in a subsequent work.

Acknowledgements This work has been performed under a research project with Corcos Industriale S.a.s.

References

1. Chen, F.Y.: Gripping mechanisms for industrial robots: an overview. Mech. Mach. Theory **17** (5), 299–311 (1982)
2. Gareth, R.S.H.S., Monkman, J., Hesse, S.: Robot Grippers. Wiley, Weinheim (2007)
3. Butterfass, J., Fischer, M., Grebenstein, M., Haidacher, S., Hirzinger, G.: Design and experiences with DLR hand II. In: Proceedings of the World Automation Congress 2004, Seville, Spain (2004)
4. Townsend, W.: The barrett hand grasper programmable flexible part handling and assembly. Ind. Robot Int. J. **27**(3), 181–188 (2000)
5. Cutkosky, M.R.: On grasp choice, grasp models, and the design of hands for manufacturing tasks. IEEE Trans. Robot. Autom. **5**(3), 269–279 (1989)

Flexible Actuator for Biomorphic Applications: Performances and Energy Consumption Evaluation

Andrea Manuello Bertetto, Carlo Ferraresi, Luigi Antonio Besalduch, Roberto Ricciu and Andrea Cadeddu

Abstract The article describes an experimental evaluation of performance of a flexible actuator specially conceived for propulsion of biomorphic robots. Static and dynamic characteristics are presented, particularly referring to the energy consumption.

Keywords Flexible pneumatic actuator · Biomimetic actuator · Biomimetic robot · Fish-like robot

1 Introduction

Industrial, agricultural, military, environmental and medical applications often require actuators able to perform special tasks. Pneumatic actuators are often able to give an effective response to these needs, especially when they are characterized by unconventional architectures: the flexible pneumatic actuators are a very interesting example [1]. One of the first specimens was the McKibben artificial muscle [2]. Flexible actuators are very useful for medical applications [3, 4], micro manipulation, grasping of delicate object, harvesting of agricultural products, and pipes net inspection [5, 6].

A.M. Bertetto (✉)
Dipartimento di Ingegneria Meccanica, Chimica e dei Materiali,
Università degli Studi di Cagliari, Cagliari, Italy
e-mail: andrea.manuello@unica.it

C. Ferraresi
Dipartimento di Ingegneria Meccanica e Aerospaziale, Politecnico di Torino, Turin, Italy
e-mail: carlo.ferraresi@polito.it

L.A. Besalduch · R. Ricciu · A. Cadeddu
Dipartimento di Ingegneria Civile, Ambientale e Architettura,
Università degli Studi di Cagliari, Cagliari, Italy
e-mail: besalduch@unica.it

T. Borangiu (ed.), *Advances in Robot Design and Intelligent Control*,
Advances in Intelligent Systems and Computing 371,
DOI 10.1007/978-3-319-21290-6_12

Experimental and theoretical studies suggested that flexible actuators may provide effective performance for fish-like propulsion. The classical Gray's studies [7–9] underline the interesting advantages of the oscillating tail instead of traditional aquatic propulsion. An innovative choice for the tail actuation system is offered by the flexible actuators described in [10–12].

The goal of this work is to evaluate the operating characteristics and the energy cost for a particular bending actuator used for aquatic fish-like robot propulsion. The activity was worked out by test rigs especially designed. Some results are referred and discussed to evaluate the advisability of this type of actuator to move the fish-like fin of an aquatic robot.

2 Experimental Set-Up and Methodology

The actuator here designed and realised is a cylindrical device having a rubber made internal double chamber divided by a longitudinal wall. The two heads allow linking to the robot body and to the tail, the sealing of the chambers and the supply. Around the cylindrical rubber body, numerous rigid rings prevent radial deformations, forcing the axial one. The work envelope of the actuator is a 2D plane region [11].

The flexible actuator is represented in Fig. 1. The pictures show the assembled actuator with the radial constraints represented by the rigid rings, the two heads and the section of the internal cylindrical rubber chamber. The figure shows also the geometry of the elastomeric body section; two chambers divided by a longitudinal wall: unbalancing the pressure drop between the chambers the actuator stretches axially and bends in a plane perpendicular to the longitudinal wall. The rubber is an

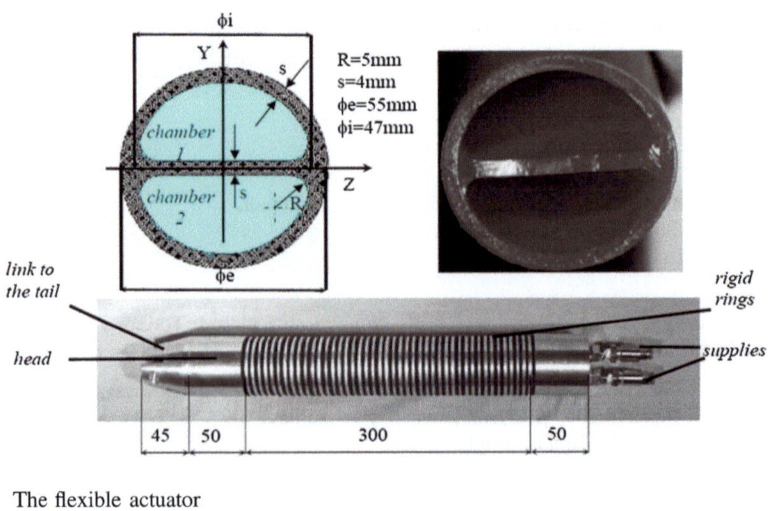

Fig. 1 The flexible actuator

elastomeric material with 70 shore A hardness; it permits high deformations preserving an elastic behaviour, and minimising the residual deformation depending on load velocity.

The actuator workspace was determined and traced by means of a specially designed experimental set up as in Fig. 2a. The actuator is vertically suspended, and the measuring of the end effector (EF) position is performed in a 3D space, by a trigonometric indirect measure. The end effector position (EF), measured at given values of the fluid pressures in the chambers defines the workspace in relation to the chamber supply pressures. The experimental set-up provides by means of wire position transducers the length variation of the edges of a pyramid having the basis on a horizontal plane and the upper peak linked to the actuator EF (Fig. 3).

To determine the energy consumption, a second test rig summarized in Fig. 2b was used. It consists of the same actuator of the first test rig, but, in order to control the air volume injected in the actuator, a pneumatic cylinder (bore diameter of 63 mm and stroke of 750 mm) was used, whose rod displacement was registered by means of a linear gauge (span: 0–1270 mm, static sensitivity of 9,8455 V/m and

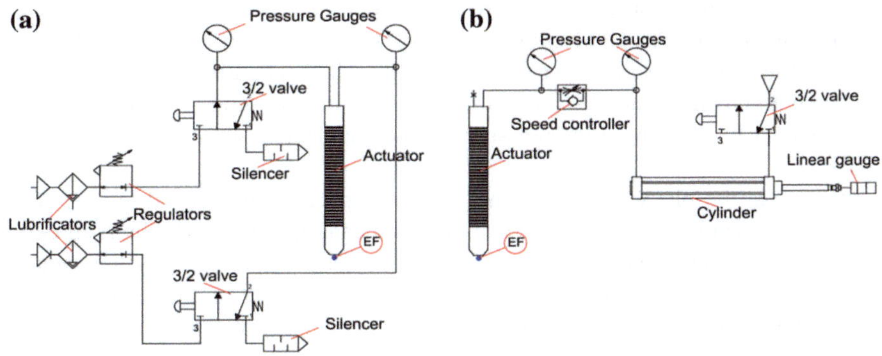

Fig. 2 Schematic representation of the first (**a**) and second (**b**) test rig

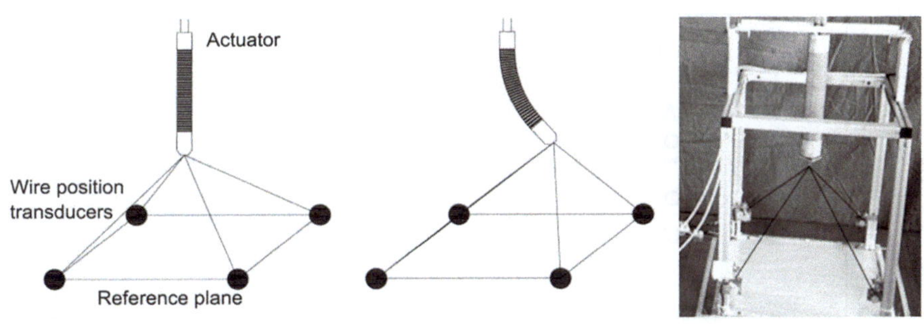

Fig. 3 The test rig to define the actuator work space

repeatability: ±0.02 % full stroke), and the pressure by means of two pressure gauges (span: 689,476 Pa, static sensitivity: 25.0 mV/100 kPa).

The position of the flexible actuator was acquired by a digital photo camera. The Data Acquisition System used was: NI USB-6221. The data was collected and processed by Matlab software.

3 Workspace of the Actuator

The actuator static workspace, represented in Fig. 4a, is a plane surface perpendicular to the plane of the longitudinal internal wall. The continuous red line has been drawn with a maximum pressure of 5 bar relative and was traced starting from A with both chambers discharged; increasing the pressure only in chamber 1, the end effector (EF) moves along the A–B curve reaching the point B for a pressure of 5 bar in chamber 1 and zero in the other one. Keeping the pressure of 5 bar in the chamber 1 and raising the pressure in the chamber 1 from zero to 5 bar the EF reaches point C. Increasing the pressure in chamber 2 the curve B–C is described by the end effector and in C the fluid pressure is 5 bar in both chambers. By discharging the chamber 1 the curve C–D is then travelled. The EF reaches again point A when both chambers are discharged. All points within the edge can be reached by the actuator EF, combining the pressures in the chambers within the range 0–5 bar relative.

In Fig. 4a, the black line represents the workspace of the second test rig for a maximum pressure of 5 bars. With this set-up we can reproduce only the A–B curve but as the two lines match quite well, the behaviour is approximately the same.

Figure 4b shows the velocity and the acceleration of the EF in the second test rig. It is evident from the velocity vectors, as the trend is very smooth and also the acceleration and the deceleration are not such abrupt as for common hydraulic

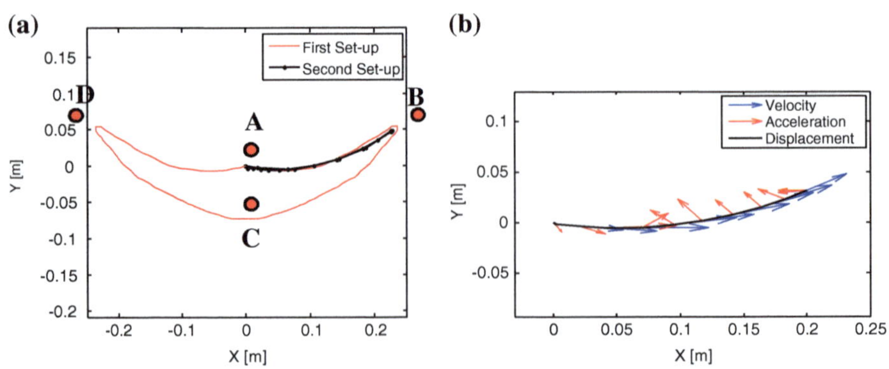

Fig. 4 **a** The actuator's 2D work space. **b** The actuator's 2D work space, velocity and acceleration of the EF in the second test rig

actuators, that they are more suitable for the reproduction of fish fins. This behaviour is shown in detail in Fig. 5. In particular in Fig. 5a it can be seen that the velocity component's behaviour is mainly represented by the horizontal component Vx, because the extension of the EF path is not so relevant. In any case, the vertical component starts in the downwards direction, due to the initial stretching of the actuator; then, after approximately half path, assumes significant values in the upwards direction. In Fig. 5b as expected the acceleration assumes significant values at the edges of the path. The acceleration is relevant at the beginning because the pressure force acts on the internal surface of the head bending the actuator at the end of the path due to deceleration.

These kinematic considerations can be highlighted in the Fig. 6, observing the different deformed geometries during the movement. In Fig. 6a, without pressure, the actuator is in the rest position. Increasing the relative pressure (B), it starts to

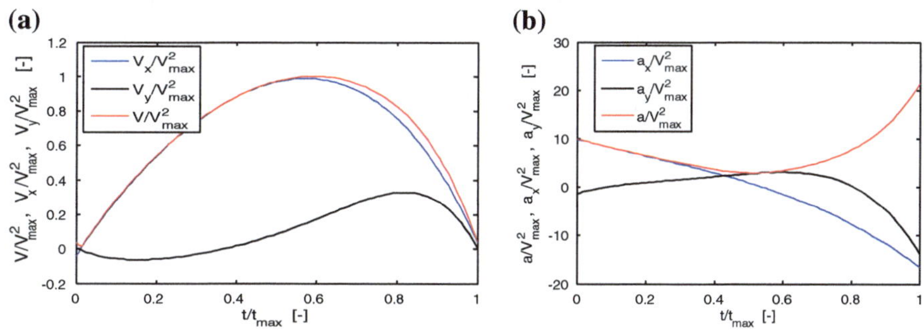

Fig. 5 The in-time normalized trends of components and resultant for the EF velocity and acceleration (**a**, **b**)

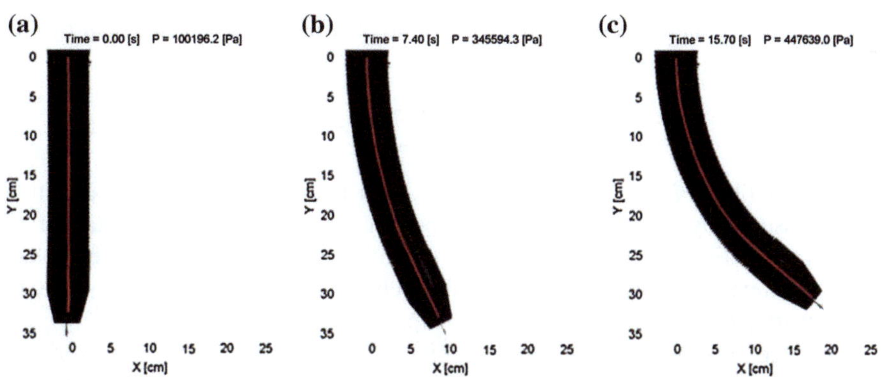

Fig. 6 Sequence of deformed geometry increasing the air supply pressure from 0 to 4.5 bars, relative in only one chamber

bend, and at approximately 4.5 bar (C) the actuator reaches its maximum curvature. From (A) to (B) there is a stretching initial phase with a small curvature; then, from (B) to (C) the stretching stops and the bending behaviour prevails.

4 Energetic Performances

Flexible actuators are suitable for propulsion of mobile robots in particular environment and for non-conventional applications; therefore to provide endurance allowing the accomplishment of a mission, an energy-saving behaviour is fundamental.

To investigate this aspect, the test rig shown in Fig. 2b has been used. The equipment can be considered as a "real" (irreversible) closed thermodynamic system, where the cylinder is used to transfer to the actuator a given volume of air at known conditions of pressure and temperature. The system allows energy transfer as heat and as mechanical work, while no matter (mass) is exchanged with the surroundings.

The experiment is composed of two steps: in the first one, the cylinder receives work from the surrounding and the gas is compressed; in the second step the cylinder returns the work previously received by expanding the internal air and reaching a new equilibrium state.

The temperature is assumed almost invariant because of the slowness of the process. The infinitesimal variation of work is:

$$\delta W = P \times \delta V + V \times \delta P \tag{1}$$

where the work (W) is negative if it is made by the surrounding.

The total work W is calculated by integrating the Eq. (1). This can be done in an approximated way considering the whole process as a sequence of small volume variations at constant pressure; therefore the total work was evaluated step by step with the following expression:

$$W = \sum Wi = \sum (Pi \times \delta Vi) \tag{2}$$

The quantities Pi and δVi, where $i = 1, \ldots, N$ is the number of the acquisition, were measured by the instruments described in Fig. 2b. The curves obtained are represented in Fig. 7; Fig. 7a, b show the evolution in time of the pressure P(t) and of the volume V(t). Figure 7c shows the behaviour in the Clapeyron plane for different frequencies of the compression/expansion cycles.

The areas subtended by the curves in Fig. 7c represent the mechanical works: in the first step the work done on the system and in the second one the work returned by the system. The work balance is represented by the net area within the cycle which represents the lost work (the energy consumption due to internal friction of the flexible actuator material and fluidic resistances).

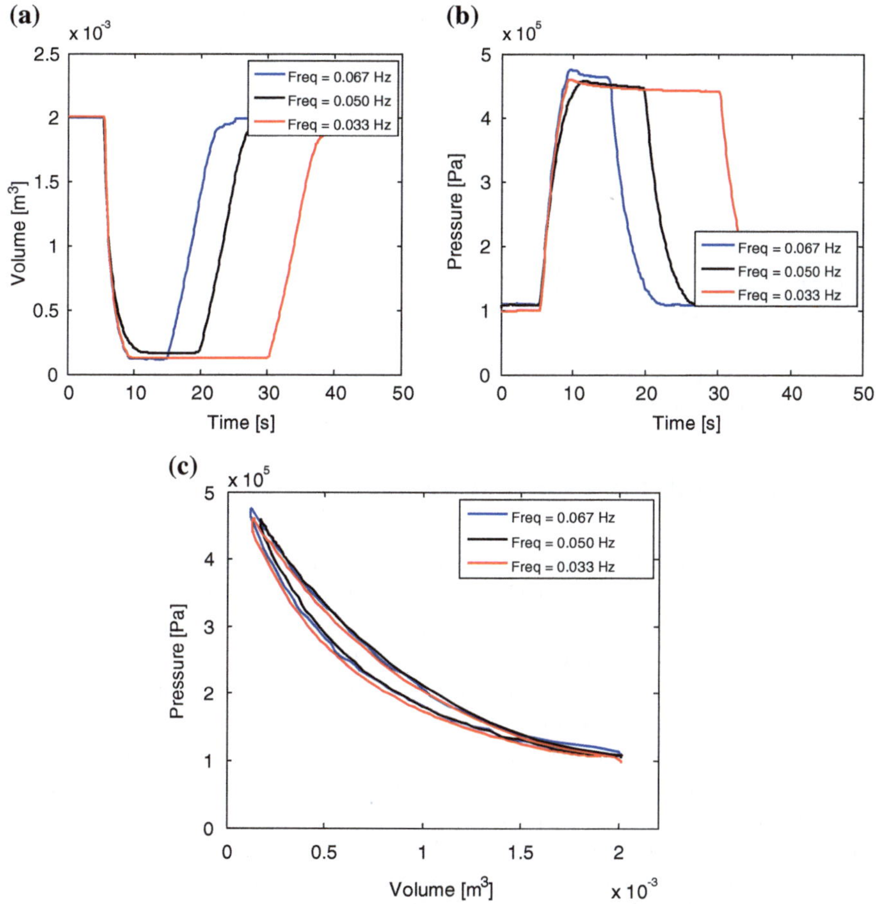

Fig. 7 Air volume injected to the flexible actuator (**a**), corresponding pressure in the actuator (**b**) and curves of pressure vs. volume (**c**) for different oscillating frequencies

The experiment highlighted an energy loss of about 47 J for each cycle, while the energy retrieved in the return stroke was 385 J. This value is approximately independent from the considered cycle; this suggests that the main losses must be due to the internal friction of the elastomeric actuator material, while other effects (heat exchange and fluidic resistances) are negligible for slow functioning like in this case. Of course, for higher working frequencies, fluidic losses would have greater influence.

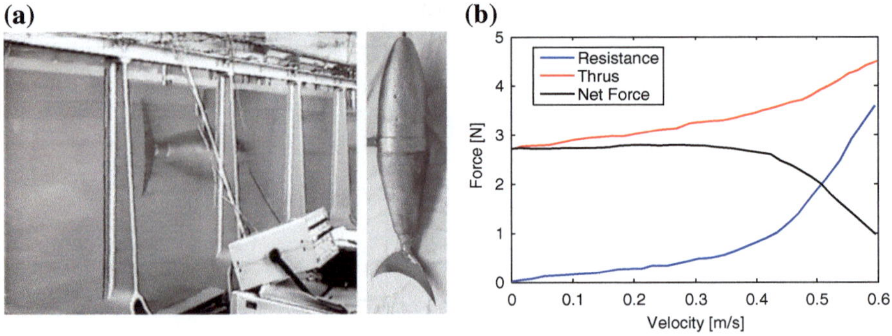

Fig. 8 **a** The fish-like robot in the hydraulic test channel. **b** Average values of forces measured during the oscillatory motion

5 A Case Study: A Fish-Like Robot

In order to verify the effectiveness of the actuator for propulsion of a fish-like robot, specific tests were performed in water, with the experimental setup shown in Fig. 8a [11]. During the actuator's motion and for a given velocity of the fluid, the net force acting on the robot was measured. In a separate test the resistance of the robot for different fluid velocities and motionless actuator, was measured too. The fin-actuator thrust is taken as the sum of resistance and net force.

The graph in Fig. 8b shows the average values of forces measured during the oscillatory motion of the fin. The oscillating tail frequency was 0.5 Hz and a pneumatic air supply pressure of 6 bar relative was given to perform arc amplitude of the tail oscillating motion of 0.5 m, referred to the thrust centre of the tail. The net force is positive up to a fish velocity of 0.6 m/s. This should indicate the maximum forward velocity of such a fish-like robot propelled by this tail at these values of frequency and amplitude tail motion.

6 Conclusions

This paper reports a first experimental investigation on a special flexible actuator, fluid operated, particularly conceived for biomimetic propulsion of fish-like robots.

The tests investigated the actuator's performances also from the energetic point of view, confirming its suitability for such an application type.

This study permitted to individuate the main problems that have to be faced and solved in future work, namely optimization of the fluid power transmission and the choice of materials with low internal friction for the actuator's body.

Acknowledgements The research reported by the authors was funded by the project INNNOVA. RE (P.O.R. of Sardinia Region, FESR 2007–2013, Axis 6.2.1a, R&T-Rescue).

References

1. Ferraresi, C., Franco, W., Manuello Bertetto, A.: Flexible pneumatic actuators: a comparison between the McKibben and the straight fibres muscle. J. Robot. Mechatron. **13**(1), 56–63 (2001)
2. Schulte, H.F.: The Characteristics of the McKibben Artificial Muscle. The Application of External Power in Prosthetics and Orthotics, Appendix H, Publ. 874, National Academy of Sciences, Washington, DC, pp. 94–115 (1961)
3. Shimizu, K., Mori, Y., Hayashi, K., Shunori, A., Kawakami, S., Hashida, M.: Gene delivery in mice using an implanted pneumatically actuated microsystem. In: IEEE 24th International Conference on Micro Electro Mechanical Systems (MEMS), pp. 181–184 (2011)
4. Manuello Bertetto, A., Meili, S., Concu, A., Crisafulli, A.: An inflatable pneumatic system for blood pressure recovery. Mech. Des. Struct. Mach. **40**(4), 506–518 (2012). (Taylor and Francis)
5. Suzumori, K., Miyagawa, T., Kimura, M., Hasegawa, Y.: Micro inspection robot for 1-in pipes. IEEE-ASME Trans. Mechatron. **4**(3), 286–292 (1999)
6. Manuello Bertetto, A., Ruggiu, M.: Pole climbing pneumatic robot. In: Proceedings of the 5th JFPS International Symposium on Fluid Power, pp. 43–48 (2002)
7. Gray, J.: Studies in animal locomotion I. The movement of fish with special reference to the eel. J. Exp. Biol. **10**, 88–104 (1933)
8. Gray, J.: Studies in animal locomotion II. The relationship between the waves of muscular contraction and the propulsive mechanism of the eel. J. Exp. Biol. **10**, 386–390 (1933)
9. Gray, J.: Studies in animal locomotion III. The propulsive mechanism of whiting. J. Exp. Biol. **10**, 391–400 (1933)
10. Shao, J., Wang, L., Yu, J.: Development of an artificial fish-like robot and its application in cooperative transportation. Control Eng. Pract. **16**, 569–584 (2008)
11. Cataudella, C., Ferraresi, C., Manuello Bertetto, A.: Flexible actuator for oscillating tail marine robot. Int. J. Mech. Control Torino **2**, 13–21 (2001)
12. Manuello Bertetto, A., Ruggiu, M.: Tail actuator propulsion device for aquatic robot. J. Robot. Mechatron. **18**(1), 80–90 (2006)

References

[illegible]

Part III
Advanced Robot Motion Control

Optimizing Trajectory Points for High Speed Robot Assembly Operations

Florin Anton, Silvia Anton, Silviu Raileanu and Theodor Borangiu

Abstract The work presented in this paper reports a research done in order to optimize the relation speed—trajectory length for a complex robotic assembly task. The assembly task consists in fixing an engine part with 8 screws, the screws being already inserted and pre-fixed but not tightened; in some cases the screws must be screwed for a length of 1 cm, and then tightened at 25 Nm. The operation duration (cycle time) should be of maximum 45 s measured from the time the pallet enters in the working area until the pallet exits the working area. Due to the conveyor operation which takes 12 s to place the pallet in the working position and to remove the pallet from the working area, only 33 s remain for the robot operation including the operation time of the screwdriver placed on the robot. The solution is based on developing an algorithm that uses the dynamics equations of the robot to compute the time needed to accomplish the task, based on the load of the robot and the stop points on the trajectory.

Keywords Robot dynamics · Industrial robot · Trajectory optimization · Assembly · Cycle time

F. Anton (✉) · S. Anton · S. Raileanu · T. Borangiu
Department of Automation and Industrial Informatics,
University Politehnica of Bucharest, Bucharest, Romania
e-mail: florin.anton@cimr.pub.ro

S. Anton
e-mail: silvia.anton@cimr.pub.ro

S. Raileanu
e-mail: silviu.raileanu@cimr.pub.ro

T. Borangiu
e-mail: theodor.borangiu@cimr.pub.ro

T. Borangiu (ed.), *Advances in Robot Design and Intelligent Control*,
Advances in Intelligent Systems and Computing 371,
DOI 10.1007/978-3-319-21290-6_13

1 Introduction

In manufacturing applications where industrial robots are involved, the *cycle time* represents an important production indicator. The cycle time represents the time required for a robot to repetitively accomplish its task; this means that the cycle time is the time spent to execute a production operation in a certain work station in the fabrication line. If the cycle time is shorter, then the productivity is increased, and this is one of the goals for every manufacturer.

The cycle time depends on different parameters, for example the complexity of the task, the trajectory length and shape, the payload of the robot and the tool performance.

This problem has been approached from multiple points of view, for example Carlson et al. [1] proposed a novel method for quality and throughput optimization based on a systematic search algorithm which exploits properties of the welding process in order to minimize the dimensional variation and robot travelling time in welding stations. Huang et al. [2] addresses the problem of realizing multi-robot coordination that is robust against pattern variation in pick-and-place tasks; therefore, they propose combining part-dispatching rules to coordinate robots, by integrating a greedy randomized adaptive search procedure (GRASP) and a Monte Carlo strategy (MCS).

Abdel-Malek and Li [3] addressed the problem of finding the robot location in a manufacturing cell that minimizes the execution time of its assigned tasks. Minimizing the robot cycle time leads to increased productivity in several industrial applications. In this approach, the robot geometric structure and specifications were considered in developing models using Inverse Kinematics to determine the travelling times between the different tasks of a particular manufacturing cell. Then, an optimization scheme is introduced to find the robot base location that minimizes its cycle time.

Nilakantan et al. [4] propose models with dual focus on time and energy to minimize the cycle time and total energy consumption simultaneously: one model (time-based model) with the primary focus to optimize cycle time and the other model (energy-based model) with the primary focus to optimize total energy consumption. The models proposed have a significant managerial implication in real assembly line systems. Depending upon the priorities of the management—primary focus on reducing cycle time or total energy consumption—suitable models could be selected. The proposed models are useful to reduce the total energy consumption and cycle time in robotic assembly lines. It can be observed that the computation time for control with time-based model is less compared to control using energy-based model.

Another approach is to reduce energy consumption without sacrificing cycle time. Björkenstam et al. [5] combined recent algorithms for collision free numerical optimal control and for optimal sequencing, and created an algorithm that was successfully applied to several industrial cases demonstrating that the proposed method can be used effectively in practical applications to find fast and energy efficient solutions. Energy optimization related to trajectory planning was also approached in [1, 6–8].

Our approach is to optimize the cycle time by minimizing the distance which the robot end effector should travel, and to maximize the robot speed to obtain a robot cycle time acceptable for the application and also avoid collisions.

2 Application Description

The application consists in an assembly task executed by an ABB IRB 2400, 6 d.o.f. industrial robot. The robot is mounted on the ceiling above a conveyor belt. On the conveyor belt, engine components are presented on pallets, the components being pre fixed in 8 screws. The time required for this operation is maximum 45 s, but the time taken by the conveyor belt to bring a pallet in the working position and to remove the pallet after the screw tightening operation is 12 s. In this case the time remaining for the robot operation is of 33 s.

The robot is equipped with a Desoutter EME38-20 J [9] automatic screwdriver with telescopic pipe wrench with a screw search feature. The entire configuration: screwdriver, wrench, telescope and robot mounting kit weights 9.3 kg. The weight was determined by executing the robot routine Load Identify from the Call Service Routine group. The screwdriver operation time is about 1 up to 1.5 s for each screw; depending on how well the screws are pre fixed (this operation is done by human operators and some time the screw must be screwed for about 1 cm by the robot before tightening). Figure 1 shows two images with the engine assembly and screw position for screws 1, 2, 3 and 4. The screws are positioned on two rows, and some of them are very close to the engine edges.

Because some of the screws are very close to the edge of the engine and because the pipe wrench has the search feature (that means that it wobbles with ±4 mm) there is the risk that the pipe wrench could enter in collision with the engine. In order to avoid this problem a constraint has been imposed regarding the way in which the wrench should approach the screw: the robot will move using a linear

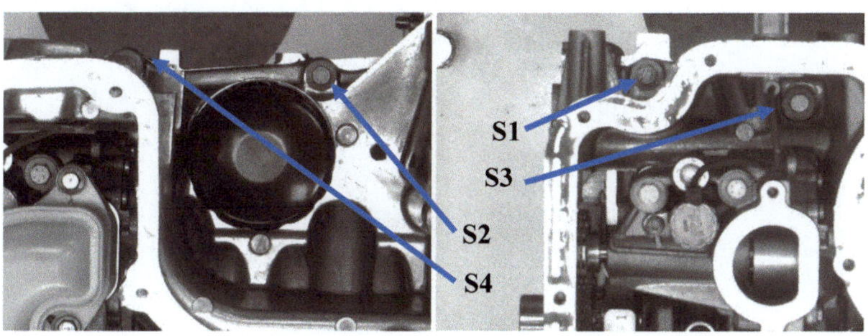

Fig. 1 The engine and screw positions

movement (in Cartesian space) from outside the engine to the screw. The approach point and the point where the wrench touches the screw define a line in space which has a distance of 12 mm from the engine edge in order to avoid collisions (the pipe wrench has a diameter of 12 mm and the wobble has 4 mm; this means that the minimum distance from the engine edge to the wrench edge will be 2 mm).

The sequence of operations in the robot assembly station is:

(a) The robot is available and the workspace does not contain any pallet.
(b) A new pallet is allowed to enter the workspace and the pallet is fixed in place.
(c) The pallet RFID tag is read and the type of engine part is identified (there are three types of engines).
(d) The robot tightens the screws (screw 1–8, and then the screws 1 and 2 are tightened again) and then moves to a safe position.
(e) The pallet is removed from the workspace.

If we compute the time required for all operations one can notice that:

(a) For conveyor operation an amount of 12 s is required.
(b) For screwdriver operation there is needed a minimum time of 1 s × 10 screws (10 s) and a maximum time of 1.5 s × 8 screws and 1 s × 2 screws (14 s).
(c) For robot movement a maximum time of 19 or 23 s remains.

Figure 2 illustrates the moment when the robot engages a screw; one can see the distance on which the screw should be screwed before tightening.

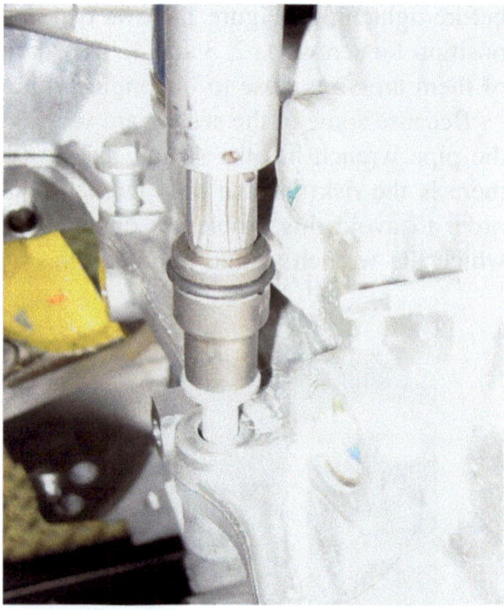

Fig. 2 The moment when the screw is engaged

3 Trajectory and Cycle Time Optimization

In order to optimize the trajectory and the cycle time we started from the constraints of the problem: the first constraint is represented by the points which are used to access the screws, which are fixed. Because the pipe wrench has a telescope with spring, after the screw tightening starts, the screw is moving down and the wrench follows without the need to move the robot.

The second constraint is the shape of the engine and its position in space. The shape of the engine was imported from a CAD file in Matlab and positioned relative to the robot base exactly in the same position as in the assembly station. The shape of the engine and its position were used in order to determine the collisions between the robot and the engine.

In Fig. 3 the optimization algorithm is schematically presented. The initial robot points are the points used to access the screws (extracted from the robot program and trained manually), see Fig. 4; the loop robot point (where the robot is waiting

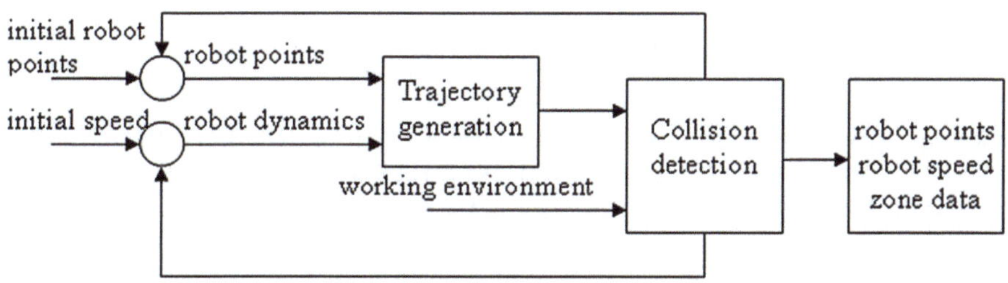

Fig. 3 The algorithm for trajectory and cycle time optimization

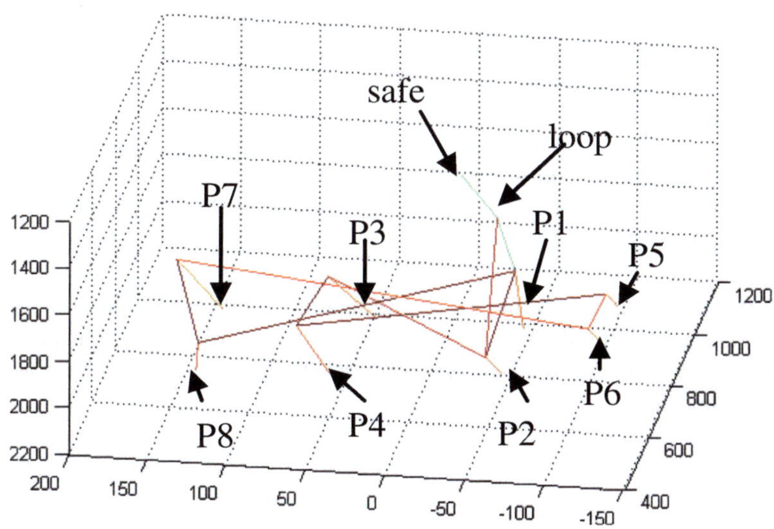

Fig. 4 The positions of the screws

the pallet with the engine) is also trained manually, and the approach points above the screws are used as intermediary points in the robot trajectory. These intermediary points are defined initially as described in Sect. 2 in a plane above the engine at 12 mm.

The initial speed is set to maximum (vmax), the robot dynamics is computed using Robotics Toolbox [10], the trajectory is computed by using the function `ctraj` and the tool weight is also added to the model.

The working environment is defined by the area where the engine is placed; the coordinates are imported in Matlab using the STL file [11] of the engine model.

The collision detection block is based on a function which computes the intersection between the engine model and the robot model (also STL) and the tool defined in the robot model.

The optimization is done in two steps; first the path is computed without taking into account the speed, then the speed and the time required to execute the motion are computed, and finally the trajectory is adjusted in order to maximize the speed.

4 Experimental Results

The trajectory is analysed in sequences: (loop, PA1, P1), (P1, PA1, PA2, P2), (P2, PA2, PA3, P3), (P3, PA3, PA4, P4), (P4, PA4, PA5, P5), (P5, PA5, PA6, P6), (P6, PA6, PA7, P7), (P7, PA7, PA8, P8), (P8, PA8, PA1, P1) where PA is the approach position for each point P1 to P8. For each sequence the zone data has been defined; for final points P1 to P8 the zone is *fine*, for the approach points the initial zone is *fine*, then the zone is increased. Once the zone is increased the PA points are translated on the segment (Px, PAx) in order to avoid the intersection of the zone with the engine. The trajectory is then generated taking into account the dynamics, the collisions are verified and the speed and time is also computed.

Figure 5 shows the simulation window where one can see the robot and the trajectory to be generated.

The zone data is increased until a zone increase is not generating a speed increase or a smaller execution time anymore. The speeds and accelerations are also plotted in order to have a visual check too. The simulation of the robot, the computation of speed and accelerations was executed using the Carlos Baraza [12] simulator.

In Fig. 6 the joint speeds and accelerations for a single execution are presented.

After executing the algorithm the zones are presented in a graphical fashion. Also, the end effector speed, the PA points and associated zone data are saved in a dedicated file. Figure 7 presents these zones.

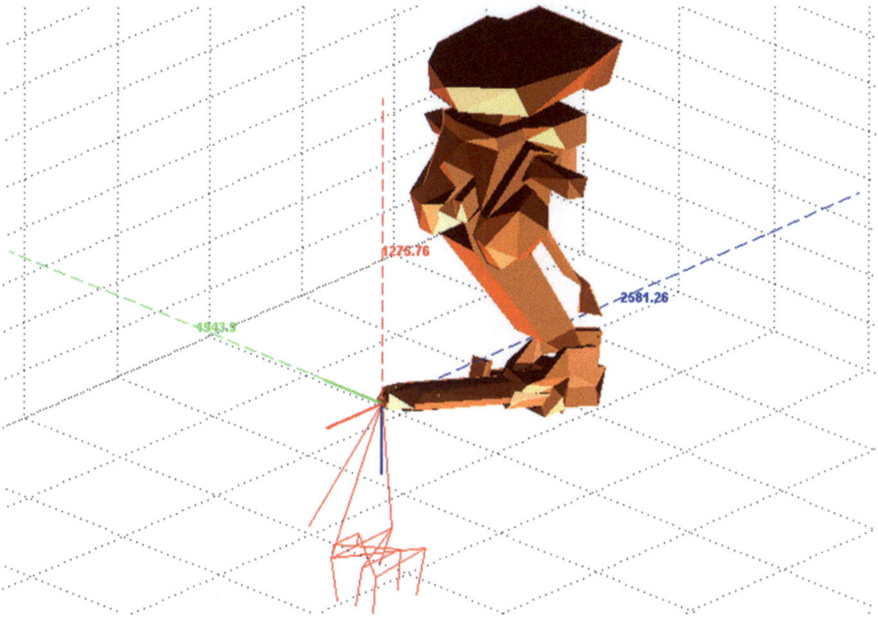

Fig. 5 The robot and the generated trajectory

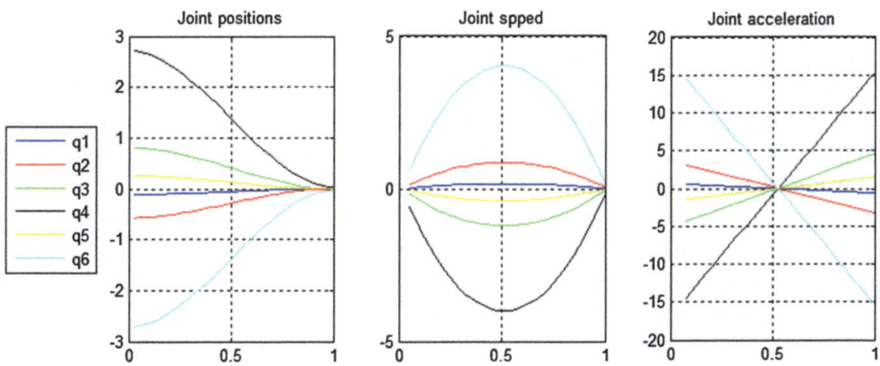

Fig. 6 The joint positions, speeds and accelerations

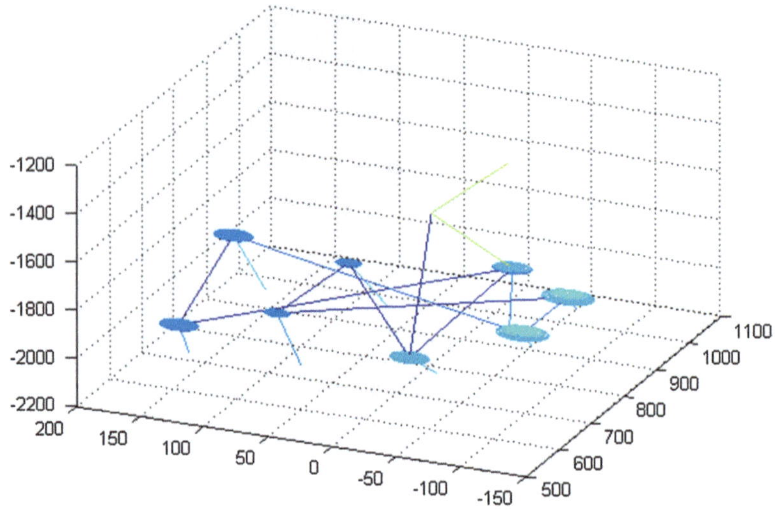

Fig. 7 The resulted zones

5 Conclusion

The resulted speed is v300: this is the maximum speed which the robot can obtain because of the important load which it handles (screwdriver and pipe wrench) and also due to the relative short distances between points.

The speed can be increased by using a set of PAx points defined at a bigger distance from the Px points but this will also increase the execution time. Another way in which the speed can be increased is by modifying the robot's accelerations, but this was not possible from the RAPID robot program, and also this could lead to premature wear of the robot.

The obtained zones were (z50, z50, z30, z25, z80, z80, z50, z40) for PA1 to PA8.

Using these settings in the simulation we obtained a cycle time of 16.53 s only for the robot motion. After the implementation on the robot we obtained a cycle time for the robot and screwdriver operation between 28 and 30 s, this is because the screwdriver operation depends on how well the screws are pre fixed and also because on the simulation the communications between the PLC and the robot were not considered.

Comparing the results with other approaches [13–16] the solution we propose can be easily integrated in manufacturing lines and adds the advantage that the trajectory can be tested offline.

Future developments will consider the possibility to define different zones for the same PA point depending on the trajectory, and also the possibility for the PA points to be tuned on different directions not only on the initial approach direction; this will allow to tune the PA position without increasing the distance that the robot will have to travel.

References

1. Carlson, J.S., Spensieri, D., Wärmefjord, K., Segeborn, J., R. Söderberg: Minimizing dimensional variation and robot traveling time in welding stations, Procedia CIRP. **23**, 77–82 (2014)
2. Huang, Y., Chiba, R., Arai, T., Ueyama, T., Ota, J.: Robust multi-robot coordination in pick-and-place tasks based on part-dispatching rules. Robot. Auton. Syst. **64**, 70–83 (2015)
3. Abdel-Malek, L.L., Li, Z.: Robot location for minimum cycle time. Eng Costs Prod Econ **17** (1–4), 29–34 (1989)
4. Nilakantan, J.M., Huang, G.Q., Ponnambalam, S.G.: An investigation on minimizing cycle time and total energy consumption in robotic assembly line systems. J. Clean. Prod. **90**(1), 311–325 (2015)
5. Björkenstam, S., Spensieri, D., Carlson, J.S., Bohlin, R., Gleeson, D.: Efficient sequencing of industrial robots through optimal control. Procedia CIRP, **23**, 194–199 (2014)
6. Fung, R.-F., Cheng, Y.-H.: Trajectory planning based on minimum absolute input energy for an LCD glass-handling. Appl Math Model **38**(11–12), 2837–2847 (2014)
7. Paes, K., Dewulf, W., Van der Elst, K., Kellens, K., Slaets, P.: Energy efficient trajectories for an industrial ABB robot. Procedia CIRP. **15**, 105–110 (2014)
8. Todtermuschke, M., Findeisen, Bauer, M.A.: Methodology for creation a reference trajectory for energetic comparability of industrial robots in body shop. Procedia CIRP. **23**, 122–126 (2014)
9. http://cadfiles.desouttertools.com/files/0003-Documentation/0100-CVI_Range_Tools_-_Controllers_-_Accessories/0140-Fixtured_Electric_Spindles_EM-EME/0141-In-Line/0001-EME_%28One_cable%29/EME38-20J_6159933813-02_pdf.pdf
10. Corke, P.I.: Robotics, vision and control, Springer, Berlin. ISBN 978-3-642-20143-1 (2011)
11. http://www.mathworks.com/matlabcentral/fileexchange/22409-stl-file-reader
12. https://www.youtube.com/watch?v=fa7GwwA3498
13. Jin, J., Gans, N.: Parameter identification for industrial robots with a fast and robust trajectory design approach. Robotics and Computer-Integrated Manufacturing **31**, 21–29 (2015)
14. Menasri, R., Nakib, A., Daachi, B., Oulhadj, H., Siarry, P.: A trajectory planning of redundant manipulators based on bilevel. Appl. Math. Comput. **250**(1), 934–947 (2015)
15. Abu-Dakka, F.J., Rubio, F., Valero, F., Mata, V.: Evolutionary indirect approach to solving trajectory planning problem for industrial robots operating in workspaces with obstacles. Eur. J. Mech. A. Solids. **42**, 210–218 (2013)
16. Kohrt, C., Stamp, R., Pipe, A.G., Kiely, J., Schiedermeier, G.: An online robot trajectory planning and programming support system for industrial use. Robotics and Computer-Integrated Manufacturing **29**(1), 71–79 (2013)

References

Cost Function-Free Optimization in Inverse Kinematics of Open Kinematic Chains

József K. Tar, László Nádai, Imre Felde and Imre J. Rudas

Abstract The traditional ways of solving various tasks "optimally" in control technology and robotics normally are based on the minimization of some cost function (or functional). On the basis of function minimization various "generalized inverse matrices" can be introduced that have special significance in the inverse kinematic tasks of redundant manipulators, where the possible solutions are ambiguous—therefore various choices are available. The solution suggested here tackles the question of optimality by the geometric interpretation of the simple and computationally efficient Gram-Schmidt algorithm. The method is presented via simulations using a redundant arm structure.

Keywords Open kinematic chain · Gram-Schmidt algorithm · Redundant robot arms · Generalized inverse · Inverse kinematic task

1 Introduction

While the solution of the *forward kinematic task* can be solved by the use of simple mathematical tools in general, that of the *inverse kinematic task* depends on the particular details of the kinematic structure under consideration. *Closed form*

J.K. Tar (✉) · I.J. Rudas
Antal Bejczy Center for Intelligent Robotics, Óbuda University, Bécsi út 96/B, H-1034, Budapest, Hungary
e-mail: tar.jozsef@nikuni-obuda.hu

I.J. Rudas
e-mail: rudas@uni-obuda.hu
URL: http://www.uni-obuda.hu

L. Nádai · I. Felde
Smartlab Knowledge Center, Óbuda University, H-1034, Budapest, Hungary
e-mail: nadai@uni-obuda.hu

I. Felde
e-mail: felde@uni-obuda.hu

T. Borangiu (ed.), *Advances in Robot Design and Intelligent Control*,
Advances in Intelligent Systems and Computing 371,
DOI 10.1007/978-3-319-21290-6_14

analytical solutions exist only in particular cases that are typical in the current industrial applications. Without the need of completeness we mention certain typical examples in the sequel.

As a classic example of 6 "*Degree of Freedom* (*DoF*)" constructions the PUMA-type robots can be mentioned, that were intensively studied even recently [1]. The existence of the closed analytical solution is guaranteed by the application of *strictly parallel* rotational axles, furthermore it is also necessary that the last three rotary axles must have a common point of intersection. The "SCARA" robots are 4 DoF constructions in which a single translational axle exists that is strictly parallel with the three also parallel rotary axles. This construction is very useful for executing assembly tasks on a plain panel with guaranteed selective compliance. As competitors of the SCARA robots the *parallel robots* can be mentioned that generally have closed kinematic chains. Clavel's "DELTA" robot is an excellent example in which parallelograms guarantee the rotation-free translation of a plateau for assembly purposes. The *hexapod robots* as the *Gough platform* from 1942 have closed kinematic chain. Its main virtue is the strong dynamic properties that enable it for working as tire test machines exerting great forces by hydraulic cylinders [2]. The above mentioned structures have the common properties as follows:

- The relatively simple solution of the inverse kinematic task is guaranteed by some geometric specialties.
- These particularities have to be guaranteed by precise manufacturing.
- When manufacturing errors occur the formal basis of the existence of the simple solution of inverse kinematics ceases. The available formulas cannot be "extrapolated" for studying the significance and effects of the modelling errors. These investigations can be done only on a more general mathematical basis.

It is worthy to note that the errors in the kinematic models mean significant problems even in our days. To reduce these errors, in [3] a laser-based calibration system was reported for the use in differential solution of the inverse kinematic task. Dealing with the kinematic uncertainties in robot control is also an actual problem, for its solving novel techniques being recently suggested in [4]. These papers considered co-operating robot arms moving some object. This task automatically leads to the problem of closed kinematic chains that is a subject area in intensive development. In [5] a new technique, based on dual quaternions, was introduced for the analysis of closed linkages with revolute joints: the theory of bonds. The authors considered that the bond structure generally comprises a lot of information on closed revolute chains with a one-parametric mobility. The usefulness of bond theory was demonstrated by giving a new and transparent proof for the well-known classification of over constrained 5R linkages.

While in the construction of *industrial robots* it is reasonable to maintain high precision and to apply special elements, in manufacturing *toy robots* neither high precision nor the application of expensive joints is reasonable. It is reasonable to assume that in this field the application of simple 1 DoF rotary joints is satisfactory without guaranteeing special restrictions as e.g. that three rotary axles must have a common point of intersection or certain axles must be strictly parallel. In these cases we cannot expect the existence of closed form analytical solutions for the inverse kinematic task. In general, the "differential tackling" of the inverse kinematics represents a viable approach. In the sequel we restrict ourselves to such constructions.

2 Formulation of the Kinematics of Open Chain Arm Structures Using Rotary Joints

It is expedient to tackle the forward kinematics by the use of the concept of "*home position*" that is illustrated by Fig. 1. In this position the arm consists of the segments $\{1, 2, \ldots, n\}$ so that to the rotation values $\{q_{1_0}, q_{2_0}, \ldots, q_{n_0}\}$ and the actual axles (as unit vectors) $\left\{e^{(1)}, e^{(2)}, \ldots, e^{(n)}\right\}$ belong together to the arm segments $\left\{r^{(1)}, r^{(2)}, \ldots, r^{(n)}\right\}$. Each of these vectors is given with respect to the workshop system of coordinates. (Without the loss of generality it is assumed that in this position the generalized coordinates take the value 0.) The vector connecting the base point of the robot and the Tool Center Point (TCP) with respect to the workshop frame can be obtained by consecutive rotations as follows: at first a rotation $O^{(n)}$ is executed around axis $e^{(n)}$ that transforms the vector connecting the base point of the last segment and the tool centre point $\{r\}$ into $O^{(n)}r$ while it leaves the other vectors invariant.

In the next step the vector connecting the base points of the previous and the current segment $\left\{r^{(n)}\right\}$ and $\left\{O^{(n)}r\right\}$ as $\left(O^{(n)}r + r^{(n)}\right)$ will be rotated by $O^{(n-1)}$ around axis $e^{(n-1)}$ $O^{(n-1)}\left(O^{(n)}r + r^{(n)}\right)$, and so on. These operations consist of a rotation and a subsequent translation that can be described by the "*homogeneous matrices*" of size $\mathbb{R}^{4\times 4}$ representing the "*Special Euclidean Matrix Group SE(3)*" as

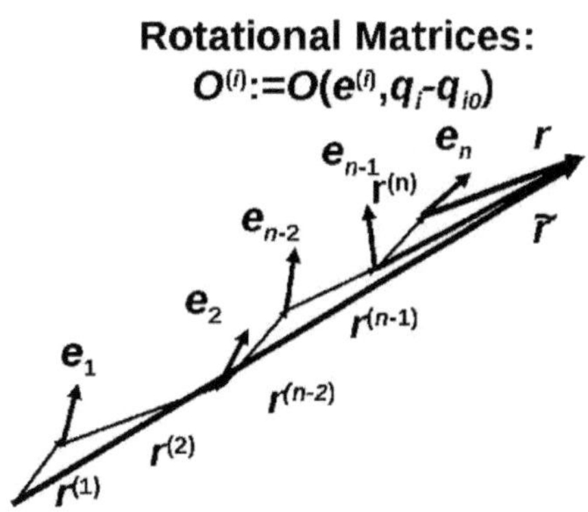

Fig. 1 The "home position" of the robot arm

$$H^{(n)}(q_n)\begin{pmatrix} r \\ 1 \end{pmatrix} = \begin{pmatrix} O^{(n)}(q_n) & r^{(n)} \\ 0^T & 1 \end{pmatrix}\begin{pmatrix} r \\ 1 \end{pmatrix} = \begin{pmatrix} O^{(n)}(q_n) + r^{(n)} \\ 1 \end{pmatrix},$$
$$\begin{pmatrix} \tilde{r} \\ 1 \end{pmatrix} = H^{(1)}(q_1)H^{(2)}(q_2)\ldots H^{(n)}(q_n)\begin{pmatrix} r \\ 1 \end{pmatrix} \tag{1}$$

where $\tilde{r}$ denotes the vector connecting the base point of the robot and the TCP, and in $H(q_1, q_2, \ldots, q_n) = H^{(1)}(q_1)H^{(2)}(q_2)\ldots H^{(n)}(q_n)$ the upper left $\mathbb{R}^{3\times3}$ block $O(q_1, q_2, \ldots, q_n) = O^{(1)}(q_1)O^{(2)}(q_2)\ldots O^{(n)}(q_n)$ corresponds to the rotational matrix that describes the rotation of the work piece moved by the robot (both are understood with respect to the workshop frame).

For giving the analytical form of the $O^{(\mathrm{i})}(q_\mathrm{i})$ matrices it is not compulsory to use the Denavit-Hartenberg conventions. The Rodrigues formula, directly using the axis vector e and the rotational angle q can be applied too [6]: $O(e,q) \stackrel{\text{def}}{=}$

$$\begin{pmatrix} c_q + (1-c_q)e_1^2 & (1-c_q)e_1e_2 - s_qe_3 & (1-c_q)e_1e_3 + s_qe_2 \\ (1-c_q)e_2e_1 + s_qe_3 & c_q + (1-c_q)e_2^2 & (1-c_q)e_2e_3 - s_qe_1 \\ (1-c_q)e_3e_1 - s_qe_2 & (1-c_q)e_3e_2 + s_qe_1 & c_q + (1-c_q)e_3^2 \end{pmatrix} \tag{2}$$

where $c_q = \cos q$ and $s_q = \sin q$. Taking into account that both the homogeneous and the rotational matrices form "*Lie groups*", the $G(t) \stackrel{\text{def}}{=} \dot{H}(t)H^{-1}(t)$ and the $\Omega(t) \stackrel{\text{def}}{=} \dot{O}(t)O^{-1}(t)$ quantities describe the elements of the tangent space of the appropriate groups at their unit elements (i.e. at the unit matrices of appropriate sizes)—in other words they are the generators of the appropriate groups.

$$H = \begin{pmatrix} O & a \\ 0^T & 1 \end{pmatrix}, \quad H^{-1} = \begin{pmatrix} O^{-1} & -O^{-1}a \\ 0^T & 1 \end{pmatrix}, \quad \dot{H} = \begin{pmatrix} \dot{O} & \dot{a} \\ 0^T & 0 \end{pmatrix},$$
$$G = \begin{pmatrix} \Omega & -\Omega a + \dot{a} \\ 0^T & 0 \end{pmatrix} \tag{3}$$

The simple physical interpretation is: the skew-symmetric Ω describes the actual rotational speed of the object rigidly attached to the tool, and $-\Omega a + \dot{a}$ describes the translational speed of the TCP with respect to the workshop frame. By taking into consideration the chain rule of differentiation in $H(q_1, \ldots, q_n)$, that is a multiplicative structure, we obtain:

$$\begin{aligned} G &= \frac{dH^{(1)}}{dq_1} H^{(1)^{-1}} \dot{q}_1 + H^{(1)} \frac{dH^{(2)}}{dq_2} H^{(2)^{-1}} H^{(1)^{-1}} \dot{q}_2 + \ldots \\ &\quad + H^{(1)} \ldots H^{(n-1)} \frac{dH^{(n)}}{dq_n} H^{(n)^{-1}} H^{(n-1)^{-1}} \ldots H^{(1)^{-1}} \dot{q}_n \\ &= G^{(1)} \dot{q}_1 + H^{(1)} G^{(2)} H^{(1)^{-1}} \dot{q}_2 + \ldots + H^{(1)} \ldots H^{(n-1)} G^{(n)} H^{(n-1)^{-1}} \ldots H^{(1)^{-1}} \\ &= \tilde{G}^{(1)} \dot{q}_1 + \ldots + \tilde{G}^{(n)} \dot{q}_n \end{aligned} \tag{4}$$

Since in the Lie groups to an *arbitrary generator G* and an *arbitrary group element W* the expression WGW^{-1} *also is a generator*, Eq. (4) has the following physical interpretation: the six independent components of the rotational and translational speed of the work piece, with respect to the workshop frame, can be described as the linear combination of group generators by the joint coordinate velocities $\{\dot{q}_1, \ldots, \dot{q}_n\}$.

By placing the six independent components of G into the rows of a column array G_{Col}, (4) will have a simple mathematical interpretation: by describing the desired velocities in G with respect to the workshop frame, these 6 components have to be constructed as the linear combination of columns with the joint coordinate velocities $\{\dot{q}_1, \ldots, \dot{q}_n\}$. By using arrays it can be written as $G_{Col} = J\dot{q}$ where J is referred to as the Jacobian matrix of the inverse kinematic task. If $n > 6$ then the system is redundant. If the rank (the number of the linearly independent columns) of J is equal to 6, we have ambiguous solutions. If this rank is less than 6, we have kinematic singularity and no exact solution of the differential inverse kinematic task exists in general (in this case it may exist only for special trajectory).

The traditional solutions as the "Moore-Penrose Pseudoinverse" [7, 8] consider this problem as optimization under constraints as for a $P > 0$ symmetric positive definite matrix one can minimize the quantity $\frac{1}{2}\dot{q}^T P \dot{q}$ under the constraints $J\dot{q} - G_{Col} = 0$.

The solution, if it exists, can be expressed as $\dot{q} = P^{-1}J^T(JP^{-1}J^T)^{-1}G_{Col}$. It is evident that if J^T has a non-empty *null space* the quadratic matrix $JP^{-1}J^T$ does not have inverse and the problem is singular. Formal possibilities for getting rid of the singularities can be regarded as some "deformations" of the original task. By introducing a small parameter $\mu > 0$ the approximate solution $\breve{\dot{q}} = P^{-1}J^T P^{-1}J^T(JP^{-1}J^T + \mu I)^{-1}G_{Col}$ always exists.

In the non-singular case when $\mu \ll min\{eigenvalues\ of\ JP^{-1}J^T\}$ the deformation of the problem is insignificant. The idea was first introduced by Levenberg in 1944 and sometimes is referred to as "*Damped Least Squares*" e.g. [9]. In 1991 Pohl [10] suggested a 2nd order approach instead of the use of the linear equations near the singularities. Pohl and Lipkin [11] in 1993 suggested complex extension of the generalized coordinates for task deformation. Other generalized inverses based on the "*Singular Value Decomposition* (*SVD*)" [12] are not related to cost function minimization but can guarantee controlled approximation errors to non-singular problems. Other formal possibility for "colouring" the problem would be the application of coordinate- and task-dependent $P(q)$ weighting in the Moore-Penrose pseudoinverse. In this approach it would not be trivial how to determine P as the function of the task's specialties. In the sequel a simple and reasonable method is suggested that drops the use of the concept of "cost function" and considers the problem through a simple geometric interpretation.

3 The Application of the Gram-Schmidt Algorithm to Avoid the Use of Generalized Inverses

Consider the following task: there is given a set of certain vectors $\{x^{(i)} \in \mathbb{R}^n | i = 1, \ldots, m \leq n\}$; the construction of a "*set of orthogonal vectors*" is desirable via the utilization of these vectors. Independently of each other, Laplace in 1820, Gram in 1883, and Schmidt [13] in 1907 suggested a simple solution. Its essence can be highlighted as follows: assume that we have two real vectors a, b of identical dimensions. Assume that instead of b we wish to use a vector that is orthogonal to a. Since orthogonality means zero scalar product, with a parameter $\lambda \in \mathbb{R}$ the $\tilde{b} \stackrel{\text{def}}{=} b + \lambda a$ can be suggested with the prescription: $\langle b + \lambda a | a \rangle = 0$ that has the solution $\lambda = \frac{-\langle b|a \rangle}{\langle a|a \rangle}$. This simple idea can be applied for the orthogonalization of an initial set of vectors as follows: (a) Choose one element of this set to be the 1st one of the new, orthogonalized set. (b) Remove its components from the remaining vectors. In this manner a subset is achieved that spans a linear subspace that is orthogonal to the 1st new vector. (c) Select an element of this subset and remove its components from the remaining vectors. The vectors obtained thus span a linear subspace that is orthogonal to the first two vectors. (d) Repeat this procedure until obtaining an orthogonal set of basis vectors. If the above reductions are also executed on an arbitrary vector b, it automatically appears as the linear combination of the new orthogonal vectors. Since during the procedure the appropriate λ values can be stored in the memory, by the use of the orthogonality of the above mentioned subspaces it is very easy to obtain b as a linear combination of the original vectors $\{x^{(i)}\}$. This procedure has the following properties: (a) if the original set of vectors cannot span the complete linear space its yields the best available approximation of b as far as the use of these basis vectors is concerned; (b) in the case of redundant original set, depending on the order of the choices, it provides one of the ambiguous solutions; (c) if the original set is not redundant it yields the unique solution, independently of the order of choosing the new basis vectors. We note that this method is very efficient computationally. A generalized inverse solves the problem for each possible vectors, but in the case of the inverse kinematic task we always need a solution only for a given, particular vector b. The suggested optimization without goal function is formulated as follows: in each step of the orthogonalization process that vector of the remaining set is chosen for which the absolute value of the scalar product with the remaining fragment of b is maximal. In this manner the rotational angle of the actual wrist joint is evidently minimized.

4 A Computational Example

In the investigations a standard 6 DoF PUMA structure has been completed with further two axles to produce a redundant, "non-canonical" 8 DoF construction with the kinematic parameters of the "home position" given in the columns of the matrices in (5) with $g_2 = 1/\sqrt{2}$, $g_3 = 1/\sqrt{3}$ (non-dimensional quantities) in the unit vectors of the rotary axles, and $L = 0.5$, $l = 0.1$, $S = 0.3$, and $h = 1$ for the arm segments in *meter* units. In the Gram-Schmidt algorithm the smallest vector not replaced by 0 had the absolute value at least 10^{-4} times that of the greatest one.

$$\left\{e^{(i)}\right\} = \begin{pmatrix} 0 & 0 & 0 & g_2 & g_2 & 0 & g_3 & g_3 \\ 0 & 1 & 1 & 0 & 0 & 1 & g_3 & -g_3 \\ 1 & 0 & 0 & -g_2 & g_2 & 0 & g_3 & g_3 \end{pmatrix}, \quad \left\{r^{(i)}\right\} = \begin{pmatrix} 0 & 0 & L & L & 0 & 0 & S & 0 \\ 0 & l & 0 & -l & 0 & 0 & 0 & S \\ h & 0 & L & -L & 0 & 0 & -S & 0 \end{pmatrix} \tag{5}$$

The interesting trajectories that can reveal the advantages of optimization are located in the vicinity of singularities. The same parameters were set in the computations, with the exception of the order of orthogonalization. The comparative Figs. 2, 3 and 4 illustrate clearly the advantages of the suggested simple optimization method.

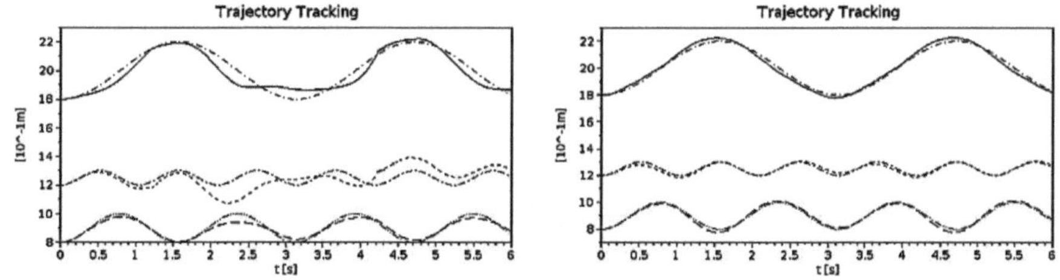

Fig. 2 Trajectory tracking in $\mathbb{R}^3$: non-optimal order (*LHS*), optimal order (*RHS*) [x_1, x_2, x_3 nominal trajectories dash-dot, long dash-dot, dash-dot-dot]

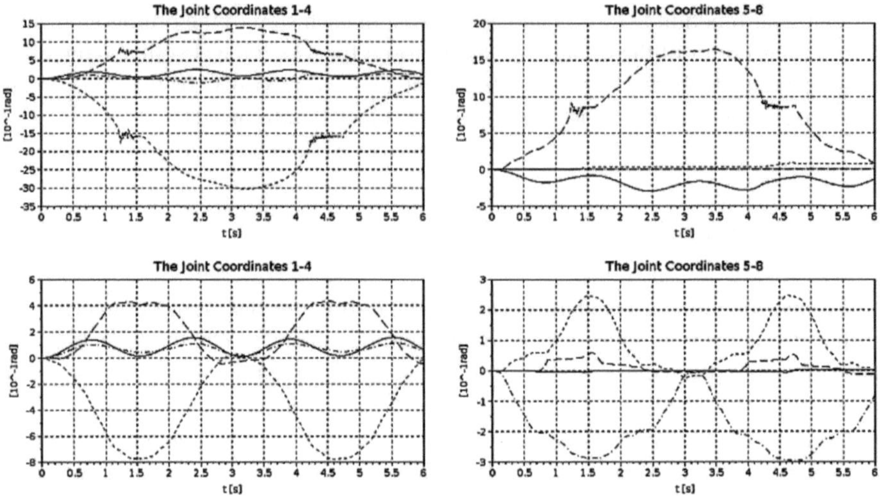

Fig. 3 The rotation of the robot axes: non-optimal order (*top*), optimal order (*bottom*) dash-dot, dash-dot-dot] (during its constant phase a given axle was not used in solving the inverse kinematic task)

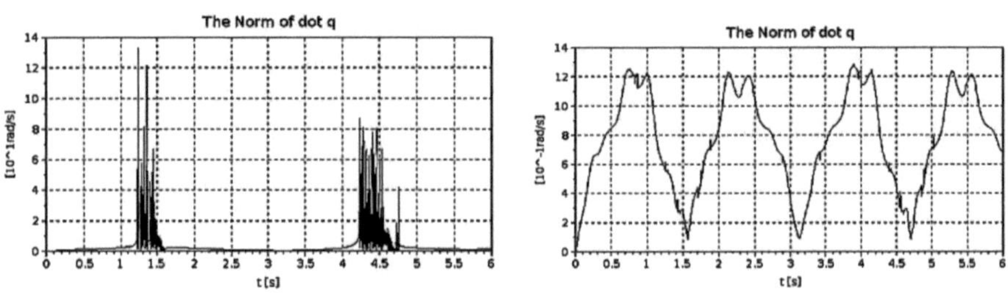

Fig. 4 The Frobenius norm of the array $\dot{q}$: non-optimal order (*LHS*), optimal order (*RHS*)

5 Conclusions

In this paper a simple, geometric interpretation-based application of the Gram-Schmidt algorithm was suggested to solve the differential inverse kinematic task of open kinematic chain robot arms with non-special construction when no closed analytical solution exists to the problem. The simple idea is choosing an "optimal" order of orthogonalization to avoid the occurrence of too great angular rotations around the robot axles.

Via completion with an error feedback term with respect to the workshop frame, this method was applied in a Kinect based system to realize hand-controlled rotation-free translation of a virtual work piece by this simulated robot.

Acknowledgments This works has been partially supported by the *Hungarian Scientific Research Fund OTKA* within the project no. K-106392.

References

1. Benitez, A., Huitzil, I., Casiano, A., Calleja, J.D.L., Medina, M.: Puma 560: robot prototype with graphic simulation environment. Adv. Mech. Eng. **2**, 15–22 (2012)
2. Merlet, J.P.: Parallel robots. In: Gladwell, G.M.L. (ed.) The series solid mechanics and its applications, vol. 128. Springer, The Netherlands (2006)
3. Park, I.W., Lee, B.J., Cho, S.H., Hong, Y.D., Kim, J.H.: Laser-based kinematic calibration of robot manipulator using differential kinematics. IEEE/ASME Trans. Mechatronics. Available: http://ieeexplore.ieee.org. doi:10.1109/TMECH.2011.2158234 (2011)
4. Aghili, F.: Adaptive control of manipulators forming closed kinematic chain with inaccurate kinematic model. IEEE/ASME Trans. Mechatron. **18**, 1544–1554 (2013)
5. Hegedűs, G., Schicho, J. and H. Schröcker: The theory of bonds: A new method for the analysis of linkages. Mech. Mach. Theory. **70**, 407–424 (2013)
6. Rodrigues, O.: Des lois géometriques qui regissent les déplacements d' un système solide dans l'espace, et de la variation des coordonnées provenant de ces déplacements considérés indpendents des causes qui peuvent les produire. J. Math. Pures Appl. **5**, 380–440 (1840)
7. Moore, E.: On the reciprocal of the general algebraic matrix. Bull. Am. Math. Soc. **26**, 394–395 (1920)
8. Penrose, R.: A generalized inverse for matrices. In: Proceedings of the Cambridge Philosophical Society, vol. **51** (1955)
9. Chiaverini, S., Egeland, O., Kanestrom, R.: Achieving user-defined accuracy with damped least squares inverse kinematics. In: IEEE International Conference on Robotics and Automation, **672**, pp. 19–22, Pisa, Italy (1991)
10. Pohl, E.: A new method of robotic rate control near singularities. In: IEEE International Conference on Robotics and Automation, vol. **405**, pp 19–22, Pisa, Italy (1991)
11. Pohl, E., Lipkin, B: Complex robotic inverse kinematic solutions. J. Mech. Des. **115** (1993)
12. Golub, G., Kahan, W.: Calculating the singular values and pseudoinverse of a matrix. SIAM J. Numer Anal **2**, 205–224 (1965)
13. Schmidt, E.: Zur Theorie der linearen und nichtlinearen Integralgleichungen. I. Teil: Entwicklung willkurlicher Funktionen nach Systemen vorgeschriebener. Matematische Annalen **63**, 442 (1907)

Control and Coordination System Supported by Biologically Inspired Method for 3D Space "Proof of Concept"

Ján Zelenka and Tomáš Kasanický

Abstract This paper presents a control strategy for a swarm of mobile/flying robots operated in 3D space. The described biologically inspired method was developed to solve exploration or monitoring tasks. We discuss a possibility to extend an algorithm [1] to 3D space in this paper. The presented model is a decentralized adaptive system with shared memory representing the environment.

Keywords 3D space exploration · Cellular automata · Biologically inspired method

1 Introduction and Motivation

One of the main issues of multi-robotics solutions is to ensure the functionality of the whole system when a malfunction of some robots occurs. Therefore it is necessary to develop systems with graceful degradation behaviour. Many works in this area were done, but most of them are based on a complicated negotiation or planning algorithms. When some of the robots cannot perform their tasks then the whole strategy must be re-planned or a complicated negotiation process starts among the remaining valid robots. Therefore a multi-robot system, capable to achieve global goals without re-planning the whole strategy was developed. This robotic system handles the area coverage problem. Let us have a space and robots equipped with sensors. Each robot can observe an assigned sector of the space (the sensors' type depends on the task). In our previous work [1] an algorithm was

J. Zelenka (✉) · T. Kasanický
Institute of Informatics, Slovak Academy of Sciences, Dúbravská Cesta 9, Bratislava, Slovakia
e-mail: jan.zelenka@savba.sk

T. Kasanický
e-mail: kasanicky@neuron.tuke.sk

T. Borangiu (ed.), *Advances in Robot Design and Intelligent Control*,
Advances in Intelligent Systems and Computing 371,
DOI 10.1007/978-3-319-21290-6_15

presented which is able to reach a global goal even when only one robot from a group remains valid. This solution is based on two dimensional cellular automata; it is sufficient for most of the real world applications, but in some cases it might be difficult to use two-dimensional approximation. For example, aerial photography task in a rugged environment will be difficult because a constant flight level is needed. Another typical task where the two-dimensional solution is useless is patrolling in a mountain area, and also patrolling in urban areas with tall buildings, etc. Following our previous work [1], we decided to develop a solution based on cellular automata, extended by a three-dimensional cellular grid.

A large body of research has investigated robot or multi-robot coverage and exploration problem. Many experimental and real applications exist in 2D [2–4, 5], 2.5D [6] and 3D [7, 8] environments. Applications of the coverage path planning in domains such as agricultural robotics [9] and unmanned aerial vehicles (UAVs) [10–13] have been reported in the literature. In [14] authors give a comprehensive explanation of the existing methods and bring a detailed classification of the existing approaches. Many approaches address optimal coverage methods. In [2] authors use a spanning tree coverage algorithm for optimal 2D space exploration. This approach also uses cellular automata as a platform and pheromone marks for coordination. Another work [15] is focused on a problem of complete coverage of an unstructured environment by a mobile robot; this paper describes a deterministic algorithm. The goal of a single robot exploration is to minimize the overall exploration time. The key problem to be solved in the context of multiple robots is to choose appropriate target points for each of the individual robots. In [16] the authors improved the path planning in 3D cellular space by D* algorithm. Investigation of efficient and inefficient coverage methods are discussed in [17]. All these methods solve the optimal coverage path for a defined swarm of robots. In case of malfunction of a single robot it is necessary to re-plan individual paths. In [18] authors use a combination of optimization methods with a virtual bird flocking for space coverage by a group of mobile robots. Multi-robots coverage problem in unknown 2D space by using the frontier-based techniques is described in [19]. These solutions were designed for 2D space and their extension for 3D was out of the investigation scope.

2 A Biologically Inspired Method for 3D Space Coverage

Our approach is based on cellular automata. Each robot in the system has its own representation of the environment (a map). The map of the robot's environment is 3D cellular grid. The robot uses the grid for navigation and as a memory. Each situation which can be observed by the robot is entered into the map. Robots in our approach cannot communicate directly; each robot communicates with the others trough pheromone marks in the environment. When a particular robot moves to a new location, the robot marks this location by its pheromone mark. Since it is difficult to create a pheromone mark in the real world (humidity, wind, evaporation, etc.) we use virtual pheromone marks. When a robot marks a cell in the 3D grid, then the robot broadcasts this information to all other robots. When a robot receives a broadcast message from another robot it subsequently changes its map. This simple communication gives robots the ability to share information. Coordination of robots is ensured by evaporating pheromone marks. Coordination is based on simple attraction/repulsion behaviour of the robots. The environment is divided into regular cubic cells.

The robot can perform one step or stay in an actual position (in case of collision) in every iteration of time; the robot can change its position from $[x_t, y_t, z_t]$ at time t to $[x_{t+1}, y_{t+1}, z_{t+1}]$ at time $t + 1$, $\{x_s, y_s, z_s\} \in N$, where N represents a set of neighbour cells (3D Moore or 3D Neuman neighbourhood) and s represent discrete time t. Each robot has a set of rules that determines its behaviour, because the designed system is based on cellular automata. The movement of robots is controlled by the following rules (taken from the 2D platform [1, 20]):

- A robot moves to a neighbour cell with the smallest pheromone value; if there exist more cells with the smallest pheromone value, the robot randomly selects one of them;
- If there exist cells which were not previously marked by robots or the pheromone value of the cell is zero, then the robot makes a move to one of these cells; the robot stores the movement direction, generates a random turn direction and random direction ratios RD, $RD \in < 1, Z^+ >$ and continues in that direction until cells in the selected direction are unmarked by pheromones;
- If more robots want to move to the same cell, then one of them will be randomly selected as the winner.
    - At time $t + 1$ the winner will move to a conflict cell and all other robots stay in their current positions.
    - Defeated robots can repeat the cell selection one time. At time $t + 1$ the winner will move to the cell and defeated robots can move to a new selected cell if no conflicts between robots occurred.

The RD parameter ensures a better distribution of robots in space at the start of the coverage process. This type of motion in the cell space was also investigated in [19]. When the robot is present in the cell it inserts a pheromone mark into the cell. Let us consider a cell $[x, y, z]$; then the value of a pheromone mark $P_{xyz}(t)$ at time t is calculate as follows:

$$R_{xyz}(t) = \begin{cases} 0 & \text{robot is not in the cell } xyz \text{ at time } t \\ 1 & \text{robot is in the cell } xyz \text{ at time } t \end{cases}; \tag{1}$$

$$P_{xyz}(t) = \begin{cases} P_{max} * R_{xyz}(t) & \text{if } P_{xyz}(t-1) = 0 \\ P_{max} * R_{xyz}(t) + P_{xyz}(t-1) - 1 & \text{if } P_{xyz}(t-1) > 0 \end{cases}. \tag{2}$$

The initialization level of the pheromone mark P_{max} is calculated at the start and depends on the exploring space size and the number of robots in the swarm:

$$P_{max} = \frac{number\ of\ cells\ in\ exploring\ space}{number\ of\ rbots\ in\ swarm} * k, \tag{3}$$

where k represents the pheromone ratio. The following simulations will be focused on verifying the pheromone ratio k in regard of mission requirements (exploration time, conflicts).

3 Simulation Results

Our program written in MATLAB software tool is able to simulate the swarm coordination based on pheromone marks in 3D space. In our simulation the visited cell is evaluated by the pheromone value according to Eqs. (1) and (2). The 3D Moore neighbourhood at the size 3 × 3 × 3 cell represents the direction selection area. The size of these areas will be tested in regard of the mission requirements (robots arrangement in space, exploration time, conflicts).

In order to verify the assumption that one robot can achieve the global goal in 3D environment (cover all space) we prepared the first simulation. Robot motion is affected by the *RD* parameter which is determined randomly. In case the movement and random direction ratio are the same, or in case 3D space the ratio is 1:1 the robot motion is rectilinear (this motion is marked as billiard ball motion, Fig. 1 [21]). The first experiment was done with one robot and the *RD* parameter was set up randomly. The robot movement was tested in space with different obstacle distribution, as illustrated in Fig. 2. Table 1 summarizes the simulation results. The

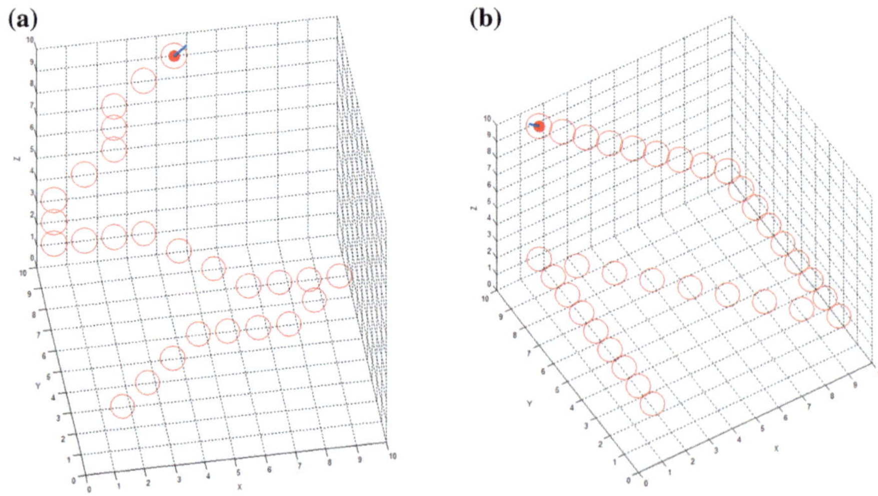

Fig. 1 **a** Motion of the robot with a random *RD* parameter and **b** billiard motion of the robot in space (*red circles* represent the pheromone mark, *red sphere* represents the robot) (color figure online)

optimal coverage method represents the situation when the exploration time equals the number of cells in the explored space (in case of one robot). Because of a small probability that the algorithm can perform optimal coverage, we decided to introduce a parameter E_a which represents the average value of space exploration time (number of iteration):

$$E_a = average\{E_1, E_2, \ldots E_n\}; \tag{4}$$

where n is the number of experiments. This value can be reduced by increasing the pheromone ratio k value or adding rules to control robot's behaviour.

The experiment with a different pheromone ratio k was done in space illustrated in Fig. 2b. Figure 3a shows how the pheromone ratio k impacts the average value of the exploration time E_a. The line has slow decreasing character. From the $k = 1.2$ one can observe minimal differences between the simulation results. In the following experiments we changed the space so that we reduced time and computing complexity of the simulation. Our experimental space represents an urban area ($90 \times 80 \times 20$ cells with obstacles). The simulation results are illustrated in Fig. 3b. In Figs. 3b and 4a, the results of the average exploration time of the swarm of one and of five robots are respectively illustrated. The graph dependency is similar in both cases. It is evident that the exploration time is dependent on the pheromone ratio (depending on the space size) and the appropriate solution for the proposed approach is from the pheromone ratio-$k > 1.1$.

Figure 4b illustrates the dependence on the number of robots in the swarm and the average exploration time consumed for the first episode. The line character is decreasing, but a high number of robots have no significant effect regarding the exploration time. Conflicts occur if there is a larger number of a robot in the swarm.

A conflict represents the situation when more robots want to go to the same cell. The raising line represents conflicts between robots in the swarm as shown in

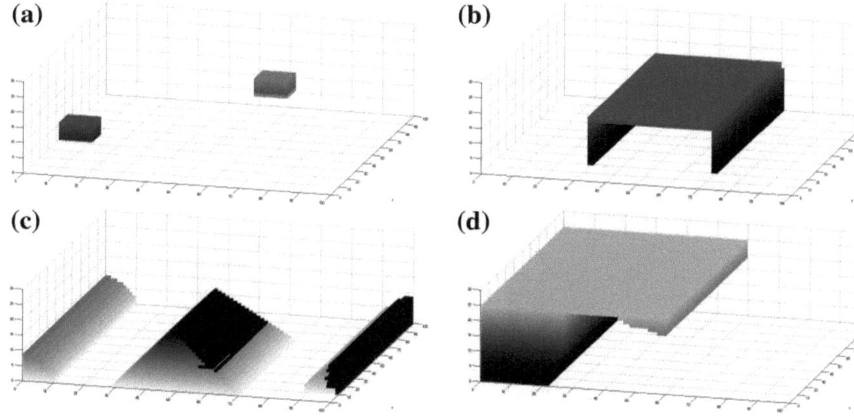

Fig. 2 Various types of explored spaces with different obstacle distribution

Table 1 Simulation results of the exploration experiment (exploration space size 100 × 100 × 30 cells, one robot in swarm, random direction ratio, $k = 1.1$, 800,000 simulated iteration, one cell as an evaluated area, 3D Moore as the direction selected area, $n = 50$ repeated simulations)

Space with obstacles	Figure 2a (1452 obstacles)	Figure 2b (11,178 obstacles)	Figure 2c (41,113 obstacles)	Figure 2d (72,500 obstacles)
Average value of the first exploration episode	525,992.7 ± 13 %	531,260.5 ± 13.2 %	441,160.8 ± 14.5 %	413,316.2 ± 23.6 %

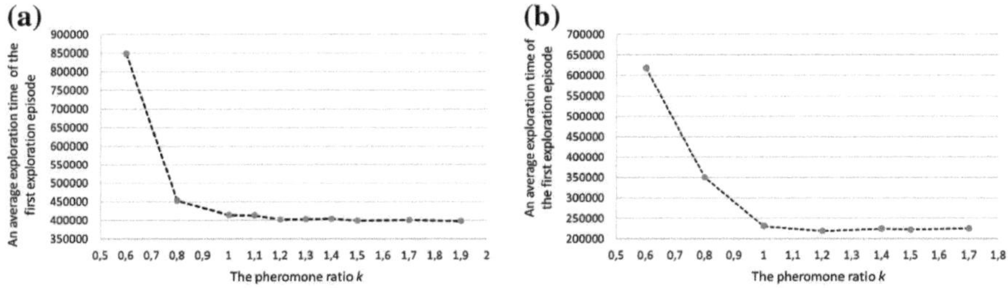

Fig. 3 The average exploration time of the first exploration episode dependent on the pheromone ratio *k*, one robot exploring space illustrated on Fig. 2b and urban area

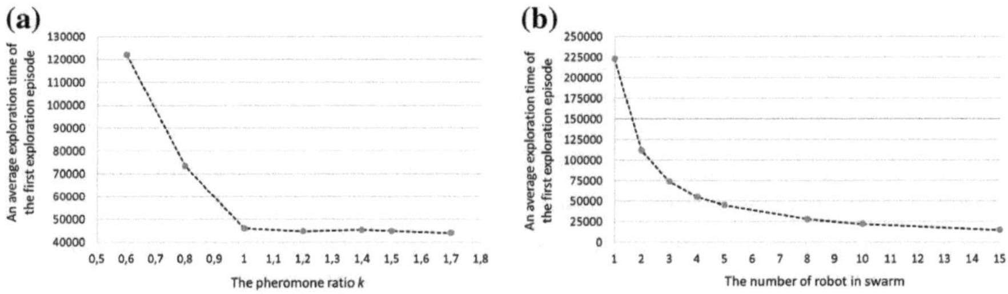

Fig. 4 The average exploration time of the first episode dependent on the pheromone ratio *k* with the swarm of 5 robots and on the number of robots in swarm, respectively

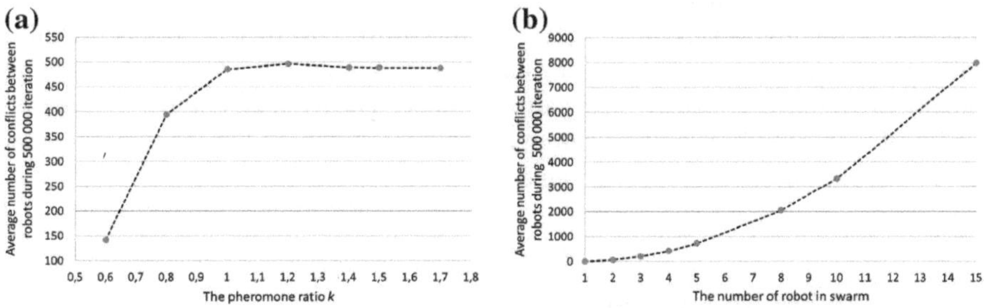

Fig. 5 An average number of conflicts between robots during 500,000 iteration

Fig. 5b. If in 10 robots-swarm the conflicts represent 1.5 % of the average exploration time (per one robot), then in 15 robots-swarm the conflicts increased to 3.5 % per each robot. Dependence on the average number of conflicts and the pheromone ratio *k* is shown in Fig. 5a. It is evident from the simulation results that the average number of conflicts is constant from $k > 1.1$. This simulation is done for the swarm of 5 robots and conflicts represent 0.23 % of the average exploration time per one robot. The upper limit of the interval in which it is optimal to choose a respective pheromone ratio was determined in the simulation where the mutual

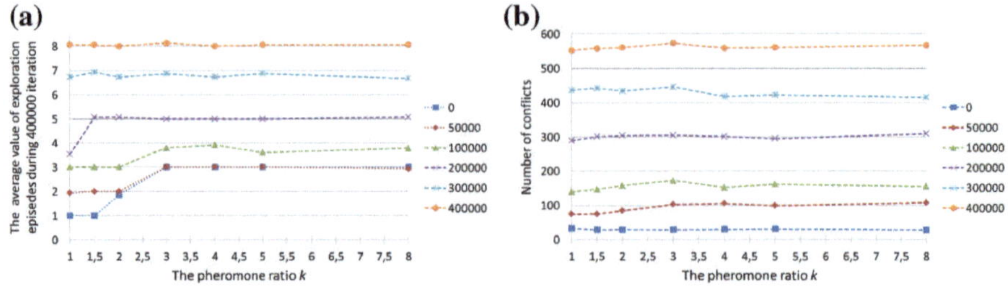

Fig. 6 The average value of exploration episodes and conflicts during 400,000 and 15 repetitions in a space with barriers with 5 robots and a variant pheromone ratio k

communication of robots was broken. Those simulations required more computation time. For this reason we illustrated the simulation results obtained from 15 repetitions for each simulation setting (pheromone ratio, time when mutual communication is switched off). Figure 6a shows the average value of exploration episodes during 400 k iterations. The individual lines represent the time when mutual communication between robots is switched off. The communication was switched off at the following iterations 0, 50, 100, 200, 300, and 400 k. The first situation (communication is switched off at iteration zero) means that the swarm of robots would not communicate from the beginning of the simulation. When the communication is switched off at 400 kth iterations, each movement of all the robots is merely based on their particular pheromone map. Each simulation will take 400 k iterations and the results represent the average value from 15 repetitions in the space representing urban area. From the simulation results (Fig. 6a) it is evident that the high pheromone value has a positive effect for robot coordination without a mutual communication. From the pheromone ratio $k > 3$, the average value of exploration episodes is partially the same. The average number of conflicts in case of a switched-off mutual communication is shown in Fig. 6b. Conflicts are not dependent on the pheromone value k.

4 Conclusion and Discussion

In this paper we propose a study on the biologically motivated method for 3D space exploration. The statement that one robot can achieve a global goal of the mission in 3D platform was verified; a robot is able to explore 3D space by the proposed method. We have investigated the pheromone ratio parameter; specifically the pheromone ratio impacts on the first exploration episode, on the number of robots in the swarm exploring area, on the increasing number of conflicts between robots and on the communication malfunction of the robots. From the simulation results it is evident that the higher pheromone value has no significant effect on the exploration time, but in the case of communication malfunctions the higher pheromone value had a positive influence on the exploration time. We assume that a correct adjustment of the pheromone neighbourhood size and of the pheromone mark quantity will decrease the number of conflicts between robots. In the above described approach, robots can move in 3D space. When a robot made changes in altitude some changes in its energy state would subsequently appear. The change is much bigger when compared to the size of a covering area. Therefore it is important to investigate the movement strategy based on efficiency of the energy management. A situation might appear when the selection of new cell would be dependent not only on the pheromone value, but also on the energy efficiency parameter. However, the investigation would be strongly dependent on the robotic platform (multi-rotor copter, flying wing, etc.). The pheromone role is not only to navigate robots in the unexplored area, but also to coordinate the swarm of robots. The problem with evaporation function can occur in space with a small and narrow area. Accordingly, once the robot explored a narrow section of space, that area would be saturated with high pheromone values. This situation creates a barrier automatically blocking the route for other robots. In the future, we would like to try to solve this problem using dynamical values of pheromone marks $P_{max}(t)$.

Simulations with more robots in a swarm or in a larger explored space require a high computing performance. With the HPC support we would like to investigate the discussed problems. The solution we proposed can be applied to heterogeneous systems consisting of different types of robots (UAV, AAV). This suggestion should also be investigated in the future.

Acknowledgment This work has been partially supported by VEGA, the national scientific grant agency under the grant number 2/0194/13.

References

1. Zelenka, J., Kasanický, T.: Insect pheromone strategy for the robots coordination. In: Applied Mechanics and Materials, vol. 613, pp. 163–171. ISSN 1660-9336 (2013)
2. Senthilkumar, K.S., Bharadwaj, K.K.: Multi-robot exploration and terrain coverage in an unknown environment. Robot. Auton. Syst. **60**(1), 123–132 (2012)
3. Agmon, N., Hazon, N., Kaminka, G.A.: The giving tree: constructing trees for efficient offline and online multi-robot coverage. Annal. Math. Artif Intell, **52**(2–4), 143–168, ISSN 1012-2443 (2008)
4. Sauter, JA., Matthews, R., Van, D., Parunak, D., Brueckner, S.: Performance of digital pheromones for swarming vehicle control. In: Proceedings of the 4th International Joint Conference on Autonomous Agents And Multiagent Systems, ACM Press, pp. 903–910. ISBN:1-59593-093-0 (2005)
5. Choset, H.: Coverage for robotics—a survey of recent results. Ann. Math. Artif. Intell. **31**, 113–126 (2001)
6. Hert, S., Tiwari, S., Lumelsky, V.: A terrain-covering algorithm for an UAV. Auton Robot **3**, 91–119 (1996)
7. Atkar, P.N., Choset, H., Rizzi, A.A., Acar, E.U.: Exact cellular decomposition of closed orientable surfaces embedded in 3. In: Proceedings of the ICRA Robotics and Automation IEEE International Conference, vol. 1, pp. 699–704 (2001)

8. Atkar, P., Greenfield, A.L., Conner, D.C., Choset, H., Rizzi, A.: Uniform coverage of automotive surface patches. Int. J. Robot. Res. **24**(11), 883–898 (2005)
9. Oksanen, T., Visala, A.: Coverage path planning algorithms for agricultural field machines. J Field Robot **26**, 558–651 (2009)
10. Cheng, P., Keller, J., Kumar, V.: Time-optimal UAV trajectory planning for 3d urban structure coverage. In: Intelligent Robots and Systems, IROS'08, pp. 2750–2757 (2008)
11. Xu, A., Virie, P., Rekleitis, I.: Optimal complete terrain coverage using an unmanned aerial vehicle. In: IEEE International Conference on Robotics and Automation (2011)
12. Barrientos, A., Colorado, J., Cerro, J., Martinez, A., Rossi, C., Sanz, D., Valente, J.: Aerial remote sensing in agriculture: a practical approach to area coverage and path planning for fleets of mini aerial robots. J Field Robot. **28**(5), 667–689 (2011)
13. Xu, A., Viriyasuthee, C., Rekleitis, I.: Optimal complete terrain coverage using an Unmanned Aerial Vehicle. In: Robotics and Automation (ICRA), pp. 2513–2519 (2011)
14. Cristobal, M.J.: Autonomous exploration and mapping of unknown environments with teams of mobile robots: http://dspace.umh.es/bitstream/11000/1370/1/Miguel%20Julia%20-%20Autonomous%20Exploration%20and%20Mapping.pdf
15. Zelinsky, A., Jarvis, R.A., Byrne, J. C., Yuta, S.: Planning paths of complete coverage of an unstructured environment by a mobile robot. In: Proceedings of International Conference on Advanced Robotics, pp. 533–538 (1993)
16. Carsten, J., Ferguson, D., Stentz, A.: 3D Field D*: improved path planning and replanning in three dimensions. In: Intelligent Robots and Systems, IROS'06, pp. 3381–3386 (2006)
17. Koenig, S., Szymanski, B., Liu, Y.: Efficient and inefficient ant coverage methods. Annal. Math. Artif. Intell. vol. 31, Issue 1–4, pp. 41–76, ISSN: 1012-2443. doi:10.1023/A:1016665115585 (2001)
18. Masár, M., Budinská, I.: Robot coordination based on biologically inspired methods. In: Advanced Materials Research, vol. 664, pp. 891–896. ISSN 1022-6680 (2013)
19. Burgard, W., Moors, M., Stachniss, C., Schneider, F.E.: Coordinated multi-robot exploration. Robot IEEE Transac on, **21**(3), 376–386 (2005)
20. Zelenka, J., Kasanický, T.: Insect pheromone strategy for the robots coordination—reaction on loss communication. In: 15th IEEE International Symposium on Computational Intelligence and Informatics, Budapest: IEEE, pp. 79–83. ISBN 978-1-4799-5338-7 (2014)
21. Margolus, N.: Physics and computation, MIT Ph.D. thesis. Reprinted as Tech. Rep. MIT/LCS/TR415, MIT Lab. for Computer Science, Cambridge MA (1987)
22. Lee, T.-S., Choi, J.-S., Lee, J.-H., Lee, B.-H.: 3-d terrain covering and map building algorithm for an UAV, In: Intelligent Robots and Systems, IROS'09, pp. 4420–4425 (2009)

Preliminary Ideas on the Odometry of an Omni-directional Mobile Robot

Ioan Doroftei, Ionel Conduraru and Vasile Horga

Abstract Omni-directional mobile robots are very attractive because they have a very good mobility, which make them appropriate when they have to move in tight areas, avoid obstacles, and find the way to the next location. To move with precision in such environments, the accurate estimation of the position is very important. The authors provide in this paper information about the design of an omni-directional robot and its control system. Also, preliminary ideas about the design of an odometer are presented.

Keywords Omni-directional mobile robot · Modified Mecanum wheel · Odometry

1 Introduction

Mobile robots are very used in industrial applications, where one of their main issues is mobility, allowing them to move in small and narrow spaces and to avoid obstacles, [1, 2]. To accomplish such tasks, an omni-directional drive mechanism is very useful because it guarantees a very good mobility. Omni-directionality mainly depends on the wheels' design, the Mecanum wheel being one of the possible solutions. These wheels are well known in the robotics community, their exceptional

I. Doroftei (✉) · I. Conduraru
Mechanical Engineering Faculty, "Gheorghe Asachi" Technical University of Iasi, Iasi, Romania
e-mail: idorofte@mail.tuiasi.ro

I. Conduraru
e-mail: conduraru.ionel@yahoo.com

V. Horga
Electrical Engineering Faculty, "Gheorghe Asachi" Technical University of Iasi, Iasi, Romania
e-mail: horga@tuiasi.ro

T. Borangiu (ed.), *Advances in Robot Design and Intelligent Control*,
Advances in Intelligent Systems and Computing 371,
DOI 10.1007/978-3-319-21290-6_16

manoeuvring capabilities attracting a lot of robot manufacturers; several classes of industrial applications use robots with such omni-directional wheels. For certain motions of the mobile platform, Mecanum wheels allow the mobile platform to change its direction of motion without changing its orientation [3, 4], which is very useful when the robot has to avoid obstacles in a small and narrow space as is the industrial environment.

When the robot is autonomous, the accurate estimation of the position is a key component of its successful operation, localization being a fundamental function for navigation. Mobile robot localization refers mainly to the precise determination of the spatial coordinates, as well as of its orientation at a certain moment of time. In many applications, the orientation and an initial estimate of the robot position are known, being supplied directly or indirectly by the user or the supervisor. During the execution of the tasks, the robot must update this estimation using measurements from its sensors; this process is known as individual localization. But using only sensors that measure relative movements, the error in the pose estimation increases over time because errors continuously accumulate. To improve the pose estimation additional odometers must be used.

Guiding mobile robots along desired trajectories is an important problem in mobile robot navigation. Often the industrial environment is changing at each moment, which makes very difficult the use of maps because they may be incomplete or inaccurate. This is why determining accurately the position and orientation of a mobile robot without external devices is a challenge. For the typical differential drive platforms used by many mobile robots, the current position and orientation are easily estimated by dead-reckoning using two odometer wheels together with two rotary encoders attached to the driving wheel axes. This positioning method is simple and cheap and easy operating in real-time. The main disadvantage of this method is the accumulation of errors from one sequence of motions to the next one [5].

Landmarks are effective in absolute locations where they are reliably detected in real time. In case of false recognition of landmarks or landmark occlusion by obstacles, a combination of landmarks and dead-reckoning is a useful method for localization of the robot and trajectory correction.

Methods like odometry, inertial navigation (gyroscopes), and magnetic compasses can be used in mobile robotics directly without the use of any maps, external marks or devices, other solutions being more complicated in terms of installation and setup of external devices. A method that uses an odometry system with good performances is presented in [6]. This method, together with a global tracking system can be a systemic solution with good performances for omni-directional mobile robot navigation. Another odometry method presented in [7] uses a sensor that combines the classical differentials and an optical encoder so that it can provide only the orientation information. The orientation sensor includes an optical encoder and three differentials. Another odometry system that eliminates the errors of asymmetry and the wheel slip is presented in [8]. An odometer attached to an omni-directional mobile platform that can give information on the position and orientation of the platform, its construction being capable to eliminate errors due to slips, is presented in this work. In [9] the authors present a very low-cost control system which can be easily interfaced with any mobile platform. The sensor system requires only two optical mice which can be placed in any position under the robot. Another project used two prototypes of sensor heads, which have been constructed [10]. Head A was designed for ground moving robots for which the distance to the ground does not change significantly. Head B was intended to extract all the possible information about self-motion that can be obtained from the optical flow.

In this work, some information is provided about the design of an omni-directional robot with original modified Mecanum wheels, and its control system. The novelty and the advantages of this wheel design have been discussed in other previous works [2, 11]. Also, preliminary ideas about the design of a simple and novel odometer are discussed. This type of odometer has not yet been implemented and no experimental tests have been carried up until now in order to prove its effectiveness. This will be the subject of future work.

2 Robot Design and Kinematics

As known, an omni-directional vehicle is able to move at any time in any direction and rotate around its own vertical axis at the same time. To get an omni-directional platform, the design of its wheels is essential, Mecanum wheel being one of the possible solutions.

Our mobile robot (see Fig. 1a) is equipped with four modified Mecanum wheels (Fig. 1b), which provide three degrees of freedom. The wheel configuration of the vehicle is shown in Fig. 2.

The direct kinematics, which gives the robot velocities [2, 11], leads to

$$\begin{bmatrix} v_x(t) \\ v_y(t) \\ \Omega_z(t) \end{bmatrix} = \frac{r_w}{4} \begin{bmatrix} -1 & 1 & -1 & 1 \\ 1 & 1 & 1 & 1 \\ \frac{2}{L+W} & -\frac{2}{L+W} & -\frac{2}{L+W} & \frac{2}{L+W} \end{bmatrix} \cdot \begin{bmatrix} \omega_1(t) \\ \omega_2(t) \\ \omega_3(t) \\ \omega_4(t) \end{bmatrix}, \quad (1)$$

and the inverse one transforms velocities in the robot frame into the wheel velocities,

$$\begin{bmatrix} \omega_1(t) \\ \omega_2(t) \\ \omega_3(t) \\ \omega_4(t) \end{bmatrix} = \frac{1}{r_w} \begin{bmatrix} -1 & 1 & \frac{L+W}{2} \\ 1 & 1 & -\frac{L+W}{2} \\ -1 & 1 & -\frac{L+W}{2} \\ 1 & 1 & \frac{L+W}{2} \end{bmatrix} \cdot \begin{bmatrix} v_x(t) \\ v_y(t) \\ \Omega_z(t) \end{bmatrix}. \quad (2)$$

Equation (1) is used in the operating system of the robot to execute odometry and Eq. (2) is used to control the velocities in the robot frame.

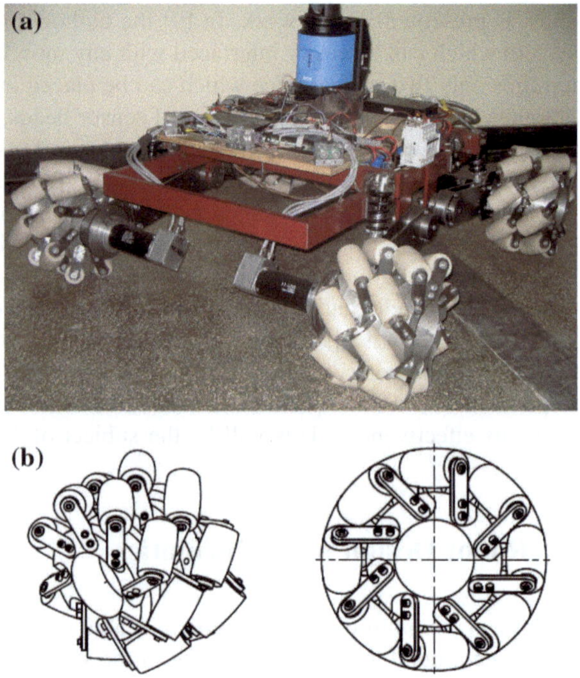

Fig. 1 Omni-directional robot with modified Mecanum wheels: **a** real robot; **b** CAD model of the proposed wheel design

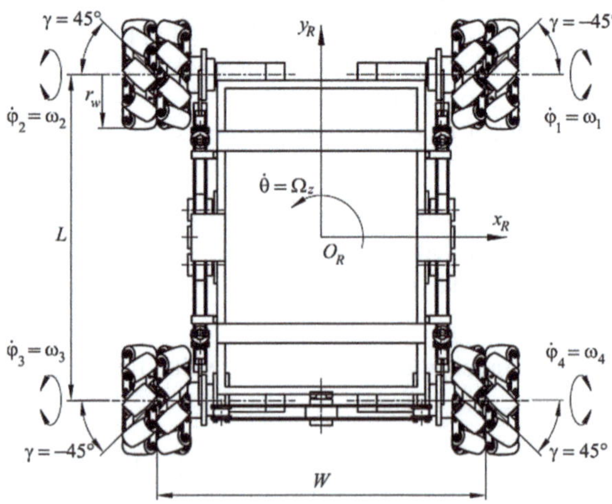

Fig. 2 Top view of the wheel configuration

3 Robot Control

The multi-processor control architecture of the omni-directional robot is presented Fig. 3. The actuating system is based on specific 4-Q-EC DES 70/10 drivers, providing commands for the four brushless AC motors (of Maxon EC60-400 type).

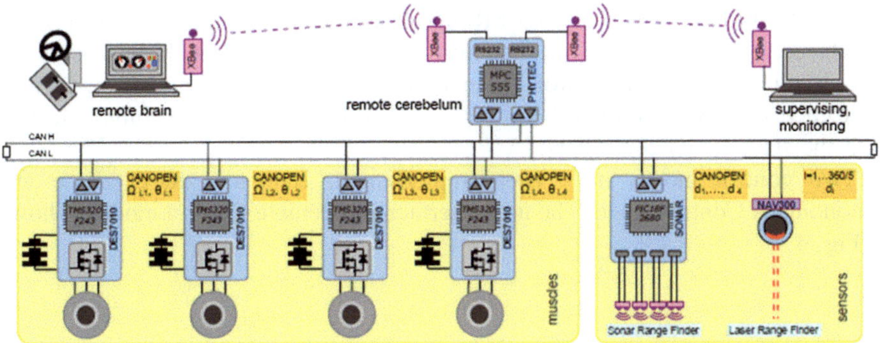

Fig. 3 Multi-microprocessor control architecture of the omni-directional robot

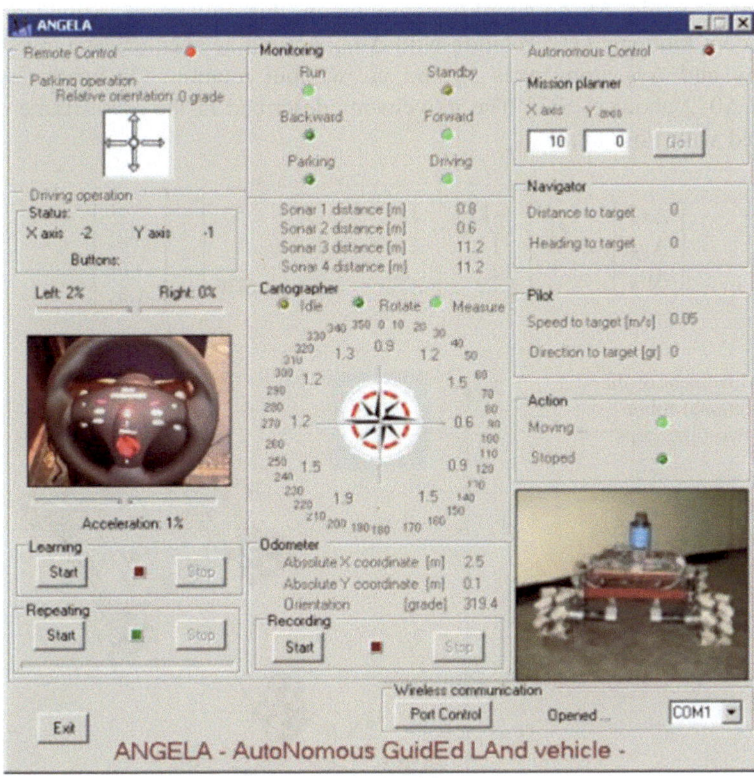

Fig. 4 User interface for Remote Control mode

One NAV300 laser sensor and six SRF05 ultrasonic sensors are used for obstacle avoidance; each motor has one incremental encoder and three Hall sensors for closed-loop displacement control. The master computing unit is implemented with a phyCORE-mpc555 controller that uses a CAN bus network for the communication with the slave (axis) controllers.

Two control modes have been implemented: Remote Control mode and Autonomous Control mode. The user interface for Remote Control mode is shown in Fig. 4.

4 Odometry Aspects

Between two time sample periods of the robot postures it is assumed that the omni-directional robot moves on a straight line while it rotates simultaneously from θ_{k-1} to θ_k (see Fig. 5). To simplify computation, this movement is divided into 3 independent motions: (1) a rotation with $\Delta\theta/2$ around z_R axis; (2) two translations with Δx_R and Δy_R along axes x_R and y_R, without rotation; (3) finally, again a rotation $\Delta\theta/2$ about z_R axis. The movements described in the robot frame can be computed with Eq. (1) as:

$$\begin{bmatrix} \Delta x_R \\ \Delta y_R \\ \Delta\theta_R \end{bmatrix} = \frac{r_w}{4} \begin{bmatrix} -1 & 1 & -1 & 1 \\ 1 & 1 & 1 & 1 \\ \frac{2}{L+W} & -\frac{2}{L+W} & -\frac{2}{L+W} & \frac{2}{L+W} \end{bmatrix} \cdot \begin{bmatrix} \Delta\varphi_1 \\ \Delta\varphi_2 \\ \Delta\varphi_3 \\ \Delta\varphi_4 \end{bmatrix}. \tag{3}$$

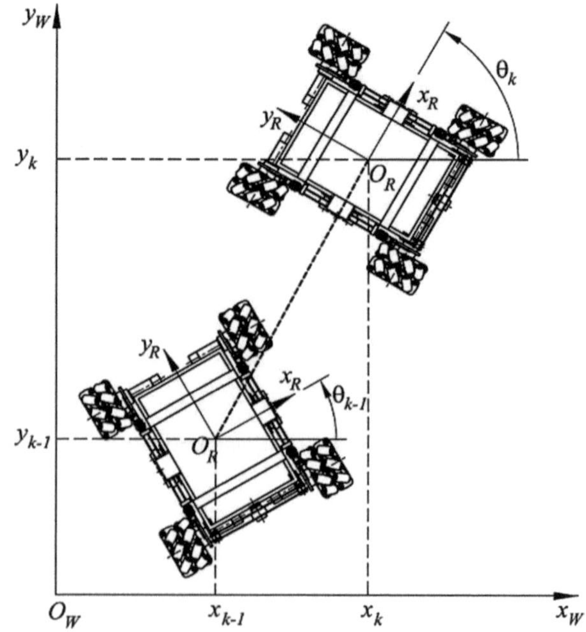

Fig. 5 Movement of the omni-directional robot between two time steps

With these movements, the new posture, (x_k, y_k, θ_k) of the omni-directional robot in the world frame $(x_W O_W y_W)$ may be computed based on the previous one $(x_{k-1}, y_{k-1}, \theta_{k-1})$:

$$\begin{cases} x_k = x_{k-1} + \Delta x_R \cos\left(\theta_{k-1} + \frac{\Delta\theta}{2}\right) - \Delta y_R \sin\left(\theta_{k-1} + \frac{\Delta\theta}{2}\right) \\ y_k = y_{k-1} + \Delta x_R \sin\left(\theta_{k-1} + \frac{\Delta\theta}{2}\right) + \Delta y_R \cos\left(\theta_{k-1} + \frac{\Delta\theta}{2}\right) \\ \theta_k = \theta_{k-1} + \Delta\theta \end{cases}. \quad (4)$$

Robot movements are corrupted by some noise caused by mechanical inaccuracies, mainly caused by slippage of the Mecanum wheels. Since the slippage of the wheels depends on the rotational speed of the free spinning rollers, the uncertainty depends on the direction of the movement in the robot frame. In order to correct positioning errors, a simple odometer solution is proposed in this work (see Fig. 6).

The odometer is based on two spherical joints, each one composed by one rubber ball 3 and six spherical bearings 4. Three universal wheels 2 are in contact with the rubber balls; on the shaft of each universal wheel an encoder was mounted. The information received from these encoders (Fig. 6) is used as displacement feedback to compute the movements of the robot:

$$\begin{cases} \Delta x_R = r\alpha_2 \\ \Delta y_R = r\alpha_1 \\ \Delta\theta_R = \frac{r}{a}(\alpha_3 - \alpha_2) \end{cases}, \quad (5)$$

where r is the radius of universal wheels, α_i (i = 1…3) are the angular strokes measured by the encoders, and a is the distance between the axes of the two rubber balls (see Fig. 6b).

To evaluate the effectiveness of this odometer, experimental tests will be carried out to confirm the effectiveness of the design.

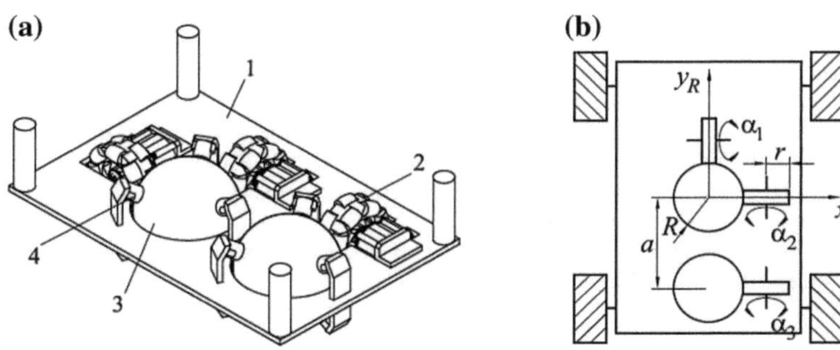

Fig. 6 Odometer: **a** 3D view; **b** top view of the robot with odometer

5 Summary

The accurate position estimation for mobile, autonomous robots represents a key function for its successful operation, because localization is the fundamental operation for correct navigation. Robot movements are corrupted by some noise caused by mechanical inaccuracies, mainly caused by slippage of the Mecanum wheels. In this work, some information about the design of an omni-directional robot using modified Mecanum wheels, and its control has been presented. Also, preliminary ideas about the design of a simple and novel odometer have been discussed. To evaluate the effectiveness of this odometer, experimental tests will be carried out to confirm the effectiveness of the design. This will be the subject of future work.

References

1. Doroftei, I.: Conceptual design of an omni-directional mobile robot. In: Visa I. (ed.) Proceedings of the 10th IFToMM International Symposium on Science of Mechanisms and Machines, pp. 115–127, Brasov, Romania, Springer (2009)
2. Doroftei, I., Stirbu, B.: Design, modeling and control of an omni-directional mobile robot. Solid State Phenom. **166–167**, 173–178 (2010)
3. Doroftei, I., Grosu, V., Spînu, V.: Omnidirectional mobile robot—design and implementation. In: Habib M.K. (ed.) Bioinspiration and Robotics: Walking and Climbing Robots, pp. 511–528. I-Tech Education and Publishing, Vienna (2007)
4. Doroftei, I., Grosu, V., Spînu, V.: Design and Control of an Omni-Directional Mobile Robot. Novel Algorithms and Techniques in Telecommunications, Automation and Industrial Electronics, pp. 105–110, Springer, Netherlands (2008)
5. Conduraru, I., Doroftei, I., Conduraru, A.: Localization methods for mobile robots—a review. Adv. Mater. Res. **837**, 561–566 (2014)
6. Ushimi, N., Yamamoto, M., Mohri, A.: Tracking control of omni-directional vehicles using two wheels caster type odometer. J Robot. Mechatron. **16**(4), 404–410 (2004)
7. Wei-Chen, L., Cong-Wei, C.: An orientation sensor for mobile robots using differentials. Int. J. Adv. Rob. Syst. **10**(134), 1–10 (2013)
8. Tehrani, A.F., Doosthosseini, A.M., Moballegh, H.R., Amini, P., Danesh Panah, M.M.: A new odometry system to reduce asymmetric errors for omnidirectional mobile robots. LNAI **3020**, 600–610 (2014)
9. Bonarini, A., Matteucci, M., Restelli, M.: Dead reckoning for mobile robots using two optical mice. In: ICINCO'04, Robotics and Automation, pp. 87–94 (2004)
10. Dahmen, H., Millers, A., Mallot, H.A.: Insect-inspired odometry by optic flow recorded with optical mouse chips. Springer, Berlin (2009)
11. Conduraru, I., Doroftei, I., Conduraru, A.: A mobile robot with modified Mecanum wheels. Adv. Mater. Res. **1036**, 775–780 (2014)

Study of Controlled Motion of Exoskeleton Moving from Sitting to Standing Position

Sergey Jatsun, Sergei Savin, Andrey Yatsun and Andrei Malchikov

Abstract In this paper we derive analytical expressions which describe the law of change of generalized coordinates of an exoskeleton, such that the mechanism performs standing up motion. Using the obtained expressions the control system has been designed and the controller's coefficients have been tuned. The behaviour of the control system with additional linear element was studied. The modified system shows significant improvement, as compared to the initial one.

Keywords Exoskeleton · Control system · Centre of mass · Control · Trajectory

1 Introduction

There are publications that study the controlled motion of walking type robots and exoskeletons. Fundamental research works in this area are [1–4]. At the same time the study of controlled stand up motion had not been yet sufficiently covered.

Research in this area presents practical interest for the following reason: stand up motion requires motor to produce greater torques than walking motion, so that the choice of motors and other design decisions for walking robots and exoskeletons should be at least partially based on the information obtained by modelling the standing up process.

This paper is focused on studying controlled motion of an exoskeleton that moves from sitting to standing position. This includes formulating performance quality measurements, synthesizing the input values of the system, synthesizing the structure of the control system and tuning the regulator.

S. Jatsun · S. Savin (✉) · A. Yatsun · A. Malchikov
Department of Mechanics, Mechatronics and Robotics,
South-West State University, Kursk, Russia
e-mail: sergey89mtkgtu@mail.ru

S. Jatsun
e-mail: teormeh@inbox.ru

T. Borangiu (ed.), *Advances in Robot Design and Intelligent Control*,
Advances in Intelligent Systems and Computing 371,
DOI 10.1007/978-3-319-21290-6_17

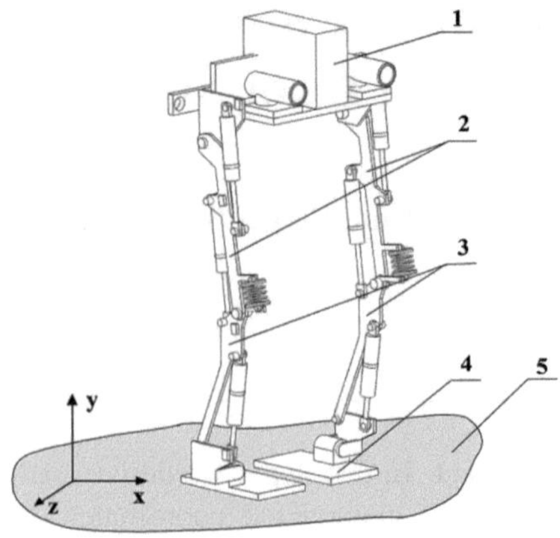

Fig. 1 3D model of lower limbs of the walking robot. *1* Torso, *2* hip, *3* shin, *4* foot, *5* bearing surface

2 Object of the Research

In this paper we study a two legged walking exoskeleton that uses linear actuators (see Fig. 1). The robot's legs consist of three bodies connected in series by means of rotational joints.

We will call those bodies: "hip", "shin" and "foot". Hips are connected to the "torso" of the robot through rotational joints. Interaction between the bearing surface and legs occurs through contact surfaces of robot's feet.

The robot has six linear actuators. Two of them are connected to the torso through rotational joints. Each hip and shin has one linear actuator attached to it in the same way. The rod of each actuator is connected to one of the mechanism's bodies (Fig. 1).

3 Synthesis of the Control System Inputs

Here we investigate a special case of standing up motion that is characterized by the fact that the velocities of all points on the mechanism bodies are coplanar. Thus to describe the motion of the mechanism one can use a model of four link planar mechanism where links are connected in series by rotational joints (see Fig. 2).

Also we assume that the foot is immobile during the whole time standing up motion takes place. We introduce a fixed coordinate system O_1xy and write down radius-vectors, which describe the position of points C_1, C_2, C_3.

In Fig. 2, φ_1, φ_2, φ_3 are the generalized coordinates which describe the orientation of the mechanism bodies. In the initial moment, the generalized coordinates have following values: $\varphi_1(0) = \varphi_1^*$, $\varphi_2(0) = \varphi_2^*$, $\varphi_3(0) = \varphi_3^*$.

Let us write down the expression for vector $\vec{r}_C$ that describes the position of the mass centre of the system (when calculating the position of the mass center we leave out the foot of the robot). To do so we introduce new constants:

Fig. 2 The schematic representation of the exoskeleton. *1* Torso, *2* hip, *3* shin, *4* foot

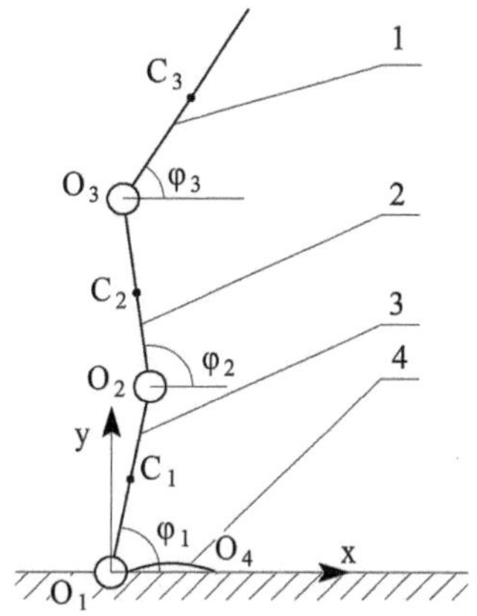

$K_1 = \frac{l_1(0.5m_1+m_2+m_3)}{m_1+m_2+m_3}$, $K_2 = \frac{l_2(0.5m_2+m_3)}{m_1+m_2+m_3}$, $K_3 = \frac{0.5l_3m_3}{m_1+m_2+m_3}$. Then the expression for $\vec{r}_C$ can be written in the following form:

$$\vec{r}_C = \begin{bmatrix} K_1\cos(\varphi_1) + K_2\cos(\varphi_2) + K_3\cos(\varphi_3) \\ K_1\sin(\varphi_1) + K_2\sin(\varphi_2) + K_3\sin(\varphi_3) \end{bmatrix} = \begin{bmatrix} x_C \\ y_C \end{bmatrix}. \quad (1)$$

Let us write down the list of conditions, which should be met during execution of standing up motion:

- Projection of the vector $\vec{r}_C$ on O_1x axis should belong to interval $(0, x_4)$;
- Projection of the vector $\vec{r}_C$ on O_1y axis should change from $y_C(0) = y_0$ to $y_C(t_1) = y_1$ during time t_1;
- Torso should rotate into the vertical position in time t_1;
- Velocities of each body at time t_1 should be equal to zero.

We can define the function that describes the change of vector $\vec{r}_C$ and angle φ_3 in the following way:

$$x_C(t) = a_1,\ y_C(t) = a_2 \cdot t^2 + a_3 \cdot t + a_4,\ \varphi_3(t) = a_5 \cdot t^2 + a_6 \cdot t + a_7. \quad (2)$$

To calculate constants a_i, $i \in \overline{1,7}$ we use the following reasoning. At the initial point of time $(t = 0)$ values $x_C(t), y_C(t)$ can be found using (1) and (2) in the following way:

$$\begin{cases} x_C(0) = K_1\cos(\varphi_1^*) + K_2\cos(\varphi_2^*) + K_3\cos(\varphi_3^*) = x_0 \\ y_C(0) = K_1\sin(\varphi_1^*) + K_2\sin(\varphi_2^*) + K_3\sin(\varphi_3^*) = y_0 \end{cases} \quad (3)$$

We assume that the initial values of the generalized coordinates are such that the condition $x_C(0) \in (0,\ x_4)$ is met. By using the list of requirements, we can write down two more equalities:$y_C(t_1) = y_1$, $\varphi_3(t_1) = 0.5\pi$. Also according to the list of conditions, the values of time derivatives $\dot{y}_C(t)$, $\dot{\varphi}_3(t)$ at the moment t_1 are equal to zero: $\dot{y}_C(t_1) = 0$, $\dot{\varphi}_3(t_1) = 0$. Thus, we obtained the system of equations, linear with respect to a_i. Solving this system we obtain values for constants a_i:

$$\begin{gathered} a_1 = x_0, \quad a_2 = \frac{1}{t_1^2}(y_0 - y_1), \quad a_3 = y_0, \quad a_4 = \frac{2}{t_1}(y_1 - y_0), \\ a_5 = \frac{1}{t_1^2}\left(\varphi_3^* - \frac{\pi}{2}\right), \quad a_6 = \varphi_3^*, \quad a_7 = \frac{2}{t_1}\left(\frac{\pi}{2} - \varphi_3^*\right). \end{gathered} \tag{4}$$

To find trajectories of the mechanism in configuration space, we solve the inverse kinematics problem. To do, we solve the following system of trigonometric equations:

$$\begin{cases} K_1\cos(\varphi_1) + K_2\cos(\varphi_2) + K_3\cos(\varphi_3) = a_1 \\ K_1\sin(\varphi_1) + K_2\sin(\varphi_2) + K_3\sin(\varphi_3) = a_2 \cdot t^2 + a_3 \cdot t + a_4 \\ \varphi_3 = a_5 \cdot t^2 + a_6 \cdot t + a_7 \end{cases} \tag{5}$$

The solutions have the following form:

$$\begin{aligned} \varphi_1 &= -\arccos\left(\frac{b_1^2 + b_2^2 + K_1^2 - K_2^2}{2K_1\sqrt{b_1^2 + b_2^2}}\right) + \alpha, \\ \varphi_2 &= \arccos\left(\frac{b_1^2 + b_2^2 + K_2^2 - K_1^2}{2K_2\sqrt{b_1^2 + b_2^2}}\right) + \alpha, \\ \varphi_3 &= a_5 \cdot t^2 + a_6 \cdot t + a_7, \end{aligned} \tag{6}$$

where $b_1 = a_1 - K_3\cos(a_5t^2 + a_6t + a_7)$, $b_2 = a_2t^2 + a_3t + a_4 - K_3\sin(a_5t^2 + a_6t + a_7)$.

4 Description of the Control System Design and Tuning of the Regulator

Derivation of dynamics equations for three link mechanisms using Lagrange equations has been done in a number of works including [5]. Here we only present a generalized form of the equations in the form that was used in papers such as [6–8]:

$$\mathrm{A}(\varphi)\ddot{\varphi} + \vec{b}(\varphi, \dot{\varphi}) + \vec{G}(\varphi) + \vec{f}(\dot{\varphi}) = \vec{\Gamma}, \tag{7}$$

where: $\mathrm{A}(\varphi) \in \mathrm{R}^{3\times3}$ is the kinetic energy matrix, $\vec{b}(\varphi, \dot{\varphi}) \in \mathrm{R}^3$ is the vector of Coriolis and centrifugal forces, $\vec{G}(\varphi) \in \mathrm{R}^3$ is the gravity force vector, $\vec{f}(\dot{\varphi}) \in \mathrm{R}^3$ is the vector of dissipative forces obtained by differentiating Rayleigh dissipation function, and $\vec{\Gamma} \in \mathrm{R}^3$ is the vector of generalized forces: $\vec{\Gamma} = [(M_1 - M_2)(M_2 - M_3)M_3]^T$ where M_1, M_2, M_3 are torques generated by actuators.

Here we consider a control system with feedback loop. The system uses multi-input multi-output proportional-integral (PI) controller. In this paper we do not look into the dynamics of the motors; we assume that we can directly and instantaneously generate torques M_1, M_2, M_3:

$$\vec{M}(t) = \begin{bmatrix} k_1 & k_1 & k_1 \\ 0 & k_1 & k_1 \\ 0 & 0 & k_1 \end{bmatrix} \begin{bmatrix} e_1 \\ e_2 \\ e_3 \end{bmatrix} + \begin{bmatrix} k_2 & k_2 & k_2 \\ 0 & k_2 & k_2 \\ 0 & 0 & k_2 \end{bmatrix} \begin{bmatrix} \int_0^t e_1 dt \\ \int_0^t e_2 dt \\ \int_0^t e_3 dt \end{bmatrix}, \tag{9}$$

where k_1, k_2 are PI controller coefficients, $\vec{M} = [M_1 \quad M_2 \quad M_3]^T$.

Let us consider a problem of tuning controller coefficients k_1, k_2 pursuing two goals: (1) minimization of oscillation amplitude, (2) minimization of error (difference between desired and real values of variables x_C, y_C, φ_3). To do that we write down expressions that relate the motion of the mechanical system (obtained as a set of numerical sequences) to two scalar values, that correspond to the quality of the control system work according to the mentioned criteria:

$$P_1 = \frac{1}{t_1} \int_0^{t_1} (\dot{\varphi}_1(t) - s_1)^2 + (\dot{\varphi}_2(t) - s_2)^2 + (\dot{\varphi}_3(t) - s_3)^2 dt, \text{ where } s_i = \frac{1}{t_1} \int_0^{t_1} \dot{\varphi}_i(t) dt, \tag{10}$$

$$P_2 = \frac{1}{t_1} \int_0^{t_1} (\varphi_{31}(t) - \varphi_1(t))^2 + (\varphi_{32}(t) - \varphi_2(t))^2 + (\varphi_{33}(t) - \varphi_3(t))^2 dt, \tag{11}$$

where P_1 is a scalar that grows quadratically with the growth of oscillation amplitude; P_2 is a scalar that grows quadratically with the growth of the error; and S_i is the average value of the generalized velocity $\dot{\varphi}_i$.

In this way, tuning the regulator means minimizing the values P_1 and P_2. Below we plot graphs which represent change of values P_1 and P_2 with respect to coefficients k_1, k_2 (see Fig. 3).

As can be seen, local minima of functions (10) and (11) do not coincide. To choose one set of coefficients k_1, k_2 let us introduce weight coefficients β_{P1} and β_{P2} which allow us to redefine the problem to minimization of a single scalar function:

$$P = \beta_{P1} P_1 + \beta_{P2} P_2 \tag{12}$$

This new function has a local minimum at $k_1 = 2500$, $k_2 = -400$ in the set of possible values of k_1, k_2 which we considered.

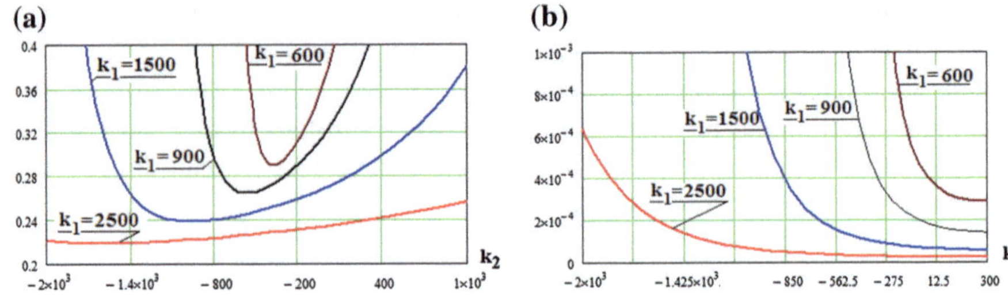

Fig. 3 **a** Change of value P_1 with respect to k_2, **b** change of value P_2 with respect to k_2

5 Introducing a Linear Compensating Element to the Control System

To improve the quality of the control system work (to minimize oscillation amplitude and error of the control system) let us introduce a linear correcting element, as shown in Fig. 4.

The correcting element does linear transformation of the regulator output such that the generalized forces vector $\vec{\Gamma}$ assumes the following form:

$$\vec{\Gamma} = \mathrm{A}\left[k_p(\vec{\varphi}_3(t) - \vec{\varphi}(t)) + k_i \int\limits_0^t (\vec{\varphi}_3(t) - \vec{\varphi}(t))dt\right], \tag{13}$$

where A is the kinetic energy matrix of the system.

Let us study Fig. 5 that provides information about how much the correcting element changes the system's dynamics (we look at the case when $k_1 = 300$, $k_2 = 0$, in Fig. 5.

By using parameters k_1, k_2 optimized in the same way as before, we can plot the trajectory of the mass centre and compare it to the best trajectory we had without the correcting element (by the best trajectory we understand the one obtained after tuning the regulator). The result is shown in Fig. 6.

It can be seen that the introduction of the correcting element leads to a significant improvement of the quality of the control system work.

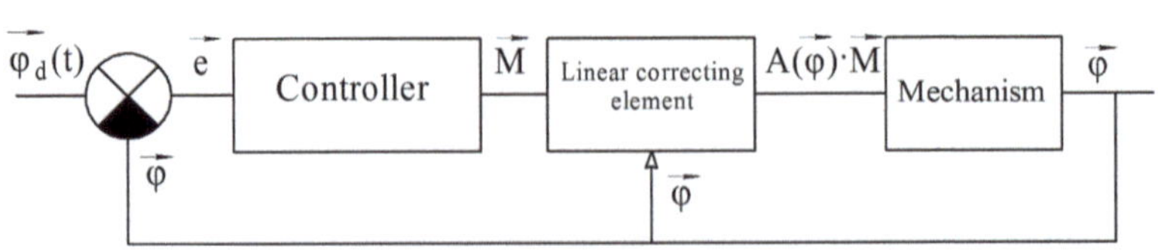

Fig. 4 Control system scheme with linear correcting element introduced

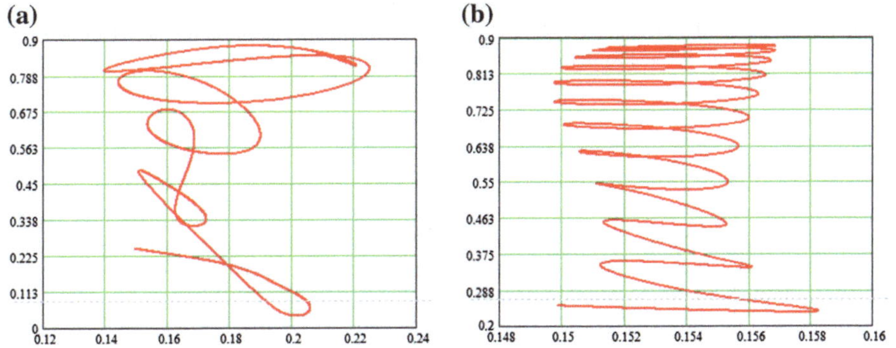

Fig. 5 Trajectory of the centre of mass **a** without the correcting element **b** with the correcting element

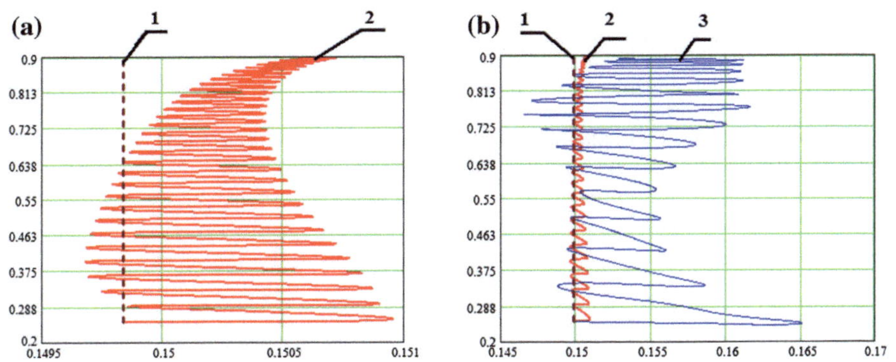

Fig. 6 Trajectory of the centre of mass (CM) of the mechanism *1*—the desired trajectory defined by Eq. (6), *2*—the trajectory obtained from control system with correcting element, *3*—the trajectory obtained with the control system without correcting element

6 Conclusions

In this paper we derived analytical expressions for laws of change of generalized coordinates that allow the mechanism to perform standing up motion. The criteria that can be used to assess the quality of control system work were presented, along with the method of the PI controller parameters tuning.

A way to improve the quality of the control system work was shown. It requires introducing a new element to the control system design in order to change the control system's structure. The linear correction element that was introduced in the control system allows one to significantly improve the quality of the control system work, but there are still oscillations.

Acknowledgements The reported research work is performed with Russian Science Foundation, in the framework of the Project No. 14-39-00008 "The establishment of the research laboratory of modern methods and robotic systems to improve the human environment".

References

1. Formalskiy, A.M.: Motion of Anthropomorphic Mechanisms. Nauka, Moscow (1982)
2. Beletskiy, V.V., Berbyuk, V.Y.: Nonlinear Model of Bipedal Walking Apparatus, Equipped with Controllable Feet. Nauka, Moscow (1982)
3. Beletskiy, V.V.: Bipedal Walking: Topics in Dynamics and Control. Nauka, Moscow (1984)
4. Vukobratovich, M.K.: Walking Robots and Anthropomorphic Mechanisms. Mir, Moscow (1976)
5. Vorochaeva, L.Y., Naumov, G.S., Yatsun, S.F.: Simulation of motion of a three link robot with controlled friction forces on a horizontal rough surface. J. Comput. Syst. Sci. Int. **54**(1), 151–164 (2015)
6. Chang, K.-S., Holmberg, R., Khatib, O.: The augmented object model: cooperative manipulation and parallel mechanism dynamics. In: IEEE International Conference on Robotics and Automation, Proceedings of ICRA'00, vol. 1 (2000)
7. Khatib, O.: Inertial properties in robotic manipulation: an object-level framework. Int. J. Rob. Res. **14**(1), 19–36 (1995)
8. Sabanovic, A., Kouhei, O.: Motion control systems. Wiley, New York (2011)

Robotic System Equipped with Catapult

Sergey Jatsun, Oksana Loktionova, Lyudmila Vorochaeva and Alexander Vorochaev

Abstract The article presents a mathematical model of motion of a robotic system that moves in jumps from the surface using the on-board catapult and selected stages of the jump. Conditions switch between stages are formulated, special attention being paid to the slideway design. A comparative analysis of the results obtained in the numerical simulation methods and the data collected from real experiments is presented.

Keywords Robotic system · On-board catapult · Acceleration · Accelerating force · Slideway · Deformation

1 Introduction

Recently, in order to accelerate a robotic system which moves by lifting-off from a surface, catapults are more and more widely used. In general we can distinguish two types of catapults: on-board and stationary. The first type has multiple acts of use, but has low efficiency because the catapult must be moved together with additional equipment, which makes the ensemble heavier. Such catapults are used for movement of robots which are designed for environment monitoring, prospecting, equipment transportation to the destination, etc. [1–5]. The second type of catapult

S. Jatsun · O. Loktionova (✉) · L. Vorochaeva · A. Vorochaev
Department of Mechanics, Mechatronics and Robotics,
South-West State University, Kursk, Russia
e-mail: oksloktionova@yandex.ru

S. Jatsun
e-mail: teormeh@inbox.ru

L. Vorochaeva
e-mail: mila180888@yandex.ru

A. Vorochaev
e-mail: sasha-vorochaev@yandex.ru

T. Borangiu (ed.), *Advances in Robot Design and Intelligent Control*,
Advances in Intelligent Systems and Computing 371,
DOI 10.1007/978-3-319-21290-6_18

provides a single acceleration of an object to reach the destination; now, the catapult remains fixed. In this case, it is easier and energetically more effective to accelerate an apparatus. Further robot movement is impossible, therefore such systems are used generally for military goals: shipping explosives or toxic substance s [6–8].

Scientists pay the greatest attention to on-board catapults, which are set on apparatuses which move with lift-off from a surface. Such catapults are used for to perform series of jumps. Computational and design methods for these devices are currently insufficiently studied. The present work is devoted to studying the influence of the apparatus acceleration in the catapult on the height and length of the jump.

2 Description of the Experimental Stand

Figure 1a presents the image of an experimental stand and Fig. 1b shows the scheme of the accelerating module of the robotic system equipped by catapult. The catapult consists of the base 1 mounted on the stand 2, where cam shaft 3 rotates. It is driven by the servo-motor 4 through the gear 5. The cam profiles interact with the pusher 6, which is set in the fasteners 7. The fasteners are connected with the support (foot) 8 by means of rods 9, which move translational in the slideway. Between the base and elements 10 of fasteners there are tension springs 11. To calculate the height and length of the jump the coordinate grid is used [9–12].

When the pusher is located on the smallest radius of the cam, the springs are not deformed, the length of springs is the smallest, and the foot must be moved from the base. During the rotation of the cam, the springs are stretched and the base extends to the foot. At the moment of transition of the largest radius of the cam at least

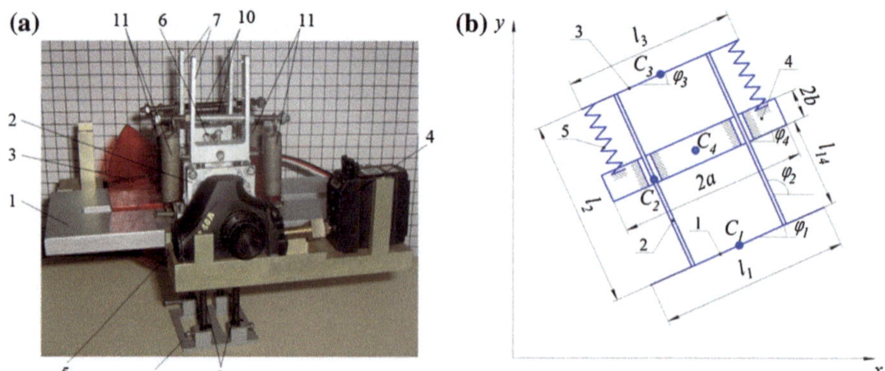

Fig. 1 Experimental stand: a photo, b settlement scheme

going to slide off of the pusher, the springs quickly compress, and the accumulated potential energy at the moment of their full compression is converted into kinetic energy. The apparatus acquires speed, being separated from the surface.

3 Mathematical Model of the Robot's Movement

To work out the mathematical model of the robotic system with catapult one should look at the scheme in Fig. 1b. The following assumptions were adopted: (a) all apparatus links are absolute solid bodies, and foot (link 1), leg (link 2), and platform (link 3) represent the rods lengths l_i, $i = 1 - 3$, and body (link 4) is a rectangle of dimensions $2a \times 2b$. The centres of mass of the links coincide with the centres of their symmetry—points C_i. The distance between the body and the foot is equal l_{14}. Between links 3 and 4 a compression spring is 5 installed [10].

The robot jumps in the vertical plane *Oxy* from the horizontal rough surface from the initial position where the link 1 is on the surface; the springs 5 are not strained, the distance $l_{14} = l_{14n}$ is greatest (Fig. 2a) [9, 10].

On the first stage takes place the preparation for accelerating. Link 1 interacts with the surface, link 4 gravitates to link 1 under the action of the force F which is generated during the rotation of the servo-motor cam, and the springs are stretched. This stage begins in initial conditions $t = t_1$, $l_{14} = l_{14n}$, $\dot{l}_{14} = 0$, and ends when the springs are completely stretched (the pusher is on the largest radius of the cam): $t = t_2$, $l_{14} = l_{14\ \min}$, $\dot{l}_{14} = 0$. The system's motion in this stage is described by one generalized coordinate—the distance l_{14}; the differential equation is:

$$m_4 \ddot{l}_{14} = F^{(1)} - F_{fr1} + m_4 g, \tag{1}$$

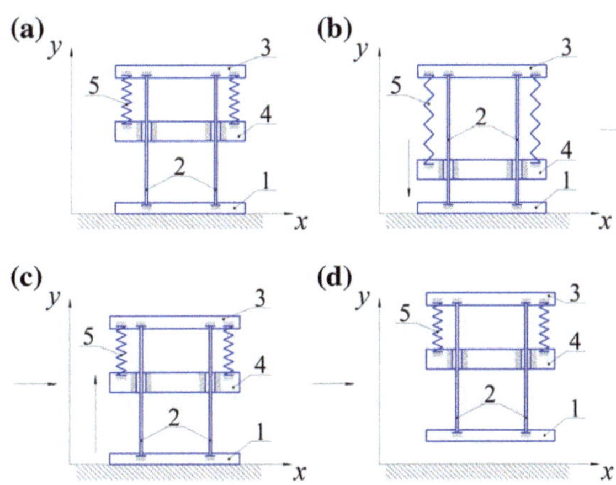

Fig. 2 Stages of the jump robotic system: **a** initial position, **b** stage 1, **c** stage 2, **d** stage 3

where: g—free fall acceleration, F_{fr1}—friction force in the progressive pair determined by the materials composing the links 2 and 4, and $F^{(1)}$—force generated by the rotation of the cam in the first stage.

The second stage starts when $t = t_2$, $l_{14} = l_{14\ \min}$, $\dot{l}_{14} = 0$, and corresponds with the apparatus acceleration because of the springs' compression after the pusher slides down from the largest radius of cam. On this stage the link 1 enters in contact with the surface, the links 2 and 3 are immovable, and the link 4 moves upward with accelerated velocity because of the progressive pair. The object action is characterized by one generalized coordinate, which is the distance l_{14}:

$$m_4\ddot{l}_{14} = F^{(2)} - F_{fr2} - m_4 g, \tag{2}$$

where F_{fr2}—the friction force in the progressive pair, and $F^{(2)}$—the force which is created by the springs in the second stage. During acceleration, due to the high speed link 4 the friction force is defined as sum of two forces: the first one equals F_{fr1}, due to materials used in the links of the progressive pair, and the second one, F_{frd}, associated with flexural deformations of links 2 moving in the slideway mounted in the body.

$$F_{fr2} = F_{fr1} + F_{frd}, \tag{3}$$

$$F_{frd} = F^{(2)} \sin \alpha, \tag{4}$$

where α is the angle between links 2 and 4 in conditions of flexural deformations of the link 2. In compliance with (4) we can see that the friction force F_{frd} increases proportionally to the value of the accelerating force $F^{(2)}$ and the angle α of link 2 is flexural.

The acceleration stage ends with the time point at completed spring compressing; the distance is $l_{14} = l_{14n}$, the link 4 speed is $\dot{l}_{14} = \dot{l}_{14}^{(2)}$, where $\dot{l}_{14}^{(2)}$ is the speed of link 4 at the end of the acceleration stage.

At time $t = t_3$ when $l_{14} = l_{14n}$ the blow between Sects. 3 and 4 happens. Their relative motion stops. The object lift-off occurs under the zeroing of the normal reaction between link 1 and the surface. After that, the third stage of the flight starts. In special cases, if the blow is totally inelastic, the initial speed $\dot{l}_{14}^{0}$ of the flight stage equals

$$\dot{l}_{14}^{0} = m_4\dot{l}_{14}^{(2)} / \sum_{i=1}^{4} m_i. \tag{5}$$

The stage of the flight starts at $t = t_3$, $y_{C4} = l_{14n} + b$, and $\dot{y}_{C4} = \dot{l}_{14}^{0}$. The position of system links, when the line of action of the accelerating force goes through the mass centre of the link 4 and is directed vertically, can be described by one generalized coordinate the projection y_{C4} position of the point C_4 on the vertical axis of the absolute coordinate system:

$$\ddot{y}_{C4}\sum_{i=1}^{4} m_i = -\sum_{i=1}^{4} m_i g. \tag{6}$$

The landing of the apparatus is not observed and described, because the aim of this work is to study the influence of the accelerating stage on the height of the jump.

The modelling of the robot motion is realized with a numerical method using a special algorithm which is based on the precise definition of the time points, when there is a transition from one stage of the jump to the next one. This leads to the change of the generalized coordinates and initial conditions of the model.

4 Results of Theoretical and Experimental Investigations

To validate the numerical simulation, the results were compared with the experimental data. Experimentally, the height of the jump which represents the distance travelled by the foot along the *Oy* axis is measured by a camera grid, located vertically over the robot. Experiments used springs with different stiffness; for each stiffness value and, accordingly, for each acceleration force created, five experiments were carried out. Then the average value of the jump was calculated using the formula:

$$H_{sr} = \sum_{i=1}^{N} H_i/N \tag{8}$$

where H_i is the height of the jump in the *i*th experiment, and $N = 5$ is the number of experiments having the same spring stiffness.

The progressive pair that consists of rods and slideway (on which the rods move) is one of the most important components of the robot system with catapult. That's why special attention was paid to modelling the slideway.

For the tested apparatus, a slideway module allowing the sliding movement was designed. This module consists of one basic component—stand 2; it has a square-topped shape with holes, where rods 9 move (Fig. 3a). The mass of such module is equal 20 g, the mass of the experimental stand is 500 g, and the spring motion is 5 cm.

Figure 4a illustrates the graphs of the experimental H_{sk} and theoretical H_{teor} evaluation of the height of the device's vertical jump, when different force values are applied to the springs, and the corresponding stiffness of the springs is considered.

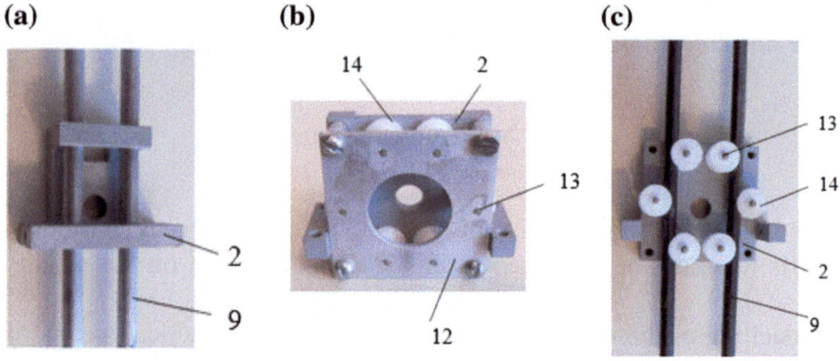

Fig. 3 Photos of the slideways: **a** sliding, **b**, **c** rolling

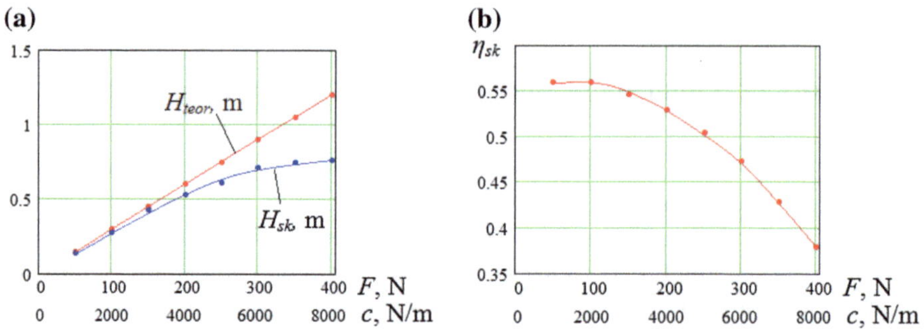

Fig. 4 Graphics dependence of sliding slideways: **a** $H_{teor}(F, c)$, $H_{sk}(F, c)$, **b** $\eta_{sk}(F, c)$

Figure 4b describes the efficiency of the progressive pair with slideway which allows the sliding motion. The efficiency was determined experimentally, leading to a value of $\eta_{sk} = 0.6$.

In Fig. 4b one can see that during the increase up to the largest value of the accelerating force (which has effect on the slideway); the energy conversion efficiency is not permanent, because it decreases from 0.6 till 0.38 on a nonlinear curve. The curve convexity has an upwards direction. This is caused by the considerable deformation of the rods, when short-time large weight bearing acts on them in the acceleration stage. This deformation augments with the increase of the accelerating force and leads to their incandescing.

To eliminate the influence of the slideway's deformation on the height of the jump a new slideway module with rolling motion, compensating the deformable condition of the rods, was designed (Fig. 3b, c). The slideway of the rolling motion consist of the stand 2 and the lid 12 where axes 13 are set, and the rollers 14 rotate about the axes 13. Three rollers were used to provide the translation of each rod 9. During the work process the rollers rotate about the fixed axes. The utilisation of rollers in slideway with rotational motion needs to assure enough distance between them to reach an efficient functioning. The length of rod 9 was therefore increased, which led to a reduction of the module's mass with 40 g—the mass of the experimental sample being 520 g. The modelling efficiency is $\eta_k = 0.85$.

Figure 5a shows the graphs of theoretical H_{teor} and experimentally H_k evaluated heights of the jump with slideways of rolling motion.

In theory the height of the jump increases proportional the increment of the force created by the springs, and their stiffness. The experimental dependence has the same character, a small discrepancy between two graphs is due to inaccuracy in the experiments. As opposed to the experimental curve with the sliding slideways are shown in Fig. 4a, the difference between curves in Fig. 5a does not exceed 5 %. It means that efficiency of the slideways of the rolling motion does not depend on the accelerating force and in average it is $\eta_{ksr} = 0.85$, which is caused by the compensating of the deformation of the punches in the designed module (Fig. 5b).

Experimental data allow concluding that robot-technical systems with on-board catapults as slideways of the translation motion should better use the slideways of the rolling motion, because of their much greater efficiency not decreasing with increasing accelerating force. Therefore, the comparison of theoretical and experimental data of the height of the jump when the variation ratio m_o/m_{zv} (mass of the body and the mass of the other links at $m_{zv} = 0.27$ kg and deformation of the springs Δ) is held to the rolling slideway, shows that the greatest force of the springs in the experiments is equal to $F = 200$ N. Combined graphs are shown in Fig. 6.

The graphs in Fig. 6a show that with the increase of the ratio m_o/m_{zv} the theoretical height of the jump decreases smoothly along a curve, the convexity of which is directed downwards. The experimental results deviate in the range of ±5 % from the theoretical ones, this discrepancy being due to measurement inaccuracy. With the increase of the springs' deformation, the height of the jump estimated theoretically by numerical method gradually increases along a law shown in Fig. 6b. The deviation of the experimental dependence does not extend beyond 10 %. Based on the analysis of these results, it was possible to draw a conclusion about the

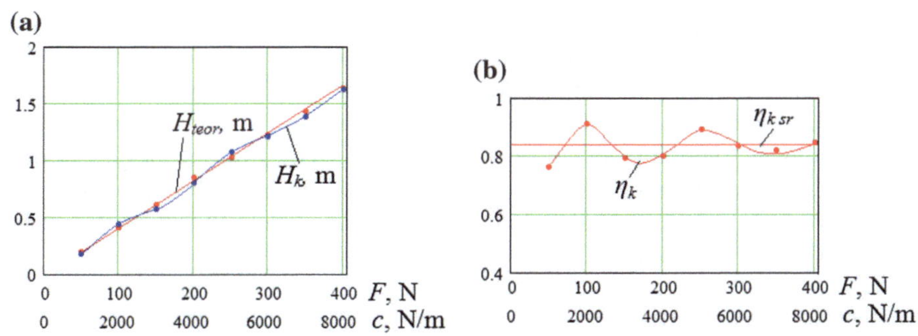

Fig. 5 Graphics dependence of rolling slideways: **a** $H_{teor}(F, c)$, $H_k(F, c)$, **b** $\eta_k(F, c)$, $\eta_{ksr}(F, c)$

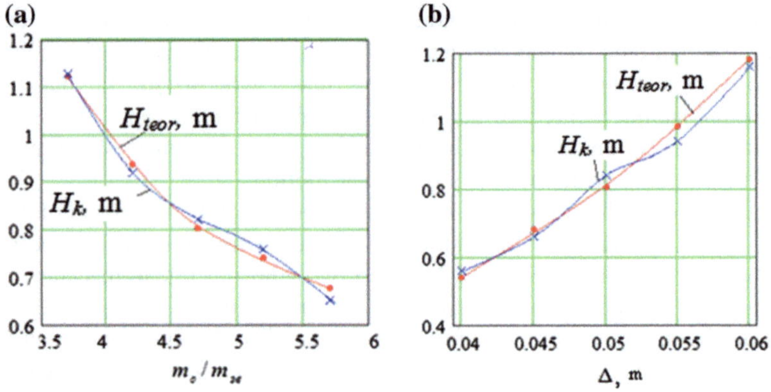

Fig. 6 Graphs of dependencies: **a** $H_{teor}(m_o/m_{zv})$, $H_k(m_o/m_{zv})$, **b** $H_{teor}(\Delta)$, $H_k(\Delta)$

adequacy of the developed mathematical model that fully and accurately describes the implementation of the jumping movement of the robotic system using on board catapult.

5 Conclusions

The paper presents the method, experimental stand and measurement tools developed for a robotic system equipped with on board catapult and moving with jumps from the surface. The sequence of jump stages and the mathematical model of the robot movement were developed; the main feature of this model is the change of generalized coordinates and initial conditions of motion at each stage.

Special attention was paid to investigating the influence of the acceleration stage on the height of the jump. Also, a model is proposed for the friction force; the model has two components, one of which depends on the materials used in the translational pair, and the second one due to the flexural deformation of the rods. The comparative analysis of results produced by numerical simulation and experimental studies confirmed the adequacy of the mathematical model. The analysis showed that the device moving in jumps from the surface, equipped with progressive accelerating pairs that use sliding slideways is impractical because deformations and jamming significantly reduce efficiency with increasing accelerating force. It is preferable the use of rolling slideways which compensate the deformation in the rods and keep constant efficiency, independent on the force generated during acceleration.

Acknowledgements The reported scientific work is performed with RNF, Project No. 14-39-00008.

References

1. Kovac, M., Schlegel, J.-C., Zufferey, D., Floreano, D.: A miniature jumping robot with self-recovery capabilities. Proceedings of the IEEE/RSJ International Conference on Robotics and Automation, Kobe, Japan, pp. 583–588 (2009)
2. Kovac, M., Schlegel, J.-C., Zufferey, D., Floreano, D.: Steerable miniature jumping robot. Auton. Rob. **28**, 295–306 (2010)
3. Hutter, M., Remy, D., Siegwart, R.: Planar hopping with a robotic leg. Eidgenössische Technische Hochschule Zurich (2011)
4. Burdick, J., Fiorini, P.: Minimalist jumping robot for celestial exploration. Int. J. Rob. Res. **22** (7), 653–674 (2003)
5. Londhe, V.N.: Energy Pumping Mechanism for Hopping Robot. Indian Institute of Technology, Bombay (2007)
6. Sapognikov, S.B., Kudriavtsev, O.A.: Compact accelerating stand for ballistic tests. Mach. Build. Ser. **20**, 139–143 (2012)
7. Zlatina, N.A., Mishina, G.I.: Ballistic devices and their use in experimental studies. Science, Moscow (1974)
8. Zelentsov, V.V., Karneychik, A.S., Vladimirov, A.: Multi-purpose ballistic laboratory: results and prospects. Eng. J. Sci. Innovation **3** (2013) http://engjournal.ru/catalog/machin/rocket/626.html
9. Volkova, L.Y., Jatsun, S.F.: Study of the influence of the point of leg attachment of hopping robot on movement of the device. Nonlinear Dyn. **9**(2), 327–342 (2013)
10. Jatsun, S.F., Volkova, L.Y.: Modeling of movement of the multi-link hopping robot and study its characteristics, Izv. RAN. Theor. Control Syst. **4**, 137–149 (2013)
11. Jatsun, S.F., Loktionova, O., Serebrovsky, V.V., Volkova, L.Y.: Modeling the jump of the four-link robots. Fundam. Res. **10**, 308–315 (2013)
12. Jatsun, S., Loktionova, O. Volkova, L.Y., Yatsun, A.: Investigation into the influence of the foot attachment point in the body on the four-link robot jump characteristics. In: Mobile Service Robotics: Proceedings of the 17th International Conference on Climbing and Walking Robots and the Support Technologies for Mobile Machines Clawar'14, pp. 101–108 (2014)

Part IV
Robot Vision and Sensory Control

A Control Predictive Framework for Image-Based Visual Servoing Applications

Corneliu Lazar and Adrian Burlacu

Abstract Model predictive control (MPC) is a very attractive control algorithm used to solve the complex problems of image-based visual servoing (IBVS) systems. Many image-based predictive controllers were reported, each being different regarding the implementing of the MPC concept. In this paper, we present a MPC framework for IBVS applications, the main contributions being a new visual predictor and the introduction of the reference trajectory.

Keywords Visual predictive control · Image based visual servoing · Robot arm motion control

1 Introduction

In order to control a robot by an image-based visual servo (IBVS) method, a set of image features must reach desired image locations. Typically, the set of features corresponds to several 3D points of the tracked target projected on the image plane [1]. Many control algorithms were used to decrease asymptotically to zero the error between the actual location of the observed features and their desired locations. Among them, the model predictive control (MPC) has been successfully used in visual servoing applications in order to control the motion of a robotic system [2–7].

In MPC, an internal model of the system is used to predict the future evolution of the system output starting at the current time, over a future prediction horizon, which is used to optimize the control signal by minimizing a constrained cost function. Distinct from the set-point trajectory, the reference trajectory is introduced in MPC to define the desired closed-loop behaviour. Along this ideal trajectory the

C. Lazar (✉) · A. Burlacu
"Gheorghe Asachi" Technical University of Iasi, Iasi, Romania
e-mail: clazar@ac.tuiasi.ro

A. Burlacu
e-mail: aburlacu@ac.tuiasi.ro

T. Borangiu (ed.), *Advances in Robot Design and Intelligent Control*,
Advances in Intelligent Systems and Computing 371,
DOI 10.1007/978-3-319-21290-6_19

system output should return to the set-point trajectory, for example when a disturbance occurs.

In recent years, many predictive controllers were used in IBVS applications to control motion of autonomous manipulators, but differ in how they have implemented the concept of MPC. The plant of the visual servo loop is considered to be composed of the robot and the camera models. Thus, in [2], the visual servo loop plant is described by an ARIMAX model including the dynamics of the velocity controlled robot (called Virtual Cartesian Motion Device -VCMD) and the dynamics introduced by the camera with image acquisition and processing. Based on this model, a Multi-Input Multi-Output (MIMO) Generalized Predictive Control (GPC) was designed to optimally control a 6 DOF manipulator. In [3], a nonlinear MPC algorithm is presented, considering the constraints in terms of robot limitations and feature visibility. The robot model is based on a set of double integrators plus the internal feedback and the camera projection model. Another nonlinear predictive algorithm for IBVS systems based on an internal model control structure was introduced in [5]. The visual servoing task is formulated as a nonlinear optimization problem over a prediction horizon which offers the advantage of taking into account the visibility and the 3D constraints.

This paper presents a review of predictive control algorithms developed by the authors for visual servoing of robot manipulators with eye-in-hand configuration. Firstly, considering point features, we introduce an internal model predictor based on the interaction matrix [4, 6], different from those used in [2, 3], but very suitable for image predictions. Two years later, the same model (called local model) was used for image prediction [5]. Secondly, we introduce the reference trajectory concept from predictive control to image based predictive controllers [6, 7]. Finally, we propose a control predictive framework for IBVS applications, which is detailed in this paper.

2 MPC Framework for IBVS

The main goal of IBVS is to find a control algorithm capable to drive the robot arm so that a set of image features $\mathbf{f}$, composed of image coordinates $[u_i \quad v_i]$ of several Cartesian points belonging to the tracked target reaches a desired value $\mathbf{f}^*$. An important role in designing the control algorithm is played by the interaction matrix $\mathbf{J}$ defined:

$$\dot{\mathbf{f}} = \mathbf{J}\mathbf{v}_c \tag{1}$$

where $\mathbf{v}_c = [\mathbf{v}^T\, \boldsymbol{\omega}^T]^T$ is the camera velocity screw with its translational ($\mathbf{v}$) and rotational ($\boldsymbol{\omega}$) components. A Cartesian point, having coordinates $[\mathrm{X} \quad \mathrm{Y} \quad \mathrm{Z}]$ related to the camera frame, can be projected using a perspective model into a point feature with image coordinates $\begin{bmatrix} u & v \end{bmatrix}$. Denoting with λ the focal length, Eq. (1) becomes:

$$\begin{bmatrix} \dot{u} \\ \dot{v} \end{bmatrix} = \begin{bmatrix} \frac{\lambda}{Z} & 0 & \frac{u}{Z} & \frac{uv}{Z} & -\left(\lambda + \frac{u^2}{\lambda}\right) & v \\ 0 & -\frac{\lambda}{Z} & \frac{v}{Z} & \lambda + \frac{v^2}{\lambda} & -\frac{uv}{\lambda} & -u \end{bmatrix} \begin{bmatrix} \mathbf{v} \\ \boldsymbol{\omega} \end{bmatrix} \tag{2}$$

where Z is the corresponding point's depth in the camera frame. The control algorithm computes the camera velocity screw which is sent to the robot velocity controller.

2.1 Image Based Predictive Control

The proposed MPC framework for IBVS is based on the interaction matrix $\mathbf{L}(\mathbf{f}(k), \mathbf{Z}(k))$. Considering m point features, an internal model for the plant of the visual servo loop, which has as input the camera velocity screw and as output point features, is built. This internal model, introduced for the first time in [4], has been used to build predictors $\mathbf{f}(k+i|k)$, $i = \overline{1, h_p}$ that calculate the future evolution of the plant output over the prediction horizon h_p.

Taking into account the general concept of predictive control, where the reference trajectory is distinct from the set-point, we define such a trajectory denoted by $\mathbf{w}(k+i|k)$, $i = \overline{1, h_p}$ in [6, 7] along which the plant should go to the set-point (desired features $\mathbf{f}^*$), starting from the current features $\mathbf{f}(k)$, over the prediction horizon. In Fig. 1, the new concept of predictive control for IBVS applications is represented. Considering the dependency of the interaction matrix of depth, we put Z as vertical axis in Fig. 1. We assume a discrete-time setting and thus, the current time is k.

At the current time, the plant output $\mathbf{f}(k)$ is measured by the image acquisition and processing and the previous history of the output trajectory is known. Also, the set-point trajectory $\mathbf{r}(k+i|k)$, $i = \overline{1, h_p}$ is known over the prediction horizon. We considered in Fig. 1 a constant set-point equal to the desired point features $\mathbf{f}^*$ obtained for the depth Z^*. Distinct from the set-point trajectory, we introduce the reference trajectory $\mathbf{w}(k+i|k)$, $i = \overline{1, h_p}$ which starts at the current output $\mathbf{f}(k)$ and ends at time $k + h_p$ with the desired point features $\mathbf{f}^*$ ($\mathbf{w}(k+h_p|k) = \mathbf{f}^*$). In this way, through the reference trajectory the behaviour of the closed loop is established and the plant output will reach the set-point trajectory as fast as possible.

Using an internal model, we predict the behaviour of the plant $\mathbf{f}(k+i|k)$, $i = \overline{1, h_p}$ over the prediction horizon. The predicted output depends on the input trajectory $\mathbf{v}_c(k+i|k)$, $i = \overline{0, h_p - 1}$ which is the future control sequence over the prediction horizon. The future control sequence is chosen such as to bring the plant output at the end of the prediction horizon and the desired point features, i.e. $\mathbf{f}(k+h_p|k) = \mathbf{f}^*$. After the computation of the optimal input trajectory by

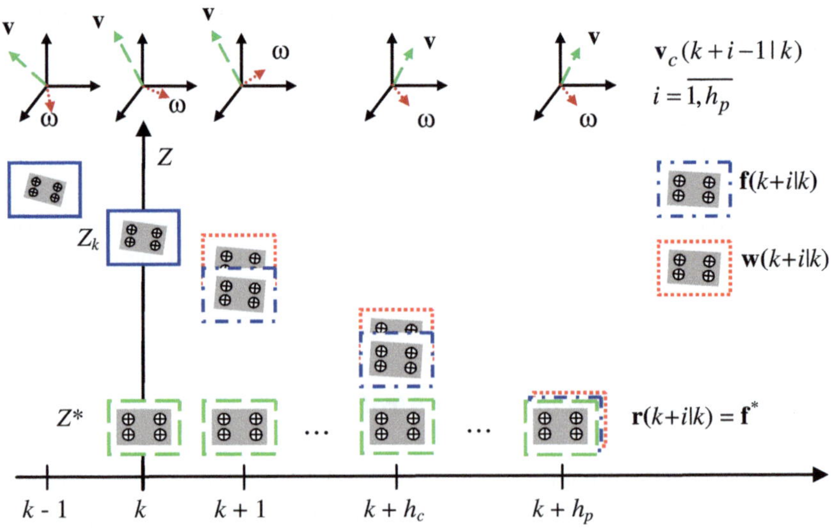

Fig. 1 Predictive control: the basic idea for IBVS applications

minimizing a sum of squares of errors $\sum_{i=1,h_p}[\mathbf{f}(k+i|k)-\mathbf{w}(k+i|k)]^2$, only the first element $\mathbf{v}_c(k|k)$ is applied to the plant and at the next sampling time the whole cycle is repeated again according to the receding horizon strategy.

2.2 *Visual Predictive Control Loop*

Using the new concept of predictive control for IBVS systems presented in Fig. 1, we developed the cascade structure from Fig. 2 for motion control of robot arms [6]. The inner loop regulates the camera velocity screw $\mathbf{v}_c$ and the outer one the robot arm motion such as to obtain a desired position described by $\mathbf{f}^*$.

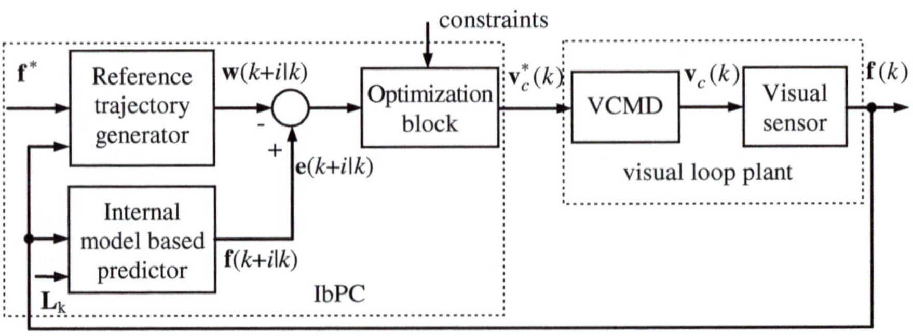

Fig. 2 Cascade structure for robot arm motion control

The inner velocity loop is considered a virtual Cartesian motion device (VCMD), usually described by a transfer matrix with camera velocity set-point $\mathbf{v}_c^*(k)$ as input and camera velocity $\mathbf{v}_c(k)$. This transfer matrix approximates the nonlinear robot dynamics using different approaches [2, 8] and typically has a diagonal form obtained through a suitable design of the multivariable inner velocity control loop.

The outer loop, based on an image based predictive controller (IbPC), computes the control signal $\mathbf{v}_c^*(k)$ so that the current point features $\mathbf{f}(k)$ should reach the desired ones $\mathbf{f}^*$. The IbPC consists of a reference trajectory generator, an internal model based predictor and an optimization block.

The internal model for image prediction was developed starting from the relation between the camera and the point features velocities given by (1) and considering the inner velocity loop as an analogue system, because of its very short sampling period (usually 1 ms), described by a diagonal transfer matrix $\mathbf{G}(s)$ [9]. Having $\mathbf{v}_c(k)$ as output of the VCMD discrete time model, the discrete time relation between camera and point features velocities is obtained by the discretization of (1) using Euler's method:

$$\mathbf{f}(k+1) = \mathbf{f}(k) + T_e \mathbf{L}_k \mathbf{v}_c(k) \tag{3}$$

The one-step ahead prediction of the image point features evolution can now be calculated using (3) and the discrete model $\mathbf{G}(z) = (1 - z^{-1})Z\{\mathbf{G}(s)/s\}$ of the VCMD, resulting:

$$\mathbf{f}(k+1|k) = \mathbf{f}(k) + T_e \mathbf{L}_k \mathbf{G}(z) \mathbf{v}_c^*(k|k) \tag{4}$$

where the notation $f(k+1|k)$ indicates that the prediction is computed at the discrete time k. The interaction matrix $\mathbf{L}_k$ is computed with the point features $(u_i(k), v_i(k))$ acquired at the current discrete time k. It is assumed that it is possible to compute at every sampling period the depth $z_i(k)$ of the current point features with respect to the camera frame. Shifting the one-step ahead prediction model (4) by recursion, the i-step ahead predictor $\mathbf{f}(k+i|k)$ is obtained:

$$\mathbf{f}(k+i|k) = \mathbf{f}(k+i-1|k) + T_e \mathbf{L}_{k+i-1} \mathbf{G}(z) \mathbf{v}_c^*(k+i-1|k) \tag{5}$$

being used as internal model based predictor.

For IBVS control systems, the set-point is the desired feature set $\mathbf{f}^*$ obtained from a reference image acquired from the desired grasping position. This image describes what the camera should see when the end-effector is correctly positioned relative to the target object. Starting from the current features $\mathbf{f}(k)$, a reference trajectory is necessary in visual predictive control to define the way how to reach the desired features $\mathbf{f}^*$ over the prediction horizon.

Beginning at the current discrete time k with the current image I_k having the point features $\mathbf{f}(k)$, the reference trajectory is designed using the image sequence $\{I_{k+i},\ i = \overline{1, h_p}\}$ with point features $\mathbf{w}(k+i|k)$ in order to obtain $\mathbf{w}(k+h_p|k) = \mathbf{f}^*$.

To generate the image plane trajectories for tracked points in an eye-in-hand system, we have chosen the 3D motion planning approach for image-based visual servoing task from [9]. Considering that the initial image is I_k with point features $\mathbf{f}(k)$, the final one is I_{k+hp} with the point features $\mathbf{f}^*$ and the object is fixed being described by four point features assumed to be coplanar but not collinear. The reference trajectory gradually varies from the current point features $\mathbf{f}(k)$ at time k to the desired point features $\mathbf{f}^*$ at time $k + h_p$. Taking into account the collineation matrix $\mathbf{C}$ representing the projective homography between the initial image I_k and the final image I_{k+hp}, the homogeneous coordinates of the four point features from the final image $\tilde{\mathbf{f}}_i^* = \left[u_i^*, v_i^*, 1\right]^T$ can be expressed with respect to the coordinates of points from the initial image $\tilde{\mathbf{f}}_i(k) = [u_i(k), v_i(k), 1]^T$, resulting $\tilde{\mathbf{f}}_i^* = \mathbf{C}\tilde{\mathbf{f}}_i(k)$, $i = \overline{1,4}$ and the possibility to compute the reference trajectory $\{\mathbf{w}(k+i|k), i = \overline{1,h_p}\}$[6].

The optimization block is developed to make future system outputs to converge to the reference trajectory. For that, an objective function J is established, generally defined as a quadratic function of predicted control error given by:

$$\mathbf{e}(k+i|k) = \mathbf{f}(k+i|k) - \mathbf{w}(k+i|k),\ i = \overline{1,h_p} \tag{6}$$

The objective function to be minimized is defined by:

$$J = \frac{1}{2}\sum_{i=1}^{h_p} \mathbf{e}^T(k+i|k)\mathbf{Q}\mathbf{e}(k+i|k) + \sum_{i=0}^{h_c-1} \mathbf{v}_c^{*T}(k+i|k)\mathbf{R}\mathbf{v}_c^*(k+i|k) \tag{7}$$

where $\mathbf{Q}$ and $\mathbf{R}$ denote positive definite, symmetric weighting matrices and h_c is the control horizon in (7). The main constraints are associated to the limits of the image called the visibility constraint, ensuring that all the features are always visible:

$$(u_i(k), v_i(k)) \in [(u_{\min}, v_{\min}), (u_{\max}, v_{\max})],\ i = \overline{1,m}. \tag{8}$$

3 Simulator for IBVS Applications

Using the MPC framework for IBVS applications, more simulators were implemented in Matlab [4, 6]. A modified version of the simulator from [6] is presented in Fig. 3, considering an object defined by 4 planar points in the Cartesian space.

The blocks 'Init pos' and 'Des conf' are used to represent the start position $\mathbf{P}_b^0$ and the desired position $\mathbf{P}_b^*$ of the object. Considering the frames attached to the robot base R_b, to the camera R_c and to the object R_o, the homogeneous matrices $\mathbf{T}_c^b$ and $\mathbf{T}_o^b$ between the frames R_c and R_b and, respectively R_o and R_b are given. Knowing the position of the desired points related to the robot $\mathbf{P}_b^*$ and the camera position related to the robot $\mathbf{T}_c^b(0)$, the points position $\mathbf{P}_c^*$ can be detected relative to

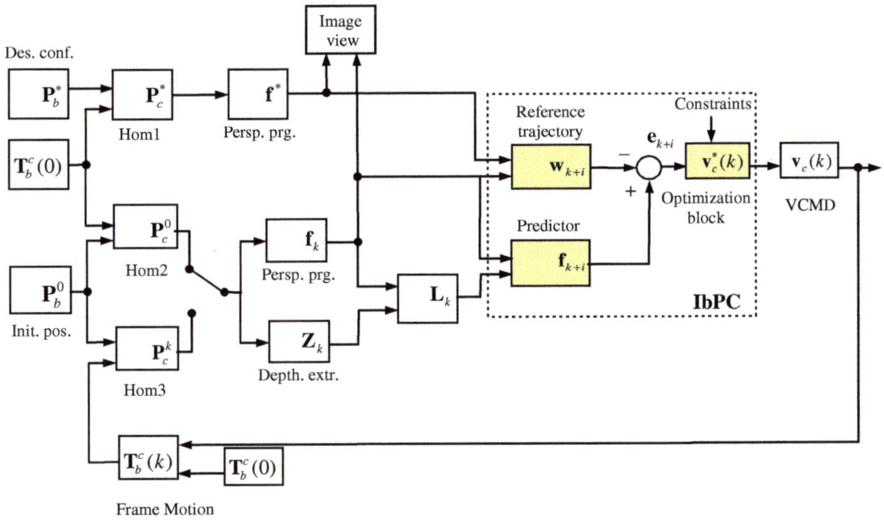

Fig. 3 Simulator for IBVS systems

the camera coordinate system R_c by using a homogeneous transformation implemented with 'Hom1' block.

By a similar procedure, the initial position $\mathbf{P}_b^0$, respectively the current positions of the points are transposed from R_b to R_c frames using 'Hom2', respectively, 'Hom3' block, resulting the initial/current position of the object points $\mathbf{P}_c^0/\mathbf{P}_c^k$ in the camera frame. The image of the object described by the points given in the Cartesian space is built using 'Persp proj' blocks, considering the intrinsic camera parameters and the image centre point coordinates [6].

The depth Z_i, necessary for the interaction matrix computation, is extracted with the 'Depth extr' block. For the visualization of the object points in the image plane, the 'image view' block is employed. Using the current feature $\mathbf{f}(k) = \mathbf{f}_k$ together with the corresponding depth $\mathbf{Z}(k) = \mathbf{Z}_k$, the interaction matrix $\mathbf{L}_k$ is computed. The 'Reference trajectory' block generates $\mathbf{w}(k+i|k) = \mathbf{w}_{k+i}$ and the 'Predictor', the output $\mathbf{f}(k+i|k) = \mathbf{f}_{k+i}$. The objective function (7) with the constraints (8) is minimized by the 'Optimization block' using the Matlab function *fmincon*.

In simulation two IbPCs were tested, with and without reference trajectory having the parameters: $h_p = 4$, $h_c = 1$, $\mathbf{Q} = e^{1-i}\mathbf{I}_8,\ \ i = \overline{1, h_p}$ and $\mathbf{R} = \mathbf{I}_6$.

The simulation results are presented in Fig. 4 and include for the two IbPCs the point features trajectories from the starting point features represented with blue circles to the desired ones depicted with red squares. Both predictive techniques fulfil the servoing task and one can observe that the reference trajectory based predictive approach has a smoother behaviour, thus making it more suitable for a real time implementation, as in [7].

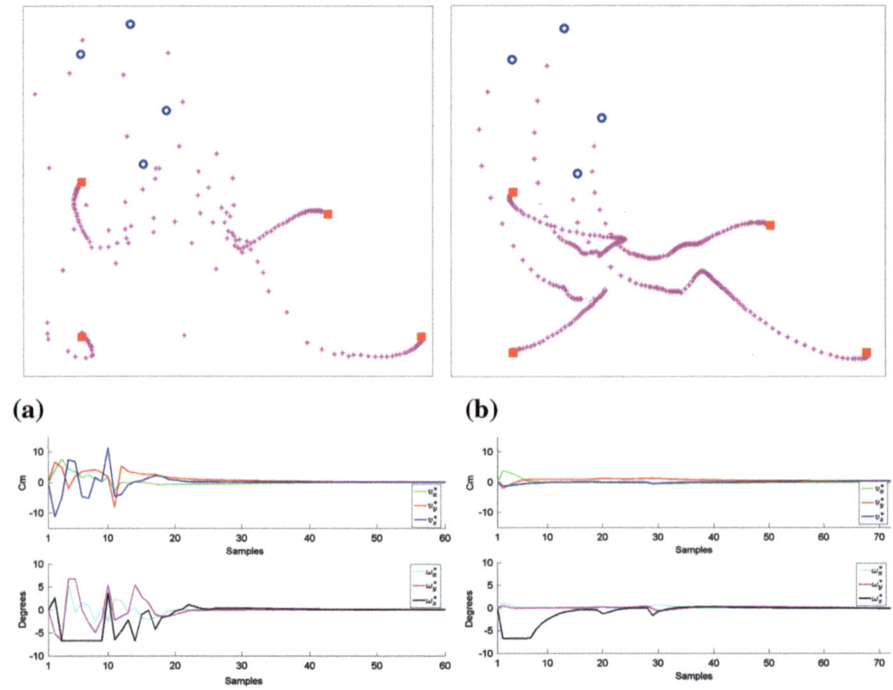

Fig. 4 Comparison of point features trajectory (*up*) and camera velocity (*down*): **a** predictive; **b** predictive with reference trajectory

4 Conclusion

The paper presented a MPC framework for IBVS applications and a review of the predictive control algorithms developed by the authors for visual servoing of robot manipulators with eye-in-hand configuration. Using a reference trajectory concept for IbVCs and an internal model based predictor the convergence and stability of robot motion have been obtained through nonlinear constraint optimization.

References

1. Chaumette, F., Hutchinson, S.: Visual servo control part I: Basic approaches. IEEE Robot. Autom. Mag. **13**(4), 82–90 (2006)
2. Gangloff, J.A., De Mathelin, M.F.: High speed visual servoing of a 6-d.o.f. manipulator using multivariable predictive control. Adv. Robot. **17**(10), 993–1021 (2003)
3. Sauvée, M., Poignet, P., Dombre, E., Courtial, E.: Image based visual servoing through nonlinear model predictive control. In: 45th IEEE Conference on Decision and Control, San Diego, pp. 1776–1781 (2006)

4. Lazar, C., Burlacu, A.: Predictive control strategy for image based visual servoing of robot manipulators. In: 9th International Conference on Automation and Information, pp. 91–97, Bucharest (2008)
5. Allibert, G., Courtial, E., Chaumette, F.: Predictive control for constrained image-based visual servoing. IEEE Trans. Robot. **26**(5), 933–939 (2010)
6. Lazar, C., Burlacu, A., Copot, C.: Predictive control architecture for visual servoing of robot manipulators. In: 18th IFAC World Congress, Milan, pp. 9464–9469, Elsevier (2011)
7. Burlacu, A., Lazar, C.: Reference trajectory based visual predictive control. Adv. Robot. **26** (8–9), 1035–1054 (2012)
8. Allotta, B., Fioravanti, D.: 3D motion planning for image-based visual servoing tasks. In: IEEE International Conference on Robotics and Automation, Barcelona, pp. 2173–2178 (2005)
9. Fujimoto, H.: Visual servoing of 6 d.o.f. manipulator by multirate control with depth identification. In: 42nd IEEE Conference on Decision and Control, Hawaii, pp. 5408–5413 (2003)

Motion Leap Compared to Data Gloves in Human Hand Tracking

Constantin Cătălin Moldovan and Ionel Starețu

Abstract The paper presents a comparative analysis of current methods for capturing human hand gestures and using then the results in virtual reality environments as a mean of direct, non-intrusive interaction between human and computer. Compared to the methods which are using different data gloves, the Motion Leap device enables new ways of interactions that are not intrusive. In this paper a system is developed that uses Motion Leap to capture human hand poses and recognize gestures, the results of the analysis being sent to a virtual reality environment for further, task-oriented high level processing.

Keywords Virtual reality · Human hand capture · Gesture recognition · Frame processing

1 Introduction

Capturing the motions of the human body and especially of human hands represents one of the most complex tasks in computer science. The complexity of the capturing process is given by the fact that human finger joints don't move perfectly in the sense that the joints of a finger move in a different way compared to the joints of another finger and moreover, one joint of a finger might move in a different manner compared to another joint of the same finger.

In time, different sensors and methods were developed and a higher accuracy and robustness were progressively obtained. The methods were categorized in invasive techniques, like those using various tracking devices and different data gloves (Cyber Glove) and non-invasive techniques based on vision techniques. At the beginning, the researchers in this field used invasive techniques which were superior from the accuracy and robustness points of view to non-invasive

C.C. Moldovan · I. Starețu (✉)
Transilvania University of Braşov, Braşov, Romania
e-mail: istaretu@yahoo.com

T. Borangiu (ed.), *Advances in Robot Design and Intelligent Control*,
Advances in Intelligent Systems and Computing 371,
DOI 10.1007/978-3-319-21290-6_20

techniques, but this handicap was progressively overcome. As the present research paper demonstrates, the accuracy given by vision-based techniques is very high, and seems even to be comparable with the invasive techniques.

The initial non-invasive sensors selected were basically RGB cameras that used markers placed on the human hand in order to accurately capture the hand's movements [1]. There were also developed methods based on contours tracking [2], edge tracking [3] or artificial intelligent methods [4]. All of them were successful to a certain degree, but, in order to obtain an accuracy compared to invasive techniques, real progress was made only recently with the development of Microsoft Kinect sensor [5] and Motion Leap sensor [6]. The latest ones provide accuracy high enough to be compared to the invasive techniques, thus indicating that they can be used in the same scope having the same kind of results but without having the same limits as the intrusive techniques.

Based on the researches made using this device [6, 7], it was demonstrated that the Kinect sensor progressively became a useful mean to capture depth data, even suggesting that it might even become an alternative solution to the intrusive techniques based on various data gloves. But, since there is a difference in the computed distance and the real distance between the Kinect Sensors and objects, using this sensor proves to be not quite efficient if the task consists in capturing complex human hands motion [6]. As an alternative to the Microsoft Kinect Sensor, a new sensor called Motion Leap was developed in 2012. Using this new sensor, the paper analyses the possibility to develop a non-invasive technique based on Motion Leap to capture various gestures from a human hand and use the resulting data to control a virtual hand, in comparison with the technique based on data glove.

The reported research is part of a more ample program that intends using a non-invasive technique to control a robotic arm and hand ensemble.

2 Motion Leap Device

The Motion Leap sensor is a newly developed sensor using the structured light concept [6]. Motion Leap allows a new way to detect human hand and recognizes human hand gestures. Compared to the Microsoft Kinect, Motion Leap constructor state that the sensor has sub-millimetre accuracy and can be considered for computer applications that require interaction in virtual reality [8]. Compared to the Microsoft Kinect sensor, the main scope for which the Motion Leap was created is to detect and track accurately and rapidly human hands with optimization of costs.

The position delivered by the Motion Leap device is relative to the centre of the device which is located in the centre of the second infrared sensor position (Fig. 1).

As can be observed in Fig. 1, the Motion Leap sensor uses three infrared emitters and two CMOS sensors in order to capture depth data. In Fig. 2 it is shown how the human hand is captured by the sensor using the internal Motion Leap viewer.

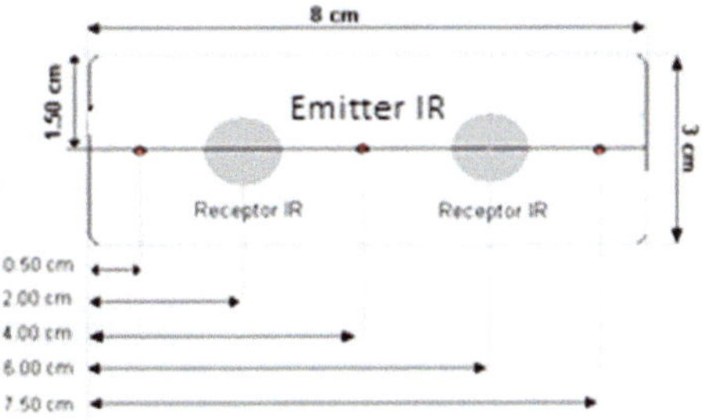

Fig. 1 Motion Leap sensor elements [6]

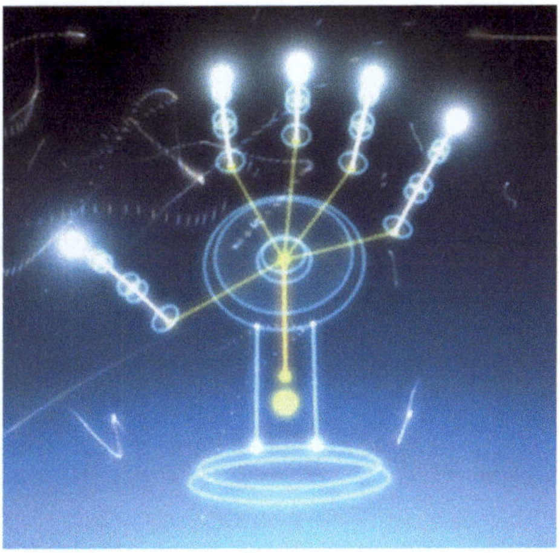

Fig. 2 Human hand capture [6]

From a functional perspective, information about the human hand, fingers, positions, joints and gestures can be recognized when the hand is in the range of 25–500 mm above the sensor (Fig. 3).

Along with the Motion Leap package, a programming interface was used; this interface, called Motion Leap SDK provides programmable access to the depth data. It can be used either by C++, Java or Microsoft .NET programming languages.

From the functional point of view, the Motion Leap device analyses a data flow which is then split by the Motion Leap SDK in frames depending on the time allowing the SDK to compute gestures or the speed which the hand is having over the sensor.

One of the most important advantages of Motion Leap is the fact that, compared to all its predecessors it can distinguish between each finger of the human hand. The internal algorithm of Motion Leap is assigning a unique identifier to every hand gesture or finger, allowing thus the algorithm to easily track each entity through a video stream. If at a certain moment of time, a finger disappears and then reappears, the internal algorithm will assign a new unique identifier.

Position data for each finger object is given with sub-millimetre accuracy on each axis X, Y and Z relative to the centre of the device (Fig. 4). Real captured

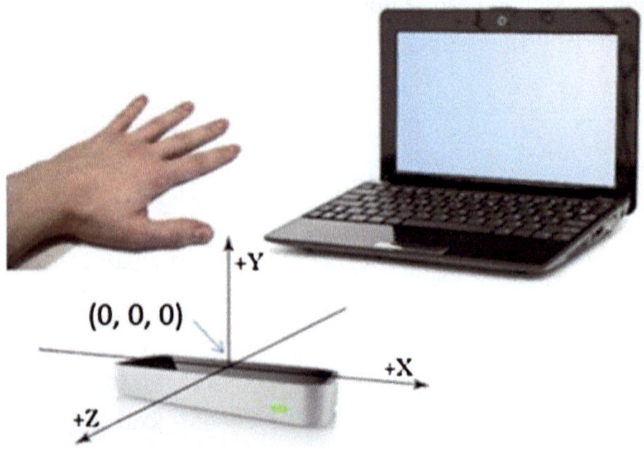

Fig. 3 Device centre and coordinate system

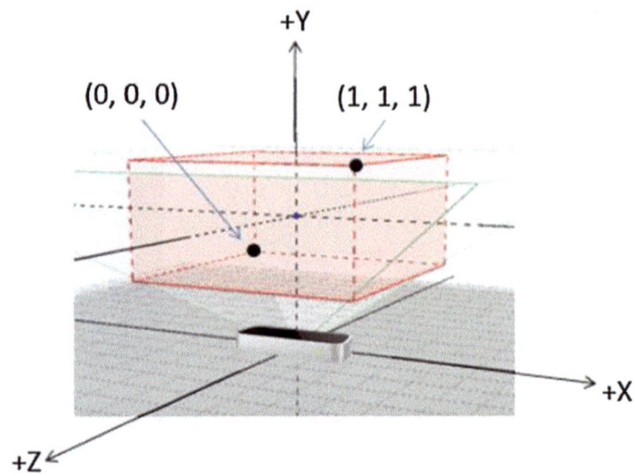

Fig. 4 Interaction cube [9]

spatial data is then translated by the internal Motion Leap SDK in different environments into the following programmable classes: *Screen*, *Interaction Box* and *Touch Zone*.

The **Screen** class describes the position and orientation of the user's monitor relatively to the coordination system used by Motion Leap. The Screen class structure includes the bottom left corner of the screen, the direction vectors of the vertical axes of the screen and the normal vectors of the screen. The same class offers intersection methods that calculate how the hand will intersect the screen. Intersection points are defined using the normalized screen coordinates (see Fig. 4). The origin point of the screen is characterized as being the 0,0 point localized in the bottom left corner with the upper right corner normalized to 1,1. That means that the intersection points x and y will be normalized between 0 and 1 (the z coordinate value will be always 0).

The **Interaction Box** class represents a rectangular space and is situated above the Motion Leap device. This rectangular space is basically the whole field of view of the device (Fig. 4). A coordinate system of the Motion Leap might be mapped easily to a point relative to the volume represented by the Interaction Box class. The value of that point is scaled in such a way that the entire volume represented by the rectangular space is treated as a cube with the edge length of 1 and with the origin in the bottom left corned the most distant from the user.

The **Touch Zone** class can be used with an object of Point-table type. This might represent a finger or a pen. The Touch Zone class is used for implementing a surface that emulates touching without actually doing touch on a physical surface. The touch zone is divided into two areas, the *hovering zone* and *touch zone* which are identified with a touch distance between –1 and +1.

3 Testing the System and Analysis of Results

In order to test the Motion Leap-based procedure and compare the results with a glove system, it was first necessary to define and then recognize different gestures of the human hand. The recognized gestures were then sent to a virtual reality environment called Grasp IT [10] using the same interface as the data gloves does.

For hand recognition several sequence frames were captured using the Motion Leap device. These sequences were captured using different light conditions (natural light or artificial light) and different visualization points. In Fig. 5 there can be seen gestures used in the recognition process.

In an empirical way it was observed that the gestures don't depend on the light conditions when Motion Leap is used. Figure 6 reproduces recognized hand gestures and displays the hand joints using the Motion Leap internal visualizer.

In order to evaluate and transmit human hand gestures to the virtual environment a module for the interaction with the virtual environment was created. The whole system is divided in five separate components: (1) hardware component, (2) frame

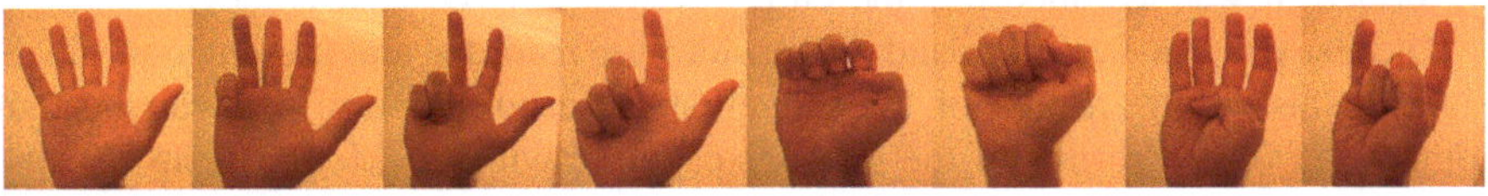

Fig. 5 Human gestures for which the recognition will be run

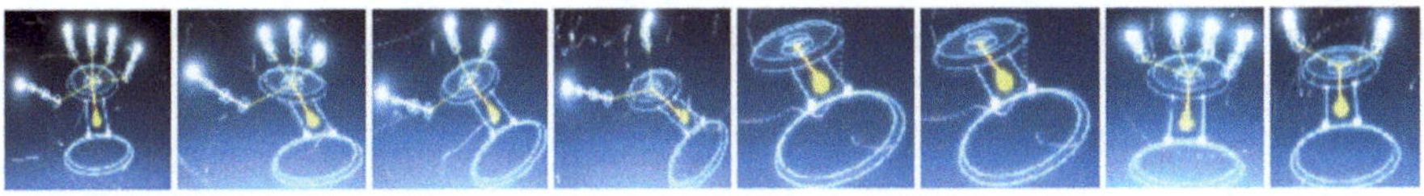

Fig. 6 Gestures recognized using Motion Leap

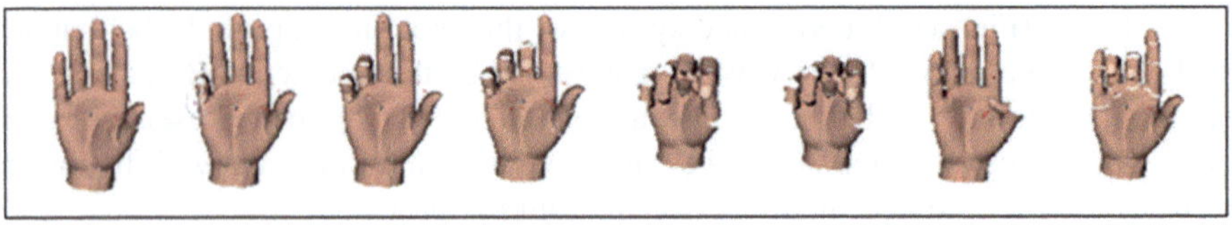

Fig. 7 Gestures transmitted to the virtual environment

processor, (3) gestures detector and (4) Hand Commander component used to transfer the data to (5) Grasp IT visual interface.

- The **Hardware component** represents the Motion Leap sensor responsible for capturing raw data.
- Once the data are available, the Motion Leap device transfers the data to a **frame processor** for a further evaluation.
- The frame processor uses the **software module** of the Motion Leap device which processes the input from the sensor and calculates the information about the current state of the hand.
- The human hand data is transmitted to the **gesture detection module** which has the main activity to detect gestures.
- Once one of gestures is detected, the detection module notifies the functional simulation component **Hand Commander** to transmit the data to the **virtual environment** for execution.

For the gestures captured above, the sequence in Fig. 7 was obtained.

4 Results and Conclusions

In order to test the accuracy of the Motion Leap device, a comparative analysis between gestures identified with the Motion Leap and with a glove system was performed. It was assumed that the Glove system has 100 % accuracy.

The test was run five times and the results were recorded and presented in Fig. 8. Figure 8 also show results of the glove-based gesture capture that can be thus compared with the results of the gestures captured with the Motion Leap sensor.

As can be seen in Fig. 8, comparing visually the results obtained with the two types of sensors, the following observations can be made:

- Using a data glove device, the precision and the accuracy are very high and the algorithm is running well in terms of repeatability.
- Using a Motion Leap sensor, the precision and the accuracy are very high too, and the algorithm is executed well in terms of repeatability with one exception for the 6th gesture.

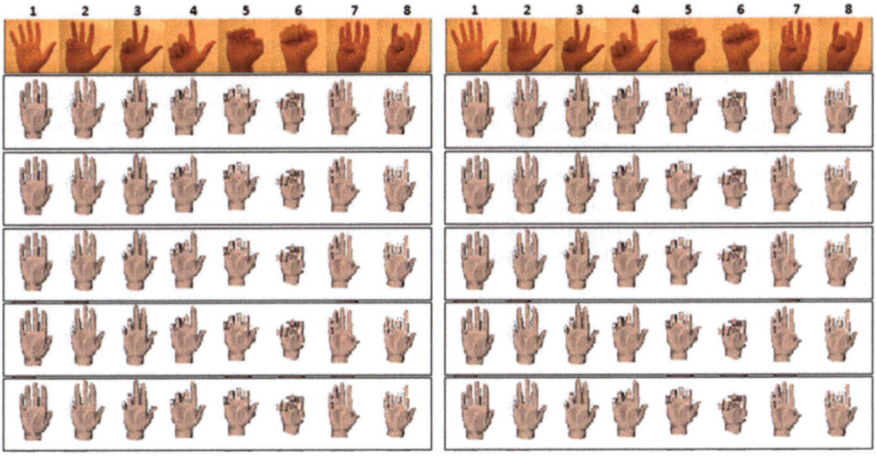

Fig. 8 Gestures transmitted to the virtual environment using the Glove device (*left*) and the Motion Leap device (*right*)

Based on the above considerations, it can be concluded that the Motion Leap sensor can be used as a means to capture data related to the human hand and the results are comparative with the methods based on data gloves.

Acknowledgments We express our gratitude to the Company Karl Cloos, Germany and its Romanian representative Robcon TM which partly sustained this research work.

References

1. Romero, J., Kjellstrom, H., Kragic, D.: Monocular real-time 3D articulated hand pose estimation. In: Proceedings of the 9th IEEE-RAS'09 International Conference on Humanoids, pp. 87–92. Paris, France (2009). ISBN 978-1-4244-4597-4
2. Asaari, M.S.M., Suandi, S.A.: Hand gesture tracking system using adaptive Kalman Filter. In: Proceedings of the 10th IEEE International Conference on Intelligent Systems Design and Applications (ISDA'10), pp. 166–171. Cairo, Egypt (2010). ISBN 978-1-4244-8134-7
3. Stenger, B., Thayananthan, A., Torr, P.H.S., Cipolla, R.: Model-based hand tracking using a hierarchical Bayesian filter. IEEE Trans. Pattern Anal. Mach. Intell. **28**(9), 1372–1384 (2006). ISSN 0162-8828
4. Moldovan, C.C., Staretu, I.: Real-time gesture recognition for controlling a virtual hand. Adv. Mater. Res., 463–464 (2012). ISSN 1662-8985
5. Moldovan, C.C., Staretu, I.: Analysis of the accuracy and robustness of Kinect sensor used with the scope of robot manipulation. Appl. Mech. Mater. **555** (2013). ISSN 1662-7482
6. Moldovan, C.C.: Theoretical and experimental research regarding the diversification of a virtual hand interaction with objects from a virtual word with application in industrial engineering, Ph.D. Thesis, defended publicly at Transilvania University of Brasov, Romania (2014)

7. Draelos, M.T.: The Kinect up close: Modifications for short-range depth imaging. In: IEEE Conference on Multisensor Fusion and Integration for Intelligent Systems (MFI'12), pp. 251–256. Hamburg, Germany (2012). ISBN 978-1-4673-2510-3
8. Chang, E.: New ways of accessing information spaces using 3D multitouch tables. In: Proceedings of the International Conference on Cyberworlds (CW'12), pp. 144–150. Birmingham, UK (2012). ISBN 0-7695-2140-1
9. Guna, J., Jakus, G., Pogačnik, M., Tomažič, S., Sodnik, J.: An analysis of the precision and reliability of the leap motion. Sensors. **14**(2) (2012). ISSN 3702-3720
10. Miller, A., Allen, P.K.: GraspIT!: A versatile simulator for robotic grasping. Rob. Autom. Mag. **11**(4), 110–122 (2004). ISSN 1070-9932

Considerations for Robot Vision in Glass Production

Anton Ružič

Abstract In this paper we present considerations for implementing and using robot vision in glass production. We concentrate specifically on the forming processes of those shell glass objects that are still produced with a considerable amount of manual labour. It is shown that every practical robot automation solution in these production phases must include the perception of the glass object's shape, implemented with computer vision. A number of particular requirements and conditions are identified, originating from high and ever changing temperature and from specific production environment conditions.

Keywords Vision systems · Glass technology requirements · Robotics

1 Introduction

The research aims at introducing robot automation for selected tasks in the glass industry. Evaluating the current situation in the European Union, we find that the largest part of glass industry production, over 60 %, is represented by container glass items (e.g. bottles, jars, and lightning covers—generally all shell and hollow glass objects).

Glass production has three main parts: first, raw glass components are prepared; then, these are melted in furnaces and objects are formed; the final part includes inspection and packaging. In this paper we consider mainly the forming processes for container/hollow glass items. These start with gathering one (or more) glass "chunks" or gobs from a furnace (or furnaces) where the glass is melted. Here the glass is in a plastic phase and has very low viscosity, at temperatures between 1050 and 1100 °C. Then a series of forming operations is carried out sequentially, so that from a solid melted gob gradually the final hollow glass object is achieved. At this

A. Ružič (✉)
Jožef Stefan Institute, Jamova cesta 39, 1000, Ljubljana, Slovenia
e-mail: anton.ruzic@ijs.si

T. Borangiu (ed.), *Advances in Robot Design and Intelligent Control*,
Advances in Intelligent Systems and Computing 371,
DOI 10.1007/978-3-319-21290-6_21

point the object becomes almost rigid and has approximately 800 °C, depending on the glass type. The chain of forming processes is somewhat different, depending on item types, mechanization and automation level, in-house technology etc. Nonetheless, hollow glass forming always includes a number of blow operations and press operations. Typically, manipulation operations and shaping operations involving respectively manipulation or contact of the glass object with some tool are also executed. Some operations include combinations of the above mentioned, applied concurrently.

Presently, two situations occur in glass container forming technology. Forming of container/hollow glass objects is automated when all of the following conditions are valid: objects have standard dimensions and shapes; the production runs are very large (from 300,000 to 500,000 on). For this purpose, specialized automated machines are used, most often so-called IS machines that use individual section technology (Fig. 1). For production with IS machines, a set of dedicated tools must be developed for each item produced. Although IS production is largely automated, it requires very meticulous settings of a large number of different parameters. These depend on each product type, tool set and other factors. Inappropriate settings result in defects. Such glass objects are rejected at the final quality inspection. While there are automated dimensional control systems utilising computer vision [1–3], the inspection of in-glass defects can still be performed only by trained operators, basically manually. They inspect object visually while manipulating them, to change the viewpoint, direction and lightning conditions; their work requires experience and previously acquired knowledge. Summarizing, for large production series and usual glass objects forming can be performed by automated IS machines.

Another situation exists when some of the following production conditions apply: the object production has medium or smaller runs (less than 300,000 objects in a run); objects have larger than usual external dimensions (e.g. larger the ordinary bottles or jars), accented non-uniform shapes; objects are composed of special glass or multiple layers of different glass. When any of these situation applies today, the production is carried out manually or semi-mechanized, (mainly manually with additional devices).

Fig. 1 Section of an automated IS machine

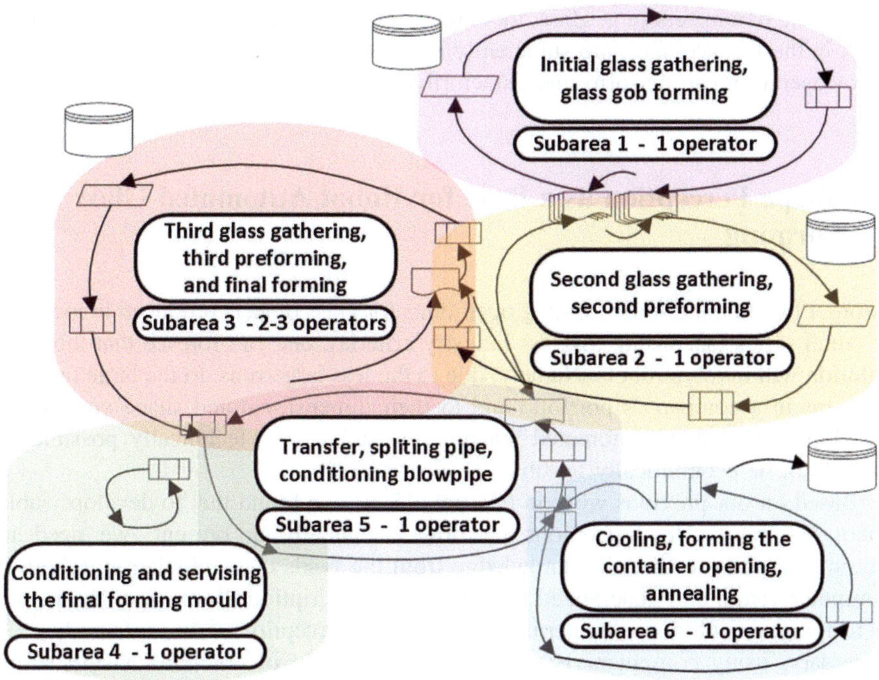

Fig. 2 Some operations in a manual glass containers forming environment

Manual forming is accomplished by a group of operators performing a series of tasks. Figure 2 shows a scheme of manual operations carried out in a real forming production cell. The operations involve, among others, a pre shaping by blowing and manipulation operation, press operations, and shaping by manual manipulation. One single forming cell requires typically 6–10 operators, depending on the items formed. Each operator is usually specialized for some operations.

One of the key operations in glass industry is the preforming operation by blowing and manipulation, where the shell is tied to a pipe or to a collar tool when the objects have larger openings. The operator achieves the desired form by a concurrent combination of manipulation and blowing of the low-viscosity shell, permanently using visual estimation of the shell's shape. Due to its high temperature, the shell is subject to deformation due to movements and to gravity. This operation is highly demanding, only few company operators being able to perform it satisfactorily for several hundreds of various items produced in this way.

Presently, in such small to medium series glass forming, the only operation where robot automation has been introduced in production, although with limited success, is glass gob gathering from furnaces [4]. Here, a rotating tool is mounted on the robot tip; the robot inserts the tool top into the furnace, gathers the adequate amount of glass and transfers it to the workplace, where a worker cuts the appropriate amount of glass. The robotized implementation is based on the assumption that all the parameters (e.g. glass, tool and environment temperatures, timing, glass level in the furnace, a.o.) are sufficiently invariant. When this is not the case, such robot operation can perform unsatisfactorily.

2 Shape Perception as a Base for Robot Automated Glass Forming

In order to automate some forming operations for glass objects produced in medium to small series (according to glass industry criteria), one can foresee that the final solution will involve robots. Namely, due to the less large runs, to the large number of items in a company's portfolio and to their dimensional and shape variety the development of fixed automated machines would not be technically possible or would not be economically feasible.

Based on our previous work in this field [5, 6], we found that to develop viable robot automation for operations performed manually at present, we need to investigate and use scientific knowledge from the fields of manipulation, robotics, computer vision and generalized trajectory task description. Because of the nature of forming, it is obvious that permanent, real-time perception of the surface shape is necessary, using computer vision. The scientific field of computer vision being quite mature and well established, a number of various methods exist that are well suited for different individual problems [7, 8]. On the other hand the majority of these methods are reliably applicable only for well determined problem domains, if the environmental conditions during image acquisition are invariant and can be appropriately controlled.

To design and develop a practical system able to perform satisfactorily in our target environment—production of glass objects, we have systematically analysed and identified all conditions and requirements that apply to the glass forming environment.

3 Specific Conditions in Manual Glass Forming Environments

3.1 Non-Rigidness and Fast Glass Object Shape Changing

The glass material is melted and conditioned in furnaces at temperatures up to 1575 °C and the temperature of the gathered glass gob is approximately 1090 °C. The object is formed in a sequence of operations in the range down to approximately 870 °C (for some glass types as low as 730 °C). The gathered object is initially a glass chunk, then it is gradually formed to become hollow, i.e. a glass shell or container. In either form (full gob or shell) the glass object has low viscosity

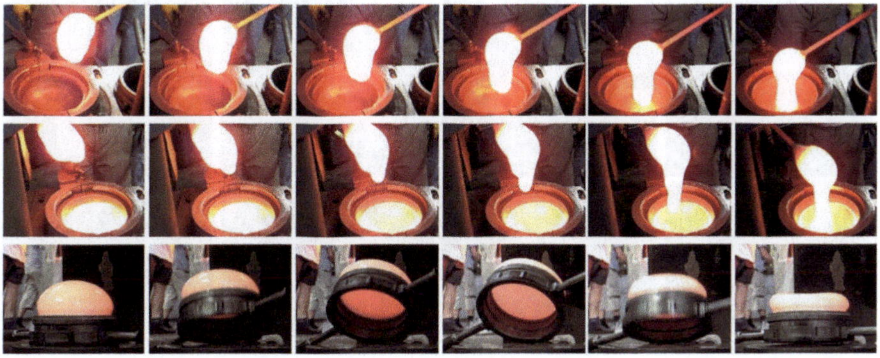

Fig. 3 Shape change speed of a high-temperature glass object: **a** 1st row—solid gob, form changes due to gravity in 500 ms; **b** 2nd row—solid gob, lower-viscosity glass, form changes in 500 ms; **c** 3rd row—glass shell, changes due to gravity in 800 ms

at these temperatures. Thus, it changes shape rapidly due to movements, when inflated and also due to gravitation, as shown in Fig. 3.

3.2 *Various Influences of Glass Radiation Depending on Temperature*

3.2.1 Glowing/Visual Radiation

The glass, heated at high temperatures, irradiates in the visual bandwidth. There are a couple of factors that must be taken into account when designing a visual subsystem.

First, as glowing glass acts as a light source, it illuminates the environment changing thus the lightning conditions of visual acquisition. Second, the radiation of heated glass can influence the characteristics of data generated by some sensors. These can function in a spectral range different from the visual one discerned by humans.

3.2.2 Radiation Change with Temperature

As glass objects change temperatures continuously during forming, the radiation and images acquired also change. Image acquisition can be additionally influenced by the shape's alteration. Figure 3 shows that the object's light radiation in the third row differs greatly from the light radiation in the first two rows, even for the very same object, just as the form reached the final aspect after approximately 8 s. A similar example is given Fig. 4, where the individual frames were taken 3 s apart. Another example of radiation change with temperature results by comparing images in Figs. 4 and 5 (the last ones show the same object after approximately 11 s).

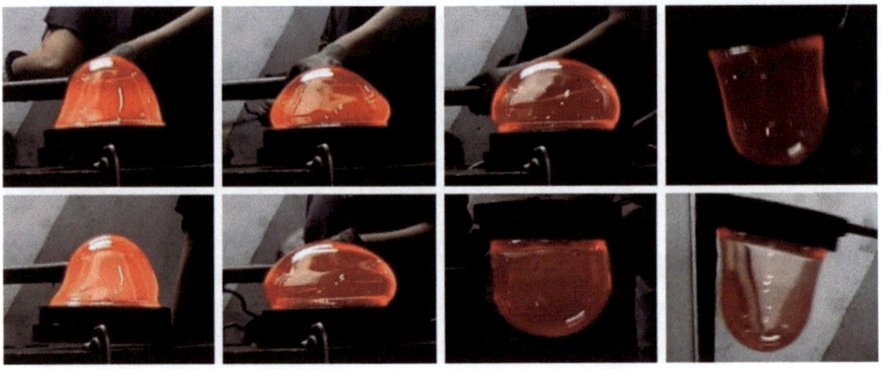

Fig. 4 Change in image characteristics of a glass object in the visual spectrum. Two rows show consecutive acquisitions; for each one, images are taken with 3 s period. The objects in the images are formed from a transparent glass; the temperature is between 800 and 900 °C

Another consideration is that in a single frame different points on the surface radiate differently due to their different temperatures.

3.3 *Influence of Different Glass Translucency and Its Change*

3.3.1 Object Background and Reflection Influence

Glass materials used for objects are often naturally translucent. This represents a problem when discriminating the object's shape from the background. For example, all images in Fig. 5 represent the same object during a very short period; nonetheless the image characteristics of the object's body appear very different, mainly because of the different background, and also due to the variable reflections.

3.3.2 Translucency Varying with Temperature and Geometry

In Fig. 4 one can notice that translucency changes with temperature, too. Thickness changes and other geometry changes can also influence considerably translucency.

3.3.3 Different Glass Materials and Multiple Glass Material Layers

Various glass materials have different basic translucency. For example, the usual soda-lime glass is transparent at room temperatures, while the opal glass is whitish, non-transparent. Furthermore, sometimes a glass objects is produced from multiple layers of different glasses.

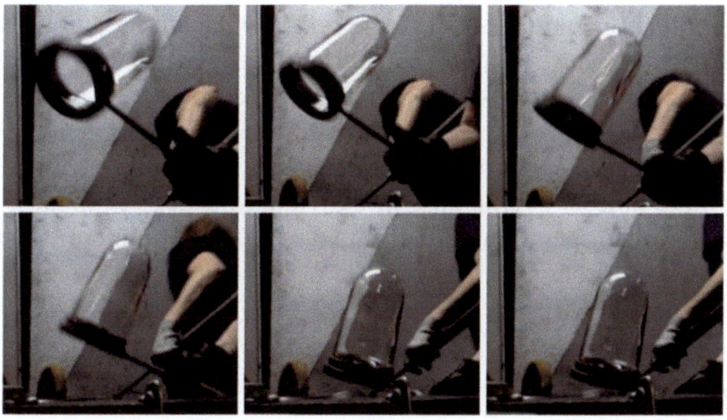

Fig. 5 Change in image characteristics of a glass object in the visual spectrum. The series of 6 consecutive images is taken during 1 s, approximately 11 s after the last series from the image in Fig. 4. The object formed has a temperature of approximately 700 °C

3.4 *Different and Varying Reflectivity*

In a similar way to translucency, a variable reflectivity also influences the acquired image. Here the reflectivity changes with temperature, too.

3.5 *Uncharacteristic Surfaces*

During forming, the surface of intermediate formed shells has no distinctive characteristics (for example edges or vertices), except along paths where the carrying tool connects to the glass shell. This makes surface characterization considerably harder.

3.6 *Limited Possibilities for Arranging Acquisition Environment*

We identified factors due to glass characteristics and high temperatures. Besides these, computer vision implementation is influenced by characteristics of real glass industry working environment. The layout of glass industry work cells is characterized by fixed positions of furnaces and crowded dispositions of machines, tools, operators and work pieces. These are dictated by operations' sequences and strict timing requirements. Due to the disposition of all components, it is often impossible to arrange the background at locations where images are acquired. The background can't be ordered also because the glass object is moved during some operations and between individual operations. Also, sometimes we cannot mount a uniform color background panel because its size could cause changes in the object's transfer trajectory and thus influence the non-rigid, low-viscosity object shape.

Another consideration is the presence of multiple glass objects in the same work cell, as different phases are performed concurrently on multiple objects. This can result in multiple objects overlapping in individual images acquired.

The glass forming operations usually involve additional intermediate reheating implemented with gas burners. The fire can also overlap with the glass objects.

3.7 Additional Production Environment Problems

3.7.1 Spontaneous Ignition

The glass object transport and manipulation devices and tools, as well as the glass object's surface may sometime spontaneously ignite in fire that degrades the acquired images. This is usually caused by lubricants applied periodically to the cell tools and devices; due to the elevated temperature, the lubricants sometimes ignite, which cannot be avoided. Such a situation is depicted in Fig. 6.

3.7.2 Pollution, Smoke

The manual glass production environment can be considerably polluted, which influences the quality of acquired images. The pollutants can be: dust resulting from production, smoke of burning or heated lubricants and other impurities. These particles obscure the image acquired. Sometimes, primary light sources are reflected from these particles. Such variable glowing reflections affect negatively the captured image stream.

Fig. 6 Influence of spontaneous ignition on images, acquired for characterizing the glass object surface. Note that image quality is degraded by both fire and smoke; **a** first row—normal camera aperture: the fire completely hides the object's image; **b** second row—small aperture: surfaces can be better determined under fire, but contrast is otherwise diminished

4 Conclusion

Several types of container glass objects are still formed mostly manually, due to technical or economic reasons. Here, robot automation would be the right solution, provided that we succeed in the design of a computer vision system able to offer permanently information on the formed object's shape (i.e., about its surface or contour geometry).

As part of automation projects carried out for a glass production company [5], we systematically investigated various requirements of the glass forming environments and real-world conditions that such a vision system have to cope with.

Although this is beyond the scope of this contribution, we anticipate that the right approach should tackle multiple points in the environment-optics-sensor-algorithms chain. For example, practical solutions would have to include: multiple cameras (to avoid clutter, to characterize non-rigid and/or deformable shapes), active vision, on-line aperture control, and concurrent acquisition with cameras in various spectra: visible, infrared, and near-infrared regions.

References

1. Laizola, E., Jimenez, A.R., Morgado, F., Calvache, M., Seco, F.: Computer-vision-based gob inspection system for monitoring and control in the glass industry. In: Proceedings of SPIE, Machine Vision Applications in Industrial Inspection XI, Santa Clara, CA, vol. 5011 (2003)
2. Rentschler, M.: Reducing risk with latest inspection technologies. Glass Worldwide (57) (2015)
3. Bierry, B.: Production monitoring solutions. Glass Worldwide (51) (2014)
4. Duplan, C.: Why ball-gatherer robots are better than slow feeders. Glass Int. (2013)
5. Ružič, A.: Instructions for installation and operation of an automated cell for a glass forming operation. Jožef Stefan Institute Report 11251 (2012)
6. Ružič, A.: Characteristics of image acquisition and processing in a high-temperature manufacturing environment. In: ROSUS 2014, Computer Processing of Images and Applications Conference, Maribor, Slovenia, pp. 42–49 (2014)
7. Szeliski, R.: Computer Vision: Algorithms and Applications. Springer, London (2010)
8. Forsyth, D.A., Ponce, J.: Computer Vision: A Modern Approach, 2nd edn. Prentice Hall, Englewood Cliffs (2011)

Rotation Angle Determination of a Rectangular Object Using an Infrared Sensorial System

Tony Stănescu, Diana Savu and Valer Dolga

Abstract Mobile robots are extensively approached as investigation topic in various research laboratories, due to their diverse applications. Currently, the interest is focused toward applications of mobile robots within unstructured working environments. The successful running of this type of applications is subject to the possibility of localizing obstacles within the working environment. This paper deals with a method to determine the position and the orientation of a rectangular object with respect to several infrared sensorial elements. Locating an object expects the determination of the reference frame coordinates of this object and of the object orientation relative to the axes of a coordinate system.

Keywords Infrared sensors · Distance measurement · Infrared data integration

1 Introduction

Infrared sensors belong to the sensor category that helps detecting obstacles within the working area of a mobile robot. Using infrared light is a simple method to determine the presence of obstacles and does not involve major costs. Bibliographical references in this field thoroughly approach the possibilities for detecting the optical radiation [1–3]. The amplitude response of infrared sensors (IR) is based on the reflected amplitude of the surrounding objects in the robot's work scene. The reflectance process depends upon the reflectance characteristics of the object's surface. The response of the infrared sensorial elements is widely used

T. Stănescu (✉) · D. Savu (✉) · V. Dolga (✉)
Department of Mechatronics, Politehnica University of Timisoara, Timişoara, Romania
e-mail: tony.stanescu@yahoo.com

D. Savu
e-mail: diana_s_ro@yahoo.com

V. Dolga
e-mail: valer.dolga@upt.ro

T. Borangiu (ed.), *Advances in Robot Design and Intelligent Control*,
Advances in Intelligent Systems and Computing 371,
DOI 10.1007/978-3-319-21290-6_22

because of its low cost and fast response time. The response time of IR elements is faster than that of ultrasonic sensors. Mohammad [4] approaches the Phong lighting model in the analysis of IR sensors when determining the distance. Korba et al. [5] used more IR sensorial elements to determine the distance but the results are not compelling. Benet et al. [6] consider that the documentation on using IR sensors in mobile robotics is still reduced. Research in this domain aims identifying new sensorial variants in IR which might be successfully integrated within the construction of mobile robots.

This paper presents an experimental method that can determine the angle with which a rectangular object is rotated with respect to a sensorial system, consisting of three IR sensors. The paper includes three sections, preceded by an "Introduction" and by the "Abstract" and ends by "Conclusions" about the approached subject. Section 2 describes theoretical notions regarding the working principle of the Sharp GP2Y0A21YK sensor. Section 3 shows how the reverse characteristic of the sensor is determined by two types of equations, and afterwards the most convenient equation is selected. Within the Sect. 4 a real study is shown, some laboratory experiments are described and the obtained results are outlined.

2 The Infrared Sensor and the Operating Principle

During the experiments mentioned in this paper, the GP2Y0A21YK Sharp sensor was used [7] (see Fig. 1), with the aim of analysing the detection of obstacles and the integration of sensorial elements in the structure of a mobile robot.

According to the device's operating principle [8], the sensorial element emits an incident ray in infrared radiation with the wavelength $\lambda = 850 \pm 70$ nm. This is reflected by the obstacle, after which it is received back by the sensor, see Fig. 2.

Novotny analyses in [10] the behaviour of the sensorial element based on the emitted energy and the received energy. The determination of the distance between the sensorial element and the obstacle is evaluated in three stages:

Fig. 1 A sharp sensor GP2Y0A21YK [7]

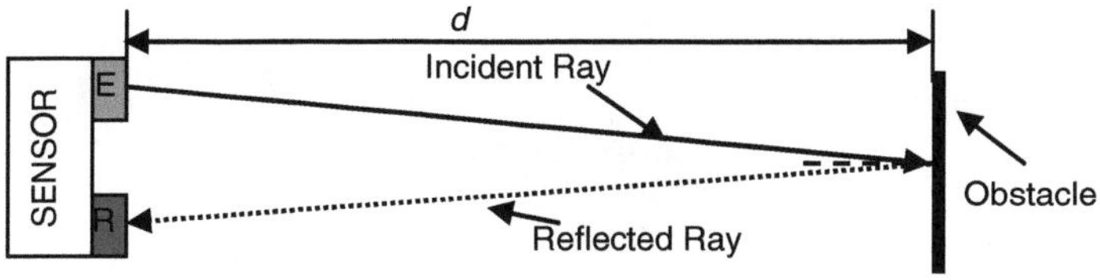

Fig. 2 The working principle of the sensorial element [9]

- Identification of the obstacle parameters;
- Calculation of the distance between the sensor and the obstacle;
- Identification of the orientation between two plans: the plan of the sensorial element and the reflection plan of the obstacle.

3 The Determination of the Reverse Characteristic for the Sensorial Element

A first set of experiments aimed to determine the distance between the IR sensor and a rectangular object with the dimensions: L = 700 mm and l = 250 mm. The reference distance was determined prior to this experiment by the laser telemeter Bosch DLE 70 Professional [11]. For this purpose, an experimental stand whose scheme is shown in Fig. 3 has been developed.

The object was placed at different distances with respect to the sensor. Figure 4 exemplifies how the object was placed in the work scene.

The authors performed several sets of measurements using the above presented stand, and the results were statistically processed. Based on these processed values, the authors mathematically approximated the inverse characteristic. The results acquired during the experiments were statistically processed and are given in Table 1, where *d* is the reference distance measured with the laser device and *U* is the average output voltage of the sensor.

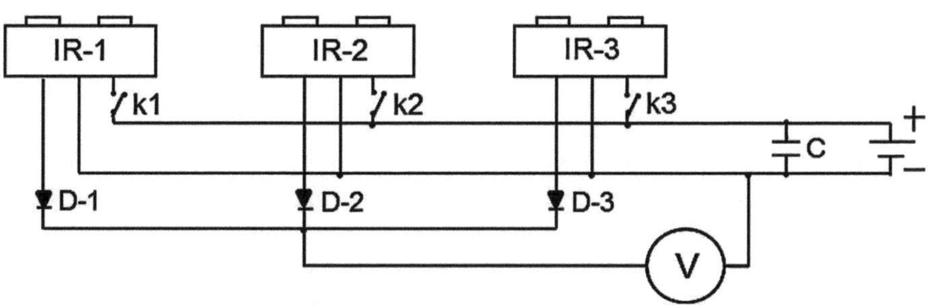

Fig. 3 The scheme of the experimental stand

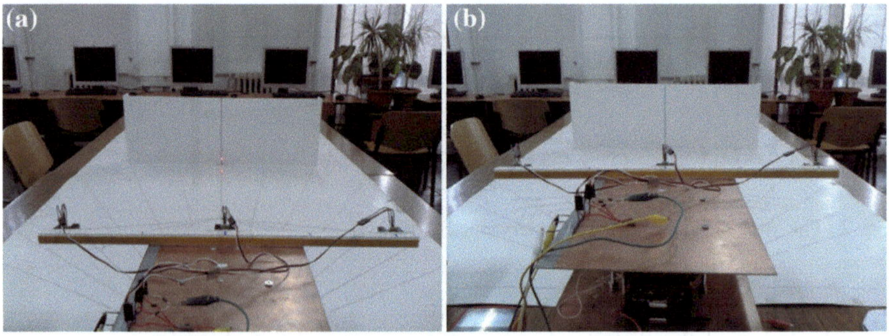

Fig. 4 Images of the object placed in the working scene: **a** far, **b** near

Table 1 Results processed after the experiment

d (mm)	U (mV)	d (mm)	U (mV)	d (mm)	U (mV)
25	1464.4	300	1711.7	1000	741.4
50	1701.5	400	1330.7	1100	702.1
75	1888.5	500	1123.3	1200	671.4
100	2486.2	600	983.9	1300	645.6
125	2553.2	700	905.4	1400	615.9
150	2472.6	800	837.6	1500	590.1
200	2248.3	900	787.3	1600	565.6

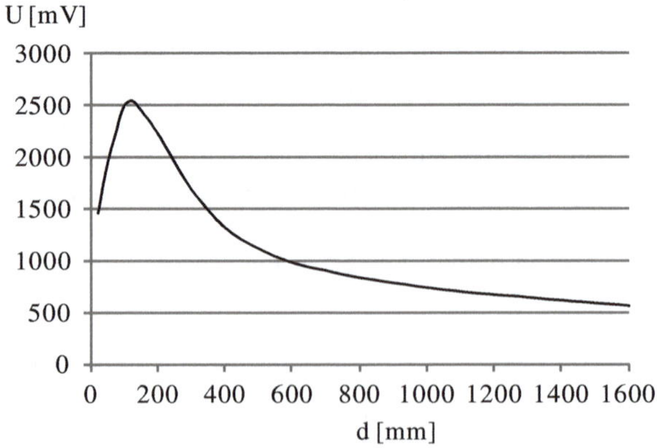

Fig. 5 The approximation curve of the sensorial element

Figure 5 illustrates the approximation curve of the sensorial element. The available data indicates a maximum for the output voltage of the infrared sensor when the object was positioned at the distance d = 125 mm.

Based on the obtained values, the reverse characteristic of the sensor was approximated. The determination of the reverse characteristic was performed in the MATLAB software environment using, during a first step, the function "polyval" (from MATLAB/Cftool), which allows the digital evaluation of the approximation polynomial for a given set of the input values. The reverse characteristic of the sensorial element was calculated over the interval [125, 1600] mm. Two approximations were applied: by an exponential equation of 2nd degree and by a polynomial equation of 4th degree.

The exponential equation of 2nd degree is:

$$d_{\text{exp.}} = 1.291 \cdot 10^{\,4} \cdot \mathrm{e}^{-0.00457 \cdot U} + 976.8 \cdot \mathrm{e}^{-0.0007403 \cdot U} \tag{1}$$

The polynomial equation of 4th degree is:

$$d_{pol.} = 5.35 \cdot 10^{-10} \cdot U^4 - 3.956 \cdot 10^{-6} \cdot U^3 + 0.01069 \cdot U^2 - 12.71 \cdot U + 6013 \tag{2}$$

Figure 6 shows the reverse characteristic represented by the exponential equation of 2nd degree, while Fig. 7 shows the reverse characteristic represented by the polynomial equation of 4th degree.

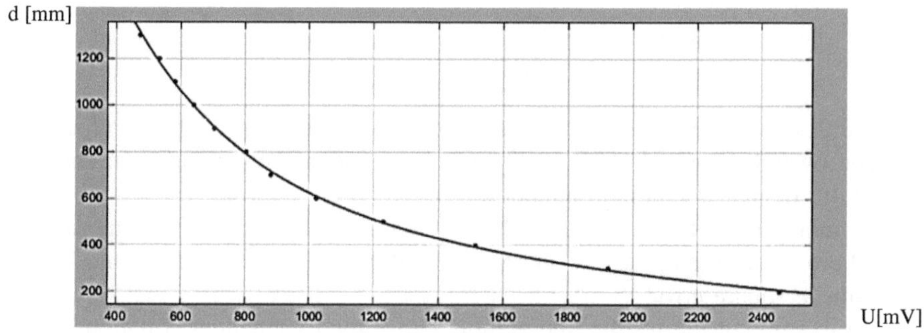

Fig. 6 Approximation of the reverse characteristic by an exponential equation of 2nd degree

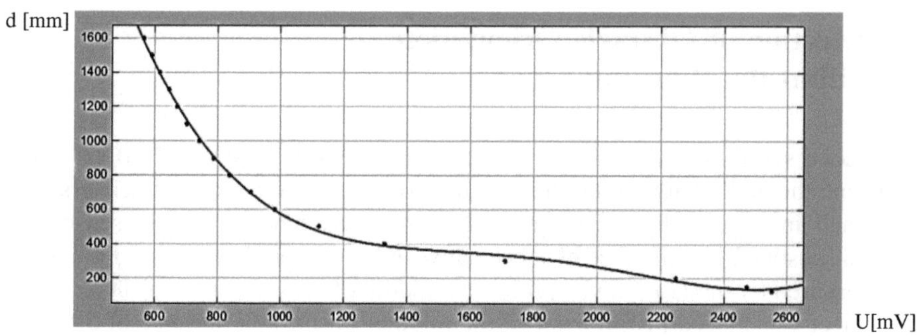

Fig. 7 Approximation of the reverse characteristic by an exponential equation of 4th degree

Table 2 Statistics indices for the approximation of the reverse characteristic

	The exponential of 2nd degree	The polynomial 4th degree
SSE	2220	4237
R—square	0.9994	0.9989
Adjusted R—square	0.9993	0.9985
RMSE	13.07	18.79

Table 3 The analysis of the results obtained with the exponential equation

U (mV)	d (mm)	$d_{exp.}$ (mm)	$d - d_{exp.}$ (mm)	U (mV)	d (mm)	$d_{exp.}$ (mm)	$d - d_{exp.}$ (mm)
2553.2	125	147.65	−22.65	787.3	900	898.82	1.17
2472.6	150	156.77	−6.77	741.4	1000	1000.17	−0.17
2248.3	200	185.35	14.64	702.1	1100	1102.60	−2.59
1711.7	300	280.26	19.73	671.4	1200	1194.53	5.46
1330.7	400	394.23	5.76	645.6	1300	1281.11	18.88
1123.3	500	501.37	−1.37	615.9	1400	1392.77	7.22
983.9	600	615.42	−15.42	590.1	1500	1501.52	−1.52
905.4	700	705.74	−5.74	565.6	1600	1616.19	−16.19
837.6	800	806.30	−6.30				

Table 2 shows the statistics indices for the approximation of the reverse characteristic over the interval [125, 1600] mm by a 2nd degree exponential equation and the 4th degree polynomial equation.

The notations in the table have the following meanings: SSE is the sum of squared errors of prediction, R_square is a coefficient of determination and RMSE is root-mean-square error.

After examining the results, the authors chose the exponential equation of 2nd degree, due to the higher statistical index. The obtained results allow also an analysis on the best results (see Table 3), where d_{exp} is the exponential equation.

4 Detection of Obstacle Orientation Within the Working Scene

The second set of experiments used three infrared sensors Sharp GP2Y0A21YK to determine the angle with which the rectangle object was rotated. The sensors were positioned in line, with a distance of 220 mm between the middle sensor and external sensors. The reference distance between the sensorial system and the object was measured by the laser device, from the middle of sensor to the object's middle. The object was placed just in front of the middle sensor, at a distance of 500 mm. The object was rotated by 10°, 20° and 30° counter clockwise respectively clockwise. Figure 8 shows images of the object positioned on the experimental stand. Processed results taken from three sensors are shown in Tables 4 and 5.

By introducing the 2nd degree exponential Eq. (1), the distances with the three sensors were obtained; the results are outlined in Tables 6 and 7. These results presented in Table 6 and in Table 7 were introduced in MATLAB/Cftool to obtain the polynomial equation of 1st degree in each case (Table 8).

It is considered that the object is at 0° when it is parallel to the axes of the three sensors. Therefore, the rotation angle is positive regardless whether the object is

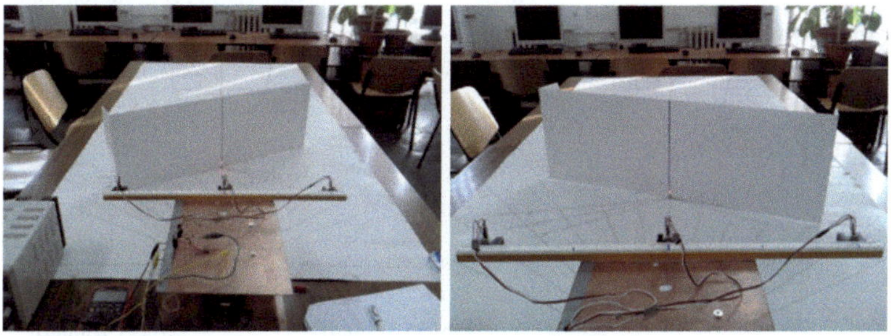

Fig. 8 Hypostases of the object placed in the working scene

Table 4 Object is rotated counter clockwise

Angle	Left sensor	Middle sensor	Right sensor
α (°)	U (mV)	U (mV)	U (mV)
10	1222.9	1130.7	1092.3
20	1327.7	1134.7	1054.6
30	1474.4	1139.2	1024.2

Table 5 Object is rotated clockwise

Angle	Left sensor	Middle sensor	Right sensor
α (°)	U (mV)	U (mV)	U (mV)
10	1085.7	1123.2	1224.6
20	1028.7	1129.6	1307.8
30	982	1116.6	1417.1

Table 6 The distance measured by the sensor when the object is rotated counter clockwise

Angle	Left sensor	Middle sensor	Right sensor
α (°)	d_1 (mm)	d_2 (mm)	d_3 (mm)
10	443.317	496.523	522.835
20	395.452	493.940	551.640
30	343.223	491.067	577.353

Table 7 The distance measured by the sensor when the object is rotated clockwise

Angle	Left sensor	Middle sensor	Right sensor
α (°)	d_1 (mm)	d_2 (mm)	d_3 (mm)
10	527.652	501.444	442.446
20	573.394	497.239	403.725
30	617.341	505.859	362.012

Table 8 The polynomial equation

	Angle (°)	Mathematical approximation equation
Object rotated counter clockwise	10	$y = 0.1807 \cdot x + 447.8$ (3)
	20	$y = 0.355 \cdot x + 402.3$ (4)
	30	$y = 0.5321 \cdot x + 353.5$ (5)
Object rotated clockwise	10	$y = -0.1936 \cdot x + 533.1$ (6)
	20	$y = -0.3856 \cdot x + 576.3$ (7)
	30	$y = -0.5803 \cdot x + 622.7$ (8)

Table 9 Results after a mathematical processing of data

Angle	Object rotated counter clockwise		Object rotated clockwise	
α (°)	tan α′	α′ (°)	tan α′	α′ (°)
10	0.1807	10.248	0.1936	10.962
20	0.355	19.554	0.3856	21.097
30	0.5321	28.031	0.5803	30.141

rotated clockwise or counter clockwise. A mathematical processing of the obtained data led to the results in Table 9, where: α is the angle by which the object was rotated, α′ is the angle determined by the sensory system and tan α is the angular coefficient of the straight lines mentioned in Table 8.

5 Conclusions

There is a good correlation between the theoretical characteristics of the sensor, as they are mentioned in the sensor's documentation [8] and the experimental ones.

The exponential equation of 2nd degree is much better than polynomial equation of 4th degree in determining the reverse characteristic of the sensor, because the statistical parameters are superior and thus lead to smaller errors.

The rotation angle of the object during the realised experiments was determined using the sensors and the reverse characteristic.

The difference between the object's rotation angle and the angle that was determined by the experimental procedure, using the IR sensors, is very low.

The method presented in this paper to determine the rotation angles of the objects can be successfully implemented in the control system of a mobile robot.

Acknowledgments This paper is supported by two Sectoral Operational Programme Human Resources Development (SOP HRD), having ID134378 and ID137070, partially financed from the European Social Fund and by the Romanian Government.

References

1. Makris, N.: Unidentified floating objects. IEEE Spectr. Int. **48**(8), 44–50 (2011)
2. Cochran, J., Kanso, E., Kelly, S.D., Xiong, H., Krstic, M.: Source seeking for two nonholonomic models of fish locomotion. IEEE Trans. Robot. **25**(5), 1166–1176 (2009). ISSN 1552-3098
3. Itineantu, A., Dolga, V.: Study of a mobile robot's behaviour in an outdoor environment (in Romanian). License thesis, University Politehnica of Timisoara (2010)
4. Mohammad, T.: Using ultrasonic and infrared sensors for distance measurement. World Acad. Sci. Eng. Technol. **3** (2009)
5. Korba, L., Elgazzar, S., Welch, T.: Active infrared sensors for mobile robots. IEEE-Trans. Instrum. Meas. **2**(43), 283–287 (1994)
6. Benet, G., Blanes, F., Simo, J.E., Perez, P.: Using infrared sensors for distance measurement in mobile robots. J. Robot. Auton. Syst. **10**, 255–266 (2002)
7. http://www.preciolandia.com/br/sensor-infravermelho-sharp-gp2y0a02yk0f-7wahv7-a.html, Feb 2015
8. https://www.sparkfun.com/datasheets/Sensors/Infrared/gp2y0a02yk_e.pdf, Feb 2015
9. Stanescu, T., Mondoc, A., Dolga, V.: Random issues in workspace analysis for a mobile robot. In: ISI Proceedings of 10th International Conference on Mathematical Problems in Engineering, Aerospace and Sciences, ICNPAA'144 Congress, 2014, Narvik, Norvegia, 15–18 July 2014, vol. 1637, pp. 1254, ISBN 978-0-7354-1276-7
10. Novotny, P.M., Ferrier, N.J.: Using infrared sensors and the Phong illumination model to measure distances. In: Proceedings of the International Conference on Robotics and Automation, Detroit, MI, USA, April 1999, vol. 2, pp. 1644–1649
11. http://www.abbaye-saint-hilaire-vaucluse.com/images_entreprises/BOSCH_telemetre.jpg, Feb 2015

Part V
Human-Robot Interaction and Collaboration

Individuals with Autism: Analysis of the First Interaction with Nao Robot Based on Their Proprioceptive and Kinematic Profiles

Pauline Chevalier, Brice Isableu, Jean-Claude Martin and Adriana Tapus

Abstract Our research aims to develop a new personalized social interaction model between a humanoid robot and an individual suffering of Autistic Spectrum Disorder (ASD), so as to enhance his/her social and communication skills. In order to define individual's profile, we posit that the individual's reliance to proprioceptive and kinematic visual cues will affect the way an individual suffering of ASD interacts with a social agent. We describe a first experiment that defines each participant's perceptivocognitive and sensorimotor profile with respect to the integration of visual inputs, thanks to the Sensory Profile questionnaire and an experimental set-up. We succeeded to form 3 groups with significant different behavioural responses inside our subject pool formed by 7 adults and 6 children with ASD. In a second experiment, we presented the Nao robot to all of our participants. We video-analysed their behaviours and compared them to the profiles we defined. In view of our results, this first interaction confirmed our hypothesis: participants with a weak proprioceptive integration and strong visual dependency had more successful interaction than participants with an overreliance on proprioceptive input and hyporeactivity to visual cues.

Keywords Autism · Personalized interaction · Socially assistive robotics · Proprioception · Kinematics

P. Chevalier · A. Tapus (✉)
Robotics and Computer Vision Lab, ENSTA-ParisTech, 91120 Palaiseau, France
e-mail: adriana.tapus@ensta-paristech.fr

P. Chevalier
e-mail: pauline.chevalier@ensta-paristech.fr

B. Isableu
CIAMS-Motor Control and Perception, Université Paris-Sud, 91405 Orsay, France
e-mail: brice.isableu@u-psud.fr

J.-C. Martin
Cognition, Perception, Use, LIMSI-CNRS, 91403 Orsay, France
e-mail: martin@limsi.fr

T. Borangiu (ed.), *Advances in Robot Design and Intelligent Control*,
Advances in Intelligent Systems and Computing 371,
DOI 10.1007/978-3-319-21290-6_23

1 Introduction

Nowadays, research in socially assistive robotics (SAR) is in expansion [1, 2]. One of the target populations is people suffering of Autistic Spectrum Disorders (ASD). Individuals with ASD have impaired skills in communication, interaction, emotion recognition, joint attention, and imitation [3]. Many studies showed that children with ASD have a great affinity with robots, computers, and mechanical components [4]. Moreover, in SAR, robots have already been used several times as tools in socialization therapies for children with ASD [5, 6]. Individuals with ASD also suffer from visual and sensory motor impairments [7, 8]. It is suggested that they manifest an exacerbated reliance on proprioceptive feedback and a deficient use of kinematic visual cues [9, 10]. A link between the individual integration of proprioceptive and visual feedbacks and communication, interactions skills, and emotion recognition had already been discussed in [10].

The goal of our research is to develop a novel personalized robot behaviour based on a social interaction model for individuals suffering of ASD. In order to define users' profiles, we make the hypothesis H1 that mitigated behavioural response (i.e., hyporeactivity) to visual motion of the scene and overreliance on proprioceptive information are linked in individuals with ASD to difficulties in integrating social cues and engaging in successful interactions. We work in collaboration with two care facilities for people suffering of ASD: MAIA Autisme (France), an association for children and adolescents with ASD, and FAM-La Lendemaine (France), a residence for adults with ASD. Our current subject pool is composed of 6 autistic children (10.9 ± 1.8 years) and 7 autistic adults (26.1 ± 7.9 years) from these two care facilities. Both of the experiments described here have been done with the same individuals.[1]

In this paper, we first present an experiment that defines each participant's perceptivo-cognitive and sensorimotor profiles with respect to the integration of visual inputs. Furthermore, we describe a second experiment where the Nao robot was presented for the first time to all our participants. Their behaviours during this first interaction with the robot, based on their profiles, are analysed and reported.

2 State of the Art

Motor, sensory, and visual impairments are present in autism, but are not taken into account in the ASD diagnosis [7, 8]. These deficits have an influence on the quality of life of individuals suffering of ASD and on their social development. An overreliance on proprioceptive information in autism has already been studied. As it has been reported in [8], individuals with autism showed normal to improved

[1]For confidentiality reasons, we encoded the participants' identity: AD# for the participants from the adults care centre and CH# for the participants from the children care centre.

integration of the proprioceptive cues in comparison to typically developed individuals. Moreover, results in [9] confirmed that postural hyporeactivity to visual information was present in the tested individuals with autism (individuals suffering from ASD with IQ comparable to that of typically developed persons). Proprioception integration in ASD was also studied so as to better understand the interaction and social capabilities. In [10], Haswell et al. asked 14 children with ASD and 13 typically developed children to perform a reaching task with their arm, while holding a robotic arm that applied a force, constraining the movement. It was found that autistic brain built a stronger than normal association between self-generated motor commands and proprioceptive feedback, confirming an overreliance on proprioceptive cues in individuals with ASD. Furthermore, it was also established that greater the reliance on proprioception was, the higher the impairments in social and imitation functions were.

The use of robots in therapy for individuals with ASD has been a great topic of interest in the last decade. Indeed, robots have been found to be great partners in learning social interaction, and imitation skills [4]. Robots are less complex in their behaviour than human peers offering more predictable and comfortable interaction behaviours [6]. Different research areas on robots for therapy for individuals with ASD have been reported in [1]: (1) developing a robot especially designed for individuals suffering of ASD, (2) design of Human-Robot Interaction, and (3) evaluation of the robot in therapies. There are many open challenges in SAR for people with ASD. However, due to small subject pools and/or short-time experiments, generalized results in improved skills are still questioned [11]. Salter et al. [12] discussed on the variability of Human-Robot Interaction set up that could biased the results and founding in this area. To the best of our knowledge, no study examined the sensorimotor and visual profiles of individuals suffering of ASD to determine and elaborate individualized and personalized scenarios for Human-Robot social interaction therapy.

3 Defining Proprioceptive and Kinematic Profiles

3.1 Methods

The first step of our work was to determine how to define each participant's perceptive-cognitive and sensorimotor profiles. We used two methodologies. First, the Sensory Profile (SP) [13] and the Adolescents/Adults Sensory Profiles (AASP) [14] were filled. They allowed evaluating individual's sensory processing preferences, being widely used in ASD research. The individual's sensory processing preferences are described in terms of the quadrants in Dunn's Model of Sensory Processing [13]: Low Registration, Sensation Seeking, Sensory Sensitivity, and Sensation Avoiding. These quadrants are formed by the junctions between the individual differences in neurological thresholds for stimulation (high-low) and self-regulation strategies (active-passive).

As we worked with two age groups, we used the appropriate questionnaire corresponding to their age range. The AASP was filled for all the +11 years old participants and the SP was filled for the youngest participants. As most of the participants did not have the cognitive level to fill the questionnaire by themselves, it has been filled with the help of their caretakers who knew well their habits and response to everyday life sensory stimuli. The caretakers were well informed of the conditions and forms of the questionnaire. In order to have homogeneous scores between age groups to assess Movement, Visual, Touch, and Auditory processing, we carefully matched and adapted the items based on their correspondence to the neurological threshold and behaviour response/self-regulation. Second, we designed an experimental setup to assess the effect of a moving virtual visual scene (VVS) on postural control, and the individual's capability to use proprioceptive inputs provided in dynamics of balance to reduce visual dependency [15]. In an instable posture, the integration of proprioceptive feedback differs among individuals. An individual integrating proprioceptive cues less than other individuals is probability visual dependent.

While exposed to a visual motion in an instable posture, his/her body sway will follow the visual stimulus. Participants were asked to stand on a force platform (FP) in front of a virtual room (Fig. 1) in three conditions:

- **C1**—*Stable position with static VVS*: each participant stood on the FP, straight ahead, with feet hip-width apart. The virtual room stays still. The recording lasted 30 s.
- **C2**—*Stable position with moving VVS*: the participant stood on the FP, straight ahead with feet hip-width apart. The virtual room has a sinusoidal movement. The recording lasted 50 s.
- **C3**—*Tandem Romberg position with moving VVS*: the participant stood on the FP, straight ahead, one foot in front of the other one (tandem Romberg stance). The virtual room has a sinusoidal movement. The recording lasted 50 s.

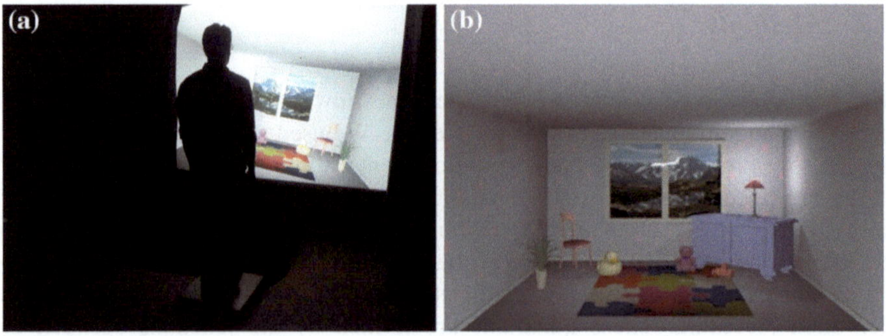

Fig. 1 Setup: **a** Experimental setup for the adults group in condition C3; **b** Capture of the virtual room used in the experiment

The virtual room was created with Blender, a free 3D rendering software. It was decorated with child toys and furniture so as to create a friendly environment. A toy plane was placed in the line of sight of the individuals in order to help them to focus on the task, and not to be distracted away. It was projected to a wall with a short focal projector in a dark room. It rolled at 0.25 Hz with an inclination of ±10°. For the adult group setup, the dimension of the projection was 2.4 m large × 1.8 m high and the participants stood at 1.3 m of the point of observation (Fig. 1b). For the children group setup, the dimension was 1.75 m large × 1.16 m high and the participants stood at 1 m.

3.2 Data Analysis

For all the sessions, a FP (AMTI OR6-5-1000) was used to record the displacement of the centre of pressure (CoP) with a sampling frequency of 1 kHz. We used a Butterworth filter with a cut-off frequency of 10 Hz on the recorded data in order to reduce the noise. We computed the Root Mean Square (RMS) of the CoP in both mediolateral and anteroposterior directions as an indicator of the individual's stability. Indeed, the RMS provides the information about the variability of the CoP in space [15]. We computed the Fast Fourier Transform (FFT) of the CoP and the correlation with a 0.25 Hz sinus so as to evaluate coupling between the postural response and the moving visual stimulus. In order to assess the correlation between sensory preferences and postural responses to our visual stimulus, we computed the correlation between the scores of the different items of the SP and the data obtained from the CoP. We computed the correlation between the CoP behaviours during different conditions to evaluate if a behaviour in a condition induced another behaviour in a different one. Then, we used clustering analysis (Dendrogram) to form behavioural groups.

3.3 Results

We observed a close relationship between the AASP patterns and the postural behaviours. Individual with higher movement sensory sensitivity score in AASP showed greater postural stability, and postural sway less driven by the moving VVS. A low visual sensation seeking behaviour also resulted in smaller postural responses to the moving VVS whereas individual with a higher visual sensitivity showed a greater postural coupling response with moving VVS. A relation between age and postural instability was found, but not between age and postural response to the moving VVS.

Clustering analyses allowed us to identify 3 groups with significant different behavioural responses (see Fig. 2): **(G1)** high scores in movement sensitivity and visual sensation seeking, low scores in visual sensory sensitivity and strong visual

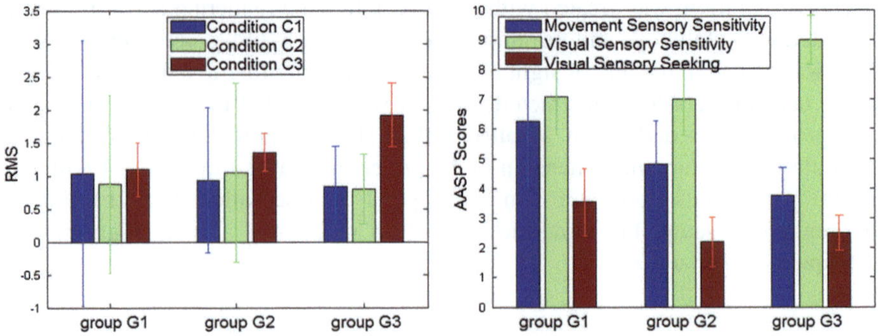

Fig. 2 *Left* Histograms of the mean RMS for the groups defined by clustering analysis for the 3 conditions. *Right* Histograms for AASP scores for the groups

independence to the moving VVS, suggesting an overreliance on proprioceptive input and hypo reactivity to visual cues; (**G2**) moderate scores in movement sensitivity and low scores in visual sensation seeking, low scores in visual sensory sensitivity and moderate reactivity to moving VVS, suggesting a reliance on both visual and proprioceptive input; and (**G3**) high scores in visual sensory sensitivity and low scores in movement sensitivity and in visual sensory seeking, and hyper reactivity to moving VVS, suggesting a weak proprioceptive integration and strong visual dependency.

4 Greetings with Nao

4.1 Method

A first Human-Robot Interaction was conducted with all our participants. The interaction purpose was to present Nao to our participants for a short duration, up to 2 min. Indeed, some of the individuals with ASD are reluctant to unusual events and change in their daily routine. The scenario was the following: after being seated in front of Nao, the robot waved to the participant and said "Hello, I am Nao. You and I, are going to be friends". If the participant was verbal, the robot asked for the participant's name. Afterwards, it continued with "Hello" followed by the name of the participant, and asked if he/she wanted it to dance for him/her, and then danced. During all the experiment, the participant was with his/her caretaker. The caretaker encouraged the participant to look at and answer to the robot. We manually analysed the video recording the participants' gaze direction and gestures towards the robot, the caretaker, and all the other directions so as to determine a match with the 3 groups described in Sect. 3.3, so as to evaluate our H1 hypothesis (Table 1).

Table 1 Mean percentage of the gaze direction for each group

Gaze direction	G1 (%)	G2 (%)	G3 (%)
Toward the robot	62.58	69.06	89.19
Toward the caretaker	13.84	1.49	1.59
Other	23.57	29.45	9.21

4.2 *Results*

The analysis of the gaze direction showed us that the individuals from G3 had gaze direction more focused on Nao and switched less their gaze than the individuals belonging to other groups. The individuals from G1 were less looking at Nao, they switched more their gaze, and looked more in other directions (nor the robot or the caretaker) than the other groups, see Table 1. Only two individuals (AD3 from G2 and AD6 from G1) looked less than 50 % of the time in other direction. The gesture analysis showed that individuals from G3 (two out of four: CH2 and CH6) had more social gestures than the other groups toward Nao robot, such as waving back to Nao while it was presenting itself. Only one of the individuals from G2 (AD1) had this behaviour and none of them was in G1. Two individuals (CH5, non-verbal, and CH6) took the initiative to touch the robot during the interaction. Overall, they were impressed by the machine. Participant AD1 recoiled when the robot moved its arms in her direction but she was the only one from G2 to wave back. The individuals from G3 showed more initiative to answer back to the robot. All of them answered to the robot after it asked the name and/or if he/she wanted it to dance. Half of the individuals from G2 had this same behaviour, and only one individual out of 4 of the G1 (CH5 from G1 is non-verbal so she is not taken into account). Moreover, participant AD4 from G1, who answered back, was really happy to interact with the robot.

5 Conclusion and Discussions

Thanks to our first experiment, we succeeded to form three groups between our participants, describing each participant's response to visual and proprioceptive inputs. With these results and our hypothesis H1, we were able to make assumptions on the behaviours each individual will have during the Human-Robot Interaction sessions. We posit from H1 that individuals from G1 will have less successful interaction than the ones from G2 and G3, and that individuals from G3 will have the most successful interactions. The interaction between the participants and the robot we conducted permitted us to analyse their behaviour toward the robots. The analysis confirmed our hypothesis made on the groups. Participants from G3 showed longer gaze direction, more speech and social gesture initiatives toward Nao and participants from G1 showed less social behaviour than the 2 other groups.

As it have been already seen in SAR almost all of our participants, children and adults, showed great interest to their new mechanical companion. We confirmed our hypothesis on the formed groups on a short greeting interaction. Moreover, the interaction was relevant because greetings are an important stage of social interaction. Presenting the robot before starting the long-term personalized interaction seemed to be an important first step, as in ASD new events can be stressful and feared.

As in [8], we did not find a relationship between age and postural behaviours, differently to typically developed individuals [16]. The groups G1 and G2 are linked to the results found in these studies. The profiles we defined gave us a group formed by individuals with a weak proprioceptive integration and strong visual dependency (G3), which is in contradiction with earlier studies showing our statement that individual with autism have an overreliance on proprioceptive cues. However, Molloy et al. [17] found an impairment of the proprioception input in autism, and that children with ASD used more the visual cues to reduce sway and maintain balance. In [18], the authors found that unlikely to typically developed individual, individuals with ASD have an impaired proprioception development. Their sensory-motor signal appear to remain at the kinaesthetic stage of typically developed 3–4 years old children, and have to rely on visual inputs. They also conjectured that the impaired proprioception of physical micro-movements of the individuals with ASD impedes as well their visual perception of micro-movements in others during real time interactions, impairing their abilities to interact with people.

These results will help us to model the customized Human-Robot Interaction sessions and adapt the robot's behaviours as a function of the participants' profile. We plan on developing a parameterized robot behaviour that will adapt its gaze, head and body posture, gestures, speech and facial expressions [11]. The collaboration with two care facilities allows us to observe the long-term effect of a therapy with humanoid robots on 13 participants, innovative in SAR for individuals with ASD. Indeed, most studies are on few days, weeks or month with subject pools up to 3–4 participants.

Acknowledgments This work is supported by IdF Doctoral Fellowship France 2013, HANDICAP theme. Many thanks to G. Lerigoleur and C. Bazile. We also thank the participants, their families and the caretakers for their participation and support.

References

1. Feil-Seifer, D., Mataric, M.J.: Defining socially assistive robotics. In: 9th International Conference on Rehabilitation Robotics, pp. 465–468. IEEE (2005)
2. Tapus, A., Mataric, M.J., Scassellati, B.: The grand challenges in socially assistive robotics. IEEE Robot. Autom. Mag. (RAM) **14**(1) (2008)

3. Charman, T., Swettenham, J., Baron-Cohen, S., Cox, A., Baird, G., Drew, A.: Infants with autism: an investigation of empathy, pretend play, joint attention, and imitation. Dev. Psychol. **33**(5), 781 (1997)
4. Hart, M.: Autism/excel study. In: Proceedings of the 7th International ACM SIGACCESS Conference on Computers and Accessibility, pp. 136–141. ACM, New York (2005)
5. Tapus, A., Peca, A., Aly, A., Pop, C., Jisa, L., Pintea, S., Rusu, A.S., David, D.O.: Children with autism social engagement in interaction with Nao, an imitative robot—a series of single case experiments. Interact. Stud. **13**(3), 315–347 (2012)
6. Kim, E.S., Berkovits, L.D., Bernier, E.P., Leyzberg, D., Shic, F., Paul, R., Scassellati, B.: Social robots as embedded reinforcers of social behavior in children with autism. J. Autism Dev. Disord. **43**(5), 1038–1049 (2013)
7. Simmons, D.R., Robertson, A.E., McKay, L.S., Toal, E., McAleer, P., Pollick, F.E.: Vision in autism spectrum disorders. Vis. Res. **49**(22), 2705–2739 (2009)
8. Gowen, E., Hamilton, A.: Motor abilities in autism: a review using a computational context. J. Autism Dev. Disord. **43**(2), 323–344 (2013)
9. Greffou, S., Bertone, A., Hahler, E.M., Hanssens, J.M., Mottron, L., Faubert, J.: Postural hypo-reactivity in autism is contingent on development and visual environment: a fully immersive virtual reality study. J. Autism Dev. Disord. **42**(6), 961–970 (2012)
10. Haswell, C.C., Izawa, J., Dowell, L.R., Mostofsky, S.H., Shadmehr, R.: Representation of internal models of action in the autistic brain. Nat. Neurosci. **12**(8), 970–972 (2009)
11. Scassellati, B., Admoni, H., Mataric, M.: Robots for use in autism research. Ann. Rev. Biomed. Eng. **14**, 275–294 (2012)
12. Salter, T., Michaud, F., Larouche, H.: How wild is wild? A taxonomy to characterize the 'wildness' of child-robot interaction. Int. J. Soc. Robot. **2**(4), 405–415 (2010)
13. Dunn, W.: Sensory Profile: User's Manual. The Psychological Corporation, USA (1999)
14. Brown, C., Dunn, W.: Adolescent-Adult Sensory Profile: User's Manual. Therapy Skill Builders, Tucson (2002)
15. Isableu, B., Fourre, B., Vuillerme, N., Giraudet, G., Amorim, M.A.: Differential integration of visual and kinaesthetic signals to upright stance. Exp. Brain Res. **212**(1), 33–46 (2011)
16. Baumberger, B., Isableu, B., Flückiger, M.: The visual control of stability in children and adults: postural readjustments in a ground optical flow. Exp. Brain Res. **159**(1), 33–46 (2004)
17. Molloy, C.A., Dietrich, K.N., Bhattacharya, A.: Postural stability in children with autism spectrum disorder. J. Autism Dev. Disord. **33**(6), 643–652 (2003)
18. Torres, E.B., Brincker, M., Isenhower, R.W., Yanovich, P., Stigler, K.A., Nurnberger, J.I., Metaxas, D.N., José, J.V.: Autism: the micro-movement perspective. Front. Integr. Neurosci. **7** (2013)

Human Robot Collaboration for Folding Fabrics Based on Force/RGB-D Feedback

Panagiotis N. Koustoumpardis, Konstantinos I. Chatzilygeroudis, Aris I. Synodinos and Nikos A. Aspragathos

Abstract In this paper, the human-robot collaboration for executing complicated handling tasks for folding non-rigid objects is investigated. A hierarchical control system is developed for the co-manipulation task of folding sheets like fabrics/cloths. The system is based on force and RGB-D feedback in both higher and lower control levels of the process. In the higher level, the perception of the human's intention is used for deciding the robot's action; in the lower level the robot reacts to the force/RGB-D feedback to follow human guidance. The proposed approach is tested in folding a rectangular piece of fabric. Experiments showed that the developed robotic system is able to track the human's movement in order to help her/him to accomplish the folding co-manipulation task.

Keywords Human robot collaboration · Folding · Cloths · Force/RGB-D control · Human intention · Co-manipulation

1 Introduction

In the industrial and craft sectors as well as in domestic and agricultural domains there are various tasks where two humans are needed to manipulate a non-rigid object. In cloth/carpet/upholstery/awning industries such tasks are transportation, handling and folding of long fabric sheets [1]. Moreover, similar tasks are found in automotive assembly such as: upholstery of seats, carpets and long cables or in robotic space applications such as blanket manipulation for satellites. Other cases

P.N. Koustoumpardis (✉) · K.I. Chatzilygeroudis · A.I. Synodinos · N.A. Aspragathos
Mechanical Engineering and Aeronautics Department, University of Patras,
Rio Patras, Greece
e-mail: koust@mech.upatras.gr

N.A. Aspragathos
e-mail: asprag@mech.upatras.gr

T. Borangiu (ed.), *Advances in Robot Design and Intelligent Control*,
Advances in Intelligent Systems and Computing 371,
DOI 10.1007/978-3-319-21290-6_24

can be found in handling of sheets for laying-up the plies of a composite material object. The folding of fabrics by people with disabilities proves to be sometimes a difficult task and especially in cases where only one hand is functional. Despite the challenges for great research opportunities in this scientific field, the automation of such tasks is still rudimentary.

The robotized manipulation of non-rigid and highly flexible sheet-like objects is a very complicated problem due to their very low bending resistance, their large deformations and their materials' non-linearity [2]. This kind of objects can change their shape by twisting, buckling, folding and wrinkling due to gravity. It is often said that they have "infinite" degrees of freedom since it is very hard to define how many are needed to define their configuration.

The approaches and the research efforts to build robotic systems for handling fabrics, considered one robot [3, 4] or two cooperative robots,[1,2,3] [5–7]. These approaches were based on fabric state recognition using mainly vision sensing, or on predefined folding motions that are based on humans' motion analysis [4], or on high speed dynamic folding without table and using cloth models [5], or on a combination of the above concepts [6, 7]. Most of these approaches, besides their needs for knowing or identifying the state/model of the fabric, are not directly applicable for folding fabrics with larger dimensions.

In an alternative viewpoint, the robots cooperate with humans which are the leaders, while the robots assist them; this could be applied in domestic or industrial fabric handling tasks. The concept of human-robot cooperation for moving a piece of fabric along one direction was presented in an earlier study [8]. In that work, the motion of a piece of fabric along a line was based on a neural network controller and compared to a PID controller. The human was the guider for this simple handling task and the robot followed him ensuring a constant tension on the fabric. Later, this was extended to the transport of the fabric along the main directions of a fixed Cartesian system [9].

In the present paper a two level hierarchical control system is proposed, based on human robot collaboration for co-manipulation of sheet like fabrics. The higher level deals with the perception of the human's motion, and the decision making for the determination of the robot's grasping point. In the lower level, the co-manipulation handling task is implemented by a hybrid force/RGB-D feedback controller. The folding of a rectangular piece of fabric is presented as a case study. In the next sections, the proposed approach for the fabric folding task is described. The two levels of the hierarchical system are presented in Sects. 2 and 3. The experimental results and the efficiency of the proposed system are presented in Sect. 4, while the prospects of the system are described through the future work in the last Sect. 5.

---

[1]Rethink Robotics, http://youtu.be/Mr7U9pQtwq8.

[2]CloPeMa research project, http://youtu.be/gK7yuPfzuD4.

[3]UC Berkeley Folding Robot, http://youtu.be/gy5g33S0Gzo.

2 Fabric Folding Based on Human-Robot Collaboration

The autonomous accomplishment of a very complex and demanding handling task, such as the folding of a fabric requires systems that should have high flexibility and intelligence to plan the motions. Thus, the human-robot collaboration, where the human acts as the guider and the robot follows, is proposed. A folding handling task according to the proposed approach is shown in Fig. 1, where a human and a robot are collaborating for folding a piece of fabric, e.g. a towel or a tablecloth laid on a table.

For the reduction of the possible fabric's configurations the well-known gravity-table pair is considered. It is also assumed that the model of the fabric is unknown due to the difficulties in fabrics' modelling. Likewise, the identification of the configuration and the state of the fabric, i.e. unfolded, wrinkled or not, folded with one or more folds etc. are not responsibilities of the control system but are duties of the human, who acts as the task leader. Our aim is to develop a robotic controller for co-manipulation of a variety of fabric types without any prior knowledge about the fabrics model and its properties, but to be able to co-manipulate only through the recognition of the intention and the actual movements of the human.

An RGB-D sensor is used to identify the configuration of the human and a vision sensor is used for localizing the fabric. A decision making system is proposed to deduce the human's intention in terms of how she/he wants to handle and fold the fabric. The system decides the appropriate robot grasping point on the fabric to accomplish the folding subtask implemented at lower level. In the co-manipulation phase the robot is synchronized with the human through a hybrid force/RGB-D controller.

3 The Two Level Hierarchical Control Scheme

A state transition diagram of the hierarchical sequence Perception—Decision making—Decision execution—Co-manipulation is designed, as shown in Fig. 2, to realize the fabric folding task, where the states of the human-fabric-robot system

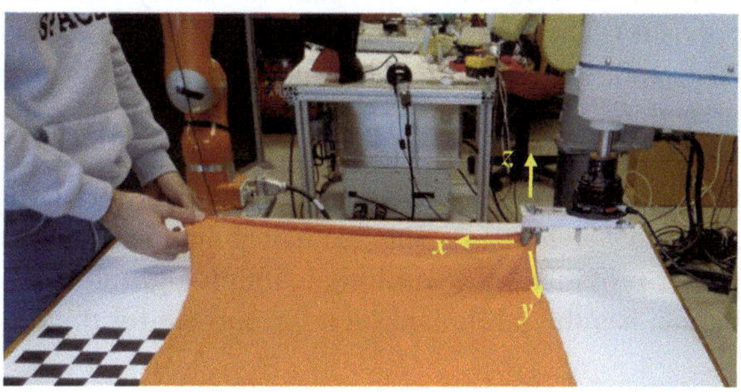

Fig. 1 Human robot collaboration for folding a fabric

are represented. The process starts with state (A) where the human's hand moves towards the fabric. The loop (H_{motion} at state A) ends and the state of the system changes to state (B) when the human grasps the fabric (H_{grasp}). At this point the intention of the human concerning the next folding sub-task is deduced, and the robot positions its end effector in order to grasp the fabric (R_{motion} at state B). The loop terminates when the robot grasps the fabric and the system's state is transferred to state (C). In state (C) the lower level of the hierarchical control is taking place. The robot follows (R_{motion} at state C) the human's actual motion (H_{motion} at state C) under force/RGB-D feedback. This action results to a folding of the fabric according to the guidance provided by the human. If the human wants many folds of the same fabric the above described procedure is repeated when the human simply releases the cloth (state transitions back to A).

3.1 *Perception and Decision Making (*Higher Level*)*

In the higher level of the control scheme, the perception of the human motion and the inference of the human intention take place. The shape of the cloth, as well as the human gripping point and his/her configuration could be a basis for correctly deciding the intended folding. Once the folding configuration is understood, the robot gripping point can be calculated and sent to the robot to start the co-manipulation.

Human pose/intention based on RGB-D sensing

Experiments were performed with humans in collaborative folding without having the ability to verbally communicate one with another so as to understand each other's intention. After that, the participants were asked to describe how they concluded what the other human intended to act. In the fabric folding process, there are two major criteria that were employed to identify the human's intention; tracking of the gripping hand position and tracking of the person's torso orientation.

This result was very close to our initial guess, since a human prefers to maintain a high dexterity across the direction of his intended motion, which is normal to his torso orientation. Also, considering the fact that psychological reasons might apply when cooperating with an industrial manipulator, the human will choose to keep his point of view towards the robot for safety reasons (keeping his eyesight always opposite). These assumptions could suffice in fabric folding tasks because the robot has a robust decision making mechanism that can understand the required folding procedure.

The localization of the fabric plays also a major role in the cooperation. The identification of the fabric configuration is a very complicated task. However, in our scenario the human's intellectual ability is much greater than the computational power provided by a machine learning algorithm. Therefore, it is assumed that the human has the fabric laid on a flat surface for the robot. The RGB-D sensor can be used for the localization process as well—if the cost is to be kept to a minimum—however an additional low cost camera is used for better results and to place the two sensors in an optimum location. Machine learning algorithms or lookup tables could be considered for decision making in complicated fabric folding tasks.

Robot action according to human intention
After the decision making process that determines the fabric location and robot grasping point, the decided robot motion is executed (intermediate level as shown in Fig. 2). The robot approaches the desired point of the fabric from the top and grasps (pinches) the fabric. For this sub-task a force-position controller is developed, where the force part (along the z axis) is responsible for approaching the table until a predefined force is measured. The gripper direction points to the corresponding point that has been grasped by the human, as shown in Fig. 1. After this level, the robot is ready to follow the human's movement controlled by a hybrid force/RGB-D controller.

3.2 *Co-manipulation Based on Hybrid Force/RGB-D Robot Control* (Lower Level)

The folding of the fabric is accomplished with a synchronized motion of the human and the robot hands. In this lower level hybrid control scheme (Fig. 3) for the co-manipulation, the RGB-D system (*Vision/Depth–feedback*) is tracking the human's hand and the force sensing system (*Force–feedback*) measures the actual forces that are applied to the fabric. The motion of the robot end-effector along the main axes is guided by the combined outputs of the force and RGB-D feedback controllers, as:

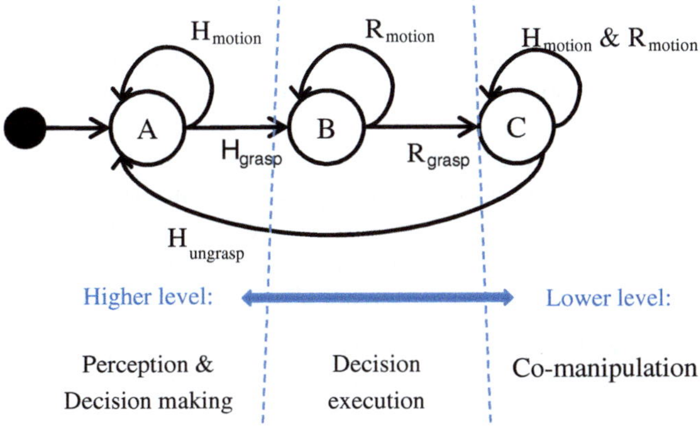

Fig. 2 The hierarchical controller and the state transition diagram for the folding task

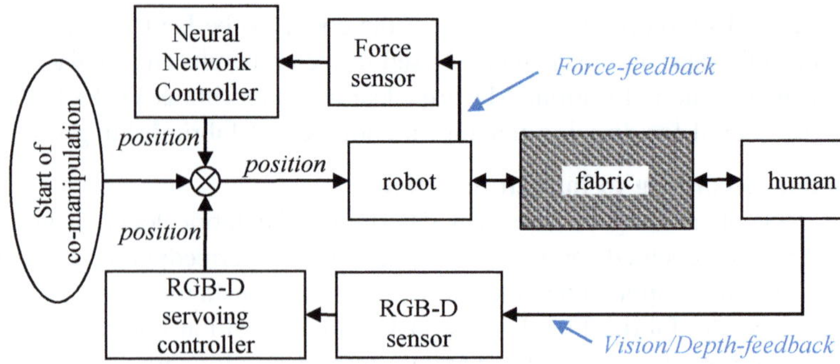

Fig. 3 Hybrid force/RGB-D control scheme

Robot position: direction, axis	+x	−x	+y	−y	+z	−z
Controller: *force* (*F*), *RGB-D* (*V/D*)	*F*	*V/D*	*V/D*	*V/D*	*F*, *V/D*	*V/D*

The x, y, z-axes are for the tool frame, i.e. the coordinate system shown in Fig. 1. The motion of the robot along the +x direction depends only on the force controller, since when the human pulls the fabric the force signal is more significant than the RGB-D one. Along the other directions the fabric resistance is negligible and therefore the measured forces are very low, while the vision/depth feedback signal is significant. Finally, for the motion along the +z direction both the force and the RGB-D controllers contribute to the end-effector motion; however, the force controller has priority to prevent collision of the robot with the table.

RGB-D feedback (V/D) for the co-manipulation during folding
The skeleton tracking library used can track the configuration of the most significant joints in the human body. However, due to the nature of the sensor as well as the intended use of this software, the accuracy of the algorithm can be compromised, leading to a significant position error that could damage the process. To overcome this problem, a moving average filter has been used that filters out large displacements over small time steps. This filter however causes an increase in the time constant of the controller, which however is acceptable due to the nature of the cooperative task.

The interface between the RGB-D servoing controller and the manipulator is for a two way communication, allowing the high-level controller to correctly identify the state changes from state C to state A (when the human releases the fabric). This is identified when the distance between the hand and the end effector increases (over a tuned threshold), and no significant change in the force is measured.

Force feedback (F) for the co-manipulation during folding.
The force feedback controller consists of a neural network controller shown in Fig. 3. A feed forward neural network with a simple topology, composed by three layers with the configuration (1–6–1) has been used as described in [8, 9].

4 Experiments and Results

The Adept Cobra s800 robot is used for the experiments. The forces are measured using the F/T system (Gamma 65/5) from ATI Industrial Automation, mounted on the robot's wrist. The RGB-D sensor is the Asus Xtion while for the fabric localization a simple low cost USB webcam is used. The perception and decision making processing of higher level is implemented in C++ and runs in ROS [10] with OpenCV [11] (for fabric localization) and NITE2 [12] with OpenNI2 drivers, publicly available at bitbucket. The high level controller is implemented in a PC and communicates with the robot controller via serial interface. The low level controller, including the force control scheme, is implemented directly on Adept SmartController CX.

To demonstrate the capabilities of the proposed controller, the task of folding a rectangular piece of fabric is tested.[4] The fabric is laid on a flat surface, and a calibration pattern is used to identify the extrinsic parameters of the camera with respect to the surface of the fabric. The coordinates of the four vertices of the fabric are calculated using a simple threshold filter and sent to the high level controller, see Fig. 4. Concurrently, the RGB-D sensor tracks the human configuration and performs a nearest neighbour query to identify whether his hand is close to any of the fabric's vertices.

The controller does not monitor the actual grasping movement for simplicity; instead it assumes that the human has performed the grasping if his hand is not moving for a short period of time, while still being close to the fabric vertex (shown in Fig. 5), for two different grasping configurations of the same fabric. The yellow sphere shows the identified vertex grasped by the human while the white sphere shows the commanded vertex that will be grasped by the robot's end effector. The green sphere shows the actual human hand position as identified by the RGB-D sensor.

In Fig. 6, the fabric is folded in 8 layers (1a–1b–2a–2b–3a–3b) in three stages. This folding is the conventional one that is followed when folding tablecloths, bed sheets and towels. The states of the higher level (perception-decision making-execution) are in the top pictures (1a, 2a, 3a), where the human and the robot grasp the fabric. At the end of the co-manipulation (lower level) the states of the fabric are shown in the bottom pictures (1b, 2b, 3b), where one fold of the fabric has been completed. The transition from one fold to the other (1 → 2, 2 → 3) is performed if the human stops her/his movement and ungrasps the fabric, while withdrawing his hand towards his torso.

[4]A demonstration of the folding task can be viewed in the Robotics Group YouTube channel.

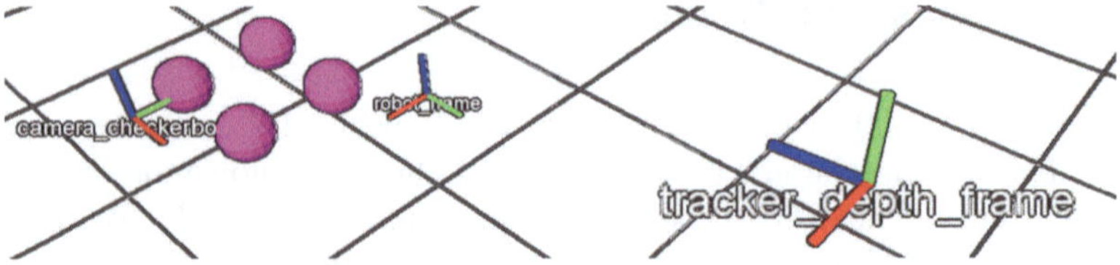

Fig. 4 The coordinate frames of the robot, RGB-D sensor and the camera extrinsic calibrated fabric coordinate frame. The *purple spheres* indicate the four identified vertices of the fabric

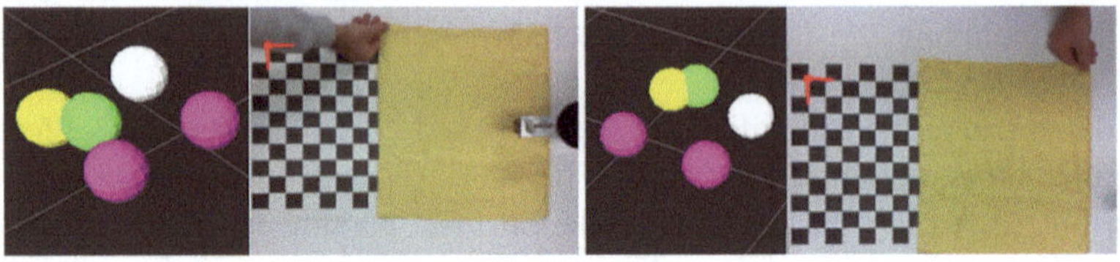

Fig. 5 Two grasping configurations for the same fabric (*left* 3D model and *right* camera feed)

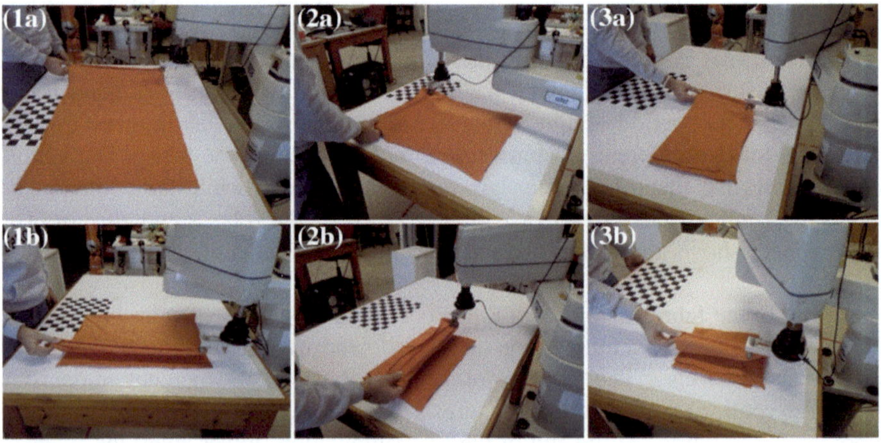

Fig. 6 Photos taken during the folding process after three consecutive folding motions

5 Conclusions and Future Work

The collaboration of a human and a robot for folding rectangular pieces of fabrics is presented in this paper. The proposed hierarchical control based approach is presented in detail and the experiments show that the co-manipulation for folding is achieved.

Our future research directions of this work are focused on the development of a sophisticated decision making system for complicated folding sequences, on the investigation of the robustness of the system and its speed response, as well as on the incorporation of the torques in order to involve the pitch/roll/yaw movements of the robot.

References

1. Saadat, M., Ping, N.: Industrial applications of automatic manipulation of flexible materials. Ind. Robot: Int. J. **29**(5), 434–442 (2002)
2. Koustoumpardis, P.N., Aspragathos, N.A.: Intelligent hierarchical robot control for sewing fabrics. Robot. Comput. Int. Manuf. **30**(1), 34–46 (2014)
3. Zoumponos, G.T., Aspragathos, N.A.: A fuzzy strategy for the robotic folding of fabrics with machine vision feedback. Ind. Robot: Int. J. **37**(3), 302–308 (2010)
4. Oshima, T., Yoshimi, T., Mizukawa, M., Ando, Y.: A study of towel folding by a robot arm—Spreading and vertex detection using image processing. In: Proceedings of 14th International Conference on Control, Automation and Systems (ICCAS'14), pp. 627–631 (2014)
5. Yamakawa, Y., Namiki, A., Ishikawa, M.: Motion planning for dynamic folding of a cloth with two high-speed robot hands and two high-speed sliders. In: Proceedings of 2011 IEEE International Conference on Robotics and Automation, pp. 5486–5491 (2011)
6. Miller, S., Van Den Berg, J., Fritz, M., Darrell, T., Goldberg, K., Abbeel, P.: A geometric approach to robotic laundry folding. Int. J. Robot. Res. **31**(2), 249–267 (2011)
7. Stria, J., Prusa, D. Hlavac, V., Wagner, L., Petrik, V., Krsek, P., Smutny, V.: Garment perception and its folding using a dual-arm robot. In: Proceedings of 2014 IEEE/RSJ International Conference on Intelligent Robots and Systems, pp. 61–67 (2014)
8. Koustoumpardis, P.N., Aspragathos, N.A.: Neural network force control for robotized handling of fabrics. In: Proceedings of International Conference on Control, Automation and Systems ICCAS 2007, COEX, Seoul, Korea, 17–20 Oct, pp. 566–571 (2007)
9. Koustoumpardis, P.N., Aspragathos, N.A.: Robot-human cooperation holding and handling a piece of fabric. In: Proceedings of 11th International Conference on Climbing and Walking Robots and Support Technologies for Mobile Machines, CLAWAR'2008, Coimbra, Portugal, 8–10 Sept, pp. 820–827 (2008)
10. Quigley, M., Conley, K., Gerkey B., et al.: ROS: An open source robot operating system. In: ICRA Workshop on Open Source Software (2009)
11. Bradski, G.: The openCV library. Dr. Dobb's J. Softw. Tools 2236121 (2000)
12. PrimeSense Inc.: Prime sensor™ NITE 1.3 algorithms notes. Online manual last viewed online at 24 Oct 2013

Case Studies for Education in Robotics: From Serious Games to "Technology to Teach Technology" Platforms

Monica Drăgoicea and Theodor Borangiu

Abstract This paper describes a specific perspective on developing case studies related to education in robotics. The proposed framework intends to support students learning how to develop distributed software applications through functionality composition. The multi-agent approach is used as a test bed for case studies development. It tries to stress the role of the application development platform in creating rich simulations, giving students the possibility to express their goals with clarity, and creating suitable application architectures to achieve their goals. The presented roadmap describes three "Technology to Teach Technology" platforms that support distributed application development. The last section of the paper gives the structure of a specific case study in mobile robotics, along with the Presage2 multi-agent platform.

Keywords Multi agent systems · Mobile robotics · Agent directed simulation

1 Introduction

Many companies are starting to develop specific software tools on a larger scale to help their users to interact with real world problems and to better develop skills for their new jobs facing a larger exposure to the Information Technology in different industry sectors. Even at the European Commission level it is recognized that ICT literacy should be encouraged and the education of tomorrow's ICT professionals should create that body of knowledge that supports creation and implementation of specific solutions to meet customers' needs and realize business opportunities [1].

M. Drăgoicea (✉) · T. Borangiu
Faculty of Automatic Control and Computers, Politehnica University of Bucharest,
313, Splaiul Independentei, 060042 Bucharest, Romania
e-mail: monica.dragoicea@acse.pub.ro

T. Borangiu
e-mail: theodor.borangiu@cimr.pub.ro

T. Borangiu (ed.), *Advances in Robot Design and Intelligent Control*,
Advances in Intelligent Systems and Computing 371,
DOI 10.1007/978-3-319-21290-6_25

The "serious games" movement is oriented today towards more educational purposes, in which players are involved in real-time interactions. These interactions are created in virtual worlds supporting decision making processes. *Innov8* [2, 3] and *CityOne* [4] are only two examples of educational games that help students evaluate their business skills and the consequences of their decisions, also addressing complex issues of the energy, water, banking and retail industries. Other examples of immersive educational simulation platforms, which are built as serious games, are *Volvo Car UK* [5], a learning solution that teaches sales people on essential topics of legislation for automobile sales, or *Darfur is Dying* [6] which teaches people about the social and economic situation in Darfur.

Besides these above mentioned names, there are many other examples of games available on the market, and several types of creators and vendors are actively playing in it. Among them there are traditional video games companies, serious games vendors, redoubtable software giants like IBM and Microsoft, or multimedia companies. Using any of these solutions implies that users have a virtual interaction with software applications or platforms that let them either entertain or solve problems, teach, investigate, and advertise in a supervised way, according to general pre-programmed game rules.

However, this paper intends to express a different perspective on specific aspects of this type of dedicated virtual worlds, both on academic perspective and personal skills' development. This perspective tries to stress the role of the application development platform, not only in creating rich simulations (see for example [7]) but also in giving students the possibility to express their goals with clarity and to create suitable application architectures to achieve these goals.

In this respect, Sect. 2 of the paper briefly describes three TTT-based platforms (*Technology to Teach Technology*) to support case studies in robotics education. The concept of self-organization in distributed software applications is further depicted in Sect. 2 along with Presage2, a programming platform dedicated to the design of multi-agent systems, and a case study in robot soccer is formulated in Sect. 3. Section 4 concludes the paper.

2 Platforms and Education in Robotics

2.1 Robocode and Object Oriented Technology

Robocode is an open source programming game in which two or more agents (embodying real world mobile robots) compete in a battle-type task [8, 9]. The simulator runs across all platforms supporting Java, being easily integrated with the Eclipse development platform [10]. Robocode evolved lately as a robot programming platform where modern AI programming and machine learning techniques can be evaluated to further education in robotics, for example [11–13].

2.2 Microsoft Robotics Developers Studio and Service-Oriented Computing

Over the last decades, new styles of writing software applications have been proposed, advancing as distributed information processing and functionality integration approaches. Service-Oriented Architecture (SOA) is being used today not only for e-commerce or business applications, but also emerged as a powerful software architecture design approach in its form of Service Oriented Computing (SOC) addressing more technologically based developments, like embedded applications [14–16].

The SOA development cycle requires specific approaches for developing service-oriented solutions like Service-Oriented Modelling and Architecture (SOMA) [17], Service-Oriented System Engineering (SOSE) [18], and dedicated environments, like SOMA-SE, the platform supporting the model-driven design of SOA solutions [19].

These specific research and development directions raised also a new evolution perspective on the educational body of knowledge necessary to acquire these new IT skills, which is different from traditional software development processes. As SOA paradigm explicitly separates software engineering tasks from programming tasks, the well trained service-oriented system engineers focus on reusable components, which they utilize to solve real-world complex problems. Therefore, the educational goal has to be shifted towards teaching students the overall application architecture and how to compose large applications using existing functionalities (software services) [20].

In the mobile robotics domain, these skills can be exposed along with the Microsoft Robotics Developer Studio (MRDS), a reference programming platform that can be used as a basis for real mobile robots integration. MRDS was proposed as a choice of expressing service-oriented computing (SOC) principles such as modularity and reusability related to applying SOA to embedded systems development with direct application to mobile robot control architecture design [21].

2.3 Presage2 and Self-Organizing Multi-Agent Systems

Considering the distributed nature of information processing in Multi-Agent Systems (MAS), Agent Oriented Computing (AOC) is used today to implement complex distributed applications and distributed intelligent systems, facing aspects like coordination [22], self-organization [23], and functionality composition [24]. At the same time, Agent Oriented Computing (AOC) and Service Oriented Computing (SOC) are proposed to be used today from an integrated exploitation perspective in which they contribute to distributed system design [25]. Agent Directed Simulation (ADS) [26] integrates different forms of agent-related research, such as agent simulation (using of simulation for agents) and agent-based simulation (using agents for simulation). Specifically, these research directions above described focus on improving distributed system design, with a strong accent on the aggregation of software components and their associated interactions in the model development. Software application functionalities are composed through interaction and they are embedded in the system model design. In a natural way, the role of the platform is also emphasised along with these new research developments, expressing *emergent phenomena* and *interactive system design* [27].

Presage2 was proposed as a simulation platform for prototyping societies of agents and developing Java-based *animation* and *simulations* of collective adaptive systems [28]. However, its usefulness proved to extend towards more complex visions, as common pool of resources management [29], along with new frameworks for resource allocation such as computational justice [30] following principles of sustainable institutions [29]. In Presage2, formal models of distributed information processing systems expressed as agent societies can be operationalized through simulation, while software components' interaction is supported by means of a powerful rule-based agent choreography mechanism.

3 Designing Case Study Robotic Applications with Presage2

With Presage2, case studies approaching education in robotics can be further designed to highlight specific educational principles to sustain the way future generations will approach real-world problems. Figure 1 presents a "**virtual build and test**" application development scenario, through which students may learn to include into the agent-based model a whole range of representations of surrounding reality. The development can be further guided following well defined mobile robotic use cases expressing *self-organizing*, *coordination*, *cooperation* and *negotiation* aspects where *behavioural assemblages* emulate a certain *group dynamics*, like *robotic foraging*, *robot soccer* and *formation maintenance* [31].

Figure 1 illustrates four main steps to follow when educational case studies in robotics are defined with Presage2 multi-agent programming platform:

- (s1) *Robotic task identification.* A requirements document is created to define overall robotic system functionalities, as well as performance measure for system validation through simulation, based on independent, local decisions and actions of the system's components;
- (s2) *Scenario definition.* A use case model is created to comply with the robotic task objectives. The agent based simulation scenarios will be further defined and the agent based simulation will be created in an environment that allows validating different aspects of interaction between task participants;
- (s3) *Formal model design of the robotic task.* The architecture of the agent based model is defined, consisting of a set of agents, the environment in which they

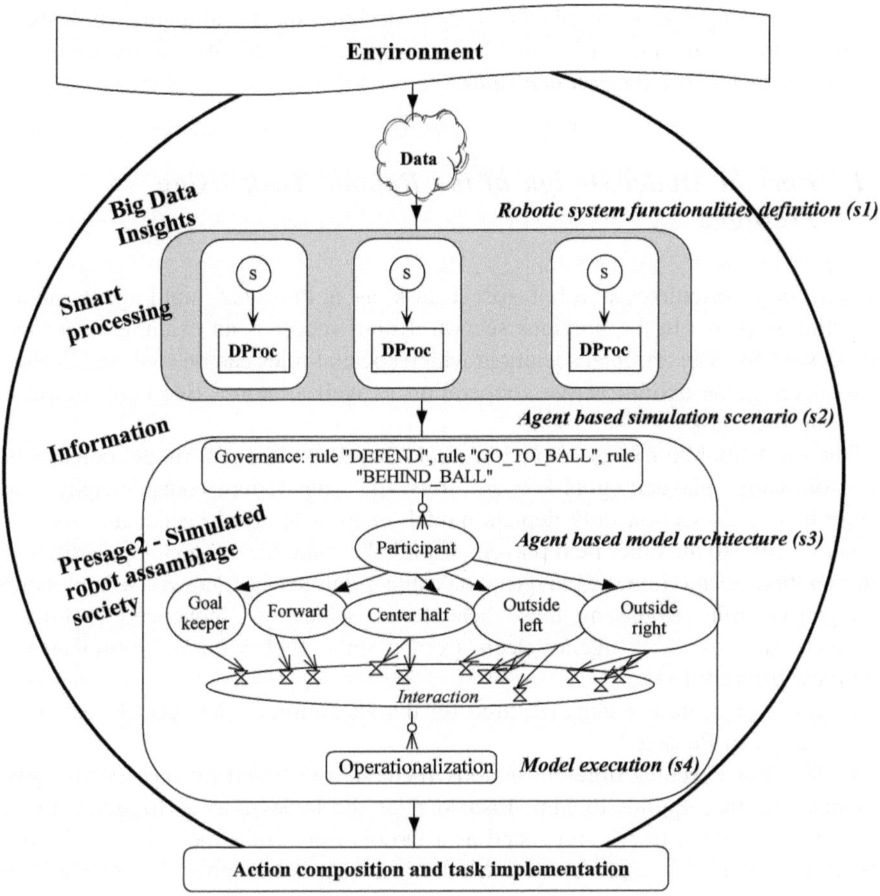

Fig. 1 Virtual build and test scenario in robot education tasks

operate, agents' communication protocol, and the set of general rules according to which they execute their actions;

- (s4) *Agent based operational modelling of the simulated robot assemblage society*. The agent based implementation model is executed in Presage2.

Fulfilling these steps leads to the creation of specific outputs in terms of models and artifacts' to be used later for action composition and task implementation on real mobile robots. As a consequence, different design and development aspects can be expressed, such as: (a) coordination of various robotic systems; (b) experimentation within a simulated environment for coordinating the interactions of the different components of the multi robotics systems; (c) modelling of the multi robotic system as a system of systems through the formal model design of the robotic task expressing coordination architectural styles; and (d) developing a set of model architectures expressing distributed system design aspects, allowing students to observe global outcomes that are the consequence of agents' *interactions* and adaptation to specific *governance rules*.

3.1 Formal Model Design of the Robotic Task Using Presage2

This section formalizes a robot soccer task as a Presage2 multi-agent model, depicting step (s3) in the previous section. Robot soccer is an example of a robot task described as an artificial society in which societal rules can be analysed. Robots in a soccer game exhibit active cooperation as well as non-active cooperation in pursuing their tasks.

The set of mobile robots defined in the Presage2 multi-agent model consists on the following players: goal-keeper, forward, outside-left, outside-right, and centre-half. This section only depicts model creation for goal-keeper and forward players, while all the other field players display the same behaviours, being different only on their position on the playground. Table 1 defines the elementary behaviours and general rules triggering these behaviours, which are sequenced to form a complete strategy; each agent selects from a set of behavioural assemblages to complete the task [31].

Figure 2 presents the steps required for the execution of the agent-based simulation model in Presage2.

Each agent executes different actions (behaviours) based on the specific governance rule that applies to him. Each one of the behaviours is triggered by an environmental condition being tested as a Drools rule. Once the condition is true, the agent action is selected to be executed in a reactive manner. For example, in

Table 1 A Presage2 formal model of a robot soccer task

Agent	Behavioural assemblage	Governance rules (Drools)	Action
Goal-keeper	DEFEND	rule "Defend" when Goal-keeper(detect_ball ()) then moveHalfDist();	Go to the half distance between ball and the middle of its team gate
	GO_TO_BALL	rule "Detect Ball" when Goal-keeper (ball_close ()) then follow_ball();	Activated when the ball is close to the gate
Forward	BEHIND_BALL	Rule "Attack" when Forward(find_ball()) then goToGate();	Activated when agent is close to the ball, positioned between the ball and the opponent gate
	GO_TO_BALL	rule "Follow Ball" when Forward (ball_close()) then follow_ball();	Activated when the ball is close to the gate

Fig. 2 Execution of the agent-based simulation model in Presage2

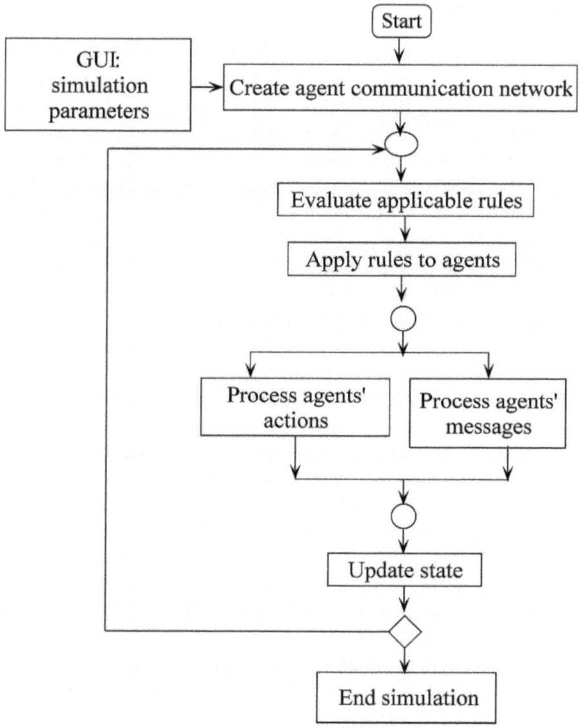

case of the forward player, once the ball position has been determined, the agent executes an approach behaviour, which should drive it into a position to kick the ball forward or even into the opponent's goal. To effectively implement actions, different methods were proposed to be used, for example the potential field method [31]. In its associated area, each robot is subjected to attractive forces (generated by the ball or the goal) and repulsive forces (generated by the opponent players).

4 Conclusions

This paper emphasises the role of the development platform in supporting teaching of specific technology-related topics. The concept of "Technology to Teach Technology" is used to highlight specific characteristics of three development platforms that support application building and testing of real world scenarios in virtual environments. A Robot Soccer task was described and its Presage2 model derived. The vision described in the paper emphasises a possible academic perspective for education in robotic engineering field that combines the SOC application development with a high-level visual modelling approach assisted by animation and simulation tools.

References

1. European Commission: The European Foundational ICT body of knowledge. E-Skills: Promotion of ICT Professionalism in Europe, Capgemini Consulting (2014)
2. IBM: Innov8: IBM introduces video game to help university students develop business skills, http://www-03.ibm.com/press/us/en/presskit/22501.wss (2007)
3. IBM: IBM serious game provides training to tackle global business challenges. IBM Press, http://www-03.ibm.com/press/us/en/pressrelease/26734.wss (2009)
4. IBM: IBM launches first smarter planet game to tackle city challenges, http://www-03.ibm.com/press/us/en/photo/32607.wss (2010)
5. Volvo: VOLVO CAR UK: Serious games replicating a real-life showroom experience. Serious game at http://www.futurelab.net/ (2007)
6. Knowlton, A.: Darfur is dying: A narrative analysis. Master of Arts Thesis, University of Nebraska at Omaha, Omaha (2009)
7. Microsoft: ESPSDK overview, http://msdn.microsoft.com/en-us/library/cc526948.aspx (2015). Accessed 6 Jan 2015
8. IBM: Robocode is the number one download on IBM alphaWorks. http://www-03.ibm.com/press/us/en/pressrelease/992.wss (2001)
9. Nelson, M.: Robocode—the official site. Available at http://robocode.sourceforge.net. on-line accessed 7 Jan 2015 (2014)
10. IBM DeveloperWorks: Secrets from the Robocode masters: A collection of hints, tips, and advice from Robocode masters. Available http://www.ibm.com/developerworks/library/j-robotips/. on-line accessed 7 Jan 2015 (2002)
11. Nielson, J. Land and B.F Jensen: Modern AI for games: Robocode. on-line at http://www.jonnielson.net/RoboReportOfficial.pdf (2010)
12. Hartness, K.: Robocode: Using games to teach artificial intelligence. J. Comput. Sci. Coll. **19** (4), 287–291 (2004)
13. Harper, R.: Evolving Robocode tanks for Evo Robocode. J. Genet. Program. Evolvable Mach. **15**(4), 403–431 (2014)
14. Karnouskos, S., Guinard, D., Savio, D., Spiess, P., Baecker, O., Trifa, V., Moreira Sa de Souza, L.: Towards the Real-Time enterprise: Service-based integration of heterogeneous SOA-ready industrial devices with enterprise applications. In: 13th IFAC Symposium on INCOM, Russia, vol. 13(1), pp. 2131–2136 (2009)
15. De Deugd, S., Carroll, R., Kelly, K.E., Millett, B., Ricker, J.: SODA: service-oriented device architecture. IEEE Pervasive Comput. **5**(3), 94–96 (2006)
16. Bohn, H., Bobek, A., Golatowski, F.: SIRENA-Service Infrastructure for Real-time Embedded Networked Devices: A service oriented framework for different domains. In: Networking, IEEE International Conference on Systems and International Conference on Mobile Communications and Learning Technologies, pp. 43–43 (2006)
17. Arsanjani, A., Ghosh, S., Allam, A., Abdollah, T., Holley, K.: SOMA: A method for developing service-oriented solutions. IBM Syst. J. vol. **47**(3), 337–396 (2008)
18. Tsai, W.T.: Service-oriented system engineering: A new paradigm. In: IEEE International Workshop on Service-Oriented System Engineering (SOSE'05), pp. 3–6 (2005)
19. Zhang, L.-J., Zhou, N., Chee, Y.-M., Jalaldeen, A., Ponnalagu, K., Sindhgatta, R. R., Arsanjani, A., Bernardini, F.: SOMA-ME: A platform for the model-driven design of SOA solutions. IBM Syst J. **47**(3), 397–413 (2008)
20. Tsai, W.T., Chen, Y., Cheng, C., Sun, X., Bitter, G., White, M.: An introductory course on service-oriented computing for high schools. J. Infor. Technol. Edu. **7**, 315–338 (2008a)
21. Tsai, W.T., Sun, X., Huang, Q., Karatza, H.: An ontology-based collaborative service-oriented simulation framework with Microsoft Robotics Studio. Simul. Model. Pract. Theor. **16**, 1392–1414 (2008b)
22. Bedrouni, A., Mittu, R., Boukhtouta, A., Berger, J.: Distributed intelligent systems. A coordination perspective. Springer Science + Business Media (2009)

23. Bernon, C., Gleizes, M.P., Migeon, F., Di Marzo Serugendo, G.: Engineering self-organizing systems. In: Di Marzo Serugendo, G., Gleizes, M.P., Karageorgos, A. (eds.) Self-organising software. From Natural to Artificial Adaptation, pp. 283–312. Springer, Berlin (2009)
24. Luo, J., Li, W., Liu, B., Zheng, X., Dong, F.: Multi-agent coordination for service composition. In: Nathan Griffiths, Kuo-Ming Chao (eds.) Agent-based service-oriented computing, pp. 47–80. Springer, Berlin (2010)
25. Griffiths, N., Chao, K.-M.: Agent-based service-oriented computing. Springer Book Series: Advanced Information and Knowledge Processing. Springer, Berlin (2010)
26. Yilmaz, L., Ören, T.: Agent-directed simulation. In: Yilmaz, L., Ören, T. (eds.) Agent-directed simulation and systems engineering. Wiley, New Jersey (2009)
27. Helbing, D., Balietti, S.: How to do agent-based simulations in the future: From modeling social mechanisms to emergent phenomena and interactive systems design. Tech. Rep. 11-06-024. Santa Fe Institute, NM, USA, Santa Fe Working Paper (2011)
28. Neville, B, Pitt, J.: PRESAGE: A programming environment for the simulation of agent societies. Programming Multi-Agent Systems, Lecture Notes in Computer Science, vol. 5442, pp. 88–103 (2008)
29. Ostrom, E.: Governing the commons. The Evolution of Institutions for Collective Action. University of Cambridge Press, Cambridge (1990)
30. Pitt, J., Busquets, D., Riveret, R.: Formal models of social processes: The pursuit of computational justice in self-organising multi-agent systems. In: 7th IEEE International Conference on Self-Adaptive and Self-Organizing Systems (SASO), pp. 269–270 (2013)
31. Dragoicea, M., Serban, T.: Behavioural diversity in cooperative multi-robot tasks. Proceedings of 16th International Workshop on Robotics in Alpe-Adria-Danube Region, pp. 412–417. Bucharest, Romania, 26–28 May 2005

Designing a Multimodal Human-Robot Interaction Interface for an Industrial Robot

Bogdan Mocan, Mircea Fulea and Stelian Brad

Abstract This paper presents a framework for multimodal human-robot interaction. The proposed framework is intended to bring important contributions to the development of human robot interaction to facilitate intuitive programming and to enable easily adapting to changes in robot task without the need of using skilled personnel. The key elements of this system are speech and hand gesture recognition, text programming, and interaction capabilities that allow the user to take over the control of the robot at any given time. Furthermore, our approach is focused on robot tasks. A user can express his/her preference for one or more modalities of interaction so that selected modalities fit user's personal needs.

Keywords Multimodal user interface · Industrial robots · Multimodal interaction

1 Introduction

Over the last decades industrial robots have become more powerful and intelligent. Thus, in many cases an investment in industrial robots is seen as a vital step that will strengthen a company's position in the market because it will increase the production rate and the process efficiency while reducing the operating costs [1]. Automation based on industrial robots represents the best solution for both productivity and flexibility [2, 3]. However, in small and medium enterprises (SMEs) robots are not commonly found. Even though the hardware cost of industrial robots

B. Mocan (✉) · M. Fulea · S. Brad
Department of Engineering Design and Robotics, Technical University of Cluj-Napoca, Cluj-Napoca, Romania
e-mail: bogdan.mocan@mur.utcluj.ro

M. Fulea
e-mail: mircea.fulea@staff.utcluj.ro

S. Brad
e-mail: stelian.brad@staff.utcluj.ro

T. Borangiu (ed.), *Advances in Robot Design and Intelligent Control*,
Advances in Intelligent Systems and Computing 371,
DOI 10.1007/978-3-319-21290-6_26

has decreased, the integration and programming costs for robots make them still unaffordable for SMEs [4, 5]. It is, thus, quite difficult to motivate a SME which is constantly under market pressure, to carry out a risky investment in robotics [2, 6, 7]. Typically for those SMEs that have frequently changing applications, it is quite expensive to afford professional programmers or technicians, and therefore a human robot interaction solution is demanded [8, 9]. Therefore, today's industrial robots do not offer rich human-robot interaction (multimodal interaction), are not simple to program for end-users, and the programming procedures are time consuming [10, 11].

The traditional online robot's programming can be done in three ways: (i) jogging an industrial robot with 6 degrees of freedom (d.o.f.) with a joystick with two d.o.f. is very time consuming and cumbersome; (ii) the operator doesn't get any visual feedback of the process result before the program has been generated and executed by the robot; (iii) many iterations are needed for even the simplest task [6, 11].

Offline programming environments like RobotStudio from ABB solve some disadvantages described above. But also off-line programming software presents several problems in many industry applications, particularly, when the robot task or the robot trajectory needs frequent change [7]. On the other side, multimodal interfaces allow users to move effortless between different modes of interaction, from visual to speech and touch, according to changes in context or user preference [8]. These interfaces have the advantage of increased usability and accessibility [12, 13]. In multimodal interfaces, the weaknesses of one modality can be offset by the strengths of another one [12]. Accessibility determines how easy it is for people to interact with the robot. Thus, multimodal interfaces can increase robot task efficiency, though perhaps not significantly, as pointed out by [14].

This paper proposes an approach that enables the development of a multimodal interface facilitating intuitive programming of an industrial robot. Our goal is to give an industrial robot the ability to communicate with its human operators in a more intelligent way, thus making the programming of industrial robots more intuitive and easy.

The paper is structured on five sections. Following this introduction, Sect. 2 describes related works regarding multimodal interfaces for programming industrial robots. Section 3 emphasizes the design process and the interface architecture that lead to an intuitive programming of an industrial robot. Section 4 describes a software demonstrator for multimodal interaction and finally, the conclusions are discussed in the Sect. 5.

2 Related Work

Even though the literature on multimodal systems approach is still scarce, various studies have shown that multimodal interfaces may be preferred by users over unimodal alternatives, because they can offer better flexibility, reliability and interaction alternatives to meet the needs of diverse users with a range of usage patterns and preferences [15–17]. Since the introduction of the "Media Room" concept by reference [18], many other systems have been developed based on multimodal interaction with the user. Researchers have established different methods for implementing such systems [19–21]. All multimodal systems are in a way similar in the sense that they receive inputs from different devices/equipment and combine the information to build a common semantic meaning from these inputs [16].

Finite-state multimodal parsing has been studied by researchers within Ref. [22]; they present a method to apply finite-state transducers for parsing inputs. In Ref. [23] speech modality alone has been used to command an industrial robot through switching between pre-programmed tasks. Reference [24] points out that coupling of speech recognition and dialog management (multimodal approach) can improve the performance of a system. A good study on incremental natural language processing and its integration with a vision system can be found in Refs. [6, 25, 26]. A multimodal interaction scheme is very convenient for robot programming because combining two or more interaction modalities can even provide improved robustness [27]. Reference [28] highlights the possibility of combining the speech and static gestures in order to program a robot for grasping and manipulate an object.

Augmented reality (AR) provides great opportunities for Human Robot Interaction (HRI) too, and has been widely used in tele-robotics because it allows the operator to work as if he is present at the remote working environment [9, 29, 30]. Through tablet PCs and head mounted displays it is possible to visualize and generate robot paths through a pointing device [30]. In their work related in Ref. [30] the authors track visually markers to define collision-free paths for the robot. Once the path is generated a virtual robot simulates the behaviour of the robot on the screen. Refs. [6] and [30] combine both augmented and virtual reality environments together with higher level voice commands to remotely operate an industrial robot.

Analyzing the available literature it can be seen that the majority of research efforts are focused on providing a suitable robotic programming method for SMEs, by improving existing programming methods (e.g. online and offline programming) or by combining different programming modalities in so called multimodal robot interaction using new concepts such as Augmented Reality.

The progress in online programming of industrial robots is largely based on sensor and control technologies to assist the operator in learning/planning complex robot motion more easily. Integrating AR with offline programming software originates from the idea of making human-robot interaction more interactive and flexible. With the development of more powerful CAD/CAM/PLM software, robotic vision, sensor technology, etc., new programming methods like multimodal interactions, suitable for SMEs, are expected to be developed in years to come.

3 Competitive Design of the Multimodal Interaction System

Designing multimodal systems for industrial robots is challenging. A set of multimodal myths have proven to be useful in designing multimodal systems, as expressed in Ref. [31]. The classic design approaches and insights from PC environments do not necessarily translate well to industrial robots' multimodal environments. Designing a multimodal interface for industrial robots must focus on robot tasks and operator needs. The set of tasks that an industrial robot can perform is directly affected by the robot pose or joint configuration at one moment [3]. Specifically, the followings are challenges of industrial robot programming: (1) *Obstacle avoidance*: in case the robot approaches an obstacle, the operator needs to modify a teaching position; (2) *Joint limit avoidance*: when an axis limit of a wrist is reached or when a singularity is entered by the robot, the operator must take out the robot from that position; (3) *Model task specific robot trajectories*: the operator must develop the logic of the task and integrate adequately target points within the robot trajectory.

Having in mind the challenges regarding classical robot programming when using competitive design principles and methods like AHP (Analytical Hierarchy Process), PUGH (Pugh Method), TRIZ and QFD (Quality Function Deployment), we developed a methodology (represented in Fig. 1) to identify the right operators' requirements, process needs, objective functions, and the best combination of the multimodal interface inputs.

The proposed methodology for selecting the inputs for the multimodal interface consists of the following steps (see Fig. 1):

Step 1 Define operator requirements, process- and robotic system needs as a set of requirements and rank them. For ranking, tools like AHP can be used.

Step 2 Define a set of target functionalities in accordance to the intuitive programming needs and deploy them against the set of need-related requirements. Thus, weights of target functionalities in relation to the multimodal

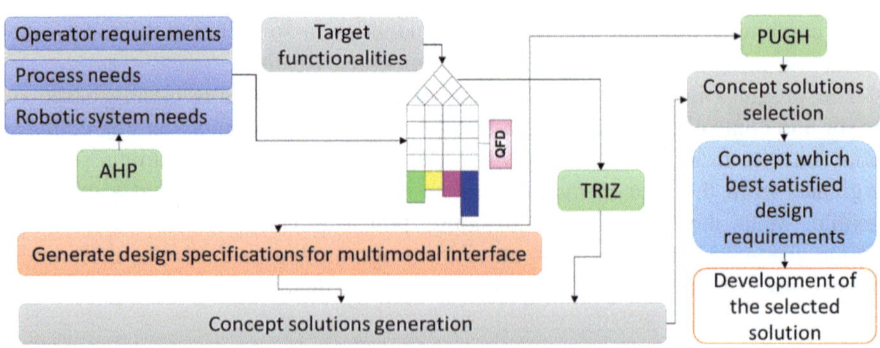

Fig. 1 Roadmap for selecting the inputs for the multimodal interface

programming needs are determined. Correlations between target functionalities have to be also established. QFD-type relationship and correlation matrices could be used to fulfil this step.

Step 3 Formulate vectors of innovation for each negative correlation between target functionalities and for each challenging target. TRIZ is a powerful tool to fulfil this process. The resulted innovation vectors represent paths towards which creativity and skills should be directed when (minimum 3) concept solutions are elaborated.

Step 4 Formulate design specifications for multimodal interface having in mind the needed target functionalities.

Step 5 Generate minimum 3 concept solutions for the multimodal interaction. Inputs from TRIZ method are expected to be integrated within the generated solutions.

Step 6 Evaluate the solutions that were generated at step 5 and select the solution that best satisfies the planned performance for the multimodal interaction interface (see Step 2). Pugh method can be used to fulfil this step.

Step 7 Results from step 6 are used for further development (detailed design and planning at component level). Use-cases, modelling languages (e.g. UML [1]) and other specific tools for software analysis and design can be used to support this step.

The identified operator requirements, process- and robotic system needs for a multimodal interface that facilitate the intuitive programming of a robot, are: (a) automatic presentation of contextually-appropriate information (18 %); (b) easy to use by an operator (25 %); (c) easy to set up (16 %); (d) graphical intuitive interface (24 %); (e) clear indication on the robot's display the logical next step (17 %). The percentages in brackets represent the rank for each requirement, identified with the help of the AHP method. The selected inputs for the multimodal interface were speech and hand gesture recognition, backed up by text programming.

Based on design guidelines and results highlighted above, the proposed framework/architecture of multimodal interface is composed of four functional modules, as illustrated in Fig. 2. The first module (*multimodal interaction*) translates hand gestures and voice command into a structured symbolic data stream. The second module (*actions interpretation*) selects the appropriate set of primitives based on user input, current state, and robot sensor data. The third module (*prioritized execution*) selects and executes primitives based on the current state, sensor inputs, and the task given by the previous step. Finally, the fourth module facilitate the *translation* of the actions, voice command and gesture *into instructions* in robot native programming language to be integrated within the robot programming task. The multimodal interaction between the human operator and robot must be supported by an intuitive graphical interface.

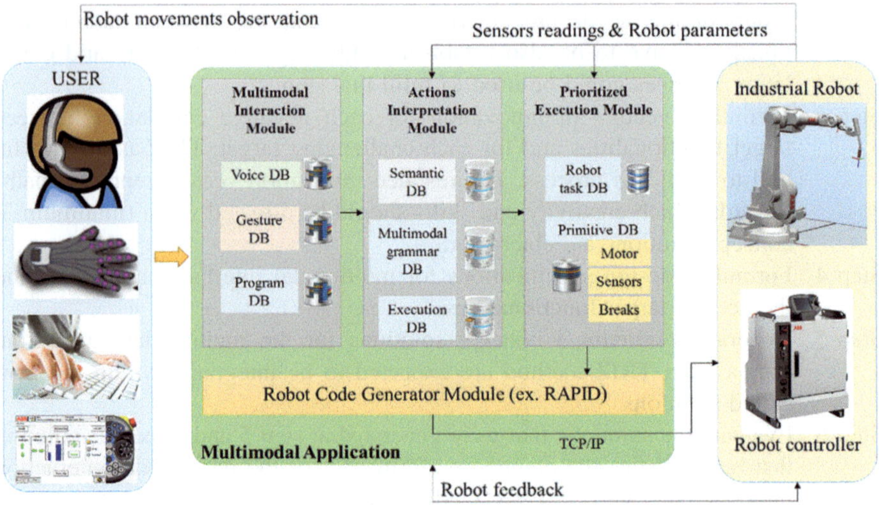

Fig. 2 The multimodal interface framework

4 Software Demonstrator for Multimodal Interface

A software demonstrator was developed (Fig. 2) to exemplify the effectiveness of the multimodal interaction in robotic systems. The demonstrator is still under development; the communication with the robot controller ABB IRB 1600 (S5) was established and we control now by hand gestures the opening and closing of the robot's gripper. The demonstrator could control the robot by text programming directly from a computer. The demonstrator (multimodal interface) has the capability to interrupt commands pre-emptively; it facilitated the verification of the operating mode through sequential programming and demonstrated effectiveness of client-server communication through the robot's TCP/IP facility. The client server application allows one user at a time to access the server in order to transmit the coordinates of a point inside the robots workspace, which the robot must reach. The client can also specify what kind of motion type the robot will perform to reach that point, having the options: linear, joint of circular.

Using the multimodal interface (software demonstrator) the operator can instantly program the robot using the graphical interface for text programming and the sensor glove for gesture commands (Fig. 3). This approach will enable production engineer to focus on manufacturing requirements rather than to the robot programming issues.

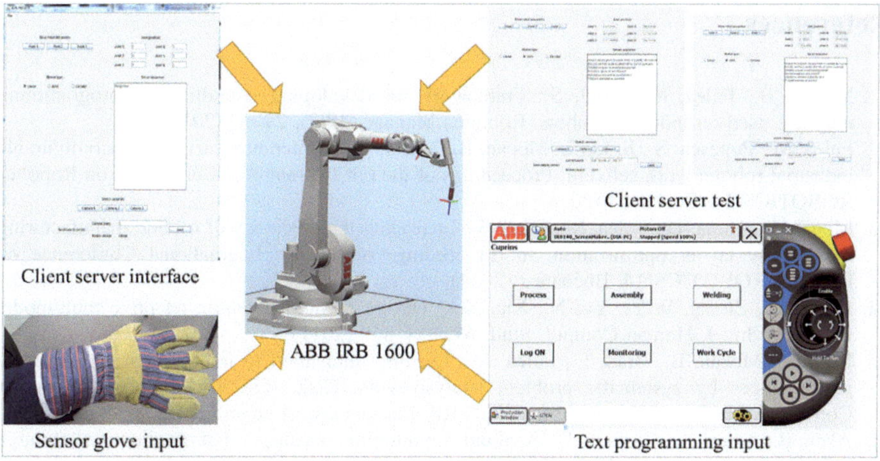

Fig. 3 Multimodal interaction demonstrator

5 Conclusions and Further Works

In this paper, we have described an overall architecture for interactive multimodal industrial robot programming interface and have illustrated the framework which uses a demonstrator. The programming approach offers, through an interface using intuitive text programming, hand gestures and speech recognition, the ability to provide interactive feedback to the industrial robot to coach it throughout the programming and execution phases. The user's intention is captured in the form of sequential robot program, and the flexibility given by the framework through real-time interaction and intuitive interface allows the captured intent to be closer to the user's real intention.

The demonstrator verifies the communication with the robot controller and the effectiveness of the text and partial hand gesture programming (till now we were able to open and close the gripper by sensor gloves) in the multimodal programming and execution approach, including the capability to interrupt commands pre-emptively.

To attain a comprehensive multimodal interface robot programming system, several elements still need to be further developed and others to be added in the future. We have to further develop the multimodal recognition module, to integrate speech interpretation module and to test additionally the actions interpretation module and the prioritized execution module. Although the programs generated by the current version can be re-executed, they are limited to simple robot task sequences. To expand the generality of the paradigm, we need to add the ability to define complex structures such as conditional logics and looping.

Acknowledgments Support within the project POSDRU/159/1.5/S/137516 PARTING is acknowledged with gratitude.

References

1. Mocan, B., Fulea, M., Brad, S.: Framework for developing a multimodal programming interface used on industrial robots. Robotica Manage. **19**(2), 29–33 (2014)
2. Fulea, M., Popescu, S., Brad, E., Mocan, B., Murar, M.: A literature survey on reconfigurable industrial robotic work cells. In: Proceedings of the 6th International Conference on Robotics ROBOTICS'14. Bucharest (2014)
3. Mocan, B., Buna, D., Fulea, M., Brad, S.: Increasing the efficiency of robotic manufacturing systems by layout optimization. In: Proceedings of the 6th International Conference on Robotics ROBOTICS'14. Bucharest (2014)
4. Kongn, J., Zhang, W.Y., Yu, N., Xia, X.J.: Design of human-centric adaptive multimodal interfaces. Int. J. Human-Comput. Stud. **69**, 854–869 (2011)
5. Brad, S., Mocan, B., Brad, E., Fulea, M.: Leading innovation to improve complex process performances by systematic problem analysis with TRIZ, TFC2014, Global Innovation Convention, EPFL, Lausanne, Procedia CIRP. Elsevier, Amsterdam (2014)
6. Akan, B., Ameri, A., Curuklu, B., Asplund, L.: Intuitive industrial robot programming through incremental multimodal language and augmented reality, 2011. In: IEEE International Conference on Robotics and Automation, pp. 3934–3939. Shanghai (2011)
7. Mocan, B., Fulea, M., Brad, E., Brad, S.: State-of-the-Art and proposals on reducing energy consumption in the case of industrial robotic systems. In: Proceedings of the 2014 International Conference on Production Research—Regional Conference Africa, Europe and the Middle East, pp. 328–334. Cluj-Napoca, Romania, 1–5 July (2014)
8. Turk, M.: Multimodal interaction: A review. Pattern Recogn. Lett. **36**, 189–195 (2014)
9. Hirzinger, G., Bals, J., Otter, M., Stelter, J.: The DLR-KUKA success story: Robotics research improves industrial robots. IEEE Robot. Autom. Mag. **12**(3), 16–23 (2005)
10. Mocan, B.: Performance planning of arc welding robotic systems using specific tools for quality planning and systematic introduction of innovation—part I, II, III. J. Calitatea—acces la success **13**(127), 80–85 (2012)
11. Mocan, B., Fulea, M.: Offline Programming of Robotic Arc Welding Systems. Acta Technica Napocensis, S: Appl. Math. Mech. **1**(54), 173–178 (2011)
12. Brick, T., Scheutz, M.: Incremental natural language processing for HRI. In: Proceedings of the ACM/IEEE International Conference on Human-Robot Interaction—HRI '07, 261–265 (2007)
13. Brad, S., Fulea, M., Brad, E., Mocan, B.: Smart deployment of demonstrators into successful commercial solutions. In: Proceedings of the 24th CIRP Design Conference, pp. 165–175. Milan, Italy, 14–16 March (2014)
14. Horst, P., Schreck, G., Willnow, C.: SME-Service networks for cooperative operation of robot installations. In: International Conference Emerging Solutions for Future Manufacturing Systems, pp. 339–346 (2006)
15. Marin, R., Sanz, P., Nebot, P., Wirz, R.: A multimodal interface to control a robot arm via the web: a case study on remote programming. IEEE Trans. Ind. Electron. **52**(6), 1506–1520 (2005)
16. Dumas, B., Lalanne, D., Oviatt, S.: Multimodal interfaces: A survey of principles, models and frameworks, Human Machine Interaction. Lecture Notes Computer Science, vol. 5440, pp. 3–26. Springer, Berlin (2009)
17. Van Wassenhove, V., Grant, K.W., Poeppel, D.: Visual speech speeds up the neural processing of auditory speech. In: Proc. Nat. Acad. Sci. **102**, 1181–1186 (2005)
18. Bolt, R.A.: Put-that-there: Voice and gesture at the graphics interface. Int. Conf. Comput. Graph. Interact. Tech. **14**, 261–264 (1980)
19. Xiao, B., Girand, C., Oviatt, S.L.: Multimodal integration patterns in children. In: International Conference on Spoken Language Processing, pp. 629–632 (2002)

20. Oviatt, S.L., Lunsford, R., Coulston, R.: Individual differences in multimodal integration patterns: What are they and why do they exist? In ACM SIGCHI Conference on Human Factors in Computing Systems, vol. 2, No. 7, 241–249 (2005)
21. Bohus, D., Horvitz, E.: Facilitating multiparty dialog with gaze, gesture, and speech. In: ACM International Conference on Multimodal Interfaces (ICMIMLMI). Beijing (2010)
22. Johnston, M., Bangalore, S: Finite-state multimodal parsing and understanding. In: Proceedings of the 18th Conference on Computational Linguistics, pp. 369–375 (2000)
23. Pires, J.N.: Robotics for small and medium enterprises: control and programming challenges, Industrial Robot (2006)
24. Fuegen, C., Holzapfel, H., Waibel, A.: Tight coupling of speech recognition and dialog management dialog-context dependent grammar weighting for speech recognition. In: 8th International Conference on Spoken Language Processing, pp. 2–5 (2004)
25. Schlangen, D., Skantze, G.: A general, abstract model of incremental dialogue processing. In: European Chapter Meeting of the ACL, 710–718 (2009)
26. Hsiao, K.-Y., Vosoughi, S., Tellex, S., Kubat, R., Roy, D.: Object schemas for responsive robotic language use. In: Proceedings of the 3rd International Conference on Human Robot Interaction—HRI '08, pp. 231–235 (2008)
27. Iba, S., Paredis, C.J.J., Khosla, P.K.: Interactive multi-modal robot programming. In: Proceedings 2002 IEEE Intetnational Conference on Robotics and Automation, pp. 161–168 (2002)
28. McGuire, P.C., Fritsch, J., Steil, J.J., Roethling, F., Fink, G.A., Wachsmuth, S., Sagerer, G., Ritter, H.: Multi-modal human-machine communication for instructing robot grasping tasks. In: International Conference on Intelligent Robots and Systems (IROS), pp. 1082–1089 (2002)
29. Fang, H., Ong, S.K., Nee, A.Y.-C.: Robot Programming Using Augmented Reality, IEEE, (2009)
30. Jara, C.A., Candelas, F.A., Gil, P., Fernandez, M., Torres, F.: An augmented reality interface for training robotics through the web. Communication, pp. 189–194 (2005)
31. Oviatt, S.L.: Ten myths of multimodal interaction. Commun. ACM **42**(11), 74–81 (1999)
32. Marin, R., Sanz, P., Sanchez, J.: A very high level interface to tele-operate a robot via Web including augmented reality. In: Proceedings of IEEE International Conference on Robotics and Automation, (May), pp. 2725–2730 (2002)
33. Quek, F.: Toward a Vision-Based Hand Gesture Interface, Virtual Reality System Technology Conference, pp. 17–29. Singapore (1994)

Part VI
Modelling and Design of Novel Mechanisms and Robotic Structures

Modelling of the Hexapod Mobile Robot Leg Using Matlab SimMechanics

Sorin Mănoiu Olaru and Mircea Niţulescu

Abstract In this paper the authors present a robot leg used to form a hexapod mobile robot structure. For the leg in question the direct kinematics, inverse kinematics and dynamic model were derived. The dynamic model was implemented using SimMechanics toolbox from Matlab. The trajectory for the leg tip was implemented using piecewise cubic spline interpolation. The algorithms and models were developed and simulated using the Matlab software suite.

Keywords Hexapod robot · Leg dynamics · Modelling · Simulation · SimMechanics

1 Introduction

In nature, most arthropods have six legs to easily maintain static stability; it has been observed that a larger number of legs do not necessarily increase walking speed [1].

There are two main types of legs configurations that are biologically inspired. The main difference is in the way the leg swings relative to body. The first type is similar to cats, humans and birds where the leg swings around a horizontal axis. The second type is more similar to insects where the legs swing around a vertical axis providing better stability.

Current vehicles we are using have wheels for locomotion. Wheeled vehicles can achieve high speed with a relative low control complexity but only on structured terrain. Since most of the earth's surface is inaccessible to regular vehicles, there is

S.M. Olaru (✉)
Hella Romania, Caracal, Romania
e-mail: manoiu@robotics.ucv.ro

M. Niţulescu
University of Craiova, 13, Craiova, Romania
e-mail: nitulescu@robotics.ucv.ro

T. Borangiu (ed.), *Advances in Robot Design and Intelligent Control*,
Advances in Intelligent Systems and Computing 371,
DOI 10.1007/978-3-319-21290-6_27

a need for mobile robots that can handle difficult terrain. The large diversity of existing walking animals offers innumerable examples of locomotion possibilities. Legged walking machines show robustness in case of leg faults.

An important drawback of legged machines is the complexity of the control required to achieve walking even on completely flat and horizontal surface in which much simpler wheeled machines work perfectly well.

Many scientists [2, 3] use animal's similarities in their mechanical design for leg configuration and leg design, leading often to simplifications of the actual biological system. In legged animal locomotion the periodical excitation of the flexor and extensor muscles is needed in order to produce effective walking movements. For a walking structure to be able to traverse rough terrain, it needs at least 2 DoF on each leg to move forward or backward or lift over objects.

Building a mechanical structure that mimics an animal leg is still in progress [4, 5]. The most common leg structure among current studies has 3 DoF of which 2 DoF are considered located in the first joint or the segment between the first and the second joint is very small. The authors propose a leg with the segment between the first and the second joint to be much longer.

The legged locomotion on natural terrain presents a set of complex problems (foot placement, obstacle avoidance, load distribution, general stability) that must be taken into account both in mechanical construction of vehicles and in the development of control strategies.

One way to handle these issues is using models that mathematically describe the different situations; therefore modelling has become a useful tool in understanding systems complexity and for testing and simulating different control approaches [6–8]. Hexapod robots have attracted considerable attention from the scientific community in recent decades because of the benefits of this type of platform.

2 Hexapod Robot Leg

The successful design of a legged robot depends on a large extent by the leg design chosen. Since all aspects of walking are ultimately governed by the physical limitations of the leg, it is important to select a leg that will allow a maximum range of motion and that will not impose unnecessary constraints on the walking.

For the leg in question there has been chosen a 3-segment revolute kinematical chain (Fig. 1a). The segments L*i* and the joint R*i* of the leg are called: coxa, femur, tibia respectively coxa joint, femur joint and tibia joint. Coxa joint also represents the attachment point on the robot. A direct geometrical model for each leg mechanism is formulated between the moving frame $O_i(x_i,y_i,z_i)$ of the leg base, where $i = 1\ldots3$, and the fixed frame $O_G(X_G,Y_G,Z_G)$.

The assignment of link frames follows the Denavi–Hartenberg (D-H) direct geometrical modelling algorithm [9]. The overall transformation from coxa frame to leg tip frame is obtained as a product between three transformation matrices:

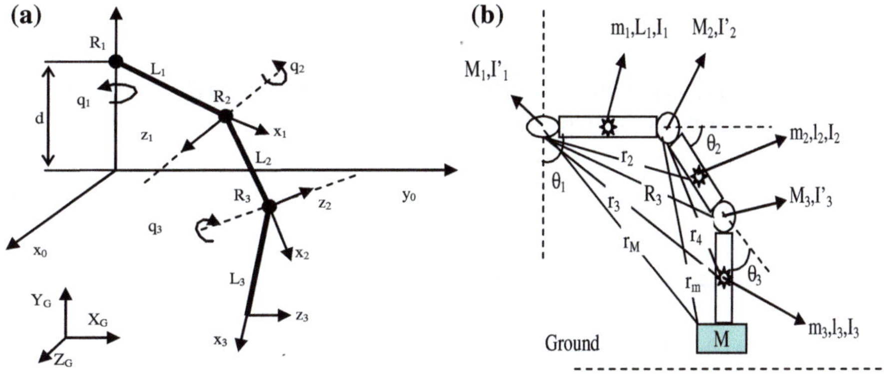

Fig. 1 Model and frame for leg kinematics (**a**) and notations for dynamic model (**b**)

$$T_{coxa}^{tip} = T_{coxa}^{femur} \cdot T_{femur}^{tibia} \cdot T_{tibia}^{tip} \quad (1)$$

Considering Fig. 1 and using D-H parameters the coordinates of each leg tip are:

$$\begin{aligned} x &= [l_1 + l_2 \cdot \cos(\theta_2) + l_3 \cdot \cos(\theta_2 - \theta_3)] \cdot \cos(\theta_1) \\ y &= [l_1 + l_2 \cdot \cos(\theta_2) + l_3 \cdot \cos(\theta_2 - \theta_3)] \cdot \sin(\theta_1) \\ z &= d_1 + l_2 \cdot \sin(\theta_2) + l_3 \cdot \sin(\theta_2 - \theta_3) \end{aligned} \quad (2)$$

where:

d_1 is the distance from global frame to coxa joint frame along Z axis;
l_i are the lengths of leg segments, i = 1…3;
θ_i are the joints angles, i = 1…3.

3 Dynamic Model

The purpose of robot dynamics is to determine the generalized forces required not only to overcome the self-inertial loads (due to the weight of the links of the robot) but also to produce the desired input motions.

In order to obtain a more precise model we divided the mass of each link in two parts (M_i-servomotor mass, m_i-link mass, $M_i > m_i$), see Fig. 1b.

Lagrange equation (Eq. 3) is an analytical approach, in which an operator (called the Euler—Lagrange) is applied on the difference between the kinetic and potential energy.

$$L(q,\dot{q}) = E_c(q,\dot{q}) - E_p(q) \tag{3}$$

Considering the generalized coordinates vector $q = [q_1, q_2, q_3]^T$, the generalized vector forces can be computed using the equation below:

$$\tau_i = \frac{d}{dt}\left(\frac{\partial L}{\partial \dot{q}_i}\right) - \frac{\partial L}{\partial q_i} \tag{4}$$

Considering Eq. 4 and Fig. 1b the generalized forces are:

$$\begin{aligned}\tau_1 = \ddot{\theta}_1\{A + B + 2Fl_1l_2\cos\theta_2 + 2El_3[l_2\cos\theta_3 + l_1\cos(\theta_2+\theta_3)]\} - \\ 2\dot{\theta}_1[Fl_1l_2\dot{\theta}_2\sin\theta_2 + El_2l_3\dot{\theta}_3\sin\theta_3 + El_1l_3\sin(\theta_2+\theta_3)(\dot{\theta}_2+\dot{\theta}_3)]\end{aligned} \tag{5}$$

$$\begin{aligned}\tau_3 = D\ddot{\theta}_3 - \{\dot{\theta}_1^2[-El_2l_3\sin\theta_3 - El_1l_3\sin(\theta_2+\theta_3)] - \dot{\theta}_2^2l_2l_3E\sin\theta_3 + gEl_3\cos(\theta_2 \\ + \theta_3)\}\end{aligned} \tag{6}$$

$$\begin{aligned}\tau_2 = \ddot{\theta}_2(C + 2El_2l_3\cos\theta_3) - 2E\dot{\theta}_2\dot{\theta}_3l_2l_3\sin\theta_3 - \{\dot{\theta}_2^2l_2l_3 2E\sin\theta_3 - \\ \dot{\theta}_1^2[Fl_1l_2\sin\theta_2 + 2El_2l_3\sin\theta_3 + 2El_1l_3\sin(\theta_2+\theta_3)] + \\ g[Fl_2\cos\theta_2 + El_3\cos(\theta_2+\theta_3)]\}\end{aligned} \tag{7}$$

where:

$I_i^{'}$ are the moments of inertia associated with the servo-motors;
$I_i^{''}$ are the moments of inertia associated with the links;
M_i is the mass of the servomotors;
m_i is the mass of the links;
M is an additional mass to approximate a part of the robot weight.

$$\begin{aligned}
A &= I_1^{'} + I_1^{''} + M_2l_1^2 \\
B &= m_2\left(l_1^2 + \frac{l_2^2}{4}\right) + M_3(l_1^2 + l_2^2) + m_3\left(l_1^2 + l_2^2 + \frac{l_3^2}{4}\right) + M(l_1^2 + l_2^2 + l_3^2) \\
C &= I_2^{'} + I_2^{''} + M_3l_2^2 + m_3\left(l_2^2 + \frac{l_3^2}{4}\right) + M(l_2^2 + l_3^2) \\
D &= I_3^{'} + I_3^{''} + Ml_3^2 \\
E &= \frac{m_3}{2} + M \\
F &= \frac{m_2}{2} + M_3 + m_3 + M
\end{aligned} \tag{8}$$

4 Hexapod Robot Leg Control

During motion the leg moves cyclically and, in order to facilitate the analysis and control the motion of the leg is split in two:

- support phase (stance) in which the robot uses the leg as support and propulsion;
- transfer phase (swing) representing the movement of the leg from one point to another.

If we consider a joint control level the leg tip coordinates are:

$$\begin{aligned} x &= d \cdot \sin(\theta_1) \\ y &= step\_length \cdot \cos(\theta_1) \\ z &= \begin{cases} step\_height, & \text{for swing} \\ 0, & \text{for stance} \end{cases} \\ d &= l_{coxa} + l_{femur} \cdot \cos(\theta_2) + l_{tibia} \cdot \cos(\theta_2 - \theta_3) \end{aligned} \tag{9}$$

For controlling the leg movement the diagram in Fig. 2 is used. Also the leg is fitted with a force sensor. The information received from the sensor is compared with a threshold and if it is bigger the leg is considered on the ground, the command to the joints being stopped.

The trajectory generator needed uses a starting point, which is the current location of the leg tip, a middle point and a finish point. Both the middle and the finish point are tested to be inside the leg workspace. The trajectory is generated using the *interp1* command from Matlab [10]. The piecewise cubic spline interpolation method was used for generating the trajectory.

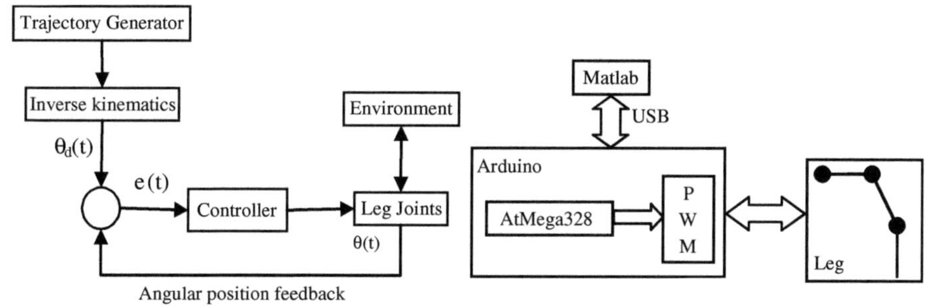

Fig. 2 Hexapod leg control diagram

5 Simulation Results

The dynamic model of the robot leg was implemented using SimMechanics, a software tool from Matlab. The model obtained using SimMechanics is shown in Fig. 3a and the simulated leg is shown in Fig. 3b.

The dynamic analysis was made in order to determine the minimum torques required for the joints to be able to maintain the leg tip on the desired trajectory. The length of the coxa, femur and tibia are respectively 8.5, 12.5 and 25 cm and the masses are 10, 25 and 50 g. The weight of each joint is 50 g.

The minimum torque graphics for femur and tibia joints, which are the most stressed ones when the robot is moving, are further presented. The following two scenarios were considered: in the first one the leg is maintained in a certain position and in the second one the leg becomes a support point.

Regarding scenario 1, the first particular imposed position for the leg is presented in Fig. 4. The motion imposed is sideways. The torque curve for femur joint is depicted in Fig. 4a and corresponds to the variation of the femur joint angle from $-\pi/2$ to 0 radians. Figure 4b shows the tibia curve necessary to maintain the femur and tibia segments collinearly. The peak values are obtained when the motion ends and the three segments are collinear.

The second imposed position from the first scenarios is shown in Fig. 5.

The femur curve is shown in Fig. 5a and like in the precedent case the joint angle varies in the range $[-\pi/2, 0]$ rad. Also, Fig. 5b depicts the minimum tibia torque curve in order to maintain the imposed position of the tibia segment during motion.

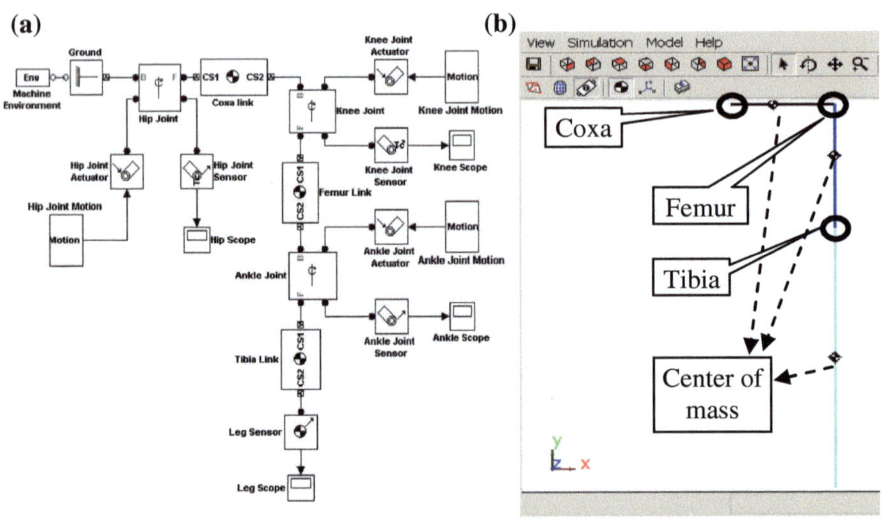

Fig. 3 SimMechanic block diagram of the leg (**a**) and the simulated model (**b**)

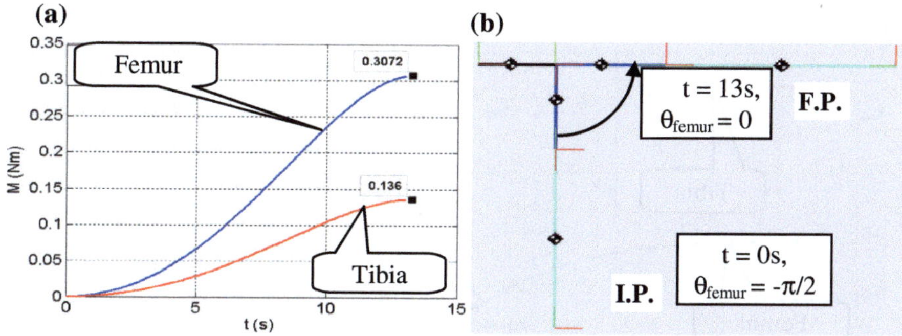

Fig. 4 Minimum torque values for femur and tibia joints (**a**) for imposed position (**b**) (case 1)

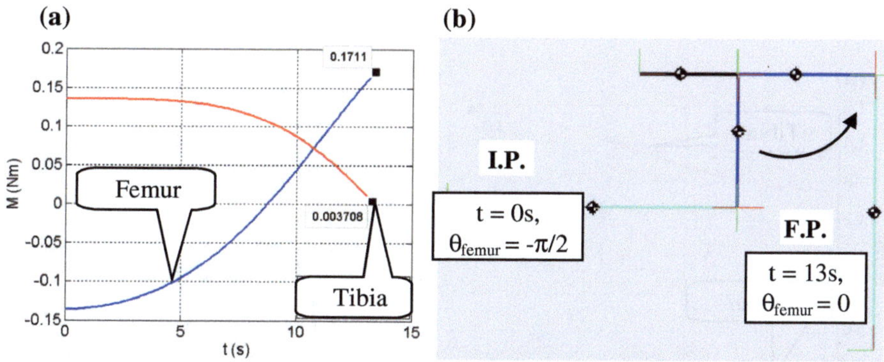

Fig. 5 Minimum torque values for femur and tibia joints (**a**) for imposed position (**b**) (case 2)

For the second scenario when the leg is considered as support point, the most disadvantageous situation is when the robot stands only on 3 legs. In this scenario the most strain is on the tibia joint.

The first analysed case is presented in Fig. 6 in which the height of the robot results from using only the tibia joint. The curve for this joint is shown in Fig. 6a and corresponds for the tibia angle from $3\pi/4$ to $\pi/2$ rad. If we compare with the previous situations we notice that the torque values that need to be developed are much greater.

Another case studied, and more important from the second scenario, is when both femur and tibia joints work together to modify the height of the robot. This situation is commonly used when the robot needs to adjust its height relative to ground or an obstacle.

For this situation the two joints are actuated. The curves for these joints are presented in Fig. 7a and correspond for a variation of femur joint angle from $\pi/4$ to 0 and for tibia joint angle from $3\pi/4$ to $\pi/2$.

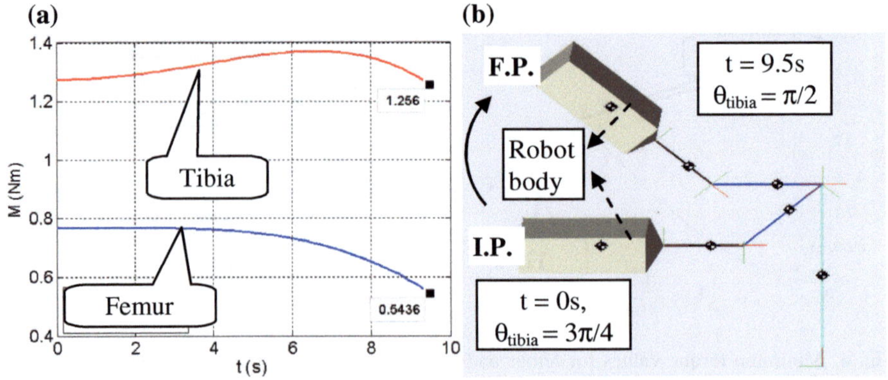

Fig. 6 Minimum torque values for femur and tibia joints (**a**) for lifting the body (**b**) (case 1)

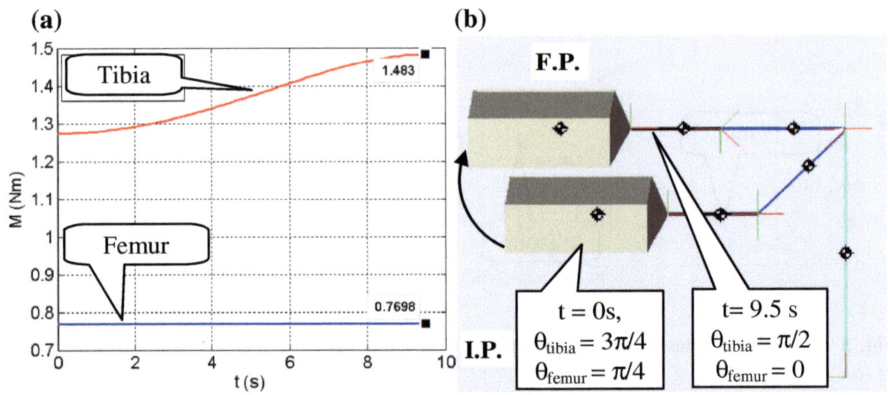

Fig. 7 Minimum torque values for femur and tibia joints (**a**) for lifting the body (**b**) (case 2)

6 Conclusions

In this paper there were examined the dynamic aspects of a hexapod robot leg that is used not only as support point but also as a mean of propelling the robot. From the analysis of the dynamics there were derived the minimum torques required mainly in the femur and tibia joints.

The movement of the leg tip along a predefined trajectory was obtained using a walking algorithm in which the dynamic analysis played a major role. The control part for the robotic leg is an experimental one and uses Matlab and Arduino Mega 2560 development board, subject to further research and improvement.

The advantage of developing the dynamic model and implementing it in SimMechanics is the possibility of viewing more types of signals such as: angles, angular acceleration, computed torque, reaction torque, and reaction force.

References

1. Bensalem, S., Gallien, M., Ingrand, F., Kahloul, I., Thanh-Hung, N.: Designing autonomous robots. IEEE Robot. Autom. Mag. **16**, 67–77 (2009)
2. Conrad, J.M., Mills, J.W.: The history and future of stiquito: A hexapod insectoid robot, Artificial Life Models in Hardware. doi:10.1007/978-1-84882-530-7_1 (2009)
3. Jakimovsk, B.: Biologically inspired approaches for locomotion of a hexapod robot OSCAR, Biologically Inspired Approaches for Locomotion, Anomaly Detection and Reconfiguration for Walking Robots. doi:10.1007/978-3-642-22505-5_5, 35–66 (2011)
4. Tao, L., Ceccarelli, M.: Additional actuations for obstacle overcoming by a leg mechanism. In: Proceedings of the 18th IFAC World Congress. doi:10.3182/20110828-6-IT-1002.00351, 6898–6903 (2011)
5. Zhou Yi, W., Jin Tong, W., Ai Hong, J., Hong Kai, L., Zhen Dong, D.: Movement behaviour of a spider on horizontal surface, Chinese Science Bulletin. doi:10.1007/s11434-011-4584-y, 2748–2757 (2010)
6. Lewinger, W.A., Quinn, R.D.: Neurobiological-based control system for an adaptively walking hexapod. Ind. Robot. Int. J. **38**(3), 258–263 (2011)
7. Yang, J.M.: Fault-tolerant gait planning for a hexapod robot walking over rough terrain. J. Intell. Rob. Syst. (2009). doi:10.1007/s10846-008-9282-x
8. Silva, M.F., Tenreiro Machado, J.A. :Fractional control of legged robots, springer proceedings in mathematics, vol. 2, pp. 647–650 (2011)
9. Schilling, R.J.: Fundamentals of robotics: analysis and control, ISBN: 0-13-344433-3 (1990)
10. Information on www.matlab.com, accessed 2014

Static Analysis of Rotary Positioning Modules for Technological Head of the Robot

Jan Semjon, Marek Vagas and Vladimir Balaz

Abstract The role of the robot actuator is transferring execution power in prescribed manner for a desired motion, so that a working mechanism can realize required positioning (in handling or technological operations) with specified accuracy and speed. A suitable choice of the actuator depends not only on its mechanical parameters, but also on sensing and control. Performance data of actuators from various manufacturers are comparable, the difference between them being mostly due to their design. The reported research is part of a project in which it was necessary to design and implement accurate reducers and actuators in various modular kinematic solutions. These modules should be possibly stacked in desired shapes, allowing them to be deployed in production machines and robotic equipment according to specific customer requirements, such as technological heads. The paper describes the design and FEM analysis of technological heads for 2-axis handling and robotic systems.

Keywords Technological head · Robot · Analysis

1 Introduction

The trend in designing new actuators for precision production equipment and for robots is to ensure high accuracy, and reduction of size and weight of the actuator. This trend can be achieved by considering mechatronic components integrating the servo motor, gear system, sensors and control technology in a compact actuator.

J. Semjon (✉) · M. Vagas · V. Balaz
Technical University of Kosice, Faculty of Mechanical Engineering, Kosice, Slovak Republic
e-mail: jan.semjon@tuke.sk

M. Vagas
e-mail: marek.vagas@tuke.sk

V. Balaz
e-mail: vladimir.balaz@tuke.sk

T. Borangiu (ed.), *Advances in Robot Design and Intelligent Control*,
Advances in Intelligent Systems and Computing 371,
DOI 10.1007/978-3-319-21290-6_28

Hence, such an actuator should benefit from the new concept of componentisation based on light weight of composite materials. The methodological validation of parameters for precision actuators is based on functional and parametric analysis of the general structure of the actuator and its recovery after breakdown. The analytical evaluation of the actuator's construction sequence relies on verifying the required parameters and properties of the actuator. Testing is done by comparing the parameters' values of sample actuators with standard values by help of measuring equipment.

Rotary axes for positioning and handling units are autonomous, functional construction modules. They possess an intelligent integration function allowing connectivity with other mechanical or control modules to obtain a more complex machinery system with higher functionality. A rotary positioning module represents an element of such a complex device; it must perform driving rotary movements with imposed positioning speed and precision [1, 2].

From the analysis of specific solutions it can be concluded that the modules deployed in technological heads are customized for the application's needs (a very small percentage of these products are standardized and described in catalogues). In consequence, standard motion functions for technological head are provided by standard motion modules, whereas multifunctional motion axes can be obtained only by customized designs of technological heads.

This leads to the need for designing appropriate dimensional series, which would be used for the creation of mutually compatible modules that can be easily integrated in ensembles of complex units. These units can be then used to compose technological heads of different shapes according to particular requirements, such as flexible and intelligent devices: grippers with rotational and translational components to manipulate objects of different shapes, etc.

Because technological heads are also used in material processing tasks such as grinding, drilling, threading and milling, their design should allow the necessary connectivity of functional modules featuring specific motions which are necessary to perform these processing tasks [3].

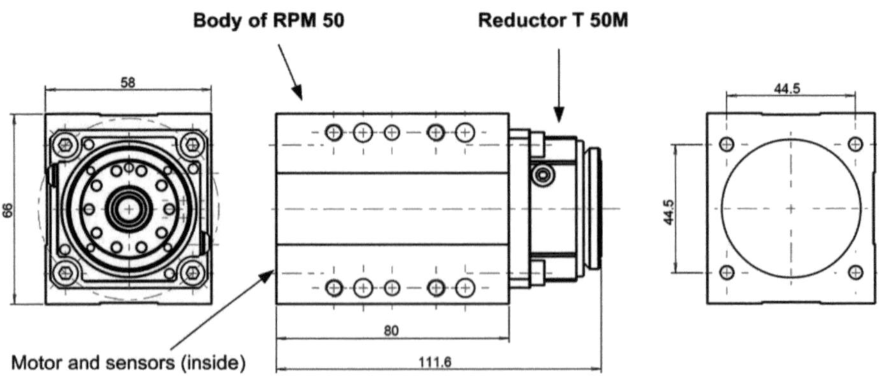

Fig. 1 The RPM 50 module

Figure 1 presents a rotary positioning module (RPM); this type of device is an electrical servomechanism capable to drive a certain type of motion (rotation or linear displacement). It consists from a gearbox, a servomotor and sensors in one compact construction [4].

2 The Basic Structure of RPM

A systemic RPM model describes the functional concept and the solution for engine construction, i.e. for the mechanism's drive that transforms forces and moments into kinematic motion parameters such as direction, acceleration, speed of the movement. The RPM model also describes the internal structure of the mechanism.

Figure 2 shows the block structure of the systemic RPM model.

The elements of the 3D systemic RPM model are:

- D—drive: provides primary energy, output of drive system;
- CB—control block: device controlling the drive signals;
- M—motor: rotary servomotor, has sensors measuring angular rotation, and safety brake. Technical versions: M1 without brake, M—B1 with integrated safety brake;
- GB—gearbox block: provides reduction of the motor M speed, higher torque at the output module, physical transfer of rotary motion (q_M) from the output of motor M to module VR (q_2); the gearbox converts the parameters of motor M (n_M—revolutions per minute, f_M—motion frequency, P_M—power performance, φ_M—path of movement/angle) to the output module VR (q_j). Technical versions: GB1 gear block with classic shaft output, GB2—gear block with hollow shaft output, GB3—gear block with flange output;

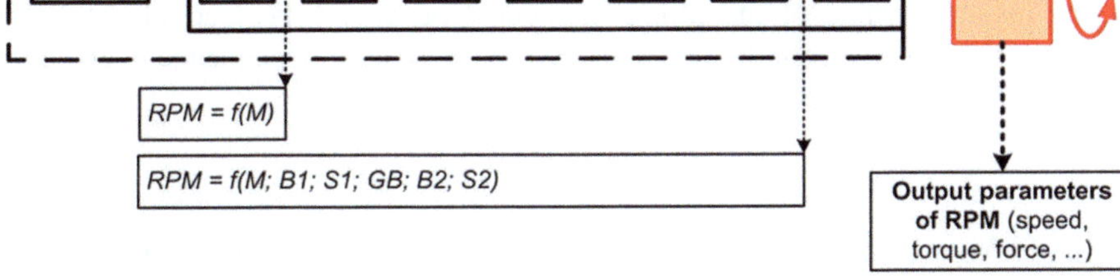

Fig. 2 Systemic RPM model—block structure

- MT—mount: guiding system assuring the contact between moving and fixed parts of the system;
- S1—ensemble of speed and position of motor M;
- S2—position sensor of RPM module;
- B1—safety brake;
- B2—positioning brake for output of module VR (at limit angular position of module RPM),
- VR—mechanical output element; represents the interface for connecting RPM at higher functional units.

The global (theoretical) model of RPM's structure RPM consists from the chain of elements: M-B1-S1-GB-B2-S2-MT-VR.

A module marked IRPM signifies an intelligent rotary positioning module containing all the components of RPM module, but it addition the drive and control blocks have intelligence, for example the drive unit is an adaptive one; it is able to react in real time to changes in the working environment during task execution. Changes are monitored based on the information retrieved from sensors (rotational speed, temperature, acceleration, torque, force). The type and number of monitored parameters depend on the current application [4–6].

In conformity with the performance objectives and technical solutions imposed by the research project, a new series of RPM modules was developed in the size range of 50–70.

3 Analysis of Carrier Bodies for RPM Modules

The FEM analysis focuses on the static analysis that aims at evaluating the appropriateness of the material to be used for the carrier body of RPM modules. The analysis considers the load of the carrier body for maximum torque values and the ensemble of forces acting on the reduction units that are part of the constructive solution of the RPM module. Based on the structure above presented, three size series of modules were selected that are most frequently used for material handling and robotic applications. They concern the modules RPM 50 and 70.

Connection holes on the module's body are located on its back and bottom sides. Therefore, a stress analysis is necessary too, that must take into account the direction in which the load is applied, see Fig. 3.

The resulting value of the tilting moment depends on the load of the RPM module which results from the radial and axial forces. This dependence is given by the equation:

$$M_{Cmax} = F_R \cdot a + F_A \cdot b \text{ [Nm]} \quad (1)$$

The FEM analysis was carried out on all three size series of RPM modules.

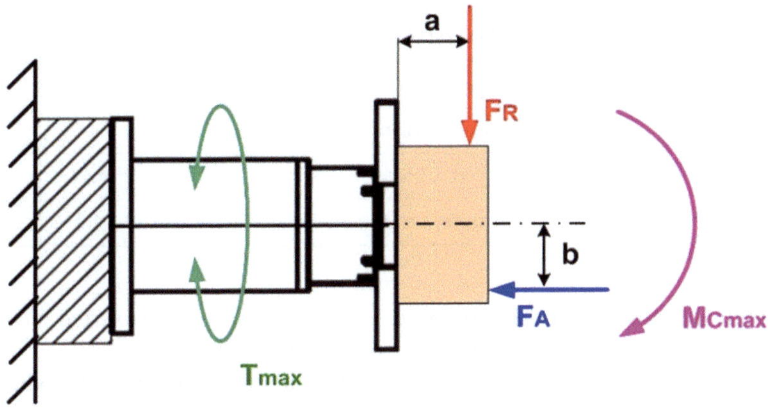

Fig. 3 Location of forces and moments

The remainder of the paper focuses on the medium size range of the module RPM 50 m. The RPM 50 module was designed with the 3D CAD program Creo 2.0 and the FEM analysis program of the Mechanical product. The material from which the module's body made is in accordance with the standard EN AW 2017, which is characterized by good machinability and good strength properties.

The proposed module body 50 RPM (2D model—sketch in Fig. 4) is also compatible with the connecting holes in which it is possible to fasten the flange through bolts.

Creating different configurations of 2 and 3-axis technological heads depends on the desired action of the applied forces and the type of application in which the heads will be used. It was therefore necessary to determine the maximum values of the load moments and forces that are applied to the output flange of the

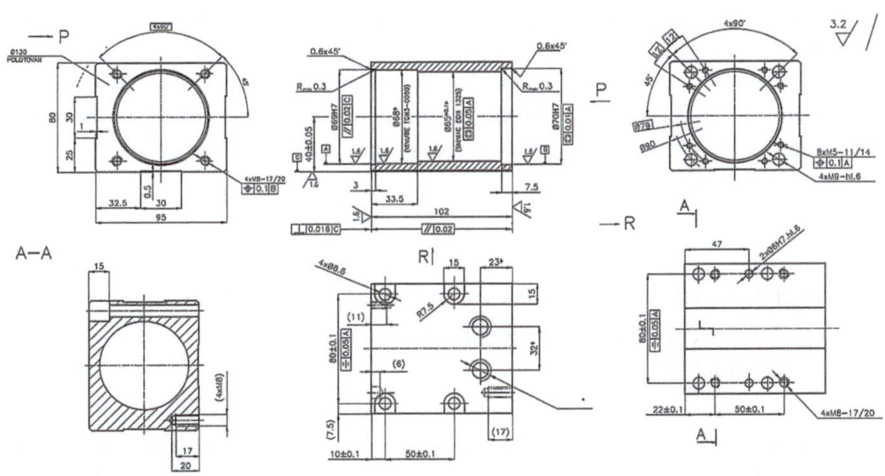

Fig. 4 Sketch of the RPM 50 module body

Maximum load output flange	Units	RPM 50	RPM 70
Rated output moment - Tmax	Nm	**36**	100
Maximal tilting moment - Mcmax	Nm	**44**	142
Rated radial force - FR	kN	**1.44**	2.6
Maximal axial force - FA	kN	**1.9**	3.7

Fig. 5 Maximum and rated load of RPM modules

technological heads. The maximum torque and forces that can be exerted on a RPM module depend on its structure and components included.

Specific value of maximum loads accepted by the RPM 50 and 70 modules are given in Fig. 5.

3.1 *Static Analysis of Stress and Strain in the Body of Module RPM 50*

This static analysis considered body loading at maximum acceleration and a torque value of 36 Nm, radial force of 1.44 kN and axial force of 1.9 kN. In order to check the interconnectivity of body modules RPM 50, there were created 3725 Tetrahedron element types, see Fig. 6. The calculation method used in Creo 2.0 / Mechanical program was QuickCheck. Wedge and Brick type elements were not used, which didn't affect the convergence of the computing algorithm. The polynomial degree was 3.

The computed of stress value (for fixation on the back of the body), had a value of 11.4 MPa, while the permissible stress for materials in accordance with EN AW 2017 must be less than 250 MPa. Comparing the calculated stress value with the allowed one, it results that the proposed construction material meets the safety requirements for a maximum torque of 36 Nm. The maximum value for the deformation of body module RPM 50 for a torque of Tmax = 36 Nm was 0.0053 mm, which represents an acceptable deformation value when the module is used in robotic applications.

3.2 *Static Analysis of Stress and Strain in the Body of Module RPM 50 at Load with Axial Force of 19 000 N*

The computed stress value (for fixation on the back of the body) had a value of 12.14 MPa. Comparing this calculated value with the maximum permissible stress of 250 MPa confirms that the proposed material for a maximum axial force FA of

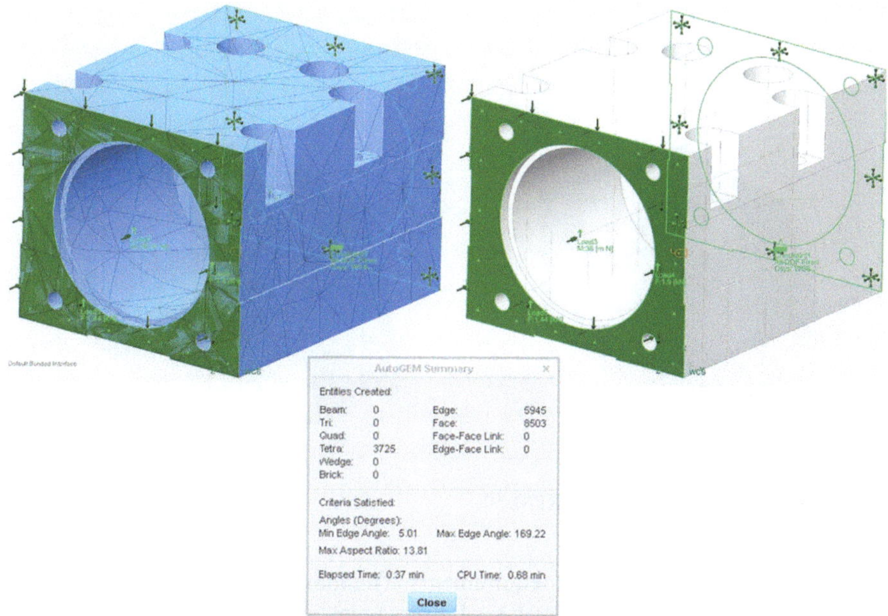

Fig. 6 Network on module for body RPM 50

1.9 kN satisfies the safety requirements. The maximum computed value for the deformation of body module RPM 50 for a load through axial force of 1.9 kN was 0.001275 mm, which represents an acceptable deformation when the module is used in robotic tasks.

3.3 *Static Analysis of Stress and Strain in the Body of Module RPM 50 at Load with Radial Force of 14 400 N*

The findings of stress value (for fixation on the back of the body) indicated a value of 13.75 MPa, much smaller than the permissible stress of 250 MPa); this results confirmed the adequacy of the proposed material for a maximum radial force FR 1.44 kN. The maximum computed value of RPM 50 deformation was 0.00837 mm (Fig. 7).

The computed stress value (for fixation on the back of the body) of the module RPM 50 under combined load of FA 1.9 kN, FR 1.44 kN and Tmax 36 Nm had a value of 6.33 MPa, much inferior to the maximum admissible one of 250 MPa). This confirms that even at a complex load exerted upon the module, the proposed

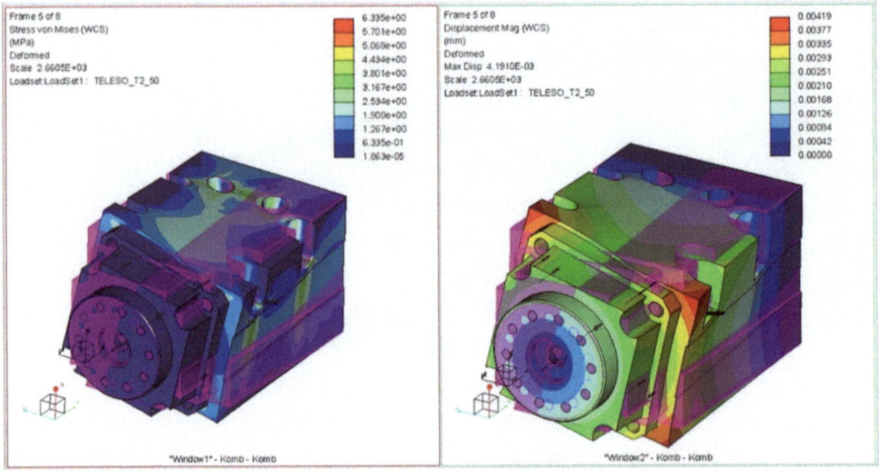

Fig. 7 Combined load for module RPM 50

construction satisfies the safety requirements. The maximum value for deformation module RPM 50 for a combined load from axial and radial forces torque was 0.00419 mm.

A similar FEM analysis was realized for the head series RPM 70.

4 Conclusion

The analysis of the properties of materials used in the construction of RPM head modules allows establishing an appropriate ratio between the mass of the module and its strength and stiffness characteristics. The development of new RPM types allows better meeting the constructive needs for automated robotized workstations. Deploying of precise technological heads on end part of the robot can improved the accuracy and repeatability of the robot's motion in space.

The FEM analysis which was carried out for the proposed modules confirmed the adequacy of using aluminium alloys in the structure of RPM modules. The utilization of such materials for the technological heads led to reduction of the module weight, while maintaining sufficient strength and rigidity. This makes possible the utilisation of such technological heads in robot manipulators with lower load capacity.

Acknowledgments The reported work represents a research contribution is the project "Aplikovaný výskum systémov inteligentnej manipulácie priemyselných robotov s neorientovanými 3D objektmi", related to the activities 1.3 and 2.1 (ITMS: 26220220164). This project was supported by the Research and Development Operational Program funded by ERDF.

References

1. Yang, G., Chen, I.M.: Task-based optimization of modular robot configurations: minimized degree-of-freedom approach. Mech. Mach. Theor. **35**(4), 517–540 (2000)
2. Semjon, J., Janos, R., Tuleja, P. and V. Balaz.: Procedure selection bearing reducer twinspin for robotic arm. Appl. Mech. Mater. **245**, 261–266, ISSN 1660-9336 (2013)
3. Semjon, J. et al.: Benchmarking analysis of the application of small bearing reducers and actuators in service robotics. Fascicle Manag Technol Eng 5.303–5.306, University of Oradea, Romania, ISBN 1583-0691 (2011)
4. Semjon, J., Varga, J. and P. Tuleja: Comparing the parameters of positioning actuators used for robots, In: DAAAM International Scientific Book 2014, Vienna : DAAAM International, 615–624, ISBN 978-3-901509-98-8, ISSN 1726-9687 (2014)
5. Wedle, A., et al.: DLRs dynamic actuator module for robotic space applications. In Proceedings of the 41st Aerospace Mechanisms Symposium, Jet Propulsion Laboratory, Robotics and Mechatronics Center (RMC), German Aerospace Center (DLR), Wessling, Germany May 16–18 (2012)
6. Hajduk, M. et al.: Trends in industrial robotics development. Appl. Mech. Mater. **282**, 1–6, ISSN 1660-9336 (2013)

A New Hyper-Redundant Arm and Control System

Viorel Stoian, Ionel Cristian Vladu and Ileana Vladu

Abstract This paper presents a new hyper-redundant robotic arm structure and its sensorial, driving and control systems. A system of cables actuated by DC motors is used for bending. The position of the robot can be obtained by bending it with the cables and by blocking the position of the needed elements using an electro pneumatic system. The major advantage of this type of actuator consists in the fact that the robot can be moved using a boundary control by cables with a single actuating unit, the position blocking system for any element being relatively simple. The main features and advantages of the sensorial system and global robot system are presented. The dynamic model of the arm is developed using Lagrange's formalism; the motion control system is based on the adaptive computed-torque method. Finally, experimental results are described.

Keywords Hyper-redundant arm · 3D angular sensor · Dynamic model · Control system

1 Introduction

Inspired by trunks, tentacles and snake backbones hyper-redundant arms are a class of manipulators that can reach any position and orientation in space. These arms have mechanical structures with continuum elements described by models with distributed parameters.

V. Stoian · I. Vladu
Faculty of Automation, Computers and Electronics, University of Craiova, Craiova, Romania
e-mail: stoian@robotics.ucv.ro

I. Vladu
e-mail: ileana.vladu@gmail.com

I.C. Vladu (✉)
Faculty of Electric Engineering, University of Craiova, Craiova, Romania
e-mail: cristian.vladu@ie.ucv.ro

T. Borangiu (ed.), *Advances in Robot Design and Intelligent Control*,
Advances in Intelligent Systems and Computing 371,
DOI 10.1007/978-3-319-21290-6_29

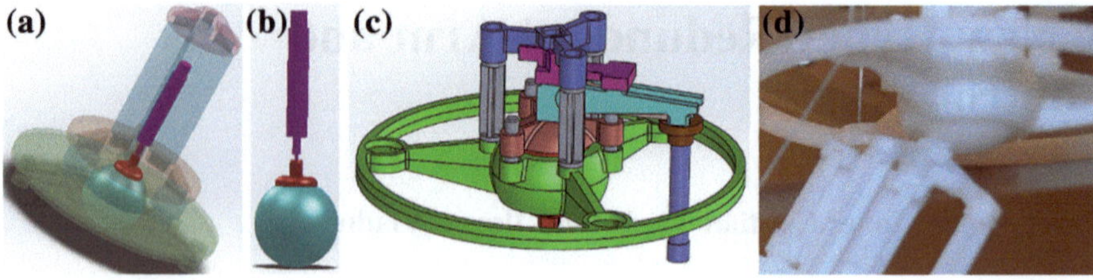

Fig. 1 Element of the hyper redundant robot (HHR), Solidworks model and implementation

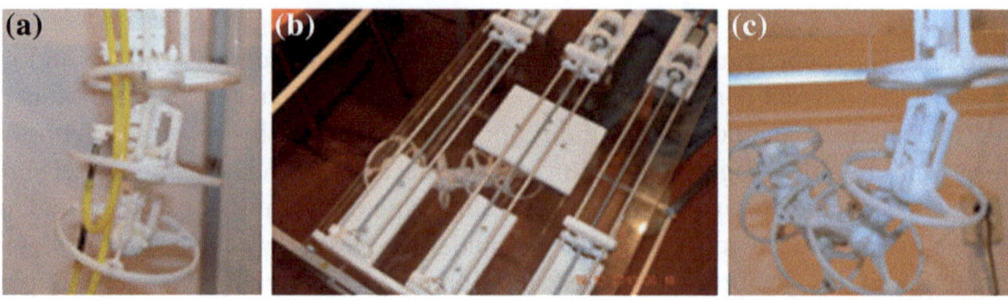

Fig. 2 The HHR robot architecture

The robotic arm is composed of identical elements connected serially. The final position of the robotic arm is obtained by changing the relative position of these elements. The structure is of backbone-type, actuated by a set of cables.

The robotic arm is considered as a single entity from the actuating point of view. The desired robot curvature is obtained using three actuating cables controlled by an electro-pneumatic system (Figs. 1 and 2).

From the constructive point of view, a robot element consists of two disks interconnected by a rigid rod. Two elements have a common disk, the mobile disk of an element representing the fixed disk for the next element. The connection of two elements is realized with a spherical joint (see Fig. 1c). The cables are connected at the terminal point of the arm, at the last mobile disk; through these cables the mobile disk rotates around the spherical articulation as against axis OX, OY or both of them. For equal cable length, the arm curvature is zero and the arm is in the relaxation position. The arm curvature is obtained by controlling the cables' length. Cables are made of very resistant plastic material. The necessity of only three actuating cables for bending the entire structure represents a major advantage.

The electro pneumatic system ensures blocking and releasing the elements. The elements that are not blocked can be bended by the cable-based actuation system. A pneumatic piston with simple action is implemented within the road. The mobile element of the piston presses on the internal sphere of spherical articulation in order to actuate (see Fig. 1a, b). The combination of actuating cables with electro pneumatic system represents a new concept for hyper-redundant robot actuating [1].

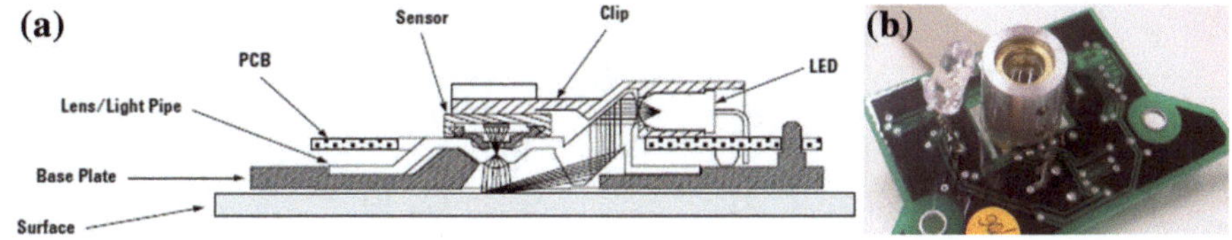

Fig. 3 Intelligent acquisition circuit for the optical sensor

The driving system is based on a DC geared motor. The driving circuit of the DC motor is realized with full bridge pulse width modulation (PWM). This system allows control of the rotation and change of direction [2].

The measurement of the spatial angular increment is realized with an optical sensor of video matrix type with LED (Fig. 3). This sensor is attached to the spherical articulation, and can effectively measure the rotation curvature described by the mobile part of the element's rotation couple, respectively the sphere. In this way the rotation spatial angular increment is measured and then decomposed by numeric computation in angular increments for each rotation axis. The implementation of the intelligent acquisition unit of the optical sensor uses an ADNS 2610 specialized circuit, described in detail in Vladu et al. [3].

The optical sensor is based on the Optical Navigation Technology and contains an image acquisition system (IAS) and a digital signal processor (DSP). The sensor detects any movement by sequential image acquisition using the IAS model, and compares successive images in real time to determine the movement. The sensor has a set of parameters which were calibrated by the manufacturer; their value must be preserved, only with minimal deviations being allowed.

The major advantage of this method consists in the possibility of measuring simultaneously the 3D angular increment and the joint error, which give the real position. Each joint of the robot has an optical sensor and an electro pneumatic system. In extreme cases each element can be considered as an arm segment, from the bending functional perspective.

2 Dynamic Model of the Hyper-Redundant Robot Arm

The hyper-redundant robot arm was modelled as follows. Each module is represented by its central axis with li length, $i = 1 \ldots$ n. Thus, a backbone architecture with n linear series segments in the plane (Fig. 4) was considered. Using the notations in Fig. 4, all computational steps of the Lagrange algorithm were performed.

The coordinates of the end tip point (x and y) are:

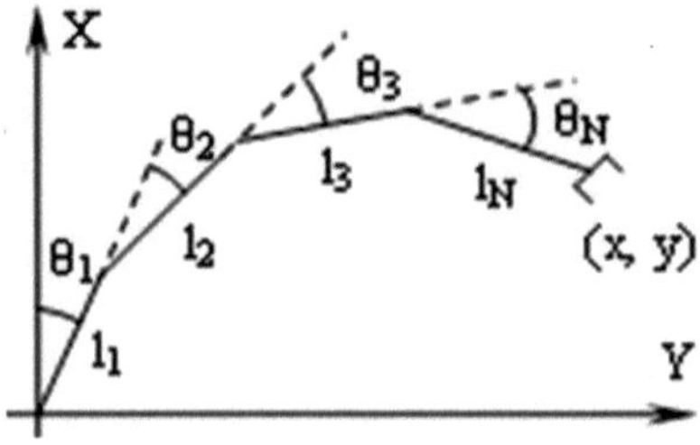

Fig. 4 The approximate model of the hyper-redundant robot arm

$$x = x_n = \sum_{i=1}^{n} l_i \cos\left(\sum_{j=1}^{i} \theta_j\right), \; y = y_n = \sum_{i=1}^{n} l_i \sin\left(\sum_{j=1}^{i} \theta_j\right) \tag{1}$$

The velocities of the x_n and y_n are:

$$\dot{x}_n = -\sum_{i=1}^{n}\left\{ l_i \left[\sin\left(\sum_{j=1}^{i} \theta_j\right)\right]\left(\sum_{j=1}^{i} \dot{\theta}_j\right)\right\} \tag{2}$$

$$\dot{y}_n = \sum_{i=1}^{n}\left\{ l_i \left[\cos\left(\sum_{j=1}^{i} \theta_j\right)\right]\left(\sum_{j=1}^{i} \dot{\theta}_j\right)\right\} \tag{3}$$

The expression of the kinetic energy of segment n is:

$$\begin{aligned} E_{Cn} &= \frac{1}{2} m_n v_n^2 + \frac{1}{2} J_{nz} \omega_n^2 = \frac{1}{2} m_n \left(\dot{x}_n^2 + \dot{y}_n^2\right) + \frac{1}{2} J_{nz} \left(\sum_{i=1}^{n} \dot{\theta}_i\right)^2 \\ &= \frac{1}{2} m_n \sum_{i=1}^{n}\left[l_i^2 \left(\sum_{j=1}^{i} \dot{\theta}_j\right)^2\right] + \frac{1}{2} m_n 2 \sum_{\substack{i=1 \\ k=1 \\ i \neq k}}^{n} \bar{l}_i \bar{l}_k \left(\sum_{j=1}^{i} \dot{\theta}_j \sum_{j=1}^{k} \dot{\theta}_j\right) + \frac{1}{2} J_{nz} \left(\sum_{i=1}^{n} \dot{\theta}_i\right)^2 \end{aligned} \tag{4}$$

The coefficient of the term $\dot{\theta}_p \dot{\theta}_k$ from expression E_{Cn} has the form:

$$K_{pk}^n = J_{nz} + m_n \sum_{\substack{i \geq p \\ j \geq k}}^{n} \bar{l}_i \bar{l}_j = K_{kp}^n \tag{5}$$

where $\bar{l}_i \bar{l}_j$ is the dot product of the vectors which are attached to the segments l_i and l_j. These vectors have the lengths and the orientation of these segments. By (5), the kinetic energy E_{Cn} can be expressed like a matrix product:

$$E_{Cn} = \frac{1}{2}\begin{bmatrix} \dot{\theta}_1 \\ \dot{\theta}_2 \\ \vdots \\ \dot{\theta}_n \end{bmatrix}^T \begin{bmatrix} K_{11}^n & K_{12}^n & \dots & K_{1n}^n \\ K_{21}^n & K_{22}^n & \dots & K_{2n}^n \\ \vdots & \vdots & \vdots & \vdots \\ K_{n1}^n & K_{n2}^n & \dots & K_{nn}^n \end{bmatrix} \begin{bmatrix} \dot{\theta}_1 \\ \dot{\theta}_2 \\ \vdots \\ \dot{\theta}_n \end{bmatrix} \tag{6}$$

Now, the total kinetic energy E_{Ct} is calculated:

$$\begin{aligned} E_{Ct}(q,\dot{q}) &= \frac{1}{2}\dot{q}^T M(q)\dot{q} = \sum_{i=1}^{n} E_{Ci} \\ &= \frac{1}{2}\begin{bmatrix} \dot{\theta}_1 \\ \dot{\theta}_2 \\ \vdots \\ \dot{\theta}_n \end{bmatrix}^T \begin{bmatrix} M_{11} & M_{12} & \dots & M_{1n} \\ M_{21} & M_{22} & \dots & M_{2n} \\ \vdots & \vdots & \vdots & \vdots \\ M_{n1} & M_{n2} & \dots & M_{nn} \end{bmatrix} \begin{bmatrix} \dot{\theta}_1 \\ \dot{\theta}_2 \\ \vdots \\ \dot{\theta}_n \end{bmatrix} \end{aligned} \tag{7}$$

where M(q) is the inertial matrix of the arm and $q = [\theta_1, \theta_2, \ldots. \theta_n]^T$ is the vector of the internal, generalized coordinates. The terms M_{pk} from (7) have the expressions:

$$M_{pk}(q) = \sum_{i=\max(p,k)}^{n} K_{pk}^{i}(q); \quad p = 1\ldots\text{n}, k = 1\ldots\text{n} \tag{8}$$

The expression of the total potential energy E_P is:

$$E_P(q) = \sum_{i=1}^{n} E_{Pi} = g\sum_{i=1}^{n} m_i\left[\sum_{j=1}^{i} l_j \sin\left(\sum_{k=1}^{j}\theta_k\right)\right] \tag{9}$$

Now, one can apply Lagrange's equation:

$$\frac{d}{dt}\left(\frac{\partial L}{\partial \dot{q}_i}\right) - \frac{\partial L}{\partial q_i} = \text{T}_\text{i} \text{ where } q_i \equiv \theta_i \tag{10}$$

where L represents Lagrange's function and has the expression:

$$L(q,\, \dot{q}) = \frac{1}{2}\sum_{i,j=1}^{n} M_{ij}(q)\dot{q}_i\dot{q}_j - E_P(q) \tag{11}$$

The two terms from Lagrange's relations become:

$$\frac{d}{dt}\frac{\partial L}{\partial \dot{q}_i} = \frac{d}{dt}\left(\sum_{j=1}^{n} M_{ij}\dot{q}_j\right) = \sum_{j=1}^{n}\left(M_{ij}\ddot{q}_j + \dot{M}_{ij}\dot{q}_j\right) \tag{12}$$

$$\frac{\partial L}{\partial q_i} = \frac{1}{2}\sum_{j,k=1}^{n} \frac{\partial M_{kj}}{\partial q_i}\dot{q}_k\dot{q}_j - \frac{\partial E_P}{\partial q_i} \quad (13)$$

Developing the derivative of M_{ij}, the relation (10) becomes:

$$\sum_{j=1}^{n} M_{ij}(q)\ddot{q}_j + \sum_{j,k=1}^{n}\left(\frac{\partial M_{ij}}{\partial q_k}\dot{q}_j\dot{q}_k - \frac{1}{2}\frac{\partial M_{kj}}{\partial q_i}\dot{q}_k\dot{q}_j\right) + \frac{\partial E_P}{\partial q_i}(q) = T_i; \quad i = 1\ldots\mathrm{n} \quad (14)$$

Rearranging the terms in (14) one obtains:

$$\sum_{j=1}^{n} M_{ij}(q)\ddot{q}_j + \sum_{j,k=1}^{n} \Gamma_{ijk}\dot{q}_j\dot{q}_k + \frac{\partial E_P}{\partial q_i}(q) = T_i \quad (15)$$

The first term involves the second derivative of the generalized coordinates. The third term is a position term representing the load due to gravity: $\partial\frac{\partial E_P}{\partial q_i}(q) = G(q)$. The quantities Γ_{ijk} are known as Christoffel symbols (of the first kind) [4]:

$$\Gamma_{ijk} = \frac{1}{2}\left(\frac{\partial M_{ij}(q)}{\partial q_k} + \frac{\partial M_{ik}(q)}{\partial q_j} - \frac{\partial M_{kj}(q)}{\partial q_i}\right) \quad (16)$$

The components of the Coriolis matrix are defined as expressed in Eq. (17):

$$C_{ij}(q, \dot{q}) = \sum_{k=1}^{n} \Gamma_{ijk}\dot{q}_k = \frac{1}{2}\sum_{k=1}^{n}\left(\frac{\partial M_{ij}(q)}{\partial q_k} + \frac{\partial M_{ik}(q)}{\partial q_j} - \frac{\partial M_{kj}(q)}{\partial q_i}\right)\dot{q}_k \quad (17)$$

where $C(q, \dot{q})\dot{q}$ is the vector of Coriolis and centrifugal forces. With the above-mentioned notations, the dynamic model of the hyper-redundant robotic structure (represented in Fig. 5) is:

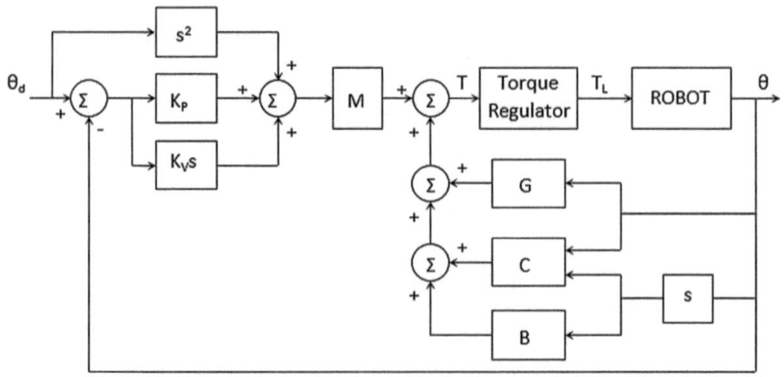

Fig. 5 The block diagram of the adaptive computed-torque controller

$$T = M(\theta)\ddot{\theta} + C\left(\theta,\, \dot{\theta}\right)\dot{\theta} + B\left(\dot{\theta}\right) + G(\theta) \tag{18}$$

3 The Control System

An adaptive control law based on the computed-torque approach (*Computed Torque Joint Servo Technique*—CTJST) is proposed for the robotic arm, ensuring asymptotic trajectory tracking [5–8].

In reality, the dynamic model (18) is not an ideal one due to parametric uncertainties like: unknown parameters (link masses, friction coefficients, inertia moments etc.) or payload disturbances. For the dynamic model (18) a computed-torque controller with some fixed estimated parameters instead of real parameters is adequate [5]:

$$T = \widehat{M}(\theta)\left(\ddot{\theta}_d + K_v\dot{e} + K_p e\right) + \hat{C}\left(\theta,\, \dot{\theta}\right)\dot{\theta} + \hat{B}\left(\dot{\theta}\right) + \hat{G}(\theta), \tag{19}$$

where $\widehat{M} = M + \Delta M$; $\hat{C} = C + \Delta C$; $\hat{B} = B + \Delta B$; $\hat{G} = G + \Delta G$ are estimates, K_v and K_p are control gain matrices, θ_d is desired trajectory and $e = \theta_d - \theta$ is the tracking error. The terms $\widehat{M}\ddot{\theta}_d + \hat{C}\dot{\theta} + \hat{B} + \hat{G}$ provide the torque value necessary to drive the system along its nominal path; the term $\widehat{M}\left(K_v\dot{e} + K_p e\right)$ provides correction torques to reduce any errors in the trajectory of the robot.

An adaptive update rule for adjusting the estimated parameters and the dynamic model (18) can be written as:

$$A\left(\theta, \dot{\theta}, \ddot{\theta}\right)\mu = M(\theta)\ddot{\theta} + C\left(\theta, \dot{\theta}\right)\dot{\theta} + B\left(\dot{\theta}\right) + G(\theta) \tag{20}$$

where $A\left(\theta, \dot{\theta}, \ddot{\theta}\right)$ is an $n \times m$ matrix of time functions known as the *regression matrix* and μ is an $m \times 1$ vector of unknown constant parameters called the *parameter error vector* [5].

In (20) robot dynamics are dissociated in unknown parameters and known time functions. From (18) and (20), the arm dynamics can be written as:

$$T = A\left(\theta, \dot{\theta}, \ddot{\theta}\right)\mu \tag{21}$$

Changing $\ddot{\theta}_d$ with $\ddot{e}$ in (19) and using (20) one obtains:

$$T = \widehat{M}(\theta)\left(\ddot{e} + K_v\dot{e} + K_p e\right) + A\left(\theta, \dot{\theta}, \ddot{\theta}\right)\hat{\mu} \tag{22}$$

where $\hat{\mu}$ is a vector which contains the time-varying estimated sizes of the unknown constant parameters.

Now, based on the previous relations, the tracking error system can be defined:

$$\ddot{e} + K_v\dot{e} + K_p e = \widehat{M}^{-1}(\theta)A\left(\theta, \dot{\theta}, \ddot{\theta}\right)\tilde{\mu} \tag{23}$$

where $\tilde{\mu}$ is the parameter error and

$$\tilde{\mu} = \mu - \hat{\mu} \tag{24}$$

The state-space representation of the tracking error system is:

$$\dot{\boldsymbol{e}} = \boldsymbol{E}\boldsymbol{e} + HM^{-1}(\theta)A\left(\theta, \dot{\theta}, \ddot{\theta}\right)\tilde{\mu} \tag{25}$$

where **e** is tracking error vector and: $\boldsymbol{e} = \begin{bmatrix} e \\ \dot{e} \end{bmatrix}$, $H = \begin{bmatrix} O_n \\ I_n \end{bmatrix}$, $E = \begin{bmatrix} O_n & I_n \\ -K_p & -K_v \end{bmatrix}$

According to [9], the Lyapunov stability analysis demonstrates that the tracking error vector is asymptotically stable if an adequate adaptive update law is used. This adaptive update law for the parameter estimate vector $\hat{\mu}$ is:

$$\dot{\hat{\mu}} = \Gamma A^T(\theta, \dot{\theta}, \ddot{\theta})\widehat{M}^{-1}(\theta)H^T \boldsymbol{P}\boldsymbol{e}, \text{ where } \Gamma = \text{diag}\,(\gamma_1, \gamma_2, \ldots\ldots\gamma_n)$$

where P is a matrix that satisfies the Lyapunov relation:

$$E^T P + PE = -Q,$$

where Q is a positive-definite, symmetric matrix.

It is also proven that the tracking error vector is asymptotically stable [9]; the proposed control method ensures that the vectors $\hat{\mu}$ and $\widehat{M}^{-1}(\theta)$ remain bounded without affecting the tracking error stability [5]. The Lyapunov function is chosen in order to eliminate the requirement for resetting the parameter estimates and the requirement for acceleration measurements in the regression matrix $A\left(\theta, \dot{\theta}, \ddot{\theta}\right)$ [10].

Figure 5 illustrates the block diagram of the adaptive computed-torque controller.

The right part of the system tries to cancel the undesired nonlinear dynamics, while the left part inserts the desired linear dynamics. In the ideal case when the estimates of the robotic arm parameters are exact, the error relation reduces to a linear 2nd-order relation independent of the robotic arm parameters and the non-linear terms are completely cancelled. If the gain matrices K_P and K_V are diagonal,

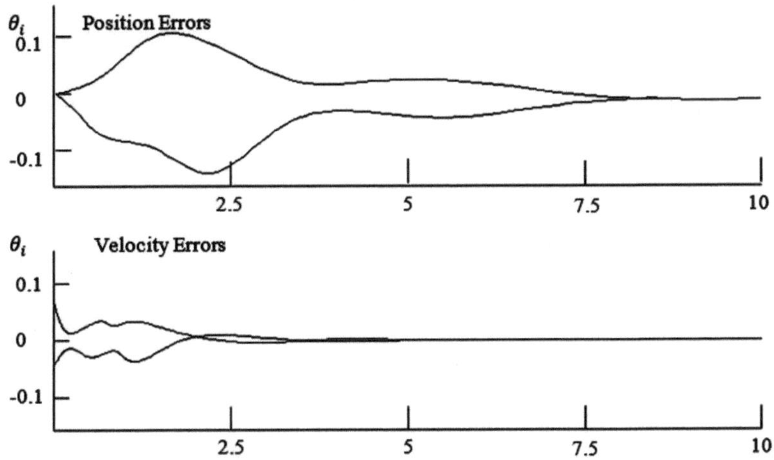

Fig. 6 Position and velocity error

then the closed-loop relations of motion are not only linear, but they are also decoupled one from another.

4 Experimental Results

For experiments, two elements of the robotic arm having the masses $m_1 = 0.04$ kg and $m_2 = 0.035$ kg were considered. The parameters of the adaptive computed-torque controller are: $k_v = 40$, $k_p = 105$, $\gamma_1 = 400$, $\gamma_2 = 400$, $\widehat{m}_1(0) = 0.05$, $\widehat{m}_2(0) = 0.45$.

The position and velocity errors are represented in Fig. 6. They decrease to zero respectively in 7 and 4 s.

5 Conclusion

The combination of actuating cable with electro-pneumatically blocking system represents a new concept for hyper-redundant robot actuating. The major advantage of the sensorial system adopted for HRR is the possibility of measuring simultaneously 3D angular increment and joint error, which are used to estimate the real position.

A dynamic model of the hyper-redundant robotic structure was established using Lagrange's formalism. The motion control system is designed according to the DSMC procedure; the fuzzy rules assure good positioning results for the robot.

The advantage of the adaptive controller is that the accuracy of manipulators carrying unknown loads improves with time because the adaptation mechanism continuously extracts tracking error data. The stability of the tracking error system is ensured by formulating an adaptive update rule. The controller performs well when the load fluctuates.

References

1. Vladu I., Ivănescu M., Vladu, I.C.: Hyper-redundant arm actuation with electro-pneumatic system. Part 1: Construction, Proceedings of 17th Int. IEEE Conference on System Theory, Control and Computing (ICSTCC'17), Sinaia, Romania (2013)
2. Vladu, I., Strîmbeanu, D., Ivănescu, M., Bîzdoacă, N., Vladu, C., Florescu, M.: Control system for a hyper-redundant robot, Proceedings of the IFAC Symposium on Information Control Problems in Manufacturing (INCOM '12), Bucharest, May 25–27 (2012)
3. Vladu I.C., Stoian V., Vladu I., Strîmbeanu, D.: Sensorial system for hyper-redundant arm, Proceedings of 18th International IEEE Conference on System Theory, Control and Computing (ICSTCC'18), Sinaia, Romania, October 17–19 (2014)
4. Murray, R.M., Li, Z., Sastry, S.S.: A Mathematical Introduction to Robotic Manipulation, CRC Press, Boca Raton, Florida, USA, ISBN: 0-8493-7981-4 (1993)
5. Craig, J.: Adaptive control of mechanical manipulators. Addison-Wesley, Reading, MA (1985)
6. Soo, Y.Y.: A robust fuzzy logic controller for robot manipulators. IEEE Trans. Syst. Man Cybernet. **27**(4), 706–713 (1997)
7. Ivanescu, M., Stoian, V.: A sequential distributed variable structure controller for a tentacle arm, Proceedings of IEEE-ICRA'96, Minneapolis, USA, Vol. 4, 3701–3706 (1996)
8. Schilling, R.J.: Fundamentals of robotics. Analysis and Control, Prentice-Hall, New Jersey, USA, ISBN: 0-13-344433-3 (1990)
9. Lewis, F., Abdallah, C., Dawson, D.M.: Control of Robot Manipulators, McMillan Publishing, New York, ISBN: 0-02-370501-9 (1993)
10. Slotine, J., Li, W.: Adaptive strategies in constrained manipulation, Proceedings IEEE International Conference on Robotics and Automation, Raleigh, NC (1985)

Dynamic Model of a Discrete Planar Tentacle Robot

Mihaela Florescu and Mircea Ivanescu

Abstract In this study, the dynamic model of a tentacle robot in two-dimensional space will be presented. We shall attempt to achieve the discretization of this type of robot, and then, by applying a nonlinear observer, the driving control system of the hyper redundant arm will be achieved. Some results obtained by simulating the robot's motion will be presented and compared.

Keywords Tentacle robot · Dynamic model · Nonlinear observer · Control

1 Introduction

In a great number of conventional engineering applications with strong constraints in the workspace of the robot, classic industrial robots do not provide a satisfactory solution. Examples of such unsuitable applications for conventional industrial robotic structures are: reaching certain positions from really narrow spaces, manipulation of objects on long and narrow trails or robot operations in indoor spaces with very low or limited accessibility.

A tentacle robot offers a much better solution for such operations. This is due to the multi redundant structure of this category of robots; because this type of mechanical structure has a theoretically infinite number of degrees of freedom, it allows acceding to the target points within complex operating spaces where a classical robot manipulator cannot have access.

In the specialized literature [1, 2], tentacle robotic systems are known as *continuous robots*, while classical robotic structures are defined as *discrete robots*.

M. Florescu (✉)
Faculty of Mechanics, University of Craiova, Craiova, Romania
e-mail: mihaelaflorescu@yahoo.com; florescu.mihaela@ucv.ro

M. Ivanescu
Faculty of Automation, Computers and Electronics, University of Craiova, Craiova, Romania
e-mail: ivanescu@robotics.ucv.ro

T. Borangiu (ed.), *Advances in Robot Design and Intelligent Control*,
Advances in Intelligent Systems and Computing 371,
DOI 10.1007/978-3-319-21290-6_30

Continuous robots are structures made by serial connection of elements whose rotary, respectively flexural motion is achieved following continuous curves. The category of continuous robots contains models of *tentacle, tube* or *coil type* (that mimics the motion carried out respectively by the octopus, elephant or snake) [3–7]. The particular structure of these models, with a large number of degrees of freedom (theoretically infinite), allows a kinematic linkage described by a continuous curve, which also ensures a very high (infinite) mobility.

The paper is structured as follows: Sect. 2 is devoted to the development of the discrete dynamic model of a tentacle planar robot. Section 3 treats a nonlinear observer for driving the robot and introduces the control laws. The conclusions are presented in the last section.

2 Dynamic Model of a Planar Tentacle Robot

In order to develop the dynamic model of a hyper redundant planar robot, any segment (denoted by i) of the arm, for which the equivalent mass of the higher segments is denoted by M, will be considered as shown in Fig. 1. It is assumed that the tentacle segment can only operate in the vertical plane xOz, having the mass uniformly distributed, with a linear density ρ, where ρ is the equivalent density of the composite materials and the electro-rheological fluid.

The spatial variable of segment's length will be denoted by s, $s \in [0, L]$. The position of a point P [1, 7] on the robot arm is written in the following form

$$x = \int_0^s \cos q' ds', \; z = \int_0^s \sin q' ds' \tag{1}$$

where q is the generalized Lagrange coordinate, in this case the absolute angle, $s' \in [0, s]$, and $q' = q(s')$.

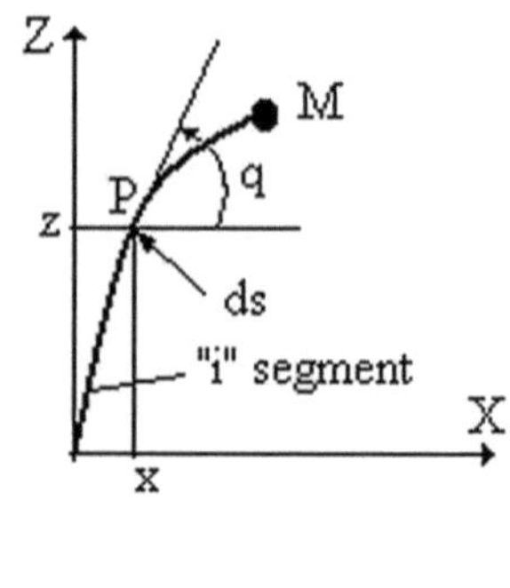

Fig. 1 Representation of *i*th segment of the arm

The generalized coordinate for the system shown in Fig. 1 will have the form

$$q = q(s,t) \tag{2}$$

The corresponding velocities are:

$$v_x = \dot{x} = \int_0^s (-\sin q')\dot{q}'ds', \; v_z = \dot{z} = \int_0^s (\cos q')\dot{q}'ds' \tag{3}$$

Considering an element with the elementary mass dm, the kinetic energy of the distributed segment will be as follows

$$dE_{c1} = \frac{v^2}{2} \cdot dm = \frac{v_x^2 + v_y^2}{2} \cdot dm \tag{4}$$

Taking into account the expressions of speeds, (3), the relation (4) becomes

$$E_{c1} = \frac{1}{2}\int_0^L \rho A \left[\left(\int_0^s (\sin q')\dot{q}'ds' \right)^2 + \left(\int_0^s (\cos q')\dot{q}'ds' \right)^2 \right] ds \tag{5}$$

The kinetic energy of the upper segments of the tentacle structure, which have the equivalent mass M, is determined in an analogous manner, in the form

$$E_{c2} = \frac{1}{2} M \left[\left(\int_0^L (\sin q')\dot{q}'ds' \right)^2 + \left(\int_0^L (\cos q')\dot{q}'ds' \right)^2 \right] \tag{6}$$

The basic potential energy for the distributed mass of the segment will be

$$dE_{p1} = g \cdot z \cdot dm, \tag{7}$$

and the total potential energy will have the form

$$E_{p1} = \rho A \int_0^L \int_0^s \sin q' ds' ds \tag{8}$$

For the upper segments of the hyper redundant arm, the potential energy is

$$E_{p2} = Mg \int_0^L \sin q ds \tag{9}$$

The deformation energy of the segment [1] is expressed by

$$E_{cL} = \int_0^L \frac{D^2}{2} E q^2 ds \tag{10}$$

where D is the cylinder's diameter, and E is the elastic modulus of the material.

The energy corresponding to the fluid's viscosity [2, 5] has the following form:

$$E_v = \int_0^L \eta q \dot{q} ds \tag{11}$$

In order to obtain the dynamic behaviour of the tentacle arm the Lagrange method developed for infinite dimensional systems will be used [1]

$$\frac{d}{dt}\left(\frac{\delta L}{\delta \dot{q}(s)}\right) - \frac{\delta L}{\delta q(s)} = T(s) \tag{12}$$

where L is the Lagrange function,

$$L = E_c + E_{nc} \tag{13}$$

in which E_c and E_{nc} is the mechanical work of the conservative forces and respectively the mechanical work of non-conservative forces.

By replacing the relations (5)–(11) in (12) and by calculating the partial derivatives, the general form of the arm's motion is obtained as

$$\begin{aligned} &\rho A \int_0^s \left(\sin(q-q')\dot{q}'^2 + \cos(q-q')\ddot{q}'\right) ds' + M \int_0^L \begin{pmatrix} \sin(q_L - q')\dot{q}'^2 + \\ + \cos(q_L - q')\ddot{q}' \end{pmatrix} ds' \\ &+ \rho A \int_0^L \cos q ds + Mg \cos q + Er^2 q + k_v \dot{q} = T \end{aligned} \tag{14}$$

where the following notations were made:

$$q' = q(s', t),\ q = q(s, t),\ q_L = q(L, t),\ s' \in [0,\ s],\ s \in [0,\ L] \tag{15}$$

The initial conditions are obtained from the initial state of the tentacle manipulator

$$q(0,s) = q_1^0(s),\ \dot{q}(0,s) = \dot{q}_1^0(s) \tag{16}$$

Equation (14) allows calculating the absolute angle q and other movement parameters such as velocity and acceleration for each point of the arm, at any time, for a control applied appropriately to the arm. Clearly, the difficulty of the problem is determined by the complexity of the equation integral—differential and its nonlinearity.

To obtain a discrete planar simplified model based on the relation (14), a spatial discretization $s_1, s_2, \ldots, s_n$, with $s_i - s_{i-1} = \Delta$ will be used with the restrictions

$$|q(s) - q(s')| < \varepsilon_1,\ |q_L(s) - q(s')| < \varepsilon_2 \tag{17}$$

where measures ε_1 and ε_2 are sufficiently low constants. Thus, one obtains

$$\ddot{q} + B\dot{q} + Cq + D = FT \tag{18}$$

where coefficients B, C, F are $n \times n$ matrices, and D is a $n \times 1$ non-linear vector,

$$B = k_v P^{-1},\ C = Er^2 P^{-1},\ D = P^{-1}Q,\ F = P^{-1} \tag{19}$$

$$P = \begin{bmatrix} (\rho A + M)\Delta & M\Delta & \ldots & M\Delta \\ (\rho A + M)\Delta & (\rho A + M)\Delta & \cdots & M\Delta \\ \vdots & \vdots & \vdots & \vdots \\ (\rho A + M)\Delta & (\rho A + M)\Delta & \cdots & (\rho A + M)\Delta \end{bmatrix} \tag{20}$$

$$Q = \begin{bmatrix} Mg\cos q_1 + \rho A\Delta \sum_{i=1}^{n} \cos q_i \\ Mg\cos q_2 + \rho A\Delta \sum_{i=1}^{n} \cos q_i \\ \vdots \\ Mg\cos q_n + \rho A\Delta \sum_{i=1}^{n} \cos q_i \end{bmatrix} \tag{21}$$

$$q_i = q(s_i),\ q = [q_1,\ q_2,\ \ldots,\ q_n]^T,\ T_i = T(s_i),\ T = [T_1,\ T_2,\ \ldots,\ T_n]^T \tag{22}$$

A linear model can be obtained when in the column vector D all the terms that depend on the force of gravity are neglected, so that Eq. (18) can be written as

$$\ddot{q} + B\dot{q} + Cq = FT \tag{23}$$

where B, C are matrices with variable coefficients determined by the electro-rheological fluid's viscosity and the electric field's intensity. From Eq. (18) a classic model can be obtained using the state vector

$$x = [q, \; q]^T \tag{24}$$

and the matrices

$$H = \begin{bmatrix} O & I \\ -C & -B \end{bmatrix}, \; g = \begin{bmatrix} O \\ -D \end{bmatrix}, \; G = \begin{bmatrix} O \\ F \end{bmatrix} \tag{25}$$

The state equations will be

$$\dot{x} = Hx + g(x) + GT, \tag{26}$$

where $g(x)$ contains the nonlinear terms from the relationship (18).

3 The Nonlinear Observer

The tentacle manipulator control law requires the acknowledgement of state variable $q(s)$, distributed along the length of the arm, with $s \in [0, \; L]$ [6]. To estimate the state variables $q(s)$ with $s \neq 0$ non-accessible, a nonlinear observer has to be found. It will be assumed that only state variable $q(0) = x_1$ can be measured. To develop a model of nonlinear observer, the nonlinear discrete model of a tentacle robot from the relation (26) will be used. The output of the system is

$$y = v^T x, \; v = [1 \quad 0 \quad 0 \quad \ldots \quad 0]^T \tag{27}$$

Theorem Nonlinear observer is defined by the relation

$$\dot{z} = Rz + h(z) + ST + Ky \tag{28}$$

where $z \in \mathbf{R}^{2n}$, R, S, K are $(2n \times 2n)$, $(2n \times 2n)$ and $(2n \times 1)$ constant matrices and $h(z)$ is a $(2n \times 1)$ nonlinear vector. The elements of the observer must verify the conditions

$$z = \Upsilon x, \; R\Upsilon + KV^T = \Upsilon H, \; h(\Upsilon x) = \Upsilon g(x), \; S = \Upsilon G \tag{29}$$

Proof The control system for the "sliding mode" method is proved by Lyapunov technique [2, 6]. A Lyapunov function V is defined as follows

$$V = \frac{1}{2}\sigma^2(e, \dot{e}) \tag{30}$$

where V is a continuous and differential function, with $V(e, \dot{e}) \geq 0$, and $V(e, \dot{e}) = 0$, if and only if $\sigma(e, \dot{e}) = 0$, in which the function $\sigma(e, \dot{e})$ has the form

$$\sigma(e,\dot{e}) = me + \dot{e} = 0 \tag{31}$$

□

The following condition is required

$$\dot{V}(e,\dot{e}) \leq 0 \tag{32}$$

and, taking into account (30) and (31), one obtains

$$\dot{V}(e,\dot{e}) = \sigma(e,\dot{e})\dot{\sigma}(e,\dot{e}) = \sigma(e,\dot{e})(m\dot{e} + \ddot{e}) \leq 0 \tag{33}$$

A reference input, denoted with $r(t)$, sufficiently smooth is considered

$$r(t) = e(t) + q(t) \tag{34}$$

which complies with the following restrictions

$$|r(t)| < \alpha_1,\ |\dot{r}(t)| < \beta_1,\ \ddot{r}(t) = 0 \tag{35}$$

From the relations (24) and (34), with $n = 1$, it is obtained

$$\dot{V}(e,\dot{e}) = \sigma(e,\dot{e})\left[m\dot{e} - \frac{1}{\delta}(T - \eta(\dot{r} - \dot{e}) - c(r - e))\right] \tag{36}$$

where coefficients δ and c are in the form

$$\delta = (\rho A + M)\Delta,\ c = E \cdot \frac{d^2}{4} \tag{37}$$

Relation (36) can be rewritten as

$$\dot{V}(e,\dot{e}) \leq \frac{1}{\delta}\sigma(e,\dot{e})T + |\sigma(e,\dot{e})|\left[|m\dot{e}| + \frac{1}{\delta}|\eta(\dot{r} - \dot{e})| + \frac{c}{\delta}|r - e|\right] \tag{38}$$

The final form of the expression from (38) is:

$$\dot{V}(e,\dot{e}) \leq \frac{1}{\delta}\left(\sigma(e,\dot{e})T + |\sigma(e,\dot{e})|\left[|\dot{e}|\left(m + \frac{\eta}{\delta}\right) + |e|\frac{c}{\delta} + \left(\frac{\eta}{\delta}\beta_1 + \frac{c}{\delta}\alpha_1\right)\right]\right) \tag{39}$$

A control law of the form in Eq. (40) below is selected:

$$T = -[k_1 \mathrm{sgn} e + k_2 \mathrm{sgn}(\sigma\dot{e})\dot{e} + k_3 \mathrm{sgn}(\sigma e)e] \tag{40}$$

By replacing (40) in (39) one obtains

$$\dot{V}(e,\dot{e}) \leq \frac{1}{\delta}(-k_1 + \eta\beta_1 + c\alpha_1)|\sigma| + \frac{1}{\delta}(-k_3 + c)|\sigma e| + \left(-\frac{k_2}{\delta} + m + \frac{\eta}{\delta}\right)|\sigma\dot{e}| \quad (41)$$

When the amplification factors of the controller, k_1, k_2, k_3, verify the restrictions

$$k_1 > \eta\beta_1 + c\alpha_1,\ k_2 > \delta\left(m + \frac{\eta}{\delta}\right),\ k_3 > c, \quad (42)$$

the function $\dot{V}(e,\dot{e})$ will be negative, and relation (41) can be rewritten as follows:

$$\dot{V}(e,\dot{e}) \leq \frac{1}{\delta}\left(-k_1^* + \eta\beta_1 + c\alpha_1\right)|\sigma(e,\dot{e})| = -\gamma|\sigma(e,\dot{e})| \quad (43)$$

where the factor k_1^* verifies the corresponding relation from (42). The switching time [1, 3] can be calculated as follows

$$t_s = \frac{1}{\gamma}|\sigma(e(0),\dot{e}(0))| \quad (44)$$

Onward, some results obtained by simulation are presented. A tentacle arm with two elements was considered, whose main parameters are shown in Table 1. The manipulator's performances are tested using a controller with a variable structure.

First, the nonlinear observer is tested. The observer's convergence for this case $(n = 2)$ is easily obtained and the model's matrices are of the form

$$R = \begin{bmatrix} -1 & 0 & 0 & -1 \\ 1{,}01 & -1 & 0 & -1 \\ 1{,}48 & 0 & -1 & -1 \\ 1{,}15 & 0 & 0 & -1 \end{bmatrix},\ K = \begin{bmatrix} 1 \\ -1 \\ -1 \\ -1 \end{bmatrix} \quad (45)$$

Figure 2 illustrates the motion performed by the first segment of the tentacle, respectively $q(1)$, between $2\pi/5$ and $\pi/2$. The estimates for nonlinear observer were shown in solid line under the conditions

$$z_2(0) = q^0(1) = 0,\ z_4(0) = q^0(1) = 0 \quad (46)$$

Next, a "bang–bang" type controller is tested, defined by

$$T = -[k_1\mathrm{sgn}(e) + k_2\mathrm{sgn}(\sigma\dot{e})\dot{e} + k_3\mathrm{sgn}(\sigma e)e] \quad (47)$$

Table 1 Parameters of the tentacle robot with 2 segments

Segment	Length (m)	Diameter (m)	ω_n	ξ_{init}
1	0.12	0.014	110	0.7
2	0.08	0.010	80	0.65

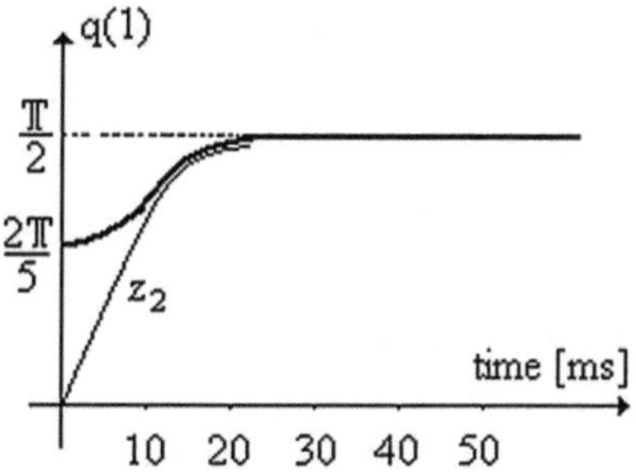

Fig. 2 Motion performed by the robot between $2\pi/5$ and $\pi/2$

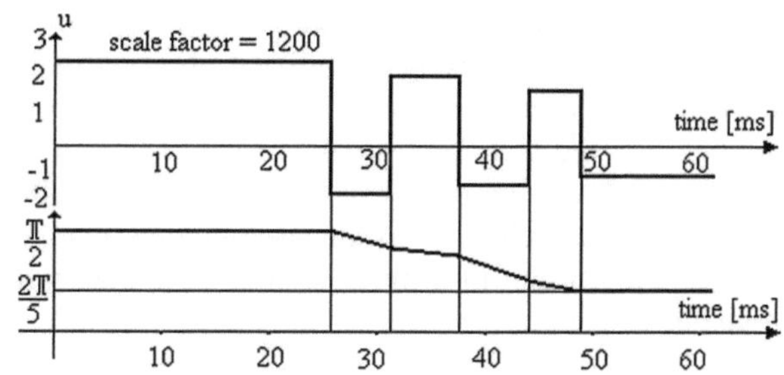

Fig. 3 Evolution of the control law u and coordinate of motion $q(1)$

A switching line for $m = 2$ was chosen. The control law u and the evolution of coordinate $q(1)$ are shown in Fig. 3.

A direct control law was used. The initial conditions of the proposed trajectory are:

$$q_1(0) = 2\pi/5,\ \dot{q}_1(0) = 0,\ q_2(0) = 2\pi/5,\ \dot{q}_2(0) = 0 \tag{48}$$

The switching times are obtained from (44) with values $t_{s1} \leq 0,026s$, $t_{s2} \leq 0,017s$

Figure 4 shows simulation results of the control system for two different types of control. The curve a in Fig. 4 is the system's trajectory for a gradual input with no

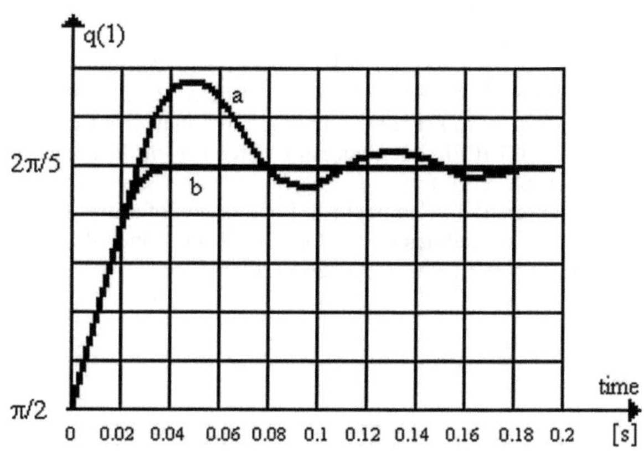

Fig. 4 The system's response to a gradual entry for different control laws: **a** control without variable structure, **b** control with variable structure

variable control structure. The second curve in Fig. 4 includes two domains, the first one having a damping coefficient $\xi = 0,45$, and the second one - the damping coefficient $\xi = 1,45$.

4 Conclusions

This paper has developed the dynamic model of a hyper redundant planar robot based on the energy relations of its arm. Then, a discretization of the model was performed by introducing a spatial meshing specified in the paper. The robot control was obtained using a nonlinear observer defined by the theorem presented. Some results describe the robot's simulated motion between the proposed limits. It can be observed that the system's response to a gradual entry for different control laws is good, but the damping coefficients have different values depending on the type of control applied. Thus, in case of the variable structure control, the robot arm reaches equilibrium in a much shorter time than when a control without variable structure is used.

Acknowledgments This work was supported by the strategic grant POSDRU/159/1.5/S/133255, Project ID 133255 (2014), co-financed by the European Social Fund within the Sectorial Operational Program Human Resources Development 2007–2013.

References

1. Ivanescu M., et al: Non conventional robot control systems, Universitaria Craiova, ISBN 973-8043-147-X (2002)
2. Ivanescu M., Florescu M.C., Popescu N., Popescu, D: Stability control of a hyper redundant arm for a grasping operation, International Conference Vibroengineering, Kaunas, 9–11 (2008)
3. Matsuno, F., Suenaga, K.: Control of redundant 3D snake robot based on kinematic model, Proceedings of the 2003 IEEE International Conference on Robotics and Automation, Taipei, Taiwan, September 14–19, 2061-2066, ISBN 0-7803-7736-2 (2003)
4. Hirose, S.: Biologically Inspired Robots (Snake-like Locomotor and Manipulator), Oxford University Press (1993)
5. Hirose, S., Fukushima, E.F.: Snakes and Strings: New Robotic Components for Rescue Operations, Tokyo Institute of Technology, 2-12-1 Ookayama Meguro-ku, Japan (2000)
6. Jones, B.A., Walker, I.D.: Limiting-case analysis of continuum trunk kinematics, Proceedings of the 2003 IEEE International Conference on Robotics & Automation, Rome, April 10–14, 1363–1368 (2007)
7. Christensen, D.J., Campbell, J.: Locomotion of miniature catom chains: scale effects on gait and velocity, Proceedings of the 2003 IEEE International Conference on Robotics and Automation, Rome,, April 10–14, 2254-2260 (2007)

How to Use 3D Printing for Feasibility Check of Mechanism Design

Marco Ceccarelli, Giuseppe Carbone, Daniele Cafolla and Mingfeng Wang

Abstract In this paper, 3D printing is presented as useful means for checking design feasibility of mechanism structures for robots. A procedure is outlined for rapid prototyping that can produce scaled prototypes for experimental validation since early stages of robot developments. An example from LARM activities shows the soundness and practical implementation of the proposed method.

Keywords Robot design · Prototyping · 3D printing · Experimental validation

1 Introduction

Prototype construction and validation testing is a fundamental activity in machine developments and particularly for robot designs. This activity is often time consuming and expensive, although necessary to achieve a proper final design as pointed out in [3]. Thus, before final production, it is required to work with prototypes that can be used for design checks and performance testing. Since early 1990s the technique of Rapid Prototyping has been developed by using scaled prototypes with easy manufacturing materials. This has led to the development of 3D printing whose practical feasibility has reached maturity only in the late 2000s. In recent years new solutions have been proposed and are available in the market

M. Ceccarelli (✉) · G. Carbone · D. Cafolla · M. Wang
LARM: Laboratory of Robotics and Mechatronics—DICEM, University of Cassino and South Latium, Via Di Biasio 43, 03043 Cassino, FR, Italy
e-mail: ceccarelli@unicas.it

G. Carbone
e-mail: carbone@unicas.it

D. Cafolla
e-mail: cafolla@unicas.it

M. Wang
e-mail: wang@unicas.it

T. Borangiu (ed.), *Advances in Robot Design and Intelligent Control*,
Advances in Intelligent Systems and Computing 371,
DOI 10.1007/978-3-319-21290-6_31

with fairly cheap printers and procedures. Thus, Rapid Prototyping with 3D printing can be considered even at early stages of the product designs.

3D printing as additive manufacturing is a process of making three dimensional solid objects from a digital file. A 3D printed object is created by laying down successive layers of material until the entire object is created. Each of these layers can be seen as a thin sliced horizontal cross-section of the eventual object. Several different 3D printing processes have been invented since the late 1970s. The printers were originally large, expensive, and highly limited in what they could produce [1]. Today a large number of additive processes are available. The main differences between processes are in the way that layers are deposited to create parts and in the materials that are used. Some of the available methods are based on melting or softening a material to produce layers, e.g. selective laser melting (SLM) or direct metal laser sintering (DMLS), selective laser sintering (SLS), fused deposition modelling (FDM), while others cure liquid materials using different sophisticated technologies such as stereo lithography (SLA). In laminated object manufacturing (LOM), thin layers are cut to shape and joined together (e.g. paper, polymer or metal). Each method has its own advantages and drawbacks. 3D printer manufacturers usually offer a choice between powder and polymer as material to print an object [8, 10, 11]. Some 3D printers can even use off-the-shelf paper sheets as material to produce a durable prototype. The main considerations for choosing a 3D printer are usually speed, cost of the printer, cost and choices of materials, and colour capabilities [5].

In this paper, the problem of using 3D printing with speedy low-cost procedure is proposed for rapid prototyping of robot mechanisms; the method can be used since early stage of mechanical design to check the feasibility of the mechanism solution regarding its functionality and feasibility.

2 Rapid Prototyping via 3D Printing

Rapid Prototyping (RP) is a term used to describe a variety of processes which are aimed at quickly creating three-dimensional physical parts from virtual 3D models by using automated printing machines. The parts are built from a 3D CAD model and match a model within the precision limits of the chosen printing process.

All RP processes are additive, since parts are built by adding layer by layer, depositing, or solidifying one or more material in a horizontal layer process [8]. In addition to additive production processes, subtractive processes such as CNC (Computer Numerical Control) machining and laser cutting can be considered for the optimization of printed prototypes by removing burrs and errors due to the chosen additive process and technology [7, 14].

3D printable models may be created with a computer aided design (CAD) package or via a 3D scanner or via a plain digital camera and photogrammetry software. The manual modelling process of preparing geometric data for 3D computer graphics is similar to plastic arts such as sculpting.

Regardless of the 3D modelling software used, a 3D model (in .skp, .dae, .3ds or similar other formats) needs to be converted to either a .STL or an .OBJ format, to allow the printing software to read it. Before printing a 3D model it must first be examined on "manifold errors". This activity is usually called the "fixup". Especially STL's that have been produced from a model obtained through 3D scanning often have many manifold errors that need to be fixed. Examples of manifold errors are surfaces that do not connect to form a solid body, gaps in the models, etc. Examples of software products that can be used to fix these errors are netfabb and Meshmixer, or even Cura or Slic3r [2, 9].

Once created, the .STL file needs to be processed by software called a "slicer". This software converts the model into a series of thin layers and produces a G-code file containing instructions tailored to a specific type of 3D printer (FDM printers). This G-code file can then be loaded on 3D printing client software, which loads the G-code, and uses it to instruct the 3D printer during the 3D printing process. It should be noted that often the client software and slicer are combined into one software program. Several open source slicer programs exist, including Skeinforge, Slic3r, and Cura as well as closed source programs including Simplify3D and KISSlicer. Examples of 3D printing clients include Repetier-Host, ReplicatorG, Printrun/Pronterface. There is one other piece of software that is often used by people using 3D printing, namely a G-Code viewer. This software lets one examine the operation path of the printer nozzle. By examining this, the user can decide to modify the G-Code to print the model in a different way, for example by selecting a different pose (position/orientation) of the object to be printed within the 3D printer workspace. On this matter one should note that the extruder's initial position and path will also affect the amount of needed support material. Examples of G-Code viewers are G-code Viewer for Blender and Pleasant3D.

A 3D printer follows the G-code instructions to lay down successive layers of liquid, powder, paper or sheet material to build the model from a series of cross sections. These layers correspond to the virtual cross sections from the CAD model, are joined or automatically fused to create the final shape. The primary advantage of this technique is its ability to create almost any shape or geometric feature including shapes that could not be feasibly manufactured by using subtractive techniques.

The printer resolution describes layer thickness and X-Y resolution in dots per inch (dpi) or micrometres (μm). Typical layer thickness is around 100 μm (250 DPI), although some machines such as the Objet Connex series and 3D Systems' ProJet series can print layers as thin as 16 μm (1600 DPI), where X-Y resolution is comparable to that of laser printers with the particles (3D dots) around 50–100 μm (510–250 DPI) in diameter [4, 6].

3D printing of a model with contemporary methods can take from several hours to several days, depending on the used method, size and complexity of the model. Additive systems can typically reduce this time to a few hours, although it varies widely depending on the type of machine, size and number of models being produced simultaneously.

Traditional techniques like injection moulding can be less expensive for manufacturing polymer products in high quantities, but additive manufacturing can be faster, more flexible and less expensive when producing relatively small quantities of parts. 3D printers give designers and concept development teams the ability to produce parts and concept models by using a desktop size printer. The printer-produced resolution is usually sufficient for many applications. However, for an improved surface finishing one can print a slightly bigger prototype and refine/polish it afterwards by using a higher resolution subtractive process [12].

Some additive manufacturing techniques are capable of using multiple materials within the same printing process. Some printers are able to print in multiple colours and colour combinations simultaneously. Support materials are removable or dissolvable upon completion of the print, and are used to support overhanging features during construction. No matter which approach a 3-D printer uses, the overall printing process is basically the same.

A procedure can be outlined for a suitable efficient 3D printing of mechanism structures through the flowchart outlined in Fig. 1.

- **Step 1**: *CAD Modelling*—Build a 3D model by using computer-aided design (CAD) software. The software may provide some hints on the structural integrity you can expect in the finished product by using data about certain materials in virtual simulations of how the object will behave under certain conditions.
- **Step 2**: *Conversion to STL*—Convert the CAD drawing to STL format. STL (Standard Tessellation Language) is a file format developed for 3D Systems in 1987 for use by stereo lithography apparatus (SLA) machines.
- **Step 3**: *Transfer to AM* (Automatic Machine)—A user copies the STL file to the computer that controls the 3D printer. There, the user can designate the size and orientation for printing choosing the best orientation.
- **Step 4**: *Parameter setup of the printer*—Printer requirements include refilling the polymers, binders and other consumables the printer will use. It also covers adding a tray to serve as a solid support base or adding the material to build temporary water-soluble supports. It is necessary to check if the parameters of the printer such as the temperature of the extruder and the platform are compatible with the chosen material and to check whether the dimensions of the designed prototype are compatible with the printer characteristic and working space.
- **Step 5**: *3D printing of elements*—The printing process is mostly automatic. Each layer is usually about 0.1 mm thick, although in some cases it can be set thinner or thicker than this value. If the quality of printing is not satisfactory, the parameters should be changed or the CAD model should be modified to adapt the printing object to the 3D printer capabilities.
- **Step 6**: *Polishing and assembly*—Safety precautions are necessary in removing the printed object from the printer platform in order to avoid injury by using wearing gloves to protect from hot surfaces or toxic chemical. Activities for polishing and assembly are required for the printed object to remove any debris or waste as well as to remove any support material.
- **Step 7**: *3D Prototype*—The prototype is ready.

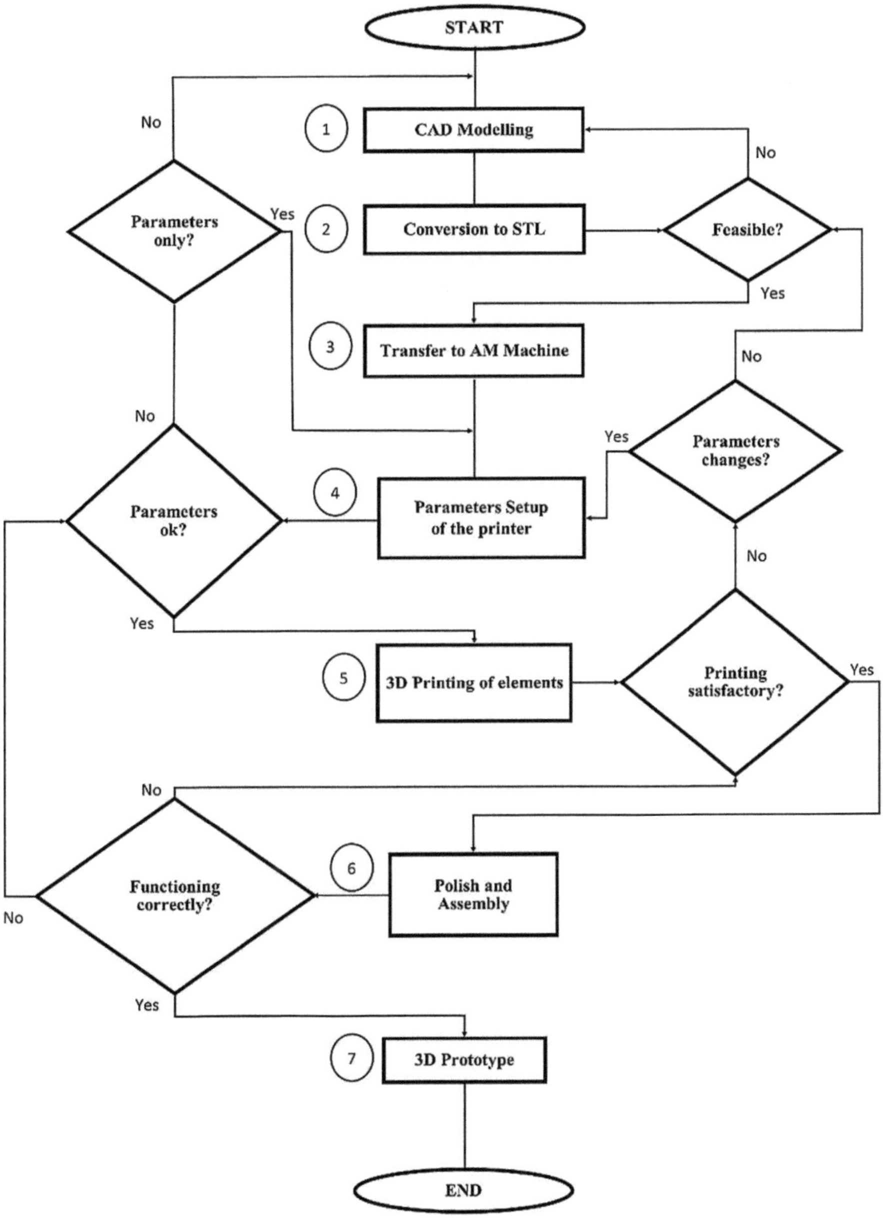

Fig. 1 A scheme of a 3D printing procedure for rapid prototyping of mechanisms

The above proposed procedure has been thought with sequential phases in order to have checks during the process for adjustments and optimizations even in the mechanical design of parts. The proposed procedure can also consider activities in a parallel sequence for carrying out further design optimization and numerical simulations of the design improvements while printing a first prototype.

3 A Case Study: LARM Tripod Locomotor

LARM tripod Locomotor is a biped locomotor based on parallel leg mechanisms [13]. The biped locomotor consists of two 3-DOF leg mechanisms and a waist. Two leg mechanisms are installed on the waist, and between the waist and feet there are two tripod mechanisms of six identical prismatic pairs with U-joints at each end.

A 3D model has been elaborated in SolidWorks® environment and the corresponding 3D prototype has been built by printing all the components and assembling them with commercial universal joints (U-joints), as shown in Fig. 2.

The 3D printer used in this case is FlashForge Creator [13], whose specification parameters are listed in Table 1. Based on the 3D CAD model, the corresponding STL format file of each component can be directly generated in SolidWorks® environment. By using the required RP program, Replicator G, and considering machine parameters like temperature, building size, printing precision and printing speed, the proper G-code can be generated and transferred to the 3D printer. G-Code is the name for the most widely used numerical control programming language. It is defined by instructions on where to move, motion speed, and motion

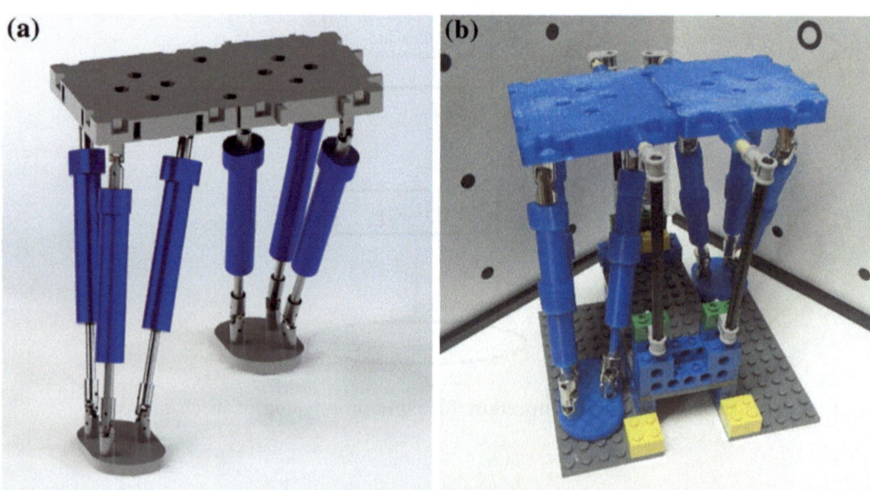

Fig. 2 The LARM tripod locomotor: **a** a 3D model in SolidWorks® environment; **b** 3D printed model

Table 1 Specification parameters of the 3D printer creator [7]

Extruder model	*Printing material*	*Software*	*Input file format*	*Compatibility*
Dual-extruder	ABS&PLA 1.75–1.8 mm	Replicator G	STL/X3G	Linux, Mac and Windows
Machine dimension	*Build size*	*Print precision*	*Positioning precision*	
320 × 467 × 381 mm	225 × 145 × 150 mm	0.1–0.2 mm	0.0025 mm on Z axis	
			0.011 mm on XY axes	
Flow velocity	*Layer thickness*	*Extruder temperature*	*Heated build platform temperature*	
Approx. 24 cc/h	0.1–0.5 mm	Maximum 230 °C	Maximum 120 °C	

path. To allow the printer to print offline disconnecting it from the computer the G-code has been converted to x3g format and transferred to a SD Card of the printer. X3g is a binary file that the machine reads, and which contains all the instructions for printing.

During parameter setting in Replicator G, the first two factors that should be considered are the workspace size of the 3D printer and the temperatures of extruder(s) and support platform. In the proposed case study, each component in the 3D CAD model has been scaled down to 25 % of the corresponding 3D CAD model size in order to fit with the printer workspace. The temperatures of extruder and support platform are set as 220 and 79 °C, respectively, since PLA has been chosen as printing material. Accordingly, travel feed rate has been set to 55 mm/s and flow rate has been set equal to 40 mm/s. The interior of the piece was filled with a rhomboidal grid with an infill of 10 %, an infill perimeter overlap ratio of 0.3 and an infill solidity of 0.1. In addition, printing orientation based on the Z axis of the 3D printer is also a key factor that affects printing quality. Generally, it is recommended to choose printing orientation based on the principle that a relative large contact area provides more support material on the support platform. However, when there is requirement of coaxially components during the assembly, the axial direction should be considered as the preferred printing orientation. For instance, each prismatic pair of the LARM Tripod locomotor consists of an upper hollow cylinder part, a lower hollow cylinder part, a ring connector and the rod part, as shown in Fig. 3. The tube is divided into two parts by considering verticality along the axial direction and convenient assembly with the rod. In order to choose the proper printing orientation, the lower tube part has been printed in two directions (Fig. 4). The print precision and the part's stiffness are is much higher when printing along the axial direction rather than along the radial one.

When all the components are printed, the next step is to polish and assemble them. In order to achieve simple polishing and convenient assembly, the support material is minimized during the parameter setting. The size of each pin is designed as being 0.2 mm smaller than the corresponding hole by considering the printing precision to avoid material overlapping. Once all the parts have been ready and

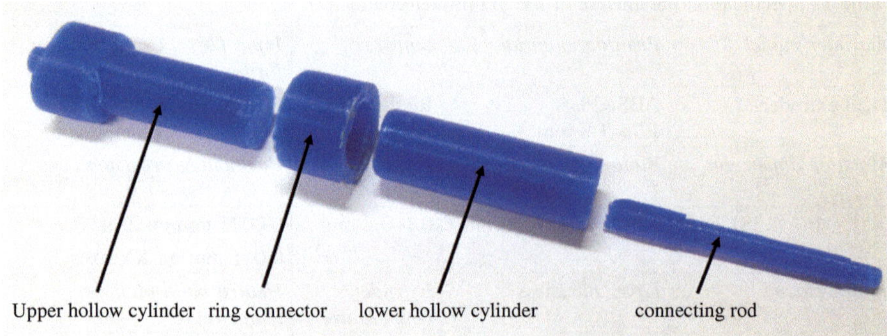

Fig. 3 A printed prismatic pair for the LARM tripod locomotor in Fig. 2

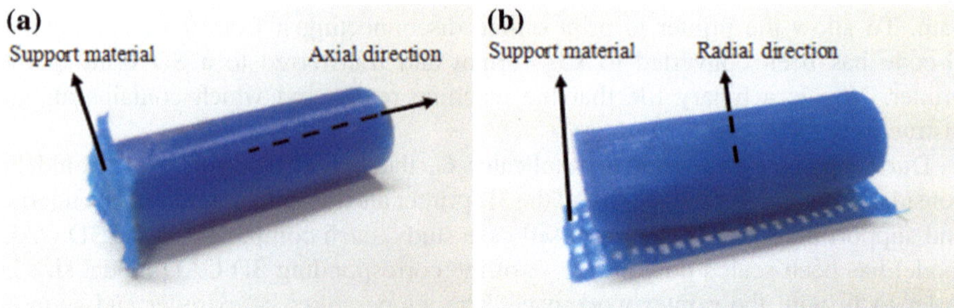

Fig. 4 The tube part of a prismatic pair when printed: **a** along the axial direction; **b** along the radial direction

Fig. 5 Surface finishing of a leg with prismatic joint

adjusted, they are assembled to obtain the 3D printed prototype of Fig. 2. Thus it was possible to check the proper mobility of the robot structure and to preliminarily validate motion efficiency. In particular, the motion range has been easily detected.

The 3D printed prototype is made from parts that can be considered as combination of main elements in the mechanical design. The parts' functionality can be evaluated by manual inspection, and, with proper actuators, even the mobility both in terms of motion ranges and constrains. The printed examples in Figs. 2, 3 and 4 show some advantages that include a better understanding of design and operation issues both on individual parts and of the entire product in a first experimental validation of the feasibility of the designed system. The surface finishing depicted in Fig. 5 shows that the printing procedure was executed correctly and the 3D printing is satisfactory.

4 Conclusions

In this paper 3D printing is recognized as a useful means in the design process with rapid prototyping of robot mechanisms through a procedure that permits a feasibility check of solutions since early stages of robot developments. Specific experiences at LARM are presented, being related to on-going development of LARM tripod Locomotor. The practical convenience and soundness of using the procedure of 3D printing prototyping for specific designs of mechanisms in humanoid robots were proved.

References

1. Amon, C.H., Beuth, J.L., Weiss, L.E., Merz, R., Prinz, F.B.: Shape deposition manufacturing with microcasting: processing, thermal and mechanical issues. J. Manuf. Sci. Eng. **120**(3) (2014)
2. Beck, J.E., Fritz, B., Siewiorek, D., Weiss, L.: Manufacturing mechatronics using thermal spray shape deposition. In: Proceedings of solid freeform fabrication symposium (1992)
3. Ceccarelli, M.: Experimental mechanics for mechanism design: an illustrated summary. In: EUROMECH colloquium 515 advanced applications and perspectives of multibody system dynamics, Blagoevgrad, paper EUM515-L1 (2010)
4. Chua, C.K., Leong, K.F., Lim, C.S.: Rapid prototyping, p. 124. World Scientific, Singapore (2003)
5. Deckard C.: Method and apparatus for producing parts by selective sintering. U.S. Patent 4863538 (1989)
6. Excell, J.: The rise of additive manufacturing. The Engineer. Retrieved (2013)
7. Flashforge 3D Printer. 3D Printer Creator. http://www.ff3dp.com/#!creator/c1txh (2015)
8. Freedman, D.H.: Layer by layer. Technol Rev **115**(1):50–53 (2012)
9. Housholder, R.: Moulding process. U.S. Patent 4247508 (1981)
10. Pham, D.T., Dimov, S.S.: Rapid manufacturing, p. 6. Springer, London (2001)
11. Prinz, F.B., Merz, R., Weiss, L., Ikawa, N.: Building parts you could not build before. In: Proceedings of 8th international conference on production engineering, pp. 40–44, Chapman & Hall, London (1997)
12. Sherman, L.M.: 3D printers lead growth of rapid prototyping. Plast. Technol. **50**(8), 43–46 (2004)
13. Wang, M.F., Ceccarelli, M.: Design and simulation of walking operation of a cassino biped locomotor. New Trend Mech. Mach. Sci. **24**, 613–621 (2015)
14. Wittbrodt, B.T., Glover A.G., Laureto, J., Anzalone, G.C., Oppliger, D., Irwin, J.L., Pearce, J. M.: Life-cycle economic analysis of distributed manufacturing with open-source 3-D printers. Mechatronics **23**(6):713 (2013)

Gibbs-Appell Equations of Motion for a Three Link Robot with MATLAB

Dan B. Marghitu and Dorian Cojocaru

Abstract The present study uses MATLAB as a tool to develop and to solve the dynamical equations of motion for an open kinematic chain. MATLAB is convenient for finding the equations of motion using Lagrange method and for solving numerically the nonlinear differential equations.

Keywords Three link robot · Equations of motion · MATLAB

1 Introduction

In this paper, MATLAB is considered for solving the mathematical equations of a three-dimensional (3D) robot with three links. The kinematic and dynamic equations of motion for the robot are deduced using a symbolical algorithm. The equations of motion are solved numerically using designed MATLAB functions.

A form of Lagrange's equations in terms of quasi-coordinates used for robots is presented in [1]. This formulation is suitable for symbolic computation and dynamic analysis of complex systems. According to Whittaker [2], the concept of quasi-coordinates was used by Lagrange and Euler, and other advanced texts in dynamics include sections on the subject [3, 4]. Baruh [5] described the equations of analytic dynamics and showed a classification based on vector principles (for example, D'Alembert's principle) or on scalar principles (for example, Hamilton's equations). Advantages and disadvantages of various approaches are analyzed. The equations of motion are the projection of the classical equations of force and

D.B. Marghitu (✉)
Auburn University, Auburn, AL 36849, USA
e-mail: marghitu@auburn.edu

D. Cojocaru
Faculty of Automatic Control, Computers and Electronics University of Craiova, Craiova, Romania
e-mail: cojocaru@robotics.ucv.ro

T. Borangiu (ed.), *Advances in Robot Design and Intelligent Control*,
Advances in Intelligent Systems and Computing 371,
DOI 10.1007/978-3-319-21290-6_32

moment balances along directions affected by the kinematic coefficients. The equations of motion for a human body model are developed in [6], using the principles of classical mechanics. The solution provides displacement and rotation of the body. Some example of motions are: simple lifting on the earth and the moon, underwater swimming, or a vertically suspended man. The achievements in the robot dynamics research starting with the development of the recursive Newton-Euler algorithm are given in [7]. The main algorithms and equations are provided, where the Kane's equations [8] can be of great value. Pan and Sharp [9] used homogeneous transformation matrices with Denavit–Hartenberg notation and the Lagrangian formulation method for a general computer program. An algebraic manipulation language, REDUCE, is employed for the open-chain structure. Equations of motion in scalar form can be automatically transferred to FORTRAN format for later numerical simulations. A recursive approach for solving snake-like robots' problems using Kane's equations is presented in [10]. The proposed method is used to derive kinematic and dynamic equations recursively. The Lagrangian method is used for the dynamics of a 3-bar deformable structure that can adapt to unconstrained environments [11]. A robot that combines the hopping movement and the wheeling movement is presented in [12]. The five-shank hopping structure was analyzed, and a non-linear spring-mass model for the robot was used. A unified method for the whole-body humanoid robot control problem in the context of multi-point multi-link contacts, constraints, and obstacles is presented in [13]. Gomes and Ruina [14] developed equations of rigid-body mechanics for an ape model with zero energy cost motions. The results are comparable with the motions of real gibbons using a 5-link model.

This paper describes the systematic computation of the equations of motion for a three link open chain using MATLAB. The software combines symbolical and numerical computations and can be applied to find and solve the equations of motions for robots, humans, and animals systems.

2 Design and Implementation

The study presented in this section is based on a three-link robot. Figure 1 is a schematic representation of a Rotation Rotation Rotation (RRR) open kinematic chain or robot arm consisting of three links 1, 2, and 3.

Let m_1, m_2, m_3 be the masses of links 1, 2, 3, respectively. Link 1 can be rotated at origin O in a "fixed" reference frame 0 (RF0) of unit vectors $[\mathbf{i}_0, \mathbf{j}_0, \mathbf{k}_0]$ about a vertical axis $\mathbf{k}_0$. The unit vector $\mathbf{k}_0$ is fixed in link 1. The link 1 is connected to link 2 at the revolute joint at point A. The element 2 rotates relative to link 1 about an axis fixed in both 1 and 2, passing through A, and perpendicular to the axis of link 1. The last link 3 is connected to 2 by means of a revolute joint at B. The mass centers of links 1, 2, and 3 are C_1, C_2, and C_3, respectively. The length of link 1 is $OA = L_1$, the length of link 2 is $AB = L_2$, and the length of link 3 is $BD = L_3$. The distance from O to C_1 is $OC_1 = L_1/2$, the distance from A to C_2 is $AC_2 = L_2/2$,

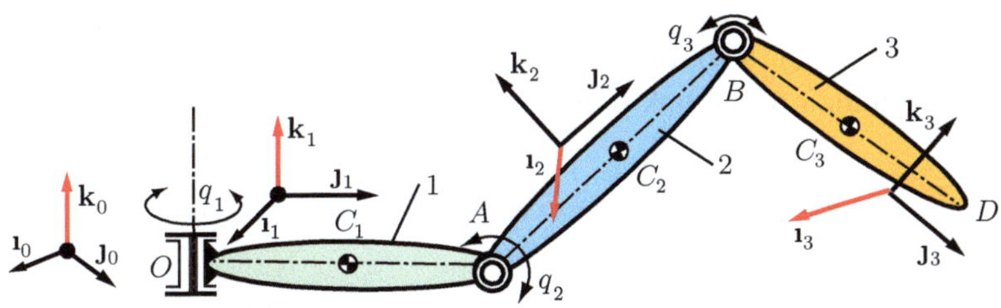

Fig. 1 RRR robot

and the distance from B to C_3 is $BC_3 = L_3/2$. The reference frame p (RFp) of unit vectors $[\mathbf{i}_p, \mathbf{j}_p, \mathbf{k}_p]$ is attached to link $p, p = 1, 2, 3$, as shown in Fig. 1.

2.1 Kinematics

The generalized coordinates (quantities associated with the the instantaneous position of the system) are $q_1(t)$, $q_2(t)$, $q_3(t)$. The first generalized coordinate q_1 denotes the radian measure of the angle between the axes of RF1 and RF0. The unit vectors $\mathbf{i}_1$, $\mathbf{j}_1$, and $\mathbf{k}_1$ can be expressed as functions of $\mathbf{i}_0$, $\mathbf{j}_0$, and $\mathbf{k}_0$

$$\begin{bmatrix} \mathbf{i}_1 \\ \mathbf{j}_1 \\ \mathbf{k}_1 \end{bmatrix} = \begin{bmatrix} c_1 & s_1 & 0 \\ -s_1 & c_1 & 0 \\ 0 & 0 & 1 \end{bmatrix} \begin{bmatrix} \mathbf{i}_0 \\ \mathbf{j}_0 \\ \mathbf{k}_0 \end{bmatrix} = R_{10} \begin{bmatrix} \mathbf{i}_0 \\ \mathbf{j}_0 \\ \mathbf{k}_0 \end{bmatrix}. \quad (1)$$

where $s_1 = \sin q_1$ and $c_1 = \cos q_1$. The transformation matrix from RF1 to RF0 is R_{10} and the transformation matrix from RF1 to RF0 is $R_{01} = R_{10}^{\mathrm{T}}$ where R_{10}^{T} is the transpose of the matrix R_{10}. The second generalized coordinate also designates a radian measure of the rotation angle between RF1 and RF2. The unit vectors $\mathbf{i}_2$, $\mathbf{i}_2$ and $\mathbf{k}_2$ can be expressed as $\mathbf{i}_2 = \mathbf{i}_1, \mathbf{j}_2 = c_2\mathbf{j}_1 + s_2\mathbf{k}_1, \mathbf{k}_2 = -s_2\mathbf{j}_1 + c_2\mathbf{k}_1$, where $s_2 = \sin q_2$ and $c_2 = \cos q_2$. The transformation matrices from RF2 to RF1 and from RF3 to RF2 (with $s_3 = \sin q_3$ and $c_3 = \cos q_3$) are

$$R_{21} = \begin{bmatrix} 1 & 0 & 0 \\ c_2 & s_2 & 0 \\ -s_2 & c_2 & 0 \end{bmatrix} \quad \text{and} \quad R_{32} = \begin{bmatrix} 1 & 0 & 0 \\ c_3 & s_3 & 0 \\ -s_3 & c_3 & 0 \end{bmatrix}. \quad (2)$$

Next, the angular velocity of the links 1, 2, and 3 will be expressed in the fixed reference frame, RF0. The angular velocity of 1 in RF0 expressed in terms of RF1 is $\omega_{10} = \dot{q}_1 \mathbf{k}_1$. The angular velocity of the link 2 with respect to link 1 expressed in terms of RF1 is $\omega_{21} = \dot{q}_2 \mathbf{i}_1$. The angular velocity of the link 2 with respect to the fixed reference frame, RF0, expressed in terms of RF1, is $\omega_{20} = \omega_{10} + \omega_{21} = \dot{q}_1 \mathbf{k}_1 + \dot{q}_2 \mathbf{i}_1$. The angular velocity of the link 3 with respect to link 2 expressed in terms of RF2 is $\boldsymbol{\omega}_{32} = \dot{q}_3 \mathbf{i}_2$. The angular velocity of the link 3 with respect to the fixed reference frame, RF0, expressed in terms of RF2, is $\boldsymbol{\omega}_{30} = \boldsymbol{\omega}_{20} + \boldsymbol{\omega}_{32}$. The angular acceleration of the link p, $p = 1, 2, 3$, in RF0 is

$$\boldsymbol{\alpha}_{p0} = \frac{d}{dt}\boldsymbol{\omega}_{p0} = \frac{^{(p)}d}{dt}\boldsymbol{\omega}_{p0} + \boldsymbol{\omega}_{p0} \times \boldsymbol{\omega}_{p0} = \frac{^{(p)}d}{dt}\boldsymbol{\omega}_{p0}, \tag{3}$$

where $\frac{^{(p)}d}{dt}$ represents the derivative with respect to time in reference frame RFp.

The position vectors of C_p, the mass center of link $p, p = 1, 2, 3$, are $\mathbf{r}_{C_1} = 0.5L_1\mathbf{j}_1$, $\mathbf{r}_{C_2} = L_1\mathbf{j}_1 + 0.5L_2\mathbf{j}_2$, and $\mathbf{r}_{C_3} = L_1\mathbf{j}_1 + L_2\mathbf{j}_2 + 0.5L_3\mathbf{j}_3$. The velocities and accelerations of $C_p, p = 1, 2, 3$, in RF0 are

$$\mathbf{v}_{C_p} = \frac{^{(p)}d}{dt}\mathbf{r}_{C_p} + \omega_{p0} \times \mathbf{r}_{C_p} \quad \text{and} \quad \mathbf{a}_{C_p} = \frac{^{(p)}d}{dt}\mathbf{v}_{C_p} + \omega_{p0} \times \mathbf{v}_{C_p}. \tag{4}$$

2.2 Gibbs-Appell Equations of Motion

The Gibbs function for the link i, $i = 1, 2, 3$ is [15]:

$$S_i = \frac{1}{2}m_i \mathbf{a}_{C_i} \cdot \mathbf{a}_{C_i} + \frac{1}{2}\alpha_{i0} \cdot \bar{I}_i \cdot \alpha_{i0} + \alpha_{i0} \cdot (\omega_{i0} \times \bar{I}_i \cdot \omega_{i0}), \tag{5}$$

where m_i is the mass of the link, $\mathbf{a}_{C_i}$ is the acceleration of the mass center of the link in RF0, $\boldsymbol{\omega}_{i0} = \boldsymbol{\omega}_{xi}\mathbf{i}_i + \boldsymbol{\omega}_{yi}\mathbf{j}_i + \boldsymbol{\omega}_{zi}\mathbf{k}_i$ is the angular velocity of the link in (0), $\boldsymbol{\alpha}_{i0} = \dot{\boldsymbol{\omega}}_{i0}$ is the angular acceleration, and $\bar{I}_i = (I_{xi}\mathbf{i}_i)\mathbf{i}_i + (I_{yi}\mathbf{j}_i)\mathbf{j}_i + (I_{zi}\mathbf{k}_i)\mathbf{k}_i$ is the central inertia dyadic of the link. The central principal axes of the link are parallel to $\mathbf{i}_i, \mathbf{j}_i, \mathbf{k}_i$ and the associated moments of inertia have the values I_{xi}, I_{yi}, I_{zi}, respectively. The dot product of the vector $\boldsymbol{\omega}$ with the dyadic $\bar{I}$ is $\boldsymbol{\omega} \cdot \bar{I} = \bar{I} \cdot \boldsymbol{\omega} = \omega_x I_x \mathbf{i} + \omega_y I_y \mathbf{j} + \omega_z I_z \mathbf{k}$. The central moments of inertia of links 1, 2, and 3 are calculated using Fig. 1. The central principal axes of link $i, i = 1, 2, 3$ are parallel to $\mathbf{i}_i$, $\mathbf{j}_i$, $\mathbf{k}_i$ and the associated moments of inertia have the values I_{ix}, I_{iy}, I_{iz}, respectively.

The Gibbs-Appell dynamical equations governing the system are:

$$\frac{\partial^2(S_1 + S_2 + S_3)}{\partial \ddot{q}_r} = Q_r, \quad r = 1, 2, 3. \tag{6}$$

To calculate the partial derivative of the function $S = S_1 + S_2 + S_3$ with respect to the variable $\ddot{q}_r$ a MATLAB function, `deriv`, was created. The function `deriv (f, g)` differentiates a symbolic expression `f` with respect to the variable `g`, where the variable `g` is a function of time `g = g(t)`. The statement `diff(f,'x')` differentiates `f` with respect to the free variable `x`. In MATLAB the free variable `x` cannot be a function of time and that is why the function `deriv` was introduced.

2.3 Generalized Active Forces

Remark: If a set of contact and/or body forces acting on a rigid body is equivalent to a couple of torque $\boldsymbol{T}$ together with force $\boldsymbol{R}$ applied at a point P of the rigid body, then the contribution of this set of forces to the generalized force, Q_r, is given by $Q_r = \frac{\partial \omega}{\partial \dot{q}_r} \cdot \mathbf{T} + \frac{\partial \mathbf{v}_P}{\partial \dot{q}_r} \cdot \mathbf{R}, r = 1, 2, \ldots,$ where $\boldsymbol{\omega}$ is the angular velocity of the rigid body in (0), $\mathbf{v}_P$ is the velocity of P in (0), and r represents the generalized coordinates.

In the case of the robotic arm, there are two kinds of forces that contribute to the generalized forces Q_1, Q_2, and Q_3 namely, contact forces applied in order to drive the links 1, 2, and 3, and gravitational forces exerted on 1, 2, and 3. The set of contact forces transmitted from 0 to 1 can be replaced with a couple of torque $\mathbf{T}_{01}$ applied to 1 at O. Similarly, the set of contact forces transmitted from 1 to 2 can be replaced with a couple of torque $\mathbf{T}_{12}$ applied to 2 at A. The law of action and reaction then guarantees that the set of contact forces transmitted from 1 to 2 is equivalent to a couple of torque $-\mathbf{T}_{12}$ to 1 at A. Next, the set of contact forces exerted by link 2 on link 3 can be replaced with a couple of torque $\mathbf{T}_{32}$ applied to 3 at B. The law of action and reaction then guarantees that the set of contact forces transmitted from 2 to 3 is equivalent to a couple of torque $-\mathbf{T}_{23}$ to 2 at B. The expressions $\mathbf{T}_{01}$, $\mathbf{T}_{12}$, and $\mathbf{T}_{23}$ are $\mathbf{T}_{01} = T_{01x}\mathbf{i}_1 + T_{01y}\mathbf{j}_1 + T_{01z}\mathbf{k}_1, \mathbf{T}_{12} = T_{12x}\mathbf{i}_2 + T_{12y}\mathbf{j}_2 + T_{12z}\mathbf{k}_2,$ and $\mathbf{T}_{23} = T_{23x}\mathbf{i}_3 + T_{23y}\mathbf{j}_3 + T_{23z}\mathbf{k}_3$. The external gravitational forces exerted on the links 1, 2, and 3 are $\mathbf{G}_i = -m_i g \mathbf{k}_1,\ i = 1, 2, 3$. The reason for replacing $\mathbf{k}_1$ in connection with the forces $\mathbf{G}_2$, and $\mathbf{G}_3$ is that they are soon to be dot-multiplied with $\frac{\partial \mathbf{v}_{C_2}}{\partial \dot{q}_r}$ and $\frac{\partial \mathbf{v}_{C_3}}{\partial \dot{q}_r}$. The contribution, $(Q_r)_1$, to the generalized active force Q_r of all the forces and torques acting on link 1 is:

$$(Q_r)_1 = \frac{\partial \omega_{10}}{\partial \dot{q}_r} \cdot \mathbf{T}_{01} + \frac{\partial \mathbf{v}_{C_1}}{\partial \dot{q}_r} \cdot \mathbf{G}_1 + \frac{\partial \omega_{10}}{\partial \dot{q}_r} \cdot (-\mathbf{T}_{12}), \quad r = 1, 2, 3. \tag{7}$$

The contribution to the generalized active force Q_r for the link 2 is

$$(Q_r)_2 = \frac{\partial \omega_{20}}{\partial \dot{q}_r} \cdot \mathbf{T}_{12} + \frac{\partial \mathbf{v}_{C_2}}{\partial \dot{q}_r} \cdot \mathbf{G}_2 + \frac{\partial \omega_{20}}{\partial \dot{q}_r} \cdot (-\mathbf{T}_{23}), \quad r = 1, 2, 3. \tag{8}$$

The contribution to the generalized active force Q_r for the link 3 is

$$(Q_r)_3 = \frac{\partial \omega_{30}}{\partial \dot{q}_r} \cdot \mathbf{T}_{23} + \frac{\partial \mathbf{v}_{C_3}}{\partial \dot{q}_r} \cdot \mathbf{G}_3, \quad r = 1, 2, 3. \tag{9}$$

The generalized active force Q_r of all the forces and torques acting on the links 1, 2, and 3 are

$$Q_r = (Q_r)_1 + (Q_r)_2 + (Q_r)_3, \quad r = 1, 2, 3. \tag{10}$$

The MATLAB results for the generalized active force Q_r, $r = 1, 2, 3$ are:

```
% Q1 = T01z
% Q2 = T12x - (L3*g*m3*cos(q2(t) + q3(t)))/2
%      - (L2*g*m2*cos(q2(t)))/2 - L2*g*m3*cos(q2(t))
% Q3 = T23x - (L3*g*m3*cos(q2(t) + q3(t)))/2
```

3 Results

For the direct dynamics the feedback control law is arbitrary selected as:

```
T01zc = -b01*diff(q1,t)-g01*(q1-q1f)-(Q1-T01z);
T12xc = -b12*diff(q2,t)-g12*(q2-q2f)-(Q2-T12x);
T23xc = -b23*diff(q3,t)-g23*(q3-q3f)-(Q3-T23x);
```

The system of equations of motion is solved numerically using the statements:

```
dx_ = [dx1; dx2; dx3; dx4; dx5; dx6];
eom_ = matlabFunction(dx_,'vars',{t,[x1;x2;x3;x4;x5;x6]});
t0 = 0; tf = 5; time = [0 tf];
q1i = 0; dq1i = 0; q2i = 0; dq2i = 0; q3i = 0; dq3i = 0;
ic0 = [q1i dq1i q2i dq2i q3i dq3i];
[t,xs] = ode45(eom_, time, ic0);
```

The `ode45` solver is used for the system of differential equations for direct dynamics. Figure 2 shows the plots of $q_1(t)$, $q_2(t)$, and $q_3(t)$, the solutions of the nonlinear differential equations, for five seconds. Figure 3 represents the successive positions of the robotic arm for the simulations.

For the inverse dynamics a desired motion of the three-link chain is specified for a time interval $0 \le t \le T_p = 5$ s. The input generalized coordinates are selected as [8]:

$$q_r(t) = q_r(0) + \frac{q_r(T_p) - q_r(0)}{T_p}\left[t - \frac{T_p}{2\pi}\sin\left(\frac{2\pi t}{T_p}\right)\right], \quad r = 1,2,3, \tag{11}$$

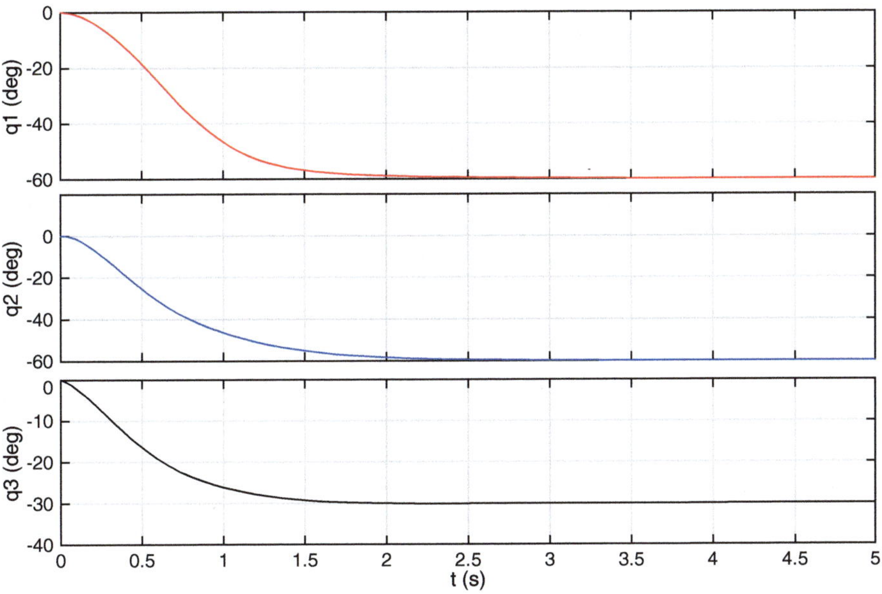

Fig. 2 Generalized coordinates $q_1(t)$, $q_2(t)$, and $q_3(t)$

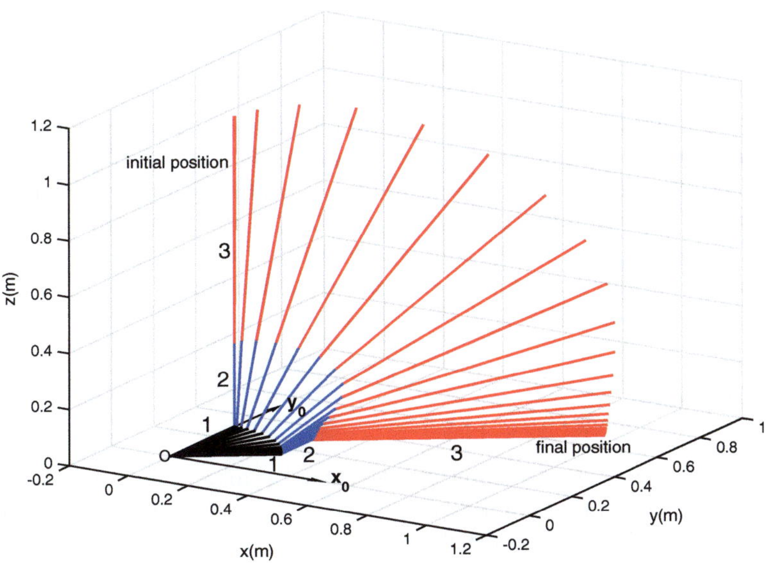

Fig. 3 Position of the robot arm

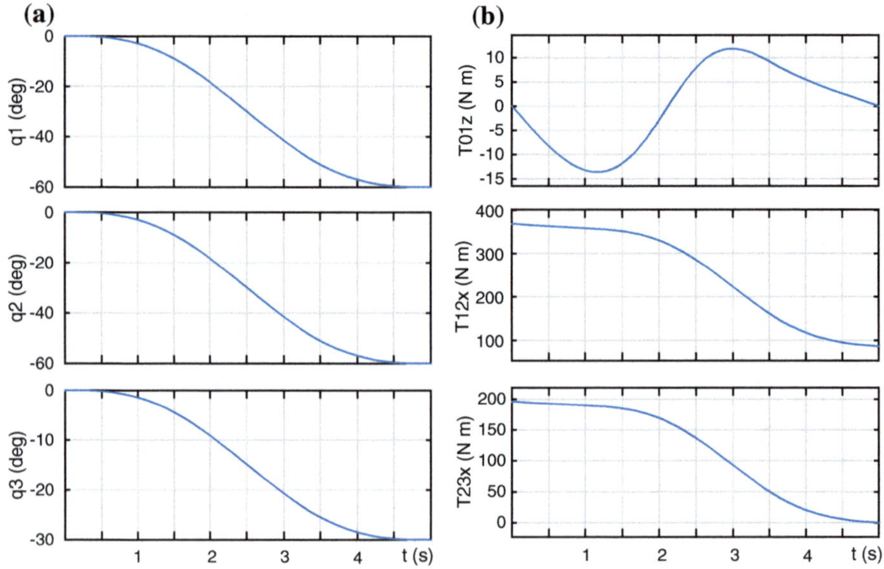

Fig. 4 Inverse dynamics: **a** input positions for the generalized coordinates and **b** calculated torques for input positions

with $q_r(T_p) = q_{rf}$. The robot arm can be brought from an initial position of rest in reference frame (0) to a final position of rest in (0) in such a way that q_1, q_2, and q_3 have specified values: $q_1(T_p) = q_{1f} = -\pi/3$ rad, $q_2(T_p) = q_{2f} = -\pi/3$ rad, and $q_3(T_p) = q_{3f} = -\pi/6$ m.

The MATLAB commands used to find the motion equations are identical with the commands presented in previous section. Figure 4a shows the input generalized coordinates. The generalized coordinates, q_1, q_2, and q_3 and their derivatives, are substituted in the expressions of T_{01x}, T_{12x}, and T_{23x}. The plots and the values of the external torques $T_{01z}(t)$, $T_{12x}(t)$, and $T_{23x}(t)$ are shown in Fig. 4b. The MATLAB programs can be found at ftp://eng.auburn.edu/pub/marghdb.

4 Conclusion

Gibbs-Appell equations of motion are developed symbolically using MATLAB. The equations are solved numerically using `matlabFunction and ode45`. Future research will focus on developing the proposed software for dynamic systems with multiple links and incorporating better numerical functions for solving the ordinary differential equations.

References

1. Quinn, R.: Equations of motion for structures in terms of quasicoordinates. J. Appl. Mech. **57** (3), 745–749 (1990)
2. Whittaker, E.: Analytical Dynamics of Particles and Rigid Bodies. Dover Publications, New York (1944)
3. Meirovitch, L.: Methods of Analytical Dynamics. McGraw Hill, New York (1970)
4. Baruh, H.: Analytical Dynamics. WCB/McGraw Hill, New York (1999)
5. Baruh, H.: Another look at the describing equations of dynamics. J. Chin. Soc. Mech. Eng. **1**, 15–24 (2000)
6. Huston, R., Passerello, C.: On the dynamics of a human body model. J. Biomech. **4**(5), 369–378 (1971)
7. Featherstone, R., Orin, D.: Robot dynamics: equations and algorithms. In: IEEE International Conference Robotics & Automation, San Francisco, CA, 826–834 (2000)
8. Kane, T., Levinson, D.: Kane's dynamical equations in robotics. Int. J. Robot. Res. **2**(3), 3–21 (1983)
9. Pan, D., Sharp, R.: Automatic formulation of dynamic equations of motion of robot manipulators. Proc. Inst. Mech. Eng. Part C: J. Mech. Eng. Sci. **202**(6), 397–404 (1988)
10. Tavakoli Nia, H., Pishkenari, H., Meghdari, A.: A recursive approach for the analysis of snake robots using kane's equations. Robotica **24**, 251–256 (2006)
11. Tur, J., Juan, S., Rovira, A.: Dynamic equations of motion for a 3-bar tensegrity based mobile robot. In: IEEE Conference on Emerging Technologies and Factory Automation, Patras, Greece, pp. 1334–1339 (2007)
12. Zhao, J., Liu, G., Yan, J., Zang, X.: Scout robot with wheeling-hopping combination locomotion. Ind. Robot: Int. J. **36**(3), 244–248 (2009)
13. Khatib, O., Sentis, L., Park, J.: A unified framework for whole-body humanoid robot control with multiple constraints and contacts. In: European Robotics Symposium 2008, STAR 44, pp. 303–312. Springer, Berlin (2008)
14. Gomes, M., Ruina, A.: A five-link 2d brachiating ape model with life-like zero-energy-cost motions. J. Theor. Biol. **237**(3), 265–278 (2005)
15. Mata, V., Provenzano, S., Cuadrado, P., Valero, F.: An o(n) algorithm for solving the inverse dynamic problem in robots by using the gibbs-appell formulation. In: Tenth World Congress IFToMM, Oulu 3, pp. 1208–1215 (1999)

Part VII
Robots in Medicine and Rehabilitation

Kinematic Behaviour of a Novel Medical Parallel Robot for Needle Placement

Bogdan Gherman, Doina Pîslă, Gabriel Kacso and Nicolae Plitea

Abstract Needle insertion procedures cover a large area of applications, like brachytherapy (BT), biopsy or fluid extraction. The paper presents the kinematics of a novel parallel robot designed for needle placement procedures. This implies that the needle will be inserted in the patient's body from the outside up to a target point following a linear trajectory, the needle tracking being achieved using: visual feedback, ultrasound if possible, otherwise computer tomography (CT). The structural synthesis, the robot kinematics, its workspace and some simulation results are presented in the paper.

Keywords Parallel robot · Needle placement · Kinematics · Workspace · Simulation

1 Introduction

Studies proved that cancer is the world's leading cause of death, followed by heart disease and stroke [1]. Cancer treatment needs to be correctly diagnosed to establish the correct type of treatment which most of the time implies a combination of radiotherapy, surgery or chemotherapy. An early and correct diagnosis allows

B. Gherman · D. Pîslă (✉) · N. Plitea
Research Centre for Industrial Robots Simulation and Testing, Technical University of Cluj-Napoca, Memorandumului 28, 400114 Cluj-Napoca, Romania
e-mail: doina.pisla@mep.utcluj.ro

B. Gherman
e-mail: bogdan.gherman@mep.utcluj.ro

N. Plitea
e-mail: nicolae.plitea@mep.utcluj.ro

G. Kacso
Iuliu Hatieganu University of Medicine and Pharmacy, Babes 8, Cluj-Napoca, Romania
e-mail: gabi.kacso@gmail.com

T. Borangiu (ed.), *Advances in Robot Design and Intelligent Control*,
Advances in Intelligent Systems and Computing 371,
DOI 10.1007/978-3-319-21290-6_33

usually better treatment options in term of life expectancy, produced trauma or quality life of the patient [2]. Prostate biopsy is still achieved today as it has been 50 years ago [3], being usually a manual procedure guided by an ultrasound probe. Brachytherapy, a relatively new approach in the fight against cancer, uses local radiation, so instead of irradiating a large area of the patient's body, in this case only the tumorous cells are irradiated, [4]. Most important, in needle placement, the needles must be precisely placed inside the tumour, at exact points chosen by the physician, [5].

Studies presented in [6] have demonstrated that a robotic device enhances the needle placement accuracy beyond the natural human capabilities. In [7] Baumann et al. proposed a system for cancer treatment using BT, named PROSPER, proving that the robot is able to reach a target in the water with a precision of less than 1 mm throughout the entire workspace. Bassan et al. described in [8] the design of a 5-DOF (degrees of freedom) hybrid robotic system that performs 3D ultrasound guided percutaneous needle insertion surgery. Hungr in [9] introduced a new 3D ultrasound robotic prostate brachytherapy system. A prototype and the 3D model of a 5-DOF hybrid robot for prostate needle insertion surgery under continuous MRI guidance has been presented in [10] by Jiang et al. A 4-DOF hybrid robot used for real-time control of transperineal prostate that works under MRI guidance and performs insertion motion manually has been introduced by Fischer in [11]. Fichtinger presented in [12] a robot assisted prostate brachytherapy system consisting of a TRUS unit, a needle insertion robot, and a software tool for treatment planning and image-registered implant system. Podder et al. showed in [13] that most of the robotic systems available for BT have been designed for prostate cancer treatment; therefore the need for new robotic solutions that target larger areas of the body is high, especially since deeply located tumours are difficult to treat using a manual procedure.

Starting from this, the authors propose a new, innovative robotic structure for general BT procedures, covering areas like the mediastinum or the paravertebral area, capable of positioning the BT needles in various organs of the patient's body and working under real-time CT monitoring.

2 The Medical Parallel Robot

In [14, 15, 16, 17] new robotic structures designed for BT have been developed, up to the experimental model level, their kinematics and workspace being presented.

The parallel robot developed in this paper targets also other needle placement applications, having a larger applicative spectrum. The medical requirements for the new robot start from the main task it has to achieve: the robot should insert rigid needles into the patient's body, having the diameter from 1.6 to 2 mm and lengths of 50–250 mm on a linear trajectory between the insertion point and the target point. According to the chosen visual feedback, the robotic structure should be either ultrasound or CT compatible, meaning that it should withstand a radiation X environment from 50 to 150 kV, have capability of teleoperation from a safe distance (over 5 m), and feature modularity and versatility in positioning the robot relative to the patient and CT. After the needles are introduced into the tumour, the needles are extracted and only the catheter will remain inside the tumour, through which the irradiation seeds will be introduced (using a specialized device).

The robotic system is of family F = 1, having a modular structure, consisting of a parallel module with M = 3 DOF (degrees of freedom) and three active joints (q_1, q_2, q_3) and a parallel module with M = 2 DOF and two active joints (a planar mechanism with the active joints q_4, q_5), see Fig. 1. The first module with three active joints consists in two translational joints, namely q_1 (along an axis parallel with the OX axis) and q_3 (along an axis parallel with the OY axis). The active joint q_2 is a rotation around an axis parallel with the OX axis, starting from the horizontal plane (the OXY plane). The second module is a parallel planar mechanism with 2 DOF, in which a bushing moves on two axes (parallel with OY and OZ). The elements 1 and 3 are connected via two Cardan joints, each having the first rotation around an axis parallel to the OZ axis (the second one being perpendicular on it).

In order to achieve the structural synthesis of the mechanism, the first module can be considered as a class 3 joint (the mobile platform 1 having 3 DOF—translations along the OX, OY and OZ axis and constant orientation), while the second module with two active joints and 2 DOF can be regarded as a class 4 joint (the final element 2, having 2 DOF). The number of moving elements of the robot is $N = 2 + 1 + 1$ (the two final elements of the two mechanisms, 1 and 2, element 3 and the needle 4). The mechanism has 1 class 5 joint (C_5, the 1 DOF revolute joint), 3 class 4 joints (C_4, the second module and the two Cardan joints) and 1 class 3 joint

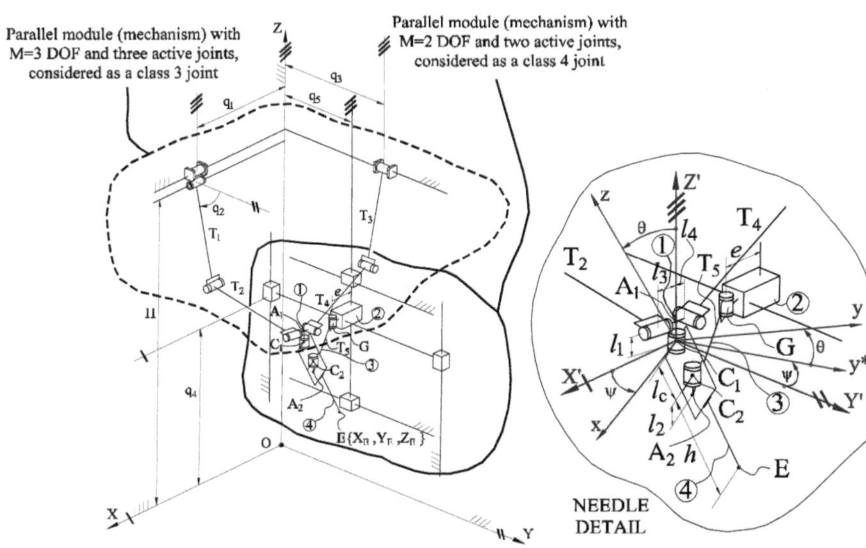

Fig. 1 The medical parallel robot—kinematic scheme

(C_3, the first module). The number of degrees of freedom can be computed using (1) [18]:

$$M = (6 - F) \cdot N - \sum_{i=1..5} (i - F) \cdot C_i = 5 \cdot N - 4 \cdot C_5 - 3 \cdot C_4 - 2 \cdot C_3 - 1 \cdot C_2 \\ = 5 \tag{1}$$

where M is the mobility of the robot (number of DOF), N is the number of moving elements, $F = 1$ is the family of the mechanism and C_i is the number of class i-th joints.

3 The Kinematic Model

To solve the geometric model of the medical robot, the following geometric parameters have been used: $T_1, T_2, T_3, T_4, T_5, l_1, l_2, l_3, l_4, h, e, l_c$ (see Fig. 1). The point $E(X_E, Y_E, Z_E)$ represents the needle tip's coordinates, while the angles ψ and θ represent the Euler's angles with:

- ψ as the rotation angle around the OZ' axis of the frame positioned in point C_1 belonging to the first Cardan joint, parallel with the OZ axis of the fixed frame in point O;
- θ as the rotation angle around the Ox axis of the frame positioned in point C_1 belonging to the first Cardan joint.

The final orientation of the needle (4) is represented by the C_1xyz frame.

3.1 The Inverse Geometric Model

The coordinates of the points C_1 and C_2 as well as A_1 and A_2 belonging to the Cardan joints are (Fig. 1—needle detail):

$$\begin{array}{ll} X_{C1} = X_E - (h + l_c) \cdot \sin(\theta) \cdot \cos(\psi) & X_{C2} = X_E - h \cdot \sin(\theta) \cdot \cos(\psi) \\ Y_{C1} = Y_E - (h + l_c) \cdot \sin(\theta) \cdot \sin(\psi) & Y_{C2} = Y_E - h \cdot \sin(\theta) \cdot \sin(\psi) \\ Z_{C1} = Z_E + (h + l_c) \cdot \cos(\theta) & Z_{C2} = Z_E + h \cdot \cos(\theta) \end{array} \tag{2}$$

$$\begin{array}{ll} X_{A1} = X_{C1} & X_{A2} = X_{C2} \\ Y_{A1} = Y_{C1} & Y_{A2} = Y_{C2} \\ Z_{A1} = Z_{C1} + l_1 & Z_{A2} = Z_{C2} - l_2 \end{array} \tag{3}$$

The two Cardan joints are not classic, meaning that the points C_1 and A_1 as well as C_2 and A_2 don't overlay, but are offset at l_1, respectively l_2, for reasons of increased motion range.

The five active joint coordinates can be computed:

$$\begin{aligned} q_1 &= X_E - (h + l_c) \cdot \sin(\theta) \cdot \cos(\psi) + l_3 \\ q_2 &= a\tan 2\left(\sin q_2, \sqrt{1 - (\sin q_2)^2}\right) \\ q_3 &= Y_{A1} = Y_E - (h + l_c) \cdot \sin(\theta) \cdot \sin(\psi) \\ q_4 &= Z_E + h \cdot \cos(\theta) - l_2 \\ q_5 &= Y_E - h \cdot \sin(\theta) \cdot \sin(\psi) - \sqrt{T_5^2 - (X_E - h \cdot \sin(\theta) \cdot \cos(\psi) - e)^2} \end{aligned} \tag{4}$$

where

$$\sin q_2 = \frac{\frac{(H - Z_{A1})}{2} + Y_{A1} \cdot \sqrt{\frac{T_1^2}{(H - Z_{A1})^2 + Y_{A1}^2} - \frac{1}{4}}}{T_1} \tag{5}$$

3.2 *The Direct Geometric Model*

In this case, the coordinates of the active joints $(q_1, q_2, q_3, q_4, q_5)$ are known, as well as the geometric parameters of the robot. The task is to determine the coordinates of the needle tip $E(X_E, Y_E, Z_E)$ and the orientation angles ψ and θ.

The coordinates of points A_1 and C_1 are determined:

$$\begin{aligned} X_{A1} &= q_1 - l_3 & X_{C1} &= X_{A1} \\ Y_{A1} &= q_3 & Y_{C1} &= Y_{A1} \\ Z_{A1} &= H - T_1 \cdot \sin\left(q_2 - \sqrt{T_2^2 - (Y_{A1} - T_1 \cdot \cos(q_2))^2}\right) & Z_{C1} &= Z_{A1} - l_1 \end{aligned} \tag{6}$$

The orientation angle θ can be determined as follows:

$$\theta = a\tan 2\left(\sqrt{l_c^2 - (Z_{C1} - q_4 - l_2)^2}, (Z_{C1} - q_4 - l_2)\right) \tag{7}$$

In order to determine the orientation angle ψ, two solutions emerge, from solving the following system of equations with the unknowns X_{A2} and Y_{A2}:

$$\begin{cases} (Y_{A2} - Y_{A1})^2 + (X_{A2} - X_{A1})^2 = l_c^2 - (Z_{C1} - Z_{C2})^2 \\ X_{A2}^2 + (Y_{A2} - q_5)^2 = c^2 \end{cases} \quad (8)$$

From (8), the two solutions of the angle ψ are:

$$\psi_i = a \tan 2(Y_{A2\_i} - Y_{A1}, X_{A2\_i} - X_{A1}) \quad (9)$$

with i = 1, 2. From (6)–(9), two solutions for the needle tip coordinates are obtained:

$$\begin{aligned} X_{E\_i} &= X_{C1} + (h + l_c) \cdot \sin(\theta) \cdot \cos(\psi_i) \\ Y_{E\_i} &= Y_{C1} + (h + l_c) \cdot \sin(\theta) \cdot \sin(\psi_i), \quad i = 1\ldots2 \\ Z_E &= Z_{C1} - (h + l_c) \cdot \cos(\theta) \end{aligned} \quad (10)$$

The inverse and direct kinematic models have been analytically obtained, with important implications in the control of the robot. Concerning the direct geometric model, the initial pose of the needle is known, thus enabling the choice of the proper solution among the two possible ones.

3.3 *Numerical Results*

The behaviour of the robotic system has been studied using the geometric model presented in previous sections as well as the developed kinematic model. A linear trajectory for the needle tip has been imposed into a simulation program created in MATLAB. The robot motion comprises two phases: first, the robot guides the needle from an initial pose (position and orientation) (point *C*) up to the insertion point into the patient's body (point *I*), achieving in the meantime also its final orientation (the angles ψ and θ); second, the robot will insert the needle following a linear trajectory from the insertion point *I* up to the target point *T*, with real-time CT control. Figure 2 illustrates the robot motion for the imposed trajectory at the level of each active joint.

The values for the geometrical parameters of the robot and the pose of the robot for the current-, insertion-, and target points are given in (11). All dimensions are in mm. For the first part of the motion, the needle's imposed maximum speed and acceleration are: $\upsilon_{\max} = 20$ mm/s and $a_{\max} = 10$ mm/s^2. After the needle is brought at the insertion point, the maximum insertion speed and acceleration are: $\upsilon_{\max} = 2.4\,\text{mm/s}$ and $a_{\max} = 1.2\,\text{mm/s}^2$.

An important issue in the control of a medical robot is to check if the imposed trajectories of the needle belong to the robot workspace. The robot workspace is generally dependent on the chosen insertion points into the patient's body. In this sense, a verification of the needle trajectories starting from the insertion up to the target point is necessary.

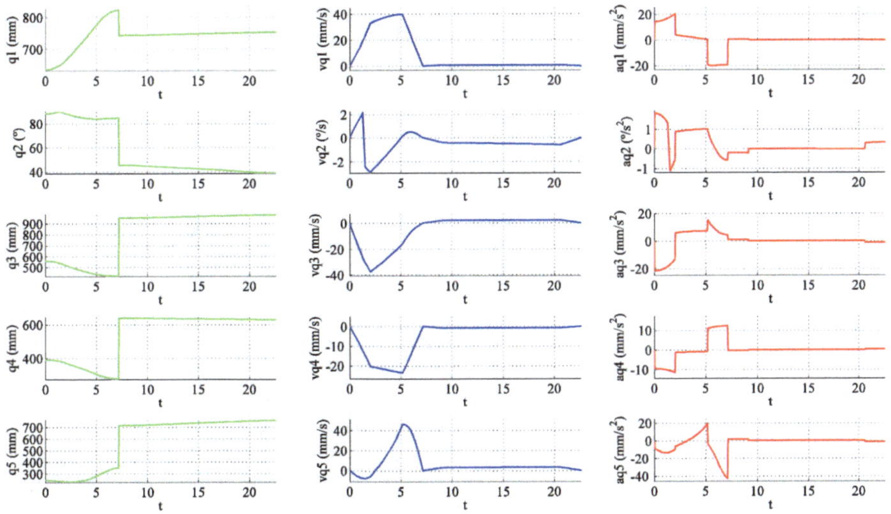

Fig. 2 Time history diagram of the active joint coordinates, speeds and accelerations

Figure 3 presents the reachable robot workspace, featuring singularity-free configurations (excluding those configurations for which the Jacobean determinant is zero or close to zero) for the indicated insertion point (dimensions in mm). It proves that the robot is capable to reach a variety of target points, in this case situated in the kidney area. Figure 4 represents the CAD model of the robot.

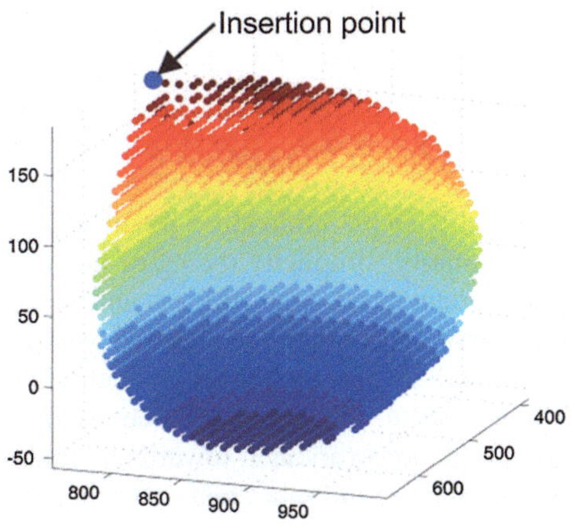

Fig. 3 Robot workspace with the insertion point $I(X_I = 525, Y_I = 780, Z_I = 185)$

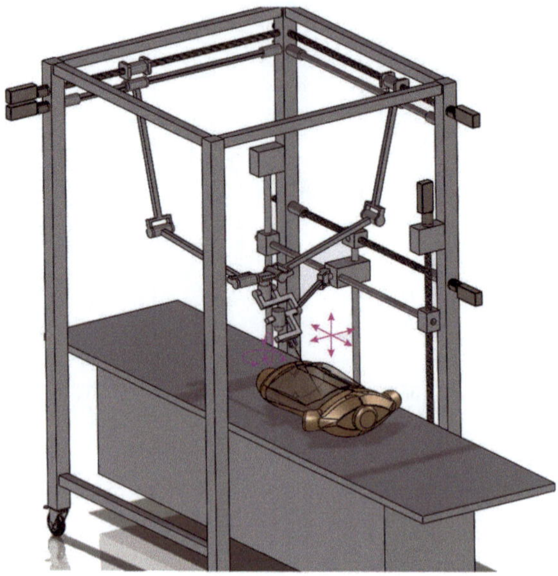

Fig. 4 CAD model of the medical robot

$$
\begin{gathered}
l_1 = 60; l_2 = 40; l_3 = 85; l_4 = 50; l_c = 112.36; h = 342.1; e = 138; \\
T_1 = T_2 = 565; T_3 = T_4 = 465; T_5 = 570.31; H = 1295; \\
\begin{cases} X_C = 550 \\ Y_C = 880 \\ Z_C = 190 \end{cases}; \quad \begin{cases} \psi_C = 90 \\ \theta_C = 45 \end{cases}; \quad \begin{cases} X_I = 525 \\ Y_I = 780 \\ Z_I = 185 \end{cases}; \quad \begin{cases} X_T = 483.01 \\ Y_T = 851 \\ Z_T = 150 \end{cases};
\end{gathered}
\tag{11}
$$

4 Conclusions

The paper presents a novel 5-DOF, innovative medical parallel robot. Its main advantages consist in: modularity, simplicity (being of family 1 and having 5 DOF for needle positioning and orientation), high stiffness and accuracy (conferred by its parallel structure). The robot kinematics has been developed and an analytical solution was presented for both inverse and direct geometric models, with important implications in the robot control. A singularity-free robot workspace has been analysed and numerical simulations (oriented towards the medical procedure) have been conducted in order to analyse the robot behaviour during the needle insertion procedures.

Acknowledgement This research was carried out in several projects: (a) the Partnership Program PN-II with financial support from MEN-UEFISCDI, Project no. 247/2014; (b) the project financed from the European Social Fond through the POSDRU program DMI 1.5, ID 137516-PARTING;

(c) the Project no. 173/2012, code PN-II-PCCA-2011-3.2-0414, entitled "Robotic assisted brachytherapy, an innovative approach of inoperable cancers—CHANCE" financed by UEFISCDI; and (d) the bilateral Austria—Romania project 745/2014 financed by UEFISCDI.

References

1. Rijo J., Ross, H.: The global economic cost of cancer. American Cancer Society, Inc., New York, No. 005444 (2010)
2. Vaida C., Gherman B., Pîsla D., Plitea, N.: A CT-scan compatible robotic device for needle placement in medical applications, In: Advanced Engineering Forum, pp. 574–583. Trans Tech Publications, Dürnten (2013). doi:10.4028/www.scientific.net/AEF.8-9.574
3. Carpenter, T.J., Forsythe, K., Kao, J., Stone, N.N., Stock, R.G.: Outcomes for patients with extraprostatic prostate cancer treated with trimodality therapy, including brachytherapy, external beam radiotherapy and hormone therapy. Brachytherapy **10**(4), 261–268 (2011)
4. Gerbaulet, A., Pötter, R., Mazeron, J.-J., Meertens, H., Limbergen, E.V.: The GEC ESTRO Handbook of Brachytherapy. European Society for Therapeutic Radiology and Oncology, Leuven (2002). ISBN 978-90-804532-6-5
5. Polo, A., Salembier, C., Venselaar, J., Hoskin, P.: Review of intraoperative imaging and planning techniques in permanent seed prostate brachytherapy. Radiother. Oncol. **94**, 12–23 (2010)
6. Mateescu, D.: Oncology Patient Guide. Bennet Publishing House, Bucharest (2010)
7. Baumann, M., et al.: Prosper: image and robot-guided prostate Brachytherapy. Aquat. Bot. **32**, 63–65 (2011)
8. Bassan, H. et al.: A novel manipulator for 3D ultrasound guided percutaneous needle insertion. In: IEEE International Conference on Robotics and Automation, pp. 617–622 (2007). ISSN 1050-4729
9. Hungr N., et al.: 3D ultrasound robotic prostate brachytherapy system with prostate motion tracking. IEEE Trans. Robot. **28**(6), 1382–1397 (2012). ISSN 1552-3098
10. Jiang, S., et al.: Kinematic analysis of a 5-DOF hybrid-driven MR compatible robot for minimally invasive prostatic interventions. Robotica **30**(7), 1147–1156 (2012)
11. Fischer, G.S., et al.: Design of a robot for transperineal prostate needle placement in MRI scanner. In: IEEE International Conference on Mechatronics, pp. 592–597. ISBN 0-7803-9712-6 (2006)
12. Fichtinger, G., et al.: Robotically assisted prostate brachytherapy with transrectal ultrasound guidance—phantom experiments. Brachytherapy **5**(1), 14–26 (2006)
13. Podder, T.K., et al.: AAPM and GEC-ESTRO guidelines for image-guided robotic brachytherapy: report of Task Group 192. Med. Phys. **41**(10), 101501 (2014). doi:10.1118/1.4895013
14. Gherman, B., Plitea, N., Galdau, B., Vaida, C., Pîsla, d.: On the kinematics of an innovative parallel robot for brachytherapy. Adv. Rob. Kinematics, 475–483. doi:10.1007/978-3-319-06698-1_49 (2014)
15. Pîsla D., Cocorean D., Vaida C., Gherman B., Pîsla A., Plitea, N.: Application oriented design and simulation of an innovative parallel robot for brachytherapy. In: Proceedings of the ASME 2014—IDETC/CIE 2014, 17–20 Aug, Buffalo, New York, USA (2014)
16. Plitea N., Szilaghyi A., Pîsla, D.: Kinematic analysis of a new 5-DOF modular parallel robot for brachytherapy. Robot. Comput. Integr. Manuf. **31**, 70–80 (2015). doi:10.1016/j.rcim.2014.07.005

17. Vaida C., Plitea N., Cocorean D., Pîsla, D.: Modeling of new spatial parallel structures with constant platform orientation using planar parallel modules. Proc. Rom. Acad. Ser. A **15**(1), 43–51 (2014). ISSN 1454-9069
18. Plitea, N., Pîsla, D., Vaida, C., Gherman, B., Szilaghyi, A., Galdau, B., Cocorean D., Covaciu, F.: On the kinematics of a new parallel robot for brachytherapy. Proc. Rom. Acad. Ser A: Math. **15**(4), 354–361 (2014)

Optimal Planning of Needle Insertion for Robotic-Assisted Prostate Biopsy

Doina Pîslă, Bogdan Gherman, Florin Gîrbacia, Călin Vaida, Silviu Butnariu, Teodora Gîrbacia and Nicolae Plitea

Abstract Robotic systems used for prostate biopsy offer important advantages compared to the manual procedures. In the robotic assisted prostate biopsy procedure, an important problem is to identify the optimal needle trajectories that allow reaching the target tissue and avoiding vital anatomical organs (major blood vessels, internal organs etc.). The paper presents an algorithm for optimal planning of the biopsy needle trajectories, based on virtual reality technologies, using as case study a novel parallel robot designed for transperineal prostate biopsy. The developed algorithm has been tested in a virtual environment for the prostate biopsy robotic-assisted procedure and results are presented.

Keywords Virtual reality · Robotic-assisted prostate biopsy · Parallel robot

D. Pîslă · B. Gherman · C. Vaida · N. Plitea
CESTER, Technical University of Cluj-Napoca, 28 Memorandumului, 400114 Cluj-Napoca, Romania
e-mail: Doina.Pisla@mep.utcluj.ro

B. Gherman
e-mail: Bogdan.Gherman@mep.utcluj.ro

C. Vaida
e-mail: Calin.Vaida@mep.utcluj.ro

N. Plitea
e-mail: Nicolae.Plitea@mep.utcluj.ro

F. Gîrbacia (✉) · S. Butnariu · T. Gîrbacia
Transilvania University of Braşov, 29 Eroilor Blvd, 500036 Braşov, Romania
e-mail: Garbacia@unitbv.ro

S. Butnariu
e-mail: Butnariu@unitbv.ro

T. Gîrbacia
e-mail: Teodora.Girbacia@unitbv.ro

T. Borangiu (ed.), *Advances in Robot Design and Intelligent Control*,
Advances in Intelligent Systems and Computing 371,
DOI 10.1007/978-3-319-21290-6_34

1 Introduction

Cancer prostate is the most commonly diagnosed cancer as well as the second most common cause of death [1]. In 2014 in the United States of America as many as 233,000 (14 %) men were estimated to be diagnosed with prostate cancer [2]. A biopsy is a procedure performed to acquire a sample of tissue from the target organ which will be used to determine whether cancer is present. In a recent study [3], it was shown that robotic-assisted prostate biopsy procedure has a higher accuracy and a better detection rate compared to biopsies performed manually by the urologist. Robot-assisted and computer-assisted surgeries are terms for technological devices using robotic systems to aid in surgical procedures [4]. Robot-assisted biopsy procedures present the following advantages compared to the manual procedures: (i) biopsy needle can be guided easier (ii) tissue deformations can be reduced because the motion is constant [3].

Recently, several different types of needle guiding robots have been developed [5–10]. At this moment, two approaches in achieving prostate biopsy are used: trans-rectal and transperineal prostate biopsy. The first approach is mainly used for manual procedures or with MRI (Magnetic Resonance Imaging) compatible robotic systems [8–10]. Other systems use the transperineal approach for less infection risks [11, 12].

The use of Virtual Reality (VR) to simulate surgery has significantly evolved during the last decade [13]. In previous researches, VR technologies have been used for learning biopsy procedure [14] and to simulate the manual biopsy procedure through haptic simulation [15, 16]. In this paper VR technologies are used for defining optimal biopsy needle trajectories for a novel parallel robotic system. Compared to other already presented VR applications for robotic needle insertion, the novelty of this approach is the possibility to automatically calculate trajectories for the proposed robotic system that allows the avoidance of the intersection with high risk areas.

The paper is structured as follows: Sect. 2 describes the new parallel robot for transperineal prostate biopsy; Sect. 3 presents the algorithm for optimal planning of needle guidance, followed by some results and conclusion.

2 The Innovative Parallel Robot Used for Prostate Biopsy

An innovative parallel robot has been designed for transperineal prostate biopsy, consisting of two modules working together [17]. Each module can be seen as a robot, the first one guiding the ultrasound probe with M = 5 DOF, while the second one guides the biopsy gun (prostate biopsy needle) having M = 5 DOF, too. Figure 1 presents the first robot (for ultrasound probe guidance) having a modular structure: it has a parallel module with M = 3 DOF with three active joints (q_1, q_2, q_3), all translational along axes parallel with the OY (q_1), respectively OZ (q_2, q_3) of

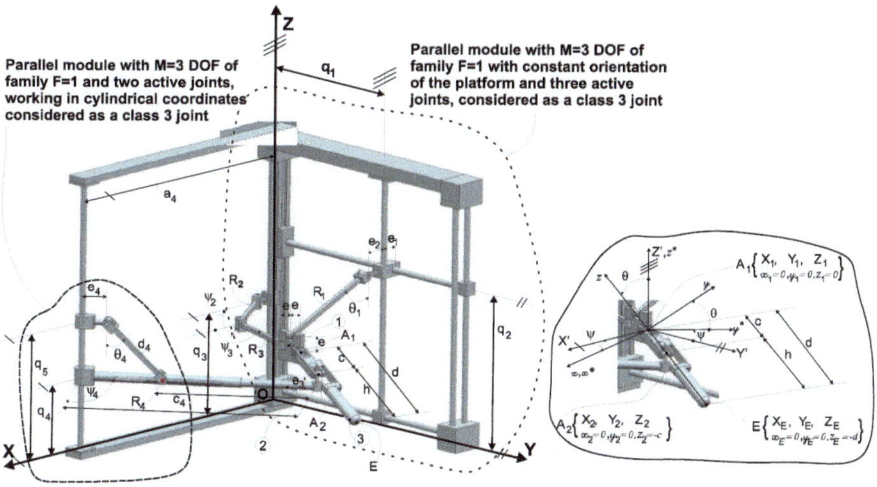

Fig. 1 BIO-PROS-1 ultrasound guidance module (robot)

the fixed frame of the robot and a second parallel module working in cylindrical coordinates with M = 3 DOF, with two active joints (q_4, q_5), both translations along an axis parallel with the robot's OZ axis. Both modules are of family F = 1, the common restriction is that the three exit elements (Fig. 1) do not execute a rotation motion, so the Euler angle $\varphi = 0$.

The second module (robot) is positioned at the distance a_4 on the OX axis from the robot fixed frame. Two Cardan joints connect the two modules, each one having the first rotational axis vertical, the second one being perpendicular on it. Between the two Cardan joints is positioned the module holding the ultrasound probe. The position and orientation of the ultrasound probe is determined by the coordinates of its tip, the point $E(X_E, Y_E, Z_E)$ and its orientation, indicated by the Euler ψ and θ angles (rotation angles around the OZ, respectively Ox of the mobile frame, placed in point A_1, Fig. 1). The second module, designed for guiding the biopsy gun, presented in Fig. 2 is similar to the one developed for ultrasound probe guidance, the difference consisting in the fact that the two modules are disposed in a single plane, at the distance X_C from the robot frame on X axis; all the active joints, $q_1', q_2', q_3', q_4', q_5'$ are translational.

The kinematic model of the robot leads to analytic solutions for both the inverse and direct models. The closure equations that describe the relation between the end-effector coordinates and the active joints from where the inverse kinematic model [the active joints coordinates: for the TRUS (Transrectal Ultrasound) guiding robot: q_1, q_2, q_3, q_4, q_5—Eq. (1) and the biopsy needle guiding robot: $q_1', q_2', q_3', q_4', q_5'$—Eq. (2)] can be determined and used in the optimal trajectory planning algorithm are:

$$\begin{cases} f_1: Y_E - d\cdot\sin(\theta)\cdot\sin(\psi) - e - q_1 = 0;\\ f_2: Z_E + d\cdot\cos(\theta) + e + \sqrt{R_1^2 - (X_E - d\cdot\sin(\theta)\cdot\cos(\psi) - e_1 - e_2)^2} - q_2 = 0;\\ f_3: Z_E + d\cdot\cos(\theta) - q_3 = 0;\\ f_4: Z_E + h\cdot\cos(\theta) - e_3 - q_4 = 0;\\ f_5: \left(c_4 + e_4 + \sqrt{d_4^2 - (q_5 - q_4)^2}\right)^2 - (a_4 - X_E + h\cdot\sin(\psi)\cdot\sin(\theta))^2\\ \quad -(Y_E - h\cdot\cos(\psi)\cdot\sin(\theta))^2 = 0; \end{cases} \tag{1}$$

$$\begin{cases} f_1: Y'_E - d'\cdot\sin(\theta')\cdot\sin(\psi') - e' - q'_1 = 0;\\ f_2: Z'_E + d'\cdot\cos(\theta') + e' + \sqrt{R_1'^2 - \left(X_C - X'_E + d'\cdot\sin(\theta')\cdot\cos(\psi') - e'_1 - e'_2\right)^2} - q'_2 = 0;\\ f_3: Z'_E + d'\cdot\cos(\theta') - q'_3 = 0;\\ f_4: Z'_E + h'\cdot\cos(\theta') - e'_3 - q'_4 = 0;\\ f_5: \left(c'_4 + e'_4 + \sqrt{d_4'^2 - (q'_5 - q'_4)^2}\right)^2 - (X_C - X_E + h\cdot\sin(\psi)\cdot\sin(\theta) - e_5)^2\\ \quad -(Y_E - h\cdot\cos(\psi)\cdot\sin(\theta))^2 = 0; \end{cases} \tag{2}$$

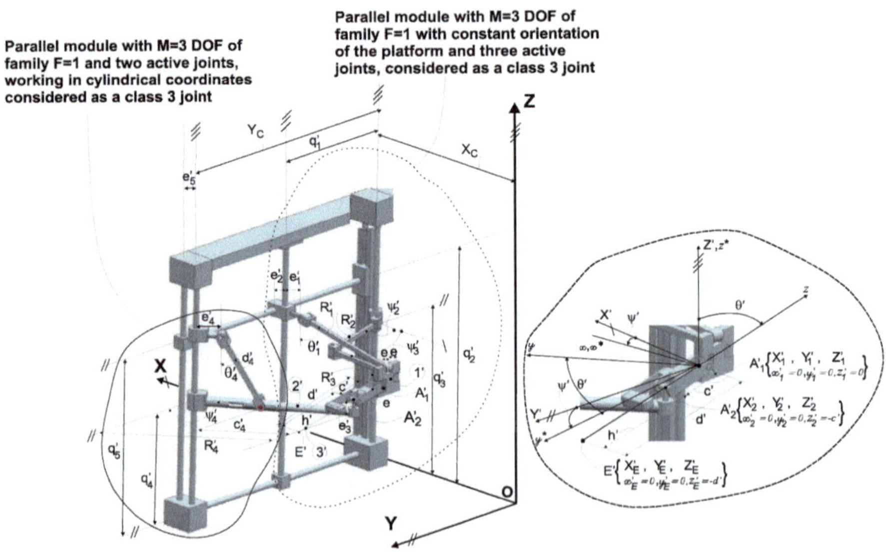

Fig. 2 BIO-PROS-1 biopsy needle guidance module (robot)

3 The Algorithm for Optimal Planning of Prostate Biopsy Needle Guidance

A proper prostate biopsy procedure will allow for a correct and quick diagnosis of the patient, leading to better life expectancy and/or life quality. A thorough analysis and simulation of the procedure will point out to the exact location and steps in the achievement of the biopsy, the system being also useful in improving the learning curve of the procedure. Using conventional TRUS images is often difficult to accurately localize the target biopsy area of the prostate and shield the organs at risk. VR technologies facilitate calculating accurate biopsy needle trajectories that avoid vital organs. The proposed algorithm for virtual planning of robotic-assisted needle insertion allows automatic definition of the needles trajectory for every target or "quadrant" of the prostate specified by the urologist. The main steps of the algorithm designed for automatic calculation of linear robot trajectories for the target point are:

- Process the 3D model of the patient reconstructed from MRI images.
- Load the 3D model of the patient into the virtual environment.
- Registration of the patient's 3D model to the robot.
- Define the target area for the biopsy procedure.
- Define parameters for needle insertion trajectories based on the target area: target area dimensions, first insertion point, target points, direction, and the vertical and horizontal displacement step.
- Successively, for each insertion point the robotic needle guidance process is conducted and the intersection between the needle tip and anatomical organs is computed. In order to determine the intersection between the linear trajectories and the high risk areas, a ray cast collision detection algorithm was used [18]. This algorithm allows detecting the contact between a linear segment and the triangles of the virtual object mesh. The algorithm returns the intersection point between these points and a line corresponding to the biopsy needle. The robot trajectories can be classified into three types: t_a, t_b, and t_c. The t_a type contains the trajectories that do not intersect the high risk areas, but intersect the target area at a point set by the urologist. The t_b type contains the trajectories that do not intersect the vital organs but the distance (d) between the needle tip and a target point set by the urologist is less than a specified threshold (Fig. 3). The t_c type contains the trajectories that intersect the vital organs. After all trajectories corresponding to the target area were checked, the optimal trajectories t_i will be considered one from t_a group. If $t_a = 0$, the new proposed trajectory t_i is selected from the t_b and is the one where the distance between the needle tip and the target point set by the urologist is minimum.
- Finally, the optimal trajectories are saved and proposed for the validation to the urologist. If $t_a = 0$ and $t_b = 0$ then it is proposed to the urologist to define other target point and then the automatic algorithm is used again.

In order to achieve the prostate biopsy procedure using this robot, the patient will be positioned in a rigid position (similar to the gynaecologic one) to the surgery

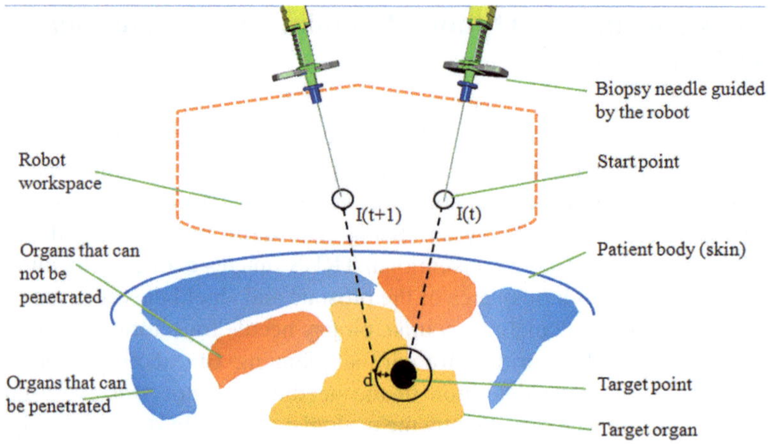

Fig. 3 Planning of biopsy needle trajectories

table, while being anesthetized from the waist down. The coordinates of the target points (possible tumours) are known and the robot is directed to them, while the needle trajectory is followed using the TRUS probe which is inserted and guided into the patient's rectum.

4 Implementation Details and Results

In the conducted test, a virtual environment containing a 3D reconstructed pelvis model of a patient has been modelled. Our test started by choosing the prostate target treatment area. The transperineal prostate biopsy procedure represents a complex process because the biopsy needle tip needs to avoid the intersection with high risk areas and accurately reach to the target position.

For the representation of the 3D model geometry of the target area in the developed VR system was used the ISO standard VRML2.0 (Virtual Reality Modeling Language). The 3D geometry model is represented using IndexedFaceSet nodes and is composed of a number of vertexes and triangles. The rendering of the 3D virtual environment was realized through the program BS Contact VRML player (http://www.bitmanagement.de). The C++ programming language has been used for the implementation of the algorithm for optimal planning of the biopsy needles. The tests have been performed using a Desktop PC with the following characteristics: Intel(R) Core(TM) i7 at 3.47 GHz and 12 GB RAM, NVidia QuadroFX 6000 GPU using Windows 7 operation system.

Figure 4 shows a robotic-assisted biopsy needle insertion for the prostate area. The needle tip reaches the target point avoiding penetration of bladder and pelvis bone.

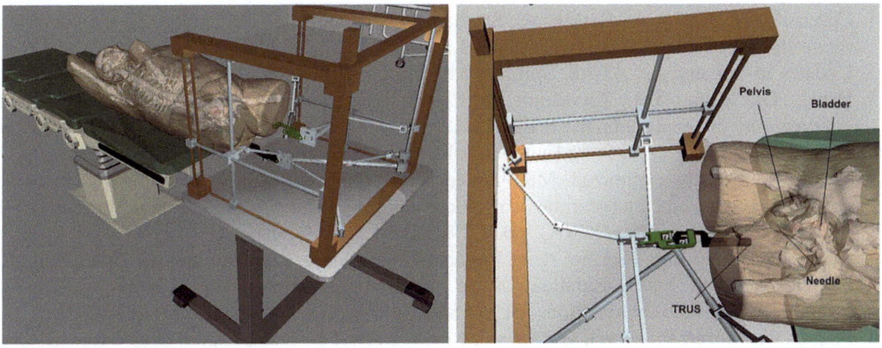

Fig. 4 Simulation of robotic-assisted prostate biopsy

5 Conclusion

This paper presents an algorithm for calculating optimal trajectories of the prostate biopsy needle applied for a novel parallel robot designed for transperineal prostate biopsy. The presented parallel robotic prostate biopsy system aims to improve the manual procedure by executing transperineal prostate biopsy with higher accuracy. The innovative parallel robotic system (consisting of two robots, each with 5 DOF) enables simultaneously the guiding of the ultrasound probe module and the biopsy gun (prostate biopsy needle). The proposed planning algorithm calculates automatically, using VR techniques, optimal linear trajectories with avoidance of high risk areas proximities for the proposed transperineal prostate biopsy parallel robot. A virtual environment has been modelled containing a virtual surgery room with accessories, the BIO-PROS-1 robot and a reconstructed 3D model of a patient. The proposed pre-planning algorithm was tested in the virtual environment and a simulation of the robotic-assisted biopsy needle insertion was presented.

Acknowledgment This paper was realized within the Partnership Programme in priority domains —PN-II, which runs with the financial support of MEN-UEFISCDI, Project no. 247/2014 and the project financed from the European Social Fund through POSDRU program, DMI 1.5, ID 137516-PARTING.

References

1. Sivaraman, A., Chauhan, S., Schatloff, O., Patel, V.: Techniques of nerve sparing in robot-assisted radical prostatectomy. In: John, H., Wiklund, P. (eds.) Robotic Urology, pp. 259–271. Springer, Berlin (2013)
2. National Cancer Institute. http://seer.cancer.gov/statfacts/html/prost.html
3. Kaye, D.R., Stoianovici, D., Han, M.: Robotic ultrasound and needle guidance for prostate cancer management: review of the contemporary literature. Curr. Opin. Urol. **24**(1), 75–80 (2014)

4. Ceccarelli, G., Patriti, A., Bartoli, A., Spaziani, A., Casciola, L.: Technology in the operating room: the robot. In: Calise, F., Casciola, L. (eds.) Minimally Invasive Surgery of the Liver. Updates in Surgery, pp. 43–48. Springer Milan, Berlin (2013)
5. Bebek, O., Myun, J.H., Cavusoglu, M.C.: Design of a parallel robot for needle-based interventions on small animals. IEEE/ASME Trans. Mechatron. **18**(1), 62–73 (2013)
6. Eslami, S., Fischer, G.S., Sang-Eun, S., Tokuda, J., Hata, N., Tempany, C.M., Iordăchiţă, I.: Towards clinically optimized MRI-guided surgical manipulator for minimally invasive prostate percutaneous interventions: constructive design. In: IEEE International Conference on Robotics and Automation (ICRA), pp. 1228–1233 (2013)
7. Krieger, A., Sang-Eun, S., Cho, N.B., Iordăchiţă, I.I., Guion, P., Fichtinger, G., Whitcomb, L. L.: Development and evaluation of an actuated MRI-compatible robotic system for MRI-guided prostate intervention. IEEE/ASME Trans. Mechatron. **18**, 273–284 (2013)
8. Poquet, C., Mozer, P., Morel, G., Vitrani, M.A.: A novel co-manipulation device for assisting needle placement in ultrasound guided prostate biopsies. In: IEEE/RSJ International Conference on Intelligent Robots and Systems (IROS), pp. 4084–4091 (2013)
9. Seifabadi, R., Iordachita, I., Fichtinger, G.: Design of a teleoperated needle steering system for MRI-guided prostate interventions. In: 4th IEEE RAS & EMBS International Conference on Biomedical Robotics and Biomechatronics (BioRob), pp. 793–798 (2012)
10. Srimathveeravalli, G., Kim, C., Petrişor, D., Ezell, P., Coleman, J., Hricak, H., Solomon, S.B., Stoianovici, D.: MRI-safe robot for targeted transrectal prostate biopsy: animal experiments. BJU Int. **113**, 977–985 (2014)
11. Ho, H., Yuen, J.S., Mohan, P., Lim, E.W., Cheng, C.W.: Robotic transperineal prostate biopsy: pilot clinical study. Urology **78**(5), 1203–1208 (2011)
12. Long, J.A., Hungr, N., Baumann, M., Descotes, J.L., Bolla, M., Giraud, J.Y., Rambeaud J.J., Troccaz, J.: Development of a novel robot for transperineal needle based interventions: focal therapy, brachytherapy and prostate biopsies. J. Urol. **188**(4), 1369–1374 (2012)
13. Rassweiler, J., Rassweiler, M.C., Müller, M., Kenngott, H., Meinzer, H.P., Teber, D., Lima, E., Petrut, B., Klein, J., Gözen, A.S., Ritter, M., Michel, M.S.: Surgical navigation in urology: European perspective. Curr. Opin. Urol. **24**, 81–97 (2014)
14. Selmi, S.Y., Fiard, G., Promayon, E., Vadcard, L., Troccaz, J.: A virtual reality simulator combining a learning environment and clinical case database for image-guided prostate biopsy. In: Proceedings of CBMS 2013—26th IEEE International Symposium on Computer-Based Medical Systems, pp. 179–184 (2013)
15. Chellali, A., Dumas, C, Milleville-Pennel, I.: Haptic communication to support biopsy procedures learning in virtual environments. Presence Teleoperators Virtual Environ. **22**, 470–489 (2013)
16. Villard, P.F., Vidal, F.P., Ap Cenydd, L., Holbrey, R., Pisharody, S., Johnson, S., Bulpitt, A., John, N.W., Bello, F., Gould, D.: Interventional radiology virtual simulator for liver biopsy. Int. J. Comput. Assist. Radiol. Surg. **9**, 255–267 (2014)
17. Plitea, N., Pîslă, D., Vaida, C., Gherman, B., Tucan, P., Govor, C., Covaciu, F.: Family of innovative parallel robots for transperineal prostate biopsy, Patent pending A/00191/13.03.2015
18. Hemmerle, J.S., Levent Gürsöz E., Prinz, F.: Rapid interference detection. In: Expert Systems and Robotics, pp. 233–241 (1991)

Towards Robot-Assisted Rehabilitation of Upper Limb Dysfunction

Irina Voiculescu, Stephen Cameron, Manfred Zabarauskas and Piotr Kozlowski

Abstract This paper proposes a robot design which combines an unobtrusive presence around the household with a patient rehabilitation tool. The robot is envisaged to spend most of its time in standby mode, springing into action at pre-set times in order to engage the patient into taking a rehabilitation program. The rehabilitation tool guides the patient through a set series of prescribed repetitive physical exercises, provides feedback and keeps track of the patient's progress, and finally summarises the feedback to a clinician who can supervise exercise uptake and effectiveness. We have achieved proof of this concept on two separate fronts which have been tested separately: an obstacle-avoidance robot which finds (and composes photographs of) people, and a stand-alone piece of software which displays and assesses physiotherapy exercises.

Keywords Mobile and autonomous robotics · Ambient assisted living · Modelling and simulation · Non-contact sensors · Kinect

1 Introduction

Robot-assisted homes, which once seemed impractical, are now emerging through the adoption of technologies such as Roomba [1], RITA (the Reliable Interactive Table Assistant) [2] and other obstacle-avoiding household robots. A potential use

I. Voiculescu (✉) · S. Cameron · M. Zabarauskas · P. Kozlowski
Department of Computer Science, Oxford University, Oxford, UK
e-mail: irina@cs.ox.ac.uk

S. Cameron
e-mail: cameron@cs.ox.ac.uk

T. Borangiu (ed.), *Advances in Robot Design and Intelligent Control*,
Advances in Intelligent Systems and Computing 371,
DOI 10.1007/978-3-319-21290-6_35

of such robots is to monitor patients while taking rehabilitation exercises. Motor disorders are associated with a variety of health conditions such as stroke, cerebral palsy or Parkinson's disease, and with patients of any age, from children to the elderly. Cases where one or more limbs are affected need clinical assessment in order to detect the extent of the limb dysfunction. Subsequently physiotherapy is prescribed for rehabilitation, and take-home patient rehabilitation programs are gradually becoming an appealing prospect to health professionals and providers, partly as they can save face-to-face clinician time and partly because patient uptake of rehabilitation exercises may be increased by a 'gadget' solution.

This paper proposes a proof-of-concept robot that combines an unobtrusive presence around the household, potential for self-deployment (for elderly or forgetful patients), and a rehabilitation assessment tool that can guide a patient through a series of prescribed repetitive physical exercises.

The main sensor proposed for the robot is an RGB-D sensor like the Kinect [3] motion sensing device. The sensor captures RGB image frames from its on-board camera and a depth image from a depth infrared camera. These devices are used for obstacle avoidance and for tracking the principal joints in a subject's upper limbs (see Fig. 1); hence our focus on rehabilitation of upper limb dysfunction. (One can also envisage extensions applicable to other parts of the body.)

We have achieved proof of this concept on two separate fronts: an obstacle-avoidance robot which finds (and frames) people, and a stand-alone piece of software which displays and assesses physiotherapy exercises. These two components of the robot have been tested [4, 5], but assembling them into a single operational unit is a task which will involve a non-trivial amount of redesign and further software development.

2 Robot Overview

2.1 Architecture

The first part of the design consists of a robot photographer. This was constructed as part of a graduate project [4] as shown in Fig. 1, and lends an appropriate hardware structure to the robot proposed here using an iClebo Kobuki base [6]. The robot photographer wanders around a room, locating people by detecting their faces [7], taking then their photographs and uploading them online. The point of taking the person's photograph is to ensure that the robot is positioned at the correct distance to observe the person's head and upper body, so as to be able to detect their upper limbs. In order to test the photography side, the robot's architecture includes a tripod and camera mounted on top of the base, both of which can be eliminated in subsequent models.

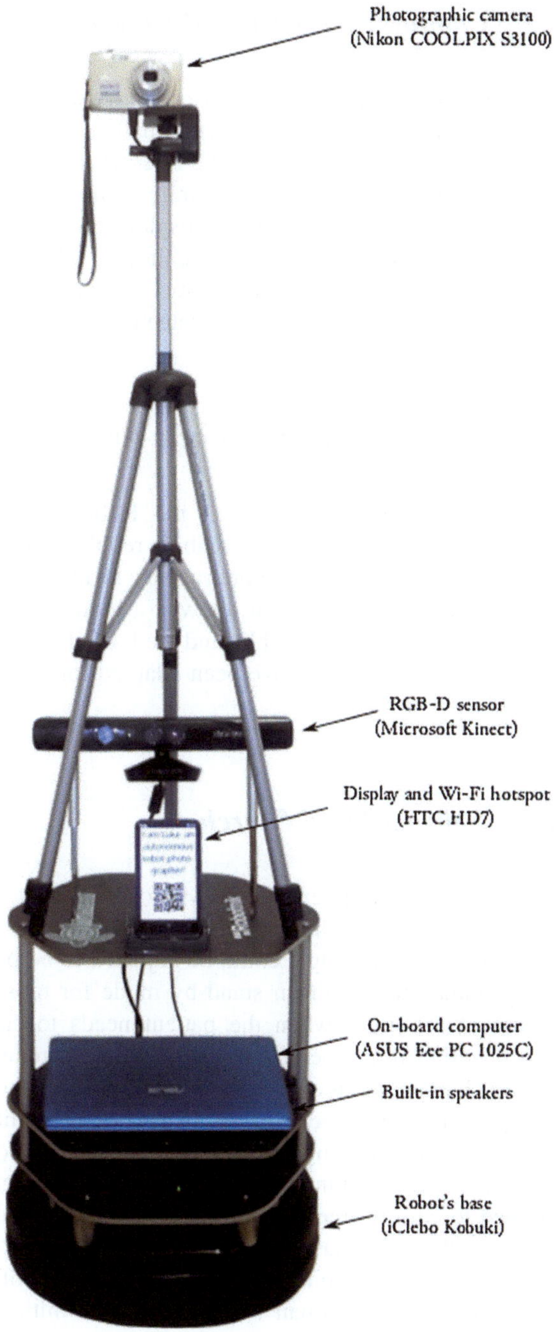

Fig. 1 Prototype photographer

2.2 Computing Power and Networking

The base is designed such that it can carry a laptop computer; wiring and sockets are provided for this purpose. Additionally, a smartphone was mounted on-board to provide wireless connection to the Internet (to upload the photographs taken and to provide debugging data to a separate computer present in the room). The smartphone also serves as the robot's display screen.

We have described in [8] the architectural design of the robot's software and the way the various components interact with each other in order to allow for navigation, obstacle avoidance, face detection and input-output operations.

2.3 Navigation and Obstacle Avoidance

Our previous experiments with the Robot Sheepdog [9] showed how simple 'potential-field' ideas, which combine repulsive and attractive fields for each object in the scene, can allow robotic agents to interact with each other and with their environment. A version that allowed a human user to drive a dog among robotic sheep was a hit with the public and the UK's Royal Society Summer Fair as early as 2001 [10]. These ideas have been adapted for the path planning component of the robot.

2.4 Sensor-Based Search

The main on-board sensor is a Microsoft Kinect, mounted on the platform at an angle of 10° to the horizontal plane, in order to point roughly to average adult human height; this angle could be adjusted to be patient-specific. The robot can be programmed to remain in stand-by mode for most of the day, and to spring into action at set times when the patient needs to carry out their exercises. At the pre-specified time, the robot self deploys, searches around the household for the nearest human and requests for them to practice their rehabilitation programme. At this point there may be scope for using face recognition software in order to find the correct human, although if the robot is designed for assisted living it is envisaged for use by people living on their own. A robot designed for use by children could be trained to confirm their identity through an entertaining set of questions.

Once the robot has navigated close enough to 'frame' its subjects, it stops and takes their photograph. A fully-fledged version of the robot would, at this point, negotiate with the human whether or not to initiate a physiotherapy session.

2.5 Gesture Assessment

The second part of the design is based on the same Kinect sensor used in the obstacle avoidance process. This is also used to monitor and score the patient's exercises. It would be easy keeping track of the patient's longitudinal progress, as well as to send regular updates to a clinician. This saves the clinician face-to-face appointment time where they would be checking whether or not the patient has been keeping on track. Additionally, the clinician can monitor the progress scores for the affected limb.

The current system provides an environment for the deployment of repetitive therapeutic tasks, but is based on a standalone application running on a conventional computer fitted with a Kinect sensor, rather than on board a robot.

We compare a series of Kinect readings, corresponding to hand actions, against pre-existing shapes such as those shown in Fig. 2. Each subject is asked to perform a range of simple arm gestures. These were chosen in consultation with researchers from the Department of Experimental Psychology in Oxford so as to be relevant to patients with apraxia following brain damage (e.g. due to stroke). Apraxic patients display disorganisation when carrying out sequential tasks, in particular being unable to plan and execute the complex sequence of motor actions [11].

Each of the subject's gestures is calibrated to a standard overall size and then assigned a score against the expected shape. The calibration is carried out relative to the Kinect sensor's detection of joint positions of the subject's body parts (head and hip), which give an estimate of the subject's upper body height, and also relative to the subject's distance from the sensor which is inferable from the depth data.

The score for each gesture is computed by comparing two sets of points using an algorithm by Catuhe [12]. Our algorithmic search traverses the list of joint positions detected by the Kinect sensor, and verifies predefined constraints. For instance, during a movement along a horizontal line from the patient's left to their right we consider the following basic checks:

- Each new hand position should be further to the patient's right.
- The height of each hand position must be within a predefined threshold of the hand's initial height.
- The first and last hand positions must be achieved within a predefined timeframe.
- The whole gesture must exceed a predefined minimum length.

Such constraints are gesture-shape-specific, which makes them difficult to be generalised. In the context of prescribed repetitive exercises the shapes can be pre-defined and so this is not a major drawback. Our application has defined a set of criteria to be used in the automatic analysis of these specific gesture shapes. These rely on the distance between consecutive positions of the hand at consecutive time steps, as well as on the change in line angle between the line of consecutive pairs

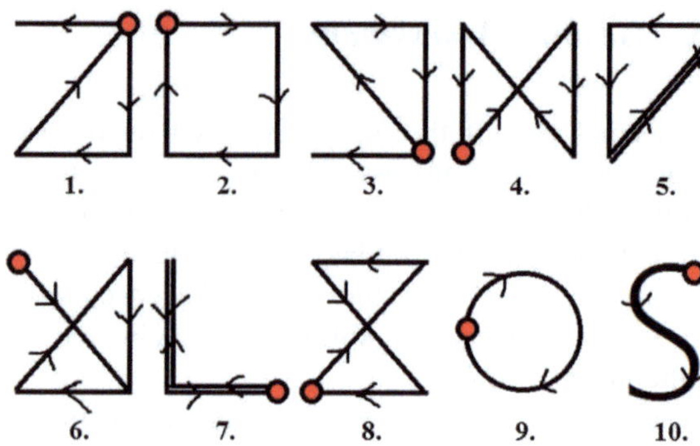

Fig. 2 A standard set of gestures

(or sets) of positions and the horizontal direction. These parameters can help computing the turning points (i.e. corners) of each gesture, and hence its centroid. The scores attributed to each gesture are based on a weighted sum of the following measurements:

1. Size and shape of actual gesture against prototype.
2. Average angular deviation of actual gesture from prototype.
3. Position of the centroid relative to the subject's own body.
4. Total speed of completing the gesture.
5. Relative speed uniformity across the figure parts (ratios of polyline segment lengths to data points per segment).

Data from healthy controls was collected and these criteria have been calibrated so as to determine the expected scoring range for a healthy individual. We also tackled the scoring using a Logistic Regression approach, but the relatively small volume of data we have collected so far did not lend itself easily to machine learning techniques. The results were comparable to those from the deterministic approach.

3 Results

3.1 Results on Human Detection and Framing

Human faces were detected successfully, and framed in appropriate proportions in the photographs. Human volunteers were asked to rate over 100 of the photographs taken in a single three-hour photo shoot, rating them on a scale from 1 ("very bad") to 5 ("very good"). More than 50 % of the photographs were rated by humans to be "good" or "very good", and more than 80 % were "neutral" or better. Examples of the face detection and overall framing are shown in Fig. 3. These results outperform other photographer robots [13].

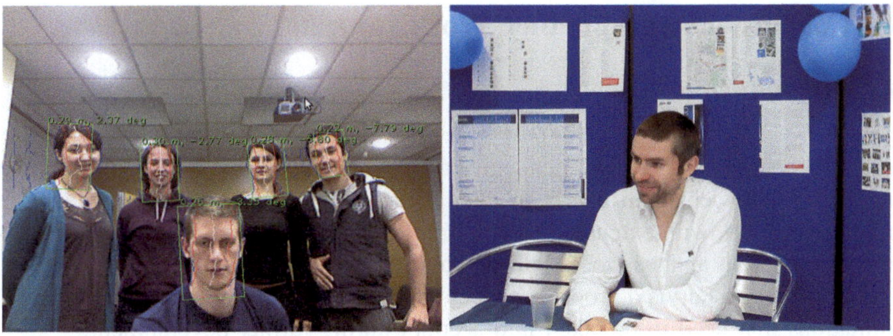

Fig. 3 Examples of face detection and picture composition

3.2 Results on Gesture Measurement

The gesture measurement has been carried out as a post-processing stage, after the tracking of a single gesture has been completed. This is mainly because the start and end points of each tracked gesture have been established using a separate process. Complexity-wise, all the elements of the processing may be run in real time.

As well as being assigned a score automatically, each gesture was also scored by a human evaluator. The human evaluation is known to be subjective, so the automatic scoring system is also likely to improve the objectivity of this assessment. The human scores did not cover evenly the scale of scores between 0 and 1. In the absence of any other objective measure, this scoring nevertheless had to be considered as a gold standard against which the machine scores were compared.

A total of 247 data points were collected for a variety of gestures. All five measurements mentioned were combined with equal weights to compute the final score. The differences between the scores allocated for each gesture by the human and by the program for each gesture yield an average deviation of 0.10324, which is around 10 % of the scale (Fig. 4). Interestingly, no significant improvement was made by manipulating the weights.

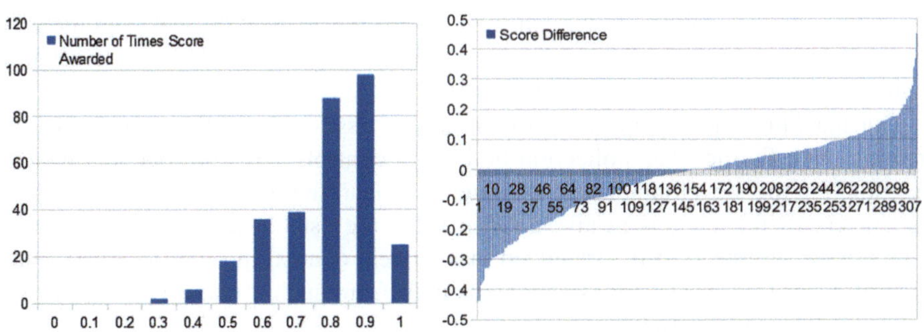

Fig. 4 Scores for gesture measurement and differences

4 Conclusions and Further Work

4.1 Conclusions

The two systems worked independently as well as could have been expected. No interaction between them has been attempted, but they have been designed in a modular fashion, with extensibility in mind. Further prototypes of the robot and its intelligent action analysis tools will be based on this work.

This robot system is a proof-of-concept and there are several (mainly hardware) limitations that need to be taken into account during further development:

- The Kinect has limited range and can be affected by heat sources (such as the Sun). However alternative RGB-D cameras can be used as they are developed, such as those being developed under Intel's RealSense initiative [14] or alternatives.
- Sideways poses as well as self-occlusion provide some challenges regarding the part of the user that is not visible to the sensor, resulting in some joints not being tracked accurately. Further Kinect releases are expected to address this.
- The information from the Kinect sensor that is processed by the software development kit is refreshed periodically at a maximum frame rate of 30 frames per second; at present there is a trade-off between smoothness and latency of movement.

4.2 Further Work

Personal care robots, such as the one outlined here, will need to be accepted by users as desirable and safe. As part of that process, ISO subcommittee TC184/SC2 has been working on a draft international standard for such devices, which has been given a standards number ISO-13482. We intend to look further at the ramifications of this work, which includes issues such as physical and electrical safety, together with standards for how the device should be operated in a non-research environment. In particular, robots are complicated electro-mechanical devices, and the way that different pieces of software can be combined together in a safe and reliable manner. They must also be consumer proof.

Our machine learning approach to the gesture assessment suffered from lack of training data. Further data collection both from controls and from patients will be essential in the development of a systematic learning mechanism for relevant parameters of the various gestures. Once these are established, randomized controlled trials will need to be organized in a clinical context in order to assess what proportion of (e.g. post-stroke) patients benefit from such a recovery program for upper limb motor function.

Some studies suggest that augmented feedback (also known as reinforced feedback in virtual environment) can be more effective than traditional rehabilitation methods [15, 16]. Our discussion with experts in post-stroke brain lesions, with hand surgeons and with physiotherapists working with children with cerebral palsy is on-going, aiming to find the most appropriate exercises to simulate in the assessment environment, as well as the most effective level of patient feedback.

Acknowledgements We are grateful to Prof Glyn Humphrey's insights on apraxia-related gestures and to Dr Mihaela Duta for system feedback and data collection.

References

1. iRobot Roomba http://www.irobot.co.uk
2. RITA—The Reliable Interactive Table Assistant http://www.assistobot.com/rita/en/
3. Kinect http://www.microsoft.com/en-us/kinectforwindows/
4. Zabarauskas, M.: Improving autonomous robot photography using RGB-D sensors. MSc thesis, Department of Computer Science, Oxford University (2013)
5. Kozlowski, P.: Measuring gesture mimicking accuracy with microsoft kinect, Fourth Year Project, Department of Computer Science, Oxford University (2013)
6. iClebo Kobuki http://kobuki.yujinrobot.com/home-en/
7. Viola, P. and M. Jones.: Rapid object detection using a boosted cascade of simple features. In: Proceedings of IEEE Computer Society Conference on Computer Vision and Pattern Recognition (2001)
8. Zabarauskas, M., Cameron: Luke, S.: An Autonomous robot photographer. In: Proceedings of ICRA (2014)
9. Vaughan, R., Sumpter, N., Henderson, J., Frost, A., Cameron, S.: Experiments in automatic flock control. Robot Auton Syst **31**, 109–117 (2000)
10. Royal Society Summer Science Exhibition https://royalsociety.org/summer-science/
11. Bickerton, W.-L., et al.: Systematic assessment of apraxia and functional predictions from the Birmingham Cognitive Screen. J Neurol Neurosurg Psychiatry **83**, 513–521 (2012)
12. Catuhe, L.: Gestures and tools for kinect (2011)
13. Ahn, D. et al.: Robot photographer with user interactivity. In: Proceedings of Intelligent Robots and Systems, pp. 5637–5643 (2006)
14. Intel RealSense, http://intel.com/RealSense
15. Kiper, P., et al.: Reinforced feedback in virtual environment for rehabilitation of upper extremity dysfunction after stroke: preliminary data from a randomized controlled trial. Biomed. Res. Int. (2014). doi:10.1155/2014/752128
16. Kwakkel, G. et al.: Effects of robot-assisted therapy on upper limb recovery after stroke: a systematic review. Clin Neurol. (2014)

Simulation and Control of a Robotic Device for Cardio-Circulatory Rehabilitation

Carlo Ferraresi, Daniela Maffiodo and Hamidreza Hajimirzaalian

Abstract In immobilized patients like paraplegic subjects, due to absence of leg muscle contraction, venous return to the heart is reduced and this may induce important diseases to the cardio-circulatory system. The application to legs of a mechanical stimulation operated by an Intermittent Pneumatic Compression (IPC) device in replacing striate muscle pump on limb veins may recover venous return to the heart, thus restoring correct cardio-circulatory performance. This paper deals with the study of an effective way to control such a device. Based on the numerical simulation of the human-machine system, the paper investigates and proposes an effective control solution.

Keywords Cardio-circulatory rehabilitation · Intermittent pneumatic compression · IPC mechatronic device · Man-machine system · Robotized rehabilitation

1 Introduction

Nowadays, the great majority of rehabilitation robotic devices are addressed to mobility. In these applications, an actively controlled robotic device applies programmed forces to induce specific loading or motion condition in a given part of the body. In a broader sense, a robotized rehabilitation can be adopted not only to maintain or improve limbs' mobility, but also to deal with diseases affecting other physiological compartments, like the cardio-circulatory system (CCS).

C. Ferraresi (✉) · D. Maffiodo · H. Hajimirzaalian
Department of Mechanical and Aerospace Engineering, Politecnico di Torino, Italy
e-mail: carlo.ferraresi@polito.it

D. Maffiodo
e-mail: daniela.maffiodo@polito.it

H. Hajimirzaalian
e-mail: hamidreza.hajimirzaalian@polito.it

T. Borangiu (ed.), *Advances in Robot Design and Intelligent Control*,
Advances in Intelligent Systems and Computing 371,
DOI 10.1007/978-3-319-21290-6_36

Actually, during any physical activity, the muscles contraction exerts important pumping action on physiological systems such as the cardiovascular or lymphatic system. In immobilized people, due to absence of leg muscle contraction, venous return to the heart is reduced and this may induce a reduction of cardiac output (CO). As a matter of fact, CO is directly influenced by venous return, which is determined primarily by the mechanical properties of the systemic circulation. The application to legs of a mechanical stimulation operated by actuators in replacing striate muscle pump on limbs veins, may recover in these patients venous return to the heart, thus restoring sufficient performance of the cardiovascular system.

Since the 1930s, it was evaluated that intermittent pneumatic compression (IPC) has a positive effect on lower extremity blood flow of calf and foot [1, 2]. Starting from the sixties, many IPC devices were developed to accommodate different medical applications: to prevent deep vein thrombosis [3, 4], for the treatment of critical limb ischemia [5–7], to compensate congestive heart failure [8] and were also recommended for lymphedema treatment [9] and sports recovery [10, 11]. Studies have shown that IPC application on legs changes central venous pressure, pulmonary artery pressure, and pulse pressure [12] and IPC application on foot increases popliteal artery blood flow [13]. Some researchers [14–17] compared different IPC devices observing positive results, even if devices were different and worked with different parameters.

Although it is possible to find in the literature the need for a more quantitative study of these devices, for example trying to model the deformation of the limb [18], the need to understand the dynamic behaviour of the device seems to be barely considered. Few researchers tried somehow to consider the effect of pressure level and inflation/deflation time [19]. Others [14] state that also further device parameters have to be considered and that further study is needed. Therefore, despite many researchers investigated on IPC effects, the lack of a general methodology to design and control an IPC device suitable for the intended application is evident. In particular, the dynamic behaviour of the device in response of the control command is practically ignored. A "robotic" approach in developing such a system is focused in particular on the appropriate control of the mechanical action exerted on the human body by the device.

Previous work [20, 21] led to the realization of different prototypes of IPC devices aimed at affecting the venous return in legs; this paper extends the results obtained in those studies. Based on the numerical simulation of the device and its interaction with the human, the paper proposes the study of an appropriate control strategy for the realization of an IPC device able to effectively act on the cardio-circulatory system in order to recover normal performance in impaired people.

2 The IPC Device

The present study is focused on a prototype previously realized in our laboratory. The main differences with respect to other IPC devices have been suggested by the application: as described in the introduction, IPC is currently used for lymphedema

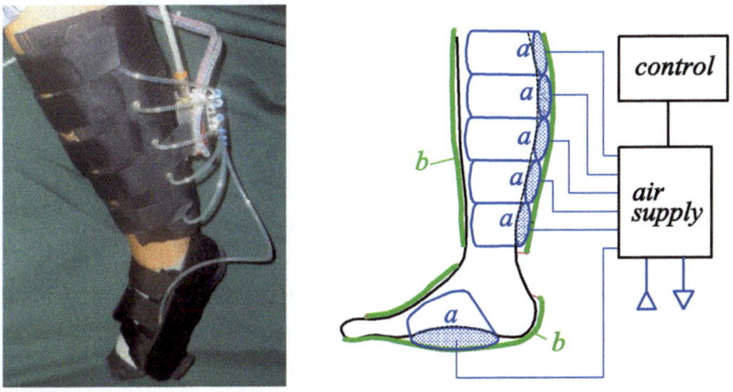

Fig. 1 Prototype of IPC device and its basic functional scheme

or limb ischemia treatment, deep vein thrombosis prevention, sport recovery, or even aesthetic treatment; all these applications are relatively low demanding as regards pressure level and response velocity. In the present application, the target was to design a device able to influence the hemodynamics of the cardio-circulatory system; therefore it is necessary to guarantee higher performance for the control of the limb/device contact pressure.

Figure 1 shows the prototype and its basic functional scheme. It includes inflatable bladders (a) supported by shells (b), a pneumatic circuit with on/off electro-pneumatic valves for the air supply and a control unit for effective control of the inner pressure in bladders. The goal is to generate on the limb surface a peristaltic and centripetal pressure wave with proper pattern.

The main shortcoming of this system was the lack of direct control of the action actually produced on the limb. The dynamic evolution of the pressure in the bladders is due not only to the time sequence imposed by the controller, but also to the characteristic of the pneumatic circuit, to physiological parameters of the subject and to the way the device is coupled with the limb.

3 PID Control of the Device

To simulate and understand the physical behaviour of the device while acting on the human an overall mathematical model was realized considering main physiological and mechanical principles, including all properties of the pneumatic circuit. A complete description of the model can be found in [20, 21]. The study of the control strategy here presented has been performed by integrating in the former numerical model all elements and logical algorithms necessary for this purpose.

Currently poor knowledge exists on a direct correlation between a pressing action on the leg and corresponding effect on the cardio-circulatory system (CCS); therefore it is impossible to define an "optimal" reference function as regards the pattern of the applied pressure. In this context, a PID controller has been considered adequate, due to its simplicity as concerns the design, the parameter-tuning and the hardware implementation [22].

To tune the PID controller and determine the values of coefficients (K_P, K_I, K_D), the mix method of Ziegler-Nichols [23] and manual tuning has been used.

In the IPC device, the contact pressure P_C is considered as the output controlled value, while for the value regulated by the PID controller either the supply pressure P_s or the mass flow rate Q can be considered. In the following, only the first method is described. Figure 2 shows the block scheme of the single-bladder system with supply pressure controlled by a pressure proportional valve (PPV).

To design the PID control, the mathematical model of the PPV has been integrated into the general model. The model of the PPV can be represented by the block diagram shown in Fig. 3 [24], where Kv is the flow coefficient, Tv is the time constant and Cv is the interior capacity of the proportional valve. P_{in} and P_{out} are the input and the output pressure of the proportional valve respectively. For a commercial proportional valve type Festo MPPE-3-1/4 which has been considered in simulations, the parameter values are $Kv = 5 \times 10^{-7}\ \mathrm{m^3/Pa/s}$, $Tv = 0.01$ s and $Cv = 1.43 \times 10^{-8}\ \mathrm{m^3/Pa}$.

With a reference contact pressure of 3.5×10^4 Pa relative to the ambient in inflating and zero in the deflating state, the best tuning was found with the coefficients: $K_P = 0.03$, $K_I = 0$, $K_D = 0.0044$, i.e. in this case the best solution points to a PD controller.

To verify the effectiveness of the method, a comparison was made between simulated and experimental results. Such a validation was made on a single-bladder system, and demonstrated the correctness of the PID coefficients.

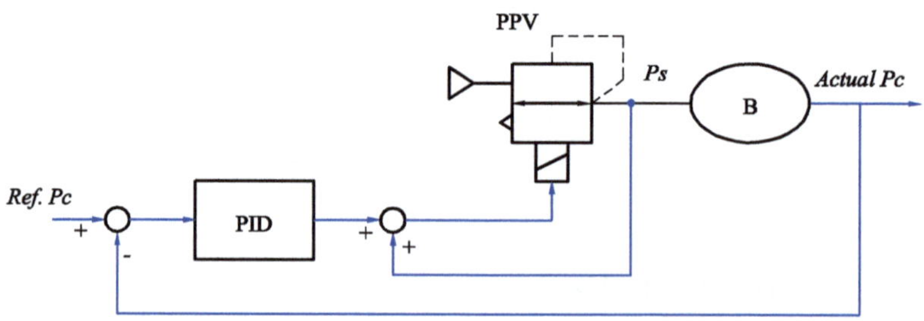

Fig. 2 Block scheme of the system with supply pressure (Ps) controlled by pressure proportional valve (PPV); contact pressure (Pc) is the outlet from bladder (B)

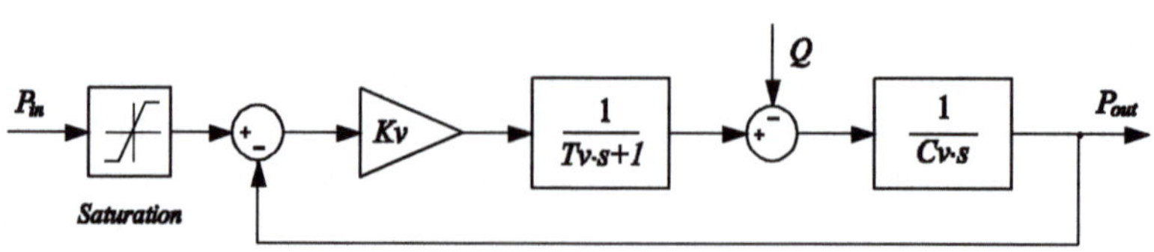

Fig. 3 Block diagram of a pressure proportional valve

Afterwards a simulation was performed on a system made up of three bladders controlled by a single proportional valve. This latter condition is particularly significant, since it represents the actual operating condition of the device: in fact, an IPC device includes a given number of bladders, which must be controlled in a coordinated way, in order to generate a definite pressure pattern on the limb surface. It is supposed that the device should generate a peristaltic and centripetal pressure wave, i.e. proceeding from distal to proximal position, in order to facilitate the venous blood return to the heart. Thus a definite reference pressure must be tracked in any bladder.

For such a multi-bladder system the controller must switch among the bladders, commanding the corresponding on-off control valve, selecting the proper contact pressure feedback and sending the proper pressure reference to the PID controller. A possible scheme of the system is presented in Fig. 4, where, posing $n = 1, 2, 3$, *Bn* are the bladders, *Vn* are the corresponding on-off control valves, *Pn* are the bladder-limb actual contact pressures, and PPV is the pressure-proportional electro-valve. Switching among bladders is performed by the controller through a specific function indicated by PLC in the scheme.

In the simulation, the three bladders were configured with different parameters, including the contact area *A*, the soft tissue stiffness K_M of the leg region where the bladder acts on, and the gap between the bladder and leg d_0. The values of these parameters were selected in such a way as to represent possible variations due to

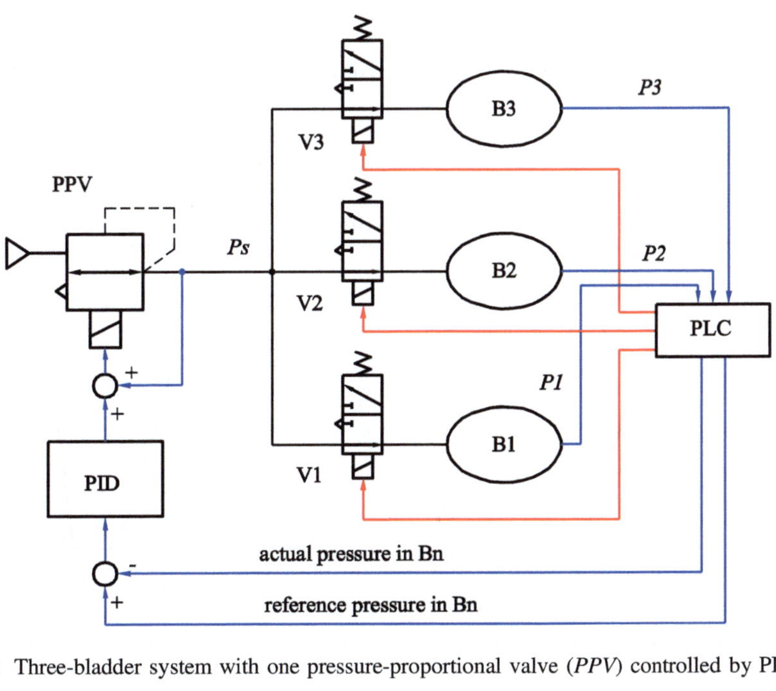

Fig. 4 Three-bladder system with one pressure-proportional valve (*PPV*) controlled by PID

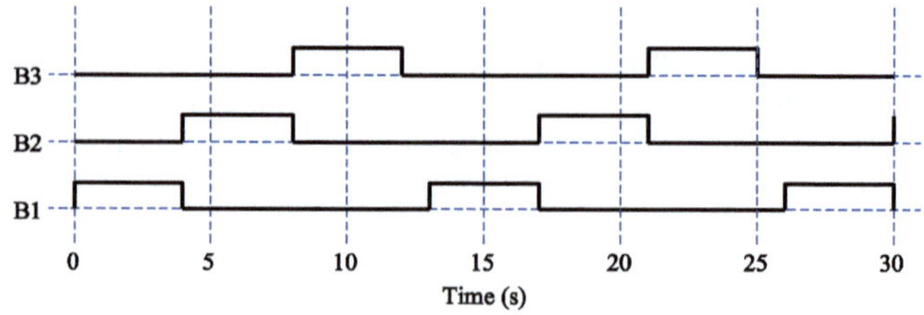

Fig. 5 Command sequence of the three bladders

muscular condition, location of the bladder on the limb and different device dressing that may occur using the real prototype.

The inflating and deflating times, imposed by the PLC function for each bladder, were 4 s and 9 s respectively and the three cycles were shifted to produce a peristaltic effect, as it is shown in the command sequence given in Fig. 5. When the command is OFF the reference pressure is *0*, when the command is ON the reference pressure is set at a defined value for all bladders.

The system was simulated in two conditions:

1. Without proportional valve and PID control: each control valve *Vn* is supplied at 5×10^4 Pa relative pressure.
2. With proportional valve and PID control: the absolute supply pressure to the valves *Vn* is regulated by the proportional valve and saturated at 2×10^5 Pa; the relative reference pressure to the PID is set to 3.5×10^4 Pa in inflating state and *0* in deflating state, for all bladders.

Figure 6 shows the simulation results without PID controller (a) and with PID controller (b).

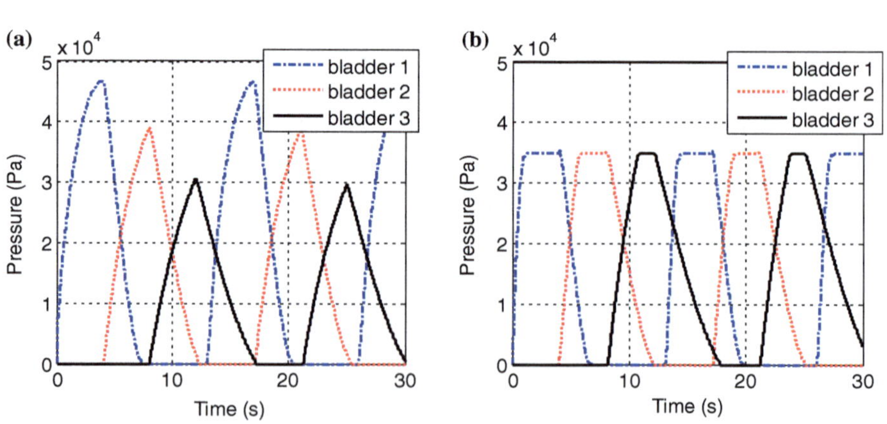

Fig. 6 Simulation of the multi-bladder system in absence (**a**) and in presence (**b**) of the PID controller

The simulation shows that an effective control of the bladder actions is almost impossible without a feedback control system, since the dynamic behaviour and the maximum level of the pressures are affected by a number of randomness that are difficult to evaluate properly. As it can be observed in Fig. 6a, the dynamic behaviour and the maximum level of the contact pressure are very different among the bladders, due to the differences among the bladder parameters. On the contrary, when the PID controller acts on the system (Fig. 6b), the contact pressure tracks much better the reference one, which is set at 3.5×10^4 Pa, in each bladder. Also the system dynamics is greatly improved in the PID solution, since the saturation of the proportional valve at 2×10^5 Pa provides the system much more energy than in the other solution where air supply must be limited at 5×10^4 Pa for bladders' safeguard.

Due to the difference in the bladder parameters, the dynamic behaviours of the three bladders are still different, as highlighted by the rise and fall times, but each bladder can reach the reference contact pressure and maintains it for a significant time. You should also consider that the tuning of the PID coefficients has been made for only one bladder, but gives acceptable results for the whole system in which the bladder characteristics were forced to exaggeratedly different values, thus verifying the ability to compensate at some extent also uncertainties like the actual value of the soft tissue stiffness and the limb/device gap.

4 Conclusion

This study concerned the effective control of an IPC device, conceived as a multi-bladder system able to realize important rehabilitation action on the cardio-circulatory system of subjects with serious mobility problems. The main issue was to individuate a proper control strategy able to create a defined pressure pattern on the limb surface, corresponding to a peristaltic and centripetal pressure wave. Such an effect could be achieved simply by using digital three-way electro-valves controlled by a programmable logic controller (PLC), but simulations showed that this solution, profitable for its low cost and weight, produces very poor results in terms of accuracy.

The method here proposed is based on the use of one electro-pneumatic pressure-proportional valve to supply the whole IPC multi-bladder device. The results show that the bladders exert to the limb the exact value of the reference pressure for about the 60 % of the desired active time (4 s). This is an excellent result, considering the natural low dynamics of such pneumatic systems and the variability of the device operating conditions. On the contrary, in absence of PID control the bladder pressure reaches a value that is practically impossible to foresee, due to the uncertainty of the physical parameters.

Future works will concern more accurate modelling of the physiological system, including also damping characteristic of the soft tissues and a complete model of the cardio-circulatory system. This will allow estimating and controlling the effect of the IPC device on the dynamic performance of the CCS, which is the final goal of the research.

References

1. Herrmann, L.G., Reid, M.R.: The conservative treatment of arteriosclerotic peripheral vascular disease. Ann Surg. **100** (1934)
2. Landis, E.M., Gibbon, J.H.: The effect of alternative suction and pressure on blood flow to the lower extremities. J. Clin. Invest. **12**(5) (1933)
3. Clagett, G.P., Reisch, J.S.: Prevention of Venus thromboembolism in general surgical patients. Results of meta-analysis. Ann. Surg. **208**, 227–240 (1988)
4. Westrich, G.H., Haas, S.B., Mosca, P., Peterson, M.: Meta-analysis of thromboembolic prophylaxis after total knee arthroplasty. J. Bone Joint Surg. Br. **82**, 795–800 (2000)
5. Louridas, G., Saadia, R., Spelay, J., Abdoh, A., Weighell, W., Arneja, A.S., Tanner, J., Guzman, R.: The art assist device in chronic lower limb ischemia. A pilot study, Int. Angiol. **21**, 28–35 (2002)
6. Delis, K.T., Nicolaides, A.N., Wolfe, J.H., Stansby, G.: Improving walking ability and ankle brachial pressure indices in symptomatic peripheral vascular disease with intermittent pneumatic foot compression: a prospective controlled study with one-year follow-up. J. Vasc. Surg. **31**, 650–661 (2000)
7. Labropoulos, N., Leon, L.R., Bhatti, A., Melton, S., Kang, S.S., Mansour, A.M., Borge, M.: Hemodynamic effect of intermittent pneumatic compression in patients with critical limb ischemia. J. Vasc. Surg. **42**(4), 710–716 (2005)
8. Nose, Y., Murata, K., Yasuaki, W., Tanaka, T., Fukagawa, Y., Yoshino, H., Susa, T., Kihara, C., Matsuzaki, M.: The impact of intermittent pneumatic compression device on deep Venus flow velocity in patients with congestive heart failure. J. Cardiol. **55**(3), 384–390 (2010
9. Johansson, K., Lie, E., Ekdahl, C., Lindfelt, J.: A randomized study comparing manual lymph drainage with sequential pneumatic compression for treatment of postoperative arm lymphedema. J. Lymphol. **31**, 56–64 (1998)
10. Wiener, A., Mizrahi, J., Verbitsky, O.: Enhancement of tibialis anterior recovery by intermittent sequential pneumatic compression of the legs. Basic Appl. Myol. **11**(2), 87–90 (2001)
11. Waller, T., Caine, M., Morris, R.: Intermittent Pneumatic Compression Technology for Sports Recovery. The Engineering of Sport, vol. 3. ISBN: 978-0-387-34680-9 (Print), 978-0-387-45951-6 (Online) (2006)
12. Unger, R.J., Feiner, J.R.: Hemodynamic effects of intermittent pneumatic compression of the legs. Anaesthesiology **67**, 266–268 (1987)
13. Morgan, R.H., Carolan, G., Psaila, J.V., Gardner, A.M.N., Fox, R.H., Woodcock, J.P.: Arterial flow enhancement by impulse compression. J. Vasc. Surg. **25**, 8–15 (1991)
14. Lurie, F., Awaya, D.J., Kinster, R.L., Eklof, B.: Hemodynamic effect of intermittent pneumatic compression and the position of the body. J. Vasc. Surg. **37**, 137–142 (2003)
15. Flam, E., Berry, S., Coyle, A., Dardik, H., Raab, L.: Blood-flow augmentation of intermittent pneumatic compression systems used for the prevention of deep vein thrombosis prior to surgery. Am. J. Surg. **171**, 312–315 (1996)
16. Christen, Y., Wutscher, R., Weimer, D., de Moerloose, P., Kruithof, E.K.O., Bounameaux, H.: Effects of intermittent pneumatic compression on venous hemodynamics and fibrinolytic activity. Blood Coagul. Fibrinolysis **8**, 185–190 (1997)
17. Froimson, M.I., Murray, T.G., Fazekas, A.F.: Venous thromboembolic disease reduction with a portable pneumatic compression device. J. Arthroplasty **24**(2), 310–316 (2009)

18. Dai, G., Gertler, J.P., Kamm, R.D.: The effects of external compression on venous blood flow and tissue deformation in the lower leg. J. Biomech. Eng. **121**, 557–564 (1999)
19. Malone, M.D., Cisek, P.L., Comerota, A.J. Jr., Holland. B., Eid, I.G., Comerota, A.J.: High-pressure, rapid inflation pneumatic compression improves venous hemodynamics in healthy volunteers and patients who are post-thrombotic. J. Vasc. Surg. **29**(4), 593–599 (1999)
20. Ferraresi, C., Hajimirzaalian, H., Maffiodo, D.: A model-based method for the design of intermittent pneumatic compression systems acting on humans. Proc. Inst. Mech. Eng. Part H J. Eng. Med. (2013). doi:10.1177/0954411913516307
21. Hajimirzaalian, H.: Robotic biomedical device for recovering cardiovascular efficiency in paraplegic patients, PhD thesis, http://porto.polito.it/2540888/, April 2014
22. Tzafestas, S., Papanikolopoulos, N.P.: Incremental fuzzy expert PID control. IEEE Trans. Ind. Electron. **37**(5), 365–371 (1990)
23. Ziegler, J.G., Nichols, N.B.: Optimum settings for automatic controllers. Trans. ASME **64**, 759–768 (1942)
24. Ferraresi, C., Velardocchia, M.: Pneumatic proportional valve with internal feedback. In: Proceedings of the 9th International Symposium on Fluid Power, Oxford, UK (1990)

Part VIII
Tracking Systems and Unmanned Aerial Vehicles

Aiming Procedure for the Tracking System

Karol Dobrovodský and Pavel Andris

Abstract The tracking system monitors and recognizes several potential objects of interest. An operator selects an object for tracking. During the tracking mode the decision is made to initiate the procedure of aiming. Three measurements of the distance are used to establish motion parameters of the tracked object. The trajectory of the motion is calculated in order to specify the future hit point position. Finally, servos drive the system into the ballistic aiming position. To increase the hit probability, the whole tracking and aiming procedure is repeated automatically.

Keywords Tracking · Prediction · Ballistic trajectory

1 Introduction

The tracking system under consideration is able to detect, recognize and continuously track moving objects of various types.

There are several specific problems which a motion detection subsystem must deal with. First of all, it is not possible to identify the movements using a simple comparison of the sequence of consecutive images [1]. A vision subsystem should eliminate motion caused by camera sensor displacements. The system must recognize an object from the moving background, track its position and report the velocity and acceleration of the tracked object. The multi-level image segmentation is discussed in [2, 3]. Potential objects to be tracked must be first recognized without specifying their identity in the context of consecutive images. Then the estimation of their identity in the context of the previous positions may be

K. Dobrovodský (✉) · P. Andris
Institute of Informatics, Slovak Academy of Sciences, Bratislava, Slovakia
e-mail: utrrdobr@savba.sk

P. Andris
e-mail: utrrandr@savba.sk

T. Borangiu (ed.), *Advances in Robot Design and Intelligent Control*,
Advances in Intelligent Systems and Computing 371,
DOI 10.1007/978-3-319-21290-6_37

performed [4]. We have presented all these image processing problems, prediction of future positions and details on processing modes in [5].

In military applications, the ballistic aiming procedure is the final phase of the activity.

Having the time and space coordinates of the predicted hit point, we look up in the ballistic tables for the aiming azimuth and elevation together with the corresponding time interval which the projectile needs to reach the hit point. This time interval defines the moment of the shot. Both servos of the system should be stopped in the shot position. The final countdown is performed up to the right moment of the shot. The goal is to hit the tracked object by the projectile shot from the gun stopped in the shot position. To increase the relatively low hit probability, the motion along the predicted trajectory during burst time is performed. The next tracking begins immediately after the shooting.

2 Engagement Procedure

The vision subsystem recognizes many potential objects for tracking. At this stage, the system works in a searching mode and recognizes the most number of objects that may be subject to tracking. A simple differential filtration is used in this mode.

The system works with day and night cameras, which are oriented in space by two degrees of freedom. In the search mode the operator manually guides the optical axis of the camera using a joystick. In this mode the extensive recognition of all potential objects for monitoring is performed. A default distance is attributed to all detected objects tentatively. After the transformation of the position coordinates, objects are placed into a common 3D model of the world (the BASE coordinate system) without specifying their identity. For new objects, pairing is performed with the objects that already exist in the model from previous observations. The pairing allocates a new object identity, i.e. identification of objects in new positions. During the extensive phase the visual system recognizes many objects; priority is given to objects near the optical axis of the camera.

The operator is doing the selection from the objects which were recognized and the system starts the tracking mode. From this moment, the object becomes the only one being processed. Online differential filtration coupled with measuring the size and the brightness of the object is employed to increase the reliability of the tracking process. Especially for flying targets, it is difficult to initiate the tracking mode by putting the optical axis of the camera on the object to be tracked [6]. Moreover, the transition from an extensive to an intensive recognition is critical at the start of the tracking mode. Finally, the only one object becomes the subject of recognition during the tracking mode. The recognition is coupled with an automatic search for the object in the predicted position in this phase.

The system allows the operator to select from the objects near the camera optical axis. The tracking mode starts after an object of tracking is selected and the number of recognized objects is reduced to one.

In detail, the operator guides the optical axis of the camera to the stain of the proposed object. A visual subsystem should recognize the stain and frame it by a green frame—an icon. If the visual subsystem does not recognize the spot automatically, i.e. does not frame it, the operator is trying to achieve framing using a correction button on the joystick. The gun barrels are always parallel to the optical axes of the cameras.

3 Hit Point Prediction

On the assumption that the tracked object's motion is performed under constant acceleration, the following formula expresses the time dependence of the predicted position

$$\mathbf{r}(t) = \mathbf{r} + \mathbf{v}t + \mathbf{a}t^2/2. \tag{1}$$

At least three consecutive vector positions are needed for the three unknown vectors **r**, **v**, **a** of the prediction formula (1). The tracking system provides only two angular coordinates (*azimuth* and *elevation*) of the tracked object position. In order to obtain all three coordinates, a multi-camera tracking system [5] or a laser rangefinder has to be used.

Let us denote by $\mathbf{r}_i$ the triples of position coordinates (transformed into the BASE) measured in times t_i, so that we have the equations $\mathbf{r}(t_i) = \mathbf{r}_i$. We need at least three vector equations for the unknown vectors **r**, **v**, **a**. Particularly, the relevant solution for $t_1 = -2T$, $t_2 = -T$, and $t_3 = 0$ is

$$\mathbf{r} = \mathbf{r}_3, \tag{2}$$

$$\mathbf{v} = (1/2T)(\mathbf{r}_1 - 4\mathbf{r}_2 + 3\mathbf{r}_3), \tag{3}$$

$$\mathbf{a} = (1/T^2)(\mathbf{r}_1 - 2\mathbf{r}_2 + \mathbf{r}_3), \tag{4}$$

where T is the time interval between two distance measurements. In general:

$$D = (t_2t_3(t_3 - t_2) + t_1t_2(t_2 - t_1) + t_3t_1(t_1 - t_3))/2 \tag{5}$$

$$\mathbf{r} = D^{-1}[t_2t_3(t_3 - t_2)\mathbf{r}_1 + t_3t_1(t_1 - t_3)\mathbf{r}_2 + t_1t_2(t_2 - t_1)\mathbf{r}_3]/2 \tag{6}$$

$$\mathbf{v} = D^{-1}\left[(t_2^2 - t_3^2)\mathbf{r}_1 + (t_3^2 - t_1^2)\mathbf{r}_2 + (t_1^2 - t_2^2)\mathbf{r}_3\right]/2 \tag{7}$$

$$\mathbf{a} = D^{-1}[(t_3 - t_2)\mathbf{r}_1 + (t_1 - t_3)\mathbf{r}_2 + (t_2 - t_1)\mathbf{r}_3]. \tag{8}$$

From the practical point of view (noise, numerical errors), it is recommended to make much more measurements and to solve an overconstrained system of equations leading to such vectors **r**, **v**, **a** that guarantee the minimum of the sum of squares of differences between the left and the right sides of equations. In such case, we have n equations:

$$\mathbf{r}(t_i) = \mathbf{r}_i, \quad i = 1, 2, \ldots n, \tag{9}$$

and the following solution takes place:

$$\mathbf{p}_k = \sum \mathbf{r}_i t_i^k, \quad k = 0, 1, 2 \tag{10}$$

$$s_k = \sum t_i^k, \quad k = 0, 1, 2, 3, 4 \tag{11}$$

$$D = s_0 s_2 s_4 + 2 s_1 s_2 s_3 - s_2^3 - s_1^2 s_4 - s_0 s_3^2 \tag{12}$$

$$\mathbf{r} = D^{-1}\left[\left(s_2 s_4 - s_3^2\right)\mathbf{p}_0 + \left(s_2 s_3 - s_1 s_4\right)\mathbf{p}_1 + \left(s_1 s_3 - s_2^2\right)\mathbf{p}_2\right] \tag{13}$$

$$\mathbf{v} = D^{-1}\left[\left(s_2 s_3 - s_1 s_4\right)\mathbf{p}_0 + \left(s_0 s_4 - s_2^2\right)\mathbf{p}_1 + \left(s_1 s_2 - s_0 s_3\right)\mathbf{p}_2\right] \tag{14}$$

$$\mathbf{a} = 2D^{-1}\left[\left(s_1 s_3 - s_2^2\right)\mathbf{p}_0 + \left(s_1 s_2 - s_0 s_3\right)\mathbf{p}_1 + \left(s_0 s_2 - s_1^2\right)\mathbf{p}_2\right] \tag{15}$$

4 Aiming Procedure

A general approach to this problem lies in geometry and timing of the predicted motion trajectory of the moving object being tracked and the ballistic trajectory of the fired projectile. Both the moving target and the flying projectile should meet in the same point in space at the same instant of time.

The graphical user interface (GUI) of the tracking system during the first shot is shown in Fig. 1. The first shot is a part of the burst that follows.

This is a space-time problem that we solved in the following way. First of all the duration of the whole aiming process is estimate. This process consists in computing the time needed for servos to reach the aiming position t_{srv}, the time needed for projectile to reach the object t_{prj}, and the time of countdown needed to make the moment of the shot precise in milliseconds t_{cnt}. Thus we can write the sum:

$$t_{hit} = t_{srv} + t_{prj} + t_{cnt}. \tag{16}$$

In order to get t_{hit} realizable, we estimate the above sum to be greater than needed with regard on the actual distance of the tracked object. Having estimated the time t_{hit}, we can estimate the hit point coordinates using the motion prediction formula (1). Let us denote the coordinates:

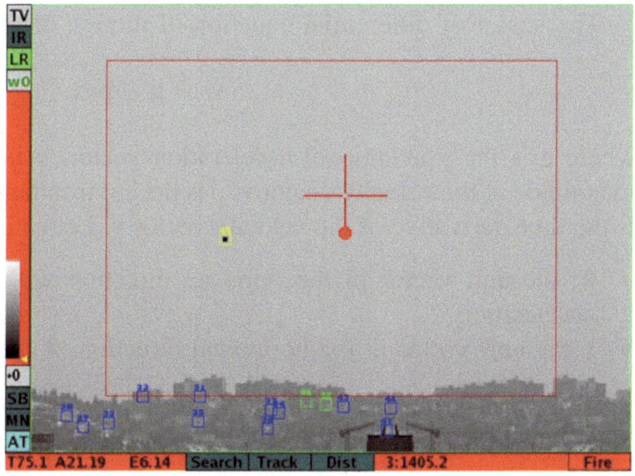

Fig. 1 Shot signalled by red strips on the left, below the picture. Symbols: aiming point (*the red cross*), predicted hit point (*the red spot*), and moving tracked object in *a yellow frame*

$$\mathbf{r}(t_{hit}) = (x_{hit},\ y_{hit},\ z_{hit}) \tag{17}$$

The next step is the calculation of the aiming azimuth:

$$\text{azimuth} = \text{atan2}(x_{hit},\ y_{hit}) \tag{18}$$

and finally, for the given horizontal distance:

$$d_{hit} = \text{sqrt}\left({x^2}_{hit} + {y^2}_{hit}\right) \tag{19}$$

and the given height z_{hit}, we look up from the ballistic tables for the aiming elevation together with the corresponding time to reach the object:

$$(d_{hit},\ z_{hit}) \Rightarrow \left(\text{elevation},\ t_{prj}\right). \tag{20}$$

Having the aiming azimuth and elevation, the control system starts positioning the servos to the aiming position in local time $t = 0$. The moment of a shot is defined by the subtraction:

$$t_{sht} = t_{hit} - t_{prj}. \tag{21}$$

Waiting for the moment of a shot is realized by the countdown in milliseconds.

5 Projectile Trajectory

This section deals with computing ballistic tables. It is an off-line numerical process that can take minutes. Data from the ballistic tables are used during the aiming procedure.

The vector of differential equation of motion of the projectile has the form:

$$\mathbf{v}' = \mathbf{g} - \mathrm{p}\mathrm{v}\mathbf{v} \tag{22}$$

where $\mathbf{g}$ is the gravitational acceleration vector, $\mathbf{v}$ is the velocity vector, v is the magnitude of the velocity vector $\mathbf{v}$, p is the environment resistance coefficient, and $\mathbf{v}'$ is the time derivative of the velocity vector $\mathbf{v}$. Let us introduce the two unit vectors:

- **k**: the unit vector in the opposite direction with respect on the gravitational acceleration
- **l**: the unit vector in the horizontal direction of the projectile.

The scalar product of the Eq. (22) with the introduced unit vectors yields the two scalar equations:

$$\mathrm{v}'_\mathrm{d} = -\mathrm{p}\mathrm{v}\mathrm{v}_\mathrm{d} \tag{23}$$

$$\mathrm{v}'_\mathrm{z} = -\mathrm{g} - \mathrm{p}\mathrm{v}\mathrm{v}_\mathrm{z}, \tag{24}$$

where $\mathrm{v_d}$ and $\mathrm{v_z}$ are respectively the horizontal and the vertical components of the velocity, the magnitude of the velocity vector being:

$$\mathrm{v} = \mathrm{sqrt}\left(\mathrm{v}_\mathrm{d}^2 + \mathrm{v}_\mathrm{z}^2\right) \tag{25}$$

The altitude dependency of the environment resistance coefficient is

$$\mathrm{p} = \mathrm{p}_0 \exp\left(-(\mathrm{z} - \mathrm{a})/\mathrm{b}\right), \tag{26}$$

where $\mathrm{p_0} = 0.00038\ \mathrm{m}^{-1}$ is the resistance at the altitude of a = 1000 m, b = 10,000 m and z is the altitude.

The discrete linear approximation of the relevant solution with the step t may be written in the form:

$$\mathrm{v(n)} = \mathrm{sqrt}\left[\mathrm{v}_\mathrm{d}^2(\mathrm{n}) + \mathrm{v}_\mathrm{z}^2(\mathrm{n})\right] \tag{27}$$

$$\mathrm{v_d(n+1) = v_d(n) - p_0 \exp[-(z(n) - a)/b] v(n) v_d(n) t} \tag{28}$$

$$\mathrm{v_z(n+1) = v_z(n) - gt - p_0 \exp[-(z(n) - a)/b] v(n) v_z(n) t} \tag{29}$$

$$\mathrm{d(n+1) = d(n) + [v_d(n+1) + v_d(n)] t/2} \tag{30}$$

$$\mathrm{z(n+1) = z(n) + [v_z(n+1) + v_z(n)] t/2.} \tag{31}$$

The initial conditions of the discrete approximation are:

$$v_d(0) = v_i \cos(e), v_z(0) = v_i \sin(e), d(0) = 0, z(0) = 0,$$

where e is the initial elevation and v_i is the magnitude of the projectile muzzle velocity.

The simulated position of the first projectile from a burst is shown in Fig. 2.

In order to reach the desired precision of the approximation, we choose the step t to be small enough (e.g. 1/100 or 1/1000 of 1 ms) and then we write only one from one hundred or one thousand points into the resulting ballistic tables.

In this way we get the precise approximation of the horizontal distance and the vertical height with the step of one millisecond.

Initial values are generated for elevation angles from $< -10, +90 >$ (in degrees) with the step of 0.0001 radian i.e. for 17,453 elevation angles. Points on the projectile trajectory using the time step of one millisecond up to 5 s are recorded for each of the elevation angles. The result is less than 88 million of the corresponding pairs:

$$(\text{elevation, time})\text{versus}(\text{horizontal distance, height}).$$

We expand the initial form of the ballistic tables by two columns—the *actual distance* and the *slope of tangent*:

$$D_i = \text{sqrt}\left(d_i^2 + z_i^2\right) \tag{32}$$

$$s_i = (z_{i+1} - z_i)/(d_{i+1} - d_i). \tag{33}$$

We shorten each trajectory of such a section in which the actual distance decreases.

The memory footprint of the ballistic tables with the range of 2.5 km is about 50 MB.

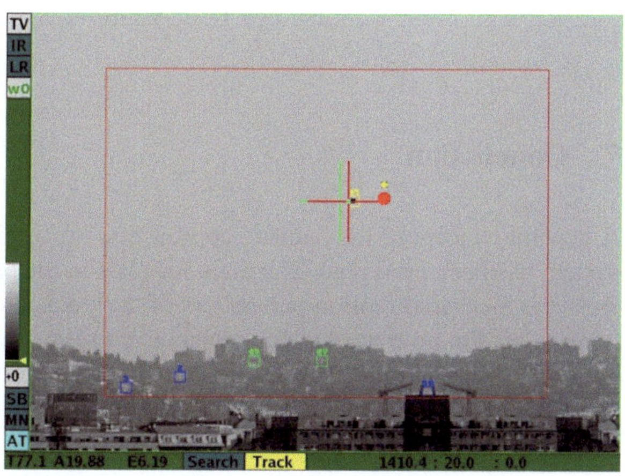

Fig. 2 The simulated projectile position (*the yellow point*) over the estimated hit point (*the red spot*) during repeated tracking (color figure online)

6 Searching in Ballistic Tables

The last point of each trajectory represents the maximum distance reachable with the corresponding elevation. Having given the hit point (z, d), all "reachable" trajectories corresponding to particular elevation angles split into two categories. There are trajectories passing above the given point and trajectories passing below the given point. The hit point is out of range otherwise. We choose the last elevation with trajectory below and the first elevation with trajectory above the given hit point using the slope of tangent between points with smaller and larger distance. We get two points for two elevations, i.e. four points. Skipping the farthest point from the given hit points there remain three equations with three unknown weights m_i for the remaining three points (z_i, d_i) and the given hit point (z, d):

$$\sum m_i = 1 \tag{34}$$

$$\sum m_i z_i = z \tag{35}$$

$$\sum m_i d_i = d. \tag{36}$$

One can write the solution in the form:

$$D = z_1(d_2 - d_3) + z_2(d_3 - d_1) + z_3(d_1 - d_2) \tag{37}$$

$$m_1 = [z_2d_3 - z_3d_2 + (d_2 - d_3)z + (z_3 - z_2)d]/D \tag{38}$$

$$m_2 = [z_3d_1 - z_1d_3 + (d_3 - d_1)z + (z_1 - z_3)d]/D \tag{39}$$

$$m_3 = [z_1d_2 - z_2d_1 + (d_1 - d_2)z + (z_2 - z_1)d]/D \tag{40}$$

The resulting product of interpolation is

$$(e, t) = \sum m_i(e_i, t_i) \tag{41}$$

7 Conclusion

A general concept of the ballistic approach to the aiming procedure for the tracking system has been presented. The main idea lies in the prediction of the static hit point from the measured motion parameters of the tracked object. The time of prediction consists of the time needed for projectile to reach the tracked object and the time needed for servos to reach the calculated ballistic aiming position. Having reached the aiming position, the final countdown is performed in milliseconds in order to hit the tracked object in the estimated time. With respect of the previous work, dealing with flying object recognition, active image filtration and various modes of the control system activity have been briefly discussed. The detailed geometric analysis of the projectile ballistic trajectory and the searching procedure in ballistic tables have been considered too.

The implementation is oriented toward the civil sector for monitoring purposes and toward the military sector for tracking and fire control purposes. The system has been developed in cooperation with a Slovak industrial partner and is based on the real-time kit [7]. It was used as semi-automatic fire control system for anti-aircraft auto cannon, being successfully tested on a shooting range.

Acknowledgments We thank the support of the Scientific Grant Agency of the Ministry of Education of the Slovak Republic and the Slovak Academy of Sciences, project number 2/0194/13.

References

1. Dobrovodský, K., Andris, P.: Automatic perception and monitoring of moving objects. In: Proceedings of the 16th International Workshop on Robotics in Alpe-Adria-Danube Region, CD, Ljubljana, Slovenia, 7–9 June 2007. ISBN 978-961-243066-5
2. Liao, P.-S., Chen, T.-S., Chung, P.-C.: A fast algorithm for multilevel thresholding. J Inf Sci Eng **17**, 713–727 (2001)
3. Arora, S., Acharya, J., Verma, A., Prasanta, K., Panigrahi, K.: Multilevel thresholding for image segmentation through a fast statistical recursive algorithm. arXiv:cs.CV/0602044 v1, Preprint submitted to Elsevier Science, May 17 (2006)
4. Dobrovodský, K., Andris, P.: Connectivity analysis with online identity estimation. In: Proceedings of the 17th International Workshop on Robotics in Alpe-Adria-Danube Region, CD, Ancona, Italy, 15–17 Sept 2008
5. Dobrovodský, K., Andris, P.: Real-time recognition and tracking of moving objects. Comput. Inf. **33**(6) (2014). ISSN 1335-9150
6. Dobrovodský, K., Andris, P.: Mechatronic vision and fire control system, In: Proceedings of the 18th International Workshop on Robotics in Alpe-Adria-Danube Region, Brasov, Romania, 25–27 May 2009. ISSN 2066-4745
7. Dobrovodský, K., Andris, P., Kurdel, P.: Computer kit for development, modeling, simulation and animation of mechatronic systems. In Proceedings of 15th International Workshop on Robotics in Alpe-Adria-Danube Region, CD, June 15–17, Balatonfüred, Hungary (2006)

Trajectory Generation with Way-Point Constraints for UAV Systems

Florin Stoican and Dan Popescu

Abstract This paper addresses some alternatives to classical trajectory generation for an unmanned aerial vehicle (UAV) which needs to pass through (or near) a priori given way-points. Using differential flatness for trajectory generation and B-splines for the flat output parameterization, the current study concentrates on flat descriptions which respect to UAV dynamics and verify way-point constraints.

Keywords Flat trajectory generation · B-spline basis functions · UAV systems

1 Introduction

UAV (unmanned aerial vehicle) systems and trajectory planning are a large part of modern robotics and motion planning strategies in general. While some simplifying assumptions regarding the kinematics and the autopilot loops of an UAV can be made, the result are dynamics which are still non-linear and difficult to control in real-time conditions. In what follows we use results from [1] to provide these dynamics and flatness constructions [2–4] to provide state and input references which are then used by a relatively simple online reference tracking controller.

This kind of construction is not unknown throughout the literature [5, 6] but several issues remain of actuality: choice of basis function for the flat-output design, validation of state and input constraints, etc. Existing results [7, 8] use B-spline basis functions to design the flat-outputs and to validate state and input constraints. Recent work of the authors [9] treats some of the above aspects and offers preliminary results for way-point relaxation and obstacle avoidance conditions.

F. Stoican (✉) · D. Popescu
Politehnica University of Bucharest, Bucharest, Romania
e-mail: florin.stoican@acse.pub.ro

D. Popescu
e-mail: dan.popescu@upb.ro

T. Borangiu (ed.), *Advances in Robot Design and Intelligent Control*,
Advances in Intelligent Systems and Computing 371,
DOI 10.1007/978-3-319-21290-6_38

The novel element of the research reported in this paper is the representation of a realistic kinematic model and the subsequent discussion on way-point relaxations via flat trajectory constructions. For the latter, in general it is required that the trajectory passes through a way-point (or a predefined neighbourhood around it) but without taking into account that a minimal time has to be spent close-by. Some preliminary results have been reported in [9] but at a significant computational cost (due to the usage of mixed integer programming). Here we propose to further explore the geometric properties of the B-spline basis functions in order to avoid an excessive computational load.

2 Preliminaries

Let us consider a nonlinear system:

$$\dot{x}(t) = f(x(t), u(t)) \tag{1}$$

where $x(t) \in \mathbf{R}^n$ is the state vector and $u(t) \in \mathbf{R}^m$ is the input vector.

The system (1) is called differentially flat if there exists $z(t) \in \mathbf{R}^m$ such that the states and inputs can be expressed in terms of $z(t)$ and its higher-order derivatives:

$$\begin{aligned} x(t) &= \Theta(z(t), \dot{z}(t), \ldots, z^{(q)}(t)), \\ u(t) &= \Phi(z(t), \dot{z}(t), \ldots, z^{(q)}(t)), \end{aligned} \tag{2}$$

where $z(t) = \gamma(x(t), u(t), \dot{u}(t), \ldots, u^{(q)}(t))$. The use of flatness reduces the problem of trajectory generation to finding an adequate flat output. This means choosing $z(t)$ such that, via mappings $\Theta(\cdot), \Phi(\cdot)$, constraints on state and inputs are verified. The flat output may be itself difficult to compute and the solution is to parameterize $z(t)$ using a set of smooth basis functions $\Lambda^i(t)$: $z(t) = \sum_{i=1}^{N} \propto_i \Lambda^i(\mathrm{t}),\ \alpha_i \in \mathbf{R}$. There are multiple choices for the basis functions $\Lambda^i(t)$ above, but many of them (e.g., polynomial basis functions) suffer numerical deficiencies: the number of functions depends on the constraints and on the degree of the derivatives appearing in the state and input parameterizations [10, 11]. Other choices are spline functions like *Bezier* [12] and *B-spline* basis functions [7, 13, 14]. In the rest of the paper we will use the latter due to their constructive properties (e.g., their complexity does not depend on the number of constraints). For further use, we recapitulate several results concerning B-splines with additional information to be found in, e.g., [14–16].

A B-spline of order d is characterized by a *knot-vector*

$$\mathbb{T} = \{\tau_0, \tau_1, \ldots, \tau_m\}, \tag{3}$$

of time instants ($\tau_0 \leq \tau_1 \leq \cdots \leq \tau_m$) that parameterizes the basis functions $B_{i,d}(t)$:

$$\begin{aligned} B_{i,1}(t) &= \begin{cases} 1, & \text{for } \tau_i \leq t < \tau_{i+1} \\ 0, & \text{otherwise} \end{cases}, \\ B_{i,d}(t) &= \frac{t - \tau_i}{\tau_{i+d-1} - \tau_i} B_{i,d-1}(t) + \frac{\tau_{i+d} - t}{\tau_{i+d} - \tau_{i+1}} B_{i+1,d-1}(t) \end{aligned} \tag{4}$$

for $d > 1$ and $i = 0, 1, \ldots, n = m - d$.

Considering a collection of *control points*:

$$\mathbf{P} = [p_0, \ldots, p_n] \tag{5}$$

we define a B-spline curve as a linear combination of the control points (5) and the B-spline basis functions $\mathbf{B}_d(t) = [B_{0,d}(t), \ldots, B_{n,d}(t)]^T$:

$$z(t) = \sum_{i=0}^{n} B_{i,d}(t) p_i = \mathbf{P}\mathbf{B}_d(t). \tag{6}$$

3 UAV Dynamics

In general, the kinematic equations governing the movement of an UAV are quite complicated (in the sense that they lead to a large-scale, coupled and nonlinear dynamics). The solution usually followed in the literature is to make simplifying assumptions (linearization around an equilibrium point, to consider particular flight conditions like coordinated turn and accelerated climb, etc.) such that reduced models are reached.

In addition, due to the complexity of the control problem, it is assumed in general that low-level flight controllers are already implemented and only higher level dynamics (the ones governing heading and attitude) are taken into account. These low-level controllers are modelled via first- (the airspeed, roll, flight-path angle and load factor) or second- (the altitude and the course angle) order dynamics and introduced in the high-order dynamics.

With these theoretical elements we can consider detailed dynamics (as taken from [1]) in flat-output representation. We take the flight-path angle γ, airspeed V_a and roll angle ϕ as being directly controlled by the autopilot through input elements γ^c, V_a^c and ϕ^c respectively. This leads to dynamics:

$$\begin{aligned} &\dot{p}_n = V_a \cos\psi \cos\gamma_a + \omega_n, && \dot{\gamma} = b_\gamma(\gamma^c - \gamma) \\ &\dot{p}_e = V_a \sin\psi \cos\gamma_a + \omega_e, && \dot{h} = V_a \sin\gamma_a - \omega_d \\ &\dot{\chi} = \tfrac{g}{V_g} \tan\phi \cos(\chi - \psi), && \dot{V}_a = b_{V_a}(V_a^c - V_a) \\ &\dot{\phi} = b_\phi(\phi^c - \phi) \end{aligned} \tag{7}$$

where h is the altitude and χ is the course angle. In all cases, the constants appearing are positive scalars whose values depend on the autopilot implementation and the state estimation scheme [1, Chap. 9].

For small aircrafts the wind can have a large influence and this is the reason we use different notation for speed and angles as measured in the aircraft and on the ground. However, whenever the wind is negligible, we can simplify some of the notation. For example, airspeed and ground speed are the same ($V_a = V_g$) and yaw angle, which in general depends on course angle, airspeed and wind speed can be relaxed to $\psi = \chi$.

We reach then the nominal dynamics:

$$\begin{aligned} &\dot{p}_n = V_a \cos\psi \cos\gamma, && \dot{p}_e = V_a \sin\psi \cos\gamma \\ &\dot{h} = V_a \sin\gamma, && \dot{\psi} = \tfrac{g}{V_a} \tan\phi \end{aligned} \tag{8}$$

where the last three equations in (9) are not repeated, as they remain the same. We choose as flat output $z = \begin{bmatrix} z_1 & z_2 & z_3 \end{bmatrix}^T = \begin{bmatrix} p_n & p_e & h \end{bmatrix}^T$. Then one can express the state elements as combinations of the flat outputs and their derivatives:

$$\begin{aligned} &\psi = \arctan\tfrac{\dot{z}_2}{\dot{z}_1}, && V_a = \sqrt{\dot{z}_1^2 + \dot{z}_2^2 + \dot{z}_3^2} \\ &\gamma = \arctan\tfrac{\dot{z}_3}{\sqrt{\dot{z}_1^2 + \dot{z}_2^2}}, && \phi = \arctan\left(\tfrac{1}{g} \cdot \tfrac{\ddot{z}_2\dot{z}_1 - \dot{z}_2\ddot{z}_1}{\sqrt{\dot{z}_1^2 + \dot{z}_2^2 + \dot{z}_3^2}}\right) \end{aligned} \tag{9}$$

Note that for input ϕ we made use of relation $\dot{\psi} = \tan\psi \cdot (1 + \tan^2\psi)^{-1}$. If in addition we consider the 1st order dynamics constraints on airspeed, roll and yaw angles we obtain the inputs:

$$\begin{aligned} V_a^c &= \sqrt{\dot{z}_1^2 + \dot{z}_2^2 + \dot{z}_3^2} + \frac{1}{b_{V_a}} \cdot \frac{\dot{z}_1\ddot{z}_1 + \dot{z}_2\ddot{z}_2 + \dot{z}_3\ddot{z}_3}{\sqrt{\dot{z}_1^2 + \dot{z}_2^2 + \dot{z}_3^2}} \\ \gamma^c &= \arctan\frac{\dot{z}_3}{\sqrt{\dot{z}_1^2 + \dot{z}_2^2}} + \frac{1}{b_\gamma} \cdot \frac{\ddot{z}_3(\dot{z}_1^2 + \dot{z}_2^2) - \dot{z}_3(\ddot{z}_1\dot{z}_1 + \ddot{z}_2\dot{z}_2)}{(\dot{z}_1^2 + \dot{z}_2^2 + \dot{z}_3^2)\sqrt{\dot{z}_1^2 + \dot{z}_2^2}} \\ \phi^c &= \arctan\left(\frac{b}{a}\right) + \frac{1}{b_\phi} \cdot \frac{\dot{b} \cdot a - a \cdot \dot{b}}{a^2 + b^2} \end{aligned} \tag{10}$$

where $a = g\sqrt{\dot{z}_1^2 + \dot{z}_2^2}$ and $b = \ddot{z}_2\dot{z}_1 - \dot{z}_2\ddot{z}_1$.

4 Way-Point Validation

Let us consider a collection of $N+1$ way-points and the time stamps associated to them:

$$\mathbb{W} = \{w_k\} \text{ and } \mathbb{T}_{\mathbb{W}} = \{t_k\}, \tag{11}$$

for any $k = 0, \ldots, N$. The goal is to construct a flat trajectory which passes 'near'[1] each way-point w_k at the time instant t_k, i.e., find a flat output $z(t)$ such that

$$x(t) = \Theta(z(t), \ldots z^{(r)}(t)) \in \{w_k\} \oplus S_k, \ \forall k = 0\ldots N \quad \text{and} \quad \forall t \in [t_k^-, t_k^+],$$

where S_k is an a priori defined *sensing region* around the k-th way-point and $[t_k^-, t_k^+]$ with $t_k^+ - t_k^- \geq T$ is an interval sufficiently large such that the UAV will 'stay' enough time around the way-point (i.e., inside region $\{w_k\} \oplus S_k$) in order to fulfil its mission).

Through the B-spline framework given in Sect. 1 we can provide a vector of control points (5) and its associated knot-vector (7) such that constraint (15) is verified:

$$\tilde{\Theta}(\mathbf{B}_d(t), \mathbf{P}) \in \{w_k\} \oplus S_k, \quad \forall k = 0\ldots N, \quad \forall t \in [t_k^-, t_k^+] \tag{12}$$

where $\tilde{\Theta}(\mathbf{B}_d(t), \mathbf{P}) = \Theta(\mathbf{P}\mathbf{B}_d(t), \ldots, \mathbf{P}M_rL_r\mathbf{B}_d(t))$ is a shorthand notation using equivalence properties between B-spline functions of different orders and their derivatives ($\mathbf{B}_d^{(r)}(t) = M_r\mathbf{B}_{d-r}(t)$ and $\mathbf{B}_{d-r}(t) = L_r\mathbf{B}_d(t)$).

Let us assume that the knot-vector is fixed ($\tau_0 = t_0$, $\tau_{n+d} = t_N$ and the intermediary points $\tau_d, \ldots, \tau_n$ are equally distributed along these extremes). Then, one can write an optimization problem with control points p_i as decision variables whose goal is to minimize a state and/or input integral cost $\Xi(x(t), u(t))$ along the interval $[t_0, t_N]$:

$$\begin{aligned} \mathbf{P} = \arg\min_{\mathbf{P}} \int_{t_0}^{t_N} ||\tilde{\Xi}(\mathbf{B}_d(t), \mathbf{P})||dt \\ \text{s.t. constraints (12) are verified} \end{aligned} \tag{13}$$

The cost $\tilde{\Xi}(\mathbf{B}_d(t), \mathbf{P}) = \Xi(\tilde{\Theta}(\mathbf{B}_d(t), \mathbf{P}), \tilde{\Phi}(\mathbf{B}_d(t), \mathbf{P}))$ can impose any penalization we deem necessary (length of the trajectory, input variation or magnitude, etc.). In general, such a problem is nonlinear (due to mappings $\tilde{\Theta}(\cdot)$ and $\tilde{\Phi}(\cdot)$) and hence

[1]Such a relaxation of the constraints makes sense in an UAV context. At a way-point w_k one needs to acquire data (photographs, information sent from the ground by a sensor network, and the like); still, we do not need to be in exactly the predefined point in order to accomplish the task, rather we can consider a certain area around the way-point.

difficult to solve. A nonlinear MPC iterative approach has been extensively studied in [13].

Condition (12) is significantly more difficult to enforce than an equality constraint (i.e., passing through the way-point w_k at time t_k). The issue comes from the fact that (12) requires the trajectory validation along a continuous time interval (i.e.$t \in [t_k^-, t_k^+]$). In what follows we propose two alternatives, both of them centred on the idea that imposing carefully some constraints in a finite number of time instants can guarantee the validation of condition (12).

Proposition 1 For a given way-point, time stamp and sensing region triple (w_k, t_k, S_k), a sufficient condition for the validation of (12) is that:

$$\tilde{\Theta}(\mathbf{B}_d(t_k), \mathbf{P}) \in \epsilon S_k \tag{14}$$

with $\varepsilon > 0$ taken such that

$$\varepsilon S_k \oplus \int_{t_k^-}^{t_k^+} ||\tilde{\Theta}(\mathbf{B}_d(t), \mathbf{P})'||dt \subseteq S_k. \tag{15}$$

Proof The integral $\int_{t_k^-}^{t_k^+} ||\tilde{\Theta}(\mathbf{B}_d(t), \mathbf{P})'||dt$ denotes the length of the trajectory over the interval $[t_k^-, t_k^+]$. If on the other fact at time t_k we verify that the state is in εS_K with ε chosen such that (15) is true, we can conclude that (14) is verified.

Remark 2 Note that (14) might not always be feasible as it may happen that the right-hand part of the inclusion is empty. Such a case denotes that the UAV cannot fulfil its mission in one pass and hence multiple passes might be required.

The second method uses the B-spline property which states that the B-spline curve always resides inside the convex hull of d consecutive control points.

Proposition 2 For a given way-point, time interval and sensing region triple $(w_k, [t_k^-, t_k^+], S_k)$ a sufficient condition for the validation of (12) is to consider $p_{r_k-d+1} \cdots p_{r_k}$ control points such that:

$$p_{r_k-i+1} \in \tilde{\Theta}^{-1}(S_k), \quad \forall i = 1 \ldots d \tag{16}$$

and $t_{r_k} = t_k^-$, $t_{r_k+1} = t_k^+$.

Proof In the interval $[t_{r_k}, t_{r_k+1}] = [t_k^-, k^+]$ the curve will stay in the convex hull of the d consecutive control points: ConvHull$\{p_{r_k-d+1} \cdots p_{r_k}\} \subseteq \tilde{\Theta}^{-1}(S_k)$.

Remark 3 This method requires a large number of control points which, for large problems and/or nonlinear cost and constraints, may turn out to be difficult to solve.

For the dynamics shown in Sect. 3, simplified to $\dot{h} = 0$ for illustration purposes we have tested the constraints specified in Proposition 1 and Proposition 2. We consider 6 way-points, each with a predefined sensing region and attached time interval. It can be seen that in both cases the trajectory will pass through the sensing

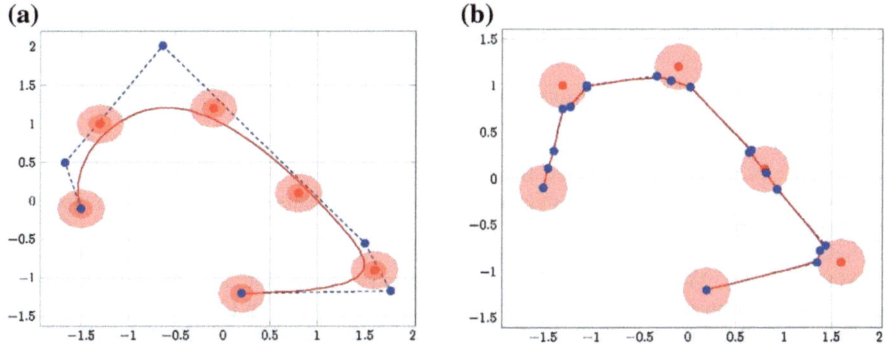

Fig. 1 Flat trajectory with sensing regions (*solid red*) and control polygon (*dashed blue*): **a** with reduced sensing region; **b** with control points convex hull

regions for the minimum required time. In Fig. 1a this is done by forcing the trajectory to be inside the deflated sensing region corresponding to the *k*-th way-point and in Fig. 1b it is done by forcing d consecutive control points to be inside the sensing region.

5 Conclusions

The present paper discusses alternatives to classic flat trajectory generation for an UAV agent which has to pass through predefined way-points. Specifically, we are interested in relaxing the constraints appearing in the construction of the trajectory in the sense that it can pass near but not necessarily through the way-points. This is done via a B-spline flat output parameterization which allows to characterize and construct efficiently the flat trajectory and to guarantee that the agent will stay a sufficient time in the sensing area attached to a way-point. Further work will consider extended simulations and more realistic working conditions (parameter variation, measurement noises, etc.). Also, additional kinematic representations will be studied in order to find more compact flat references descriptions and the B-spline representations will be employed for additional topics of interest, e.g., obstacle avoidance.

Acknowledgement The work has been partially funded by the Sectorial Operational Programme Human Resources Development 2007–2013 of the Ministry of European Funds through the Financial Agreement [grant number POSDRU/159/1.5/S/132395] and by National Research Programme STAR, project 71/2013: Multisensory robotic system for aerial monitoring of critical infrastructure systems—MUROS.

References

1. Beard, R., McLain, T.: Small unmanned aircraft: theory and practice. Princeton University Press, Princeton (2012)
2. Fliess, M., Lévine, J., Martin, P., Rouchon, P.: Flatness and defect of non-linear systems: introductory theory and examples. Int. J. Control **61**(6), 1327–1361 (1995)
3. Lévine, J.: Analysis and Control of Nonlinear Systems: A Flatness-Based Approach, Springer, Berlin (2009)
4. Sira-Ramírez, H., Agrawal, S.: Differential Flatness. Marcel Dekker, New York (2004)
5. Mellinger, D., Michael, N., Kumar, V.: Trajectory generation and control for precise aggressive manoeuvres with quadrotors. Int. J. Robot. Res. 27–36 (2012)
6. Prodan, I., Bencatel, S., Olaru, R., Sousa, J., Stoica, C., Niculescu, S.-I.: Receding horizon flight control for trajectory tracking of autonomous aerial vehicles. Control Eng. Practice **21** (10), 1334–1349 (2013)
7. Suryawan, F.: Constrained Trajectory generation and fault tolerant control based on differential flatness and B-splines. PhD thesis, University of Newcastle, Australia (2012)
8. Suryawan, F., De Dona, J., Seron, M.: Methods for trajectory generation in a magnetic-levitation system under constraints. In: 18th Mediterranean Conference on Control and Automation, pp. 945–950 (2010)
9. Stoican, F., Prodan, I., Popescu, D.: Flat trajectory generation for way-points relaxations and obstacle avoidance. Submitted for publication at MED 2015
10. Daniel, M., Daubisse, J.C.: The numerical problem of using Bézier curves and surfaces in the power basis. Comput. Aided Geom. Des. **6**(2), 121–128 (1989)
11. Farouki, R.T., Rajan, V.T.: On the numerical condition of polynomials in Bernstein form. Comput. Aided Geom. Des. **4**(3), 191–216 (1987)
12. Schumaker, L.L.: Spline Functions: Basic Theory. Cambridge University Press, Cambridge (2007)
13. De Doná, J.A., Suryawan, F., Seron, M.M., Lévine, J.: A flatness-based iterative method for reference trajectory generation in constrained NMPC. In: Nonlinear Model Predictive Control, pp. 325–333, Springer, Berlin (2009)
14. Gordon, W., Riesenfeld, R.: B-spline curves and surfaces. Comput. Aided Geom. Des. **167**, 95 (1974)
15. Patrikalakis, N., Maekawa, T.: Shape Interrogation for Computer Aided Design and Manufacturing. Springer, Berlin (2009)
16. Piegl, L., Tiller, w.: Curve and Surface Basics. Springer, Berlin (1995)

Kinematics-Based Localization of a Skid-Steer Vehicle

Rocco Galati, Ivan Giannoccaro, Arcangelo Messina and Giulio Reina

Abstract The paper addresses the localization issue for skid-steer vehicles. The large extent of slippage incurred by this type of vehicle in a turn heavily degrades the accuracy of their position estimation systems. Based on the kinematics modelling of skid-steering, the concept of equivalent track is presented. During field experiments with an off-road unmanned vehicle, this new concept proved to be effective in reducing localization errors up to two orders of magnitude.

Keywords Unmanned ground vehicles · Kinematics modelling · Localization

1 Introduction

During recent years, a fast growing interest in the area of unmanned vehicles has been observed [1]. In field robotics, wheeled vehicles take advantage of differential drive mechanism mainly because it offers a good mobility and easy control methods. Typically, it is possible to identify two different types of mobile platforms, i.e. platforms for which no-slip and pure-rolling conditions may be assumed and platforms where this effect is requires changing the current heading.

R. Galati (✉) · I. Giannoccaro · A. Messina · G. Reina
Department of Engineering for Innovation, University of Salento, Via per Monteroni, 73100 Lecce, Italy
e-mail: rocco.galati@unisalento.it

I. Giannoccaro
e-mail: ivan.giannoccaro@unisalento.it

A. Messina
e-mail: arcangelo.messina@unisalento.it

G. Reina
e-mail: giulio.reina@unisalento.it

T. Borangiu (ed.), *Advances in Robot Design and Intelligent Control*,
Advances in Intelligent Systems and Computing 371,
DOI 10.1007/978-3-319-21290-6_39

Although significant studies have been focused on motion control and localization for the first kind of platforms, very few studies that examine and accommodate them for skid-steering effects have been made [2–5]. Currently, the skid-steering principle is based on controlling the relative velocities of both tracks and wheels, much in the same way as for differential drive wheeled vehicles. However, control of tracked locomotion poses a more complex problem because variation of the relative velocity of the two tracks results in slippage as well as soil deformation in order to achieve steering [6]. Moreover, kinematics is not straightforward, since it is not possible to predict the exact motion of the vehicle only from its control inputs (i.e. optical encoders).

This means that motion control methods suitable for differential wheeled robots cannot be directly used for tracked vehicles with skid-steering driving system and a kind of slip estimation should be carried out for performing an acceptable path-following [7].

2 Kinematics of Skid-Steering

Skid-steer systems are widely used on tracked vehicles such as excavators and tanks, but are also commonly used on some four- and six-wheeled vehicles. On these vehicles, the tracks (or wheels) on each side can be driven at various speeds in forward and reverse motion. There is no explicit steering mechanism since the steering operation is accomplished by actuating each side at a different velocity or in a different direction, causing the tracks (or wheels) to slip or skid on the ground.

Because of the complex wheel-ground interactions, it is considered a challenging task to describe an accurate kinematic model for skid-steer mobile vehicles. Wheel slip has an important role in kinematic and dynamic modelling of skid-steer mobile vehicles. The slip information relates the wheel angular velocity to the linear motion of the vehicle platform. Understanding the slip information is important for skid-steer vehicle localization applications, such as dead reckoning [8, 9].

Figure 1 illustrates the kinematics of a skid-steer vehicle during a clockwise turn with reference to a right-hand vehicle-fixed coordinate system, whose origin is assumed to be in the vehicle's centre of mass. Using similar triangle properties, the turning radius equation results using the proportion between each edge of the triangles ABC, ADE:

$$\frac{v_o}{v_i} = \frac{R + \frac{B}{2}}{R - \frac{B}{2}}; \quad R = \frac{\frac{B}{2}\left(\frac{v_o}{v_i} + 1\right)}{\left(\frac{v_o}{v_i} - 1\right)} = \frac{B}{2}\left(\frac{v_o + v_i}{v_o - v_i}\right) \tag{1}$$

$$\omega_z = \frac{v_o + v_i}{2R} = \frac{v_i\left(\frac{v_o}{v_i} - 1\right)}{B} \tag{2}$$

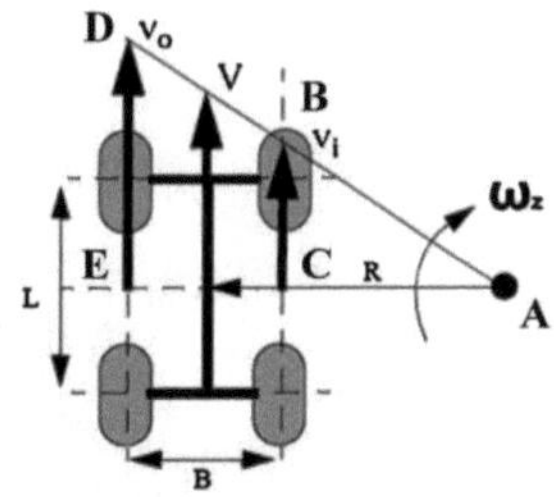

Fig. 1 Kinematics of a skid-steering vehicle

It is worth noting that the turning radius in Eq. (2) is estimated under the assumption that no slippage occurs between wheels and ground during the turning manoeuver. Anyway, in the real world, skidding and slipping effects between wheels and ground surfaces can be observed for all skid-steering vehicles where skidding is necessary to change the vehicle heading. For this reason, even at low turning speed, the traditional kinematics approach cannot describe correctly the vehicle position along a specified path. Difference between the wheel hub linear velocity and the wheel angular velocity can be measured as a result of longitudinal skid effect, well described by:

$$i = (1 - \frac{V}{r \cdot \omega}) \cdot 100 \tag{3}$$

$$R' = \frac{B}{2}\left(\frac{v_o(1 - i_o) + v_i(1 - i_i)}{v_o(1 - i_o) - v_i(1 - i_i)}\right) \tag{4}$$

while the new yaw rate estimate will be:

$$\omega_z{}' = \frac{v_o(1 - i_o) + v_i(1 - i_i)}{2R'} = \frac{v_i\left(\frac{v_o(1-i_o)}{v_i} - (1 - i_i)\right)}{B} \tag{5}$$

Even if most research activities focus on skidding estimation for vehicle localization and path planning, a correct relationship between the longitudinal skidding effect and the vehicle behaviour has not been achieved yet. This is mainly because the skidding and slipping phenomena can be considered as a function of the interaction between both left and right side wheels and the ground surface and also because they affect the encoder readings that become not useful to describe the vehicle position.

In this paper, the equivalent track concept is introduced as a parameter describing the track that a vehicle should have during a turning operation according to the encoder readings.

Different field experiments were performed to evaluate the equivalent track method, using the 4 × 4 skid-steer vehicle Husky A200, shown in Fig. 2a, while its dimensions are shown in Fig. 2b.

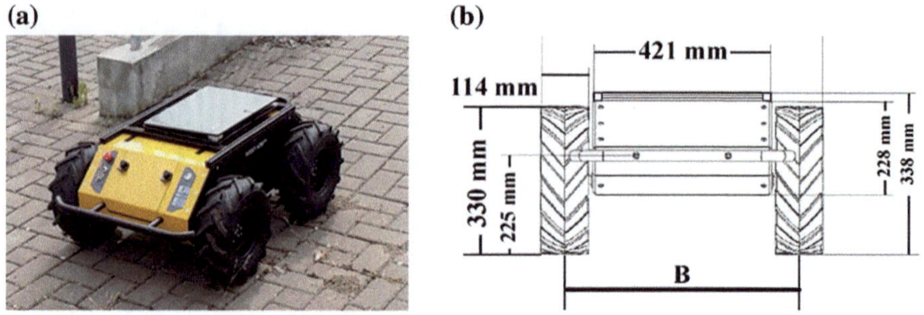

Fig. 2 **a** The Husky A200 vehicle. **b** Geometric properties

Equations provided by the manufacturer are based on the assumption of no-slippage and describe the relationships between wheel velocities and platform linear and angular velocities v and ω, respectively:

$$\mathrm{v} = \frac{v_r + v_l}{2}, \quad \omega = \frac{v_r - v_l}{B} \tag{6}$$

where the subscripts r and l indicate the right and left side respectively, and B is the geometric track of the vehicle (0.56 m for the considered robot). Moreover, since Husky A200 includes optical quadrature encoders assembled on the motor shafts measuring the rotation angles $\Delta\phi_r$ and $\Delta\phi_l$, it is possible to estimate the linear displacement and the angular displacement $\Delta\theta$.:

$$\Delta S = \frac{r}{2}(\Delta\phi_r + \Delta\phi_l) \quad \Delta\theta = \frac{r}{d}(\Delta\phi_r - \Delta\phi_l) \tag{7}$$

Finally, it is possible to update the vehicle's position at time t_{k+1} using classical odometric equations [10]:

$$x_{k+1} = x_k + \Delta S_\mathrm{k} \cos\left(\theta_k + \frac{\Delta\theta_\mathrm{k}}{2}\right) \tag{8}$$

$$y_{k+1} = y_k + \Delta S_\mathrm{k} \sin\left(\theta_k + \frac{\Delta\theta_\mathrm{k}}{2}\right) \tag{9}$$

$$\theta_{k+1} = \theta_k + \Delta\theta_\mathrm{k} \tag{10}$$

Clearly, these equations can well describe the vehicle's position during straight line paths, but they are not accurate to describe steering manoeuvres even at low speeds.

3 Equivalent Track for Localization

Figure 3 illustrates the kinematics of a skid-steering vehicle during a clockwise turn where left side wheels have positive velocity and right wheels are blocked. Under this assumption, the instantaneous centre of rotation (ICR) should lie on the right side and the diameter of the circumference drawn by the left side wheels should be twice the track length. In the real case, this consideration is not accurate since blocked wheels on the right side are dragged due to the driving wheels action on the left side. Therefore, it is possible to assume that the diameter of the drawn circumference depends on the difference between left and right velocities where the velocity of blocked wheels side cannot be considered null even if no encoder counters increment is registered.

The equivalent track method tries to describe the track that the vehicle should have during a turning manoeuver according to the encoder readings in order to bring in coincidence the velocity reported by the encoders with the actual wheel velocities, as shown in Fig. 3. This method is still valid, where both wheel sides have different velocities that can introduce significant skidding effects. Interestingly, the higher the difference between wheels sides, the lower the equivalent track.

Different tests were performed in order to map correctly the equivalent track depending on the differential velocity at the opposite sides of the vehicle under the assumption of balanced output for both left and right steering manoeuvres.

Table 1 shows average values for differential velocities and turning radius obtained by making the vehicle follow multiple circular paths with a maximum differential velocity of 0.8 m/s. As a result, it is possible to build by interpolation the relationship between the input velocity, ΔV, and the equivalent track.

Fig. 3 Equivalent track method

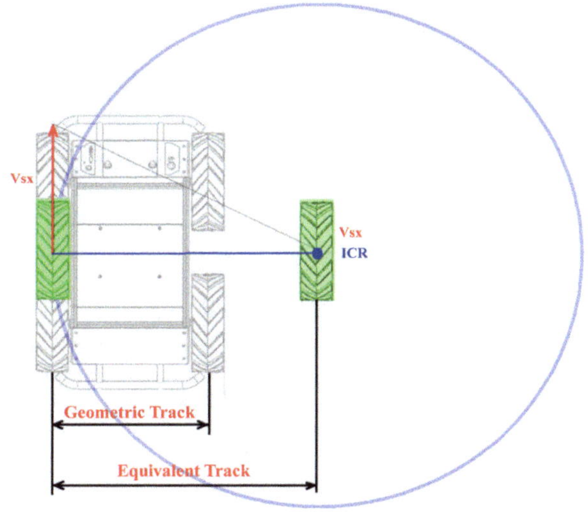

Table 1 Equivalent track mapping (average values)

$\Delta V = \lvert V_{dx} - V_{sx} \rvert$ (m/s)	B (m)
0.1	0.50
0.2	0.55
0.3	0.60
0.4	0.65
0.5	0.75
0.6	0.85
0.7	1.20
0.8	1.60

3.1 Experiments on Quasi-planar Ground

In order to evaluate the proposed method for skid-steer localization, different field experiments were performed on approximately flat ground. In particular, Fig. 4 shows a rectangular closed path (red line) covered by the vehicle while it was remotely controlled by joystick.

This test was repeated four times on the same surface and the results are shown in Fig. 5, where the blue lines refer to conventional kinematics and the red lines denote the equivalent track method.

The absolute error is defined as the distance between the starting position and the final one in the kinematic reconstruction of the vehicle's path over four repeated experiments, as summarized in Table 2. The maximum error obtained by the equivalent track method is 0.84 m that corresponds to 1.56 % of the total travel length of 54 m.

Fig. 4 Rectangular closed path

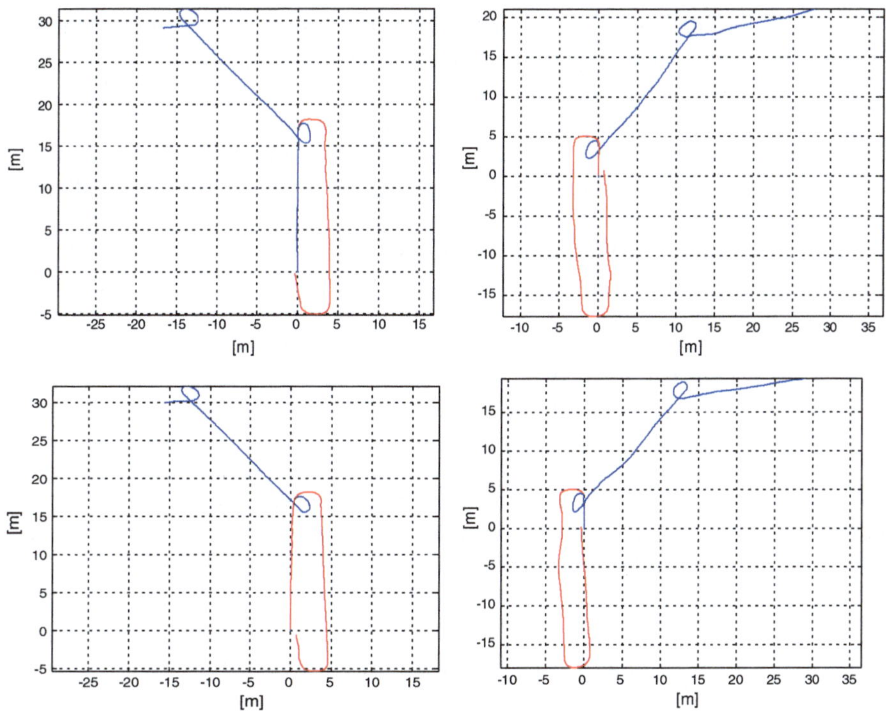

Fig. 5 Vehicle position estimation, as obtained over four closed-loop tests. *Red* position estimation using the equivalent track. *Blue* position estimation using standard kinematics

Table 2 Results of the experiments shown in Fig. 5

	Equivalent Track		Standard kinematics	
Test	Absolute error (m)	Relative error %	Absolute error (m)	Relative error %
1	0.37	0.69	36.07	67.55
2	0.28	0.52	36.44	68.26
3	0.27	0.50	36.15	67.69
4	0.84	1.56	35.24	65.99

3.2 Extension to 3D Environments

The equivalent track method can accurately describe the vehicle localization on two-dimensional planes since encoders cannot recognize ground slopes, which are very common in many outdoor environments. In order to characterize correctly the vehicle motion in the 3D world, it is necessary to detect terrain slope. An inertial sensor IMU300CC was used to measure linear acceleration, rotation rates and ground slope.

The three angular rate sensors, the MEMS accelerometers and the magnetometers on IMU300CC inertial sensor are aligned with the body frame of the sensor, so that if inertial vector data is needed, the sensor outputs are converted from the body frame to the inertial frame. This can be accomplished by introducing the Euler angles method that provides a way to give a representation of the 3D orientation of an object using a combination of three rotations about different axes (e.g., roll, pitch and yaw).

The transformation matrix used to convert body-frame angular rates to Euler angular rates is described by:

$$D(\phi, \theta, \psi) = \begin{bmatrix} 1 & \cos(\phi)\tan(\theta) & \sin(\phi)\tan(\theta) \\ 0 & \cos(\phi) & -\sin(\phi) \\ 0 & \sin(\phi)/\cos(\theta) & \cos(\phi)/\cos(\theta) \end{bmatrix} \tag{11}$$

where ϕ, θ, ψ express roll, pitch and yaw angles in the inertial frame, respectively. Let p represent the body-frame x-axis angular velocity output, q represent the body-frame y-axis angular velocity output, and r represent the body-frame z-axis output. Then, the angular velocities components related to the inertial frame are given by

$$\left(\dot{\phi}, \dot{\theta}, \dot{\psi}\right)^{\mathrm{T}} = D(\phi, \theta, \psi) * (p, q, r)^{\mathrm{T}} \tag{12}$$

Different tests were performed using these relationships to evaluate the method; the results are listed below.

Figure 6 depicts a test on an uneven surface; the closed-loop rectangular trajectory (red line in Fig. 6a) covers routes having different height profiles. Nevertheless, the proposed strategy is able to reconstruct accurately the path (Fig. 6b) with a final error of 0.86 % respect to the total length of about 70 m.

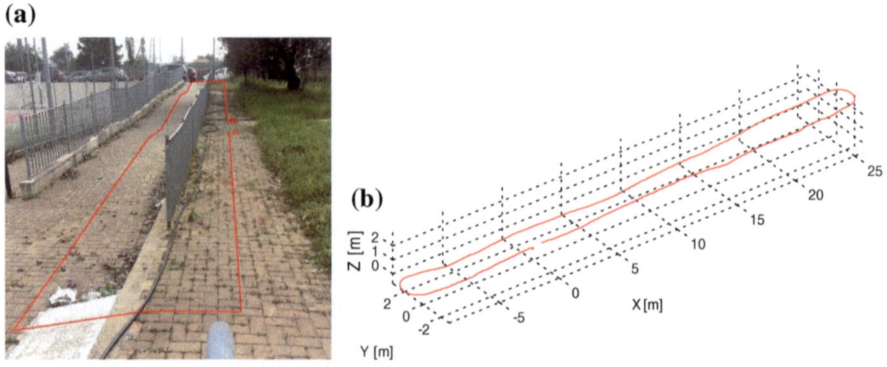

Fig. 6 **a** Real path. **b** Path reconstruction

Table 3 Errors in position estimation for the experiments shown in Fig. 6

Test	Absolute error (m)	Relative error %
1	0.31	0.86
2	0.29	0.81

In Table 3, the error analysis shows a small error obtained using the proposed method. Moreover, it is possible to appreciate the good accuracy of this strategy in estimating the slope of the surface compared to the real one.

4 Conclusions

This paper described a method for the localization of a skid-steer vehicle by using encoders and IMU sensors to define an equivalent track, instead of a fixed geometric track that can dynamically change depending on the interaction between the wheels and the terrain surface. Experimental results obtained from different paths were presented to validate the method, showing its effectiveness to localize the vehicle along the path.

References

1. Siciliano, B., Khatib, O.: Springer Handbook of Robotics. Springer, Berlin (2008)
2. Martinez, J., Mandow, A., Morales, J., Pedraza, S. and A. Garca-Cerezo: Approximating kinematics for tracked mobile robots. Int. J. Robot. Res. **24**(10), 867–878 (2005)
3. Pentzer, J., Brennan, S., Reichard, K.: Model-based prediction of skid-steer robot kinematics estimation of track instantaneous centers of rotation. J. Field Robot. **31**(3), 455–476 (2014)
4. Tuan Le, A., Rye, D., Durrant-Whyte, H.F.: Estimation of back-soil interactions for autonomous tracked vehicles. In: Proceedings of the IEEE International Conference on Robotics and Automation, Albuquerque (1997)
5. Wang, D., Low, C.B.: Modeling and analysis of skidding and slipping in wheeled mobile robots: control design perspective. IEEE Trans. Rob. **24**(3), 676–687 (2008)
6. Wong, J.: Theory of Ground Vehicles. Wiley, New York (2008)
7. Burke, M.: Path-following control of a velocity constrained tracked vehicle incorporating adaptive slip estimation. In: Proceedings of the IEEE International Conference on Robotics and Automation, pp. 97–102 (2012)
8. Reina, G., Ojeda, L., Milella, A., Borenstein, J.: Wheel slippage and sinkage detection for planetary rovers. Trans. Mechatron. **11**(2), 185–195 (2006)
9. Reina, G., Ishigami, G., Nagatani, K., Yoshida, K.: Vision-based estimation of slip angle for mobile robots and planetary rovers. In: IEEE International Conference on Robotics and Automation, Pasadena, 19–23 May 2008
10. Ojeda, L., Reina, G., Cruz, D., Borenstein, J.: The FLEXnav precision dead-reckoning System. Int. J. Veh. Auton. Syst. **4**(2), 173–195 (2006)

Dynamic Task Planning of Aerial Robotic Platforms for Ground Sensor Data Collection and Processing

Grigore Stamatescu, Dan Popescu and Cristian Mateescu

Abstract The adoption of wireless sensor network systems is becoming wide-spread in critical large-scale monitoring applications. These include but are not limited to pipeline infrastructures for oil and water, border areas, roads and railway systems. In such scenarios, airborne robotic platforms like unmanned aerial vehicles (UAVs) can provide valuable services for data collection, communication relaying and higher level supervision. This is the case for both single UAV deployment as well as for swarms of UAVs collaboratively integrated into monitoring systems. The paper discusses the opportunity for in-network pre-processing of sensor data for local UAV task planning in order to increase the efficiency of data collection processes. A gradient scheme is introduced for decision support of the UAV task planning. Results are validated by simulation.

Keywords Wireless sensor networks (WSN) · Unmanned aerial vehicles (UAV) · Information processing · Data collection · Large scale monitoring

1 Introduction

As dense monitoring and control applications become wide-spread through networks of embedded computing communication devices, significant challenges arise for effective data management at node and network levels. The integration of tens to

G. Stamatescu (✉) · D. Popescu
Department of Automation and Industrial Informatics, University Politehnica of Bucharest, Bucharest, Romania
e-mail: grigore.stamatescu@upb.ro

D. Popescu
e-mail: dan.popescu@upb.ro

C. Mateescu
Teamnet International, 22 Tudor Vladimirescu Blvd, 050883 Bucharest, Romania
e-mail: cristian.mateescu@teamnet.ro

T. Borangiu (ed.), *Advances in Robot Design and Intelligent Control*,
Advances in Intelligent Systems and Computing 371,
DOI 10.1007/978-3-319-21290-6_40

thousands of wireless sensor nodes, which have the task of periodically sampling process (or set of process) variables, and communicating the data via low-power links towards a central control gateway, demands data reduction mechanisms which are usually implemented by means of aggregation and sensor fusion [1]. The goal afterwards is to reliably relay the high level resulting pieces of information back to a central server for automatic or assisted decision support. The long distance links and storage buffers required for redundancy lead to an increase in the complexity and cost of the individual nodes which suggests the use of layered monitoring architectures. These are strongly dependent on the application specific requirements for data collection and control and must offer the possibility of dynamic parameter adjustment [2].

We focus on the essential support role that can be played by UAV in effectively supervising large scale distributed sensor systems. The WSN-UAV symbiosis for security perimeter and critical infrastructure systems monitoring has become an interesting new research direction over the last few years. Several global frameworks have been proposed and classified, among which:

- *Hierarchical structure* where the fixed wireless sensor network operates in the designated target area, and collects ground data according to predefined or dynamic behaviour. UAVs act upon several alerting and alarm levels generated by embedded algorithms running on a network control centre for data collection, processing and decision. This can be showcased by event detection such as an intruder entering a protected area which leads to on-demand dispatching of an UAV capable of video surveillance of the target area signalled by the WSN;
- *Collaborative heterogeneous structure* in which UAVs are able to support the sensor network by acting as mobile communication relays in order to restore WSN connectivity in case of a fault, like is the case of reduced network throughput or even network partitioning if a critical node fails. This can also be extended to soft constraints such as effective load balancing of the communication channel by offering alternative routes through the mobile relays. Evaluating the issue from an opposite perspective, the sensor network can operate as a fixed reference infrastructure for localization and UAV communication;
- *Mobile wireless sensor network* where the nodes are implemented as mini- and micro- UAVs, equipped with: scalar and multidimensional/multimedia sensors, enhanced local processing and storage capacity, collaborative swarm type algorithms for multi-UAV coordination and planning and various communications interfaces;
- *Hybrid application-specific systems* which include elements from all of the above.

Current research issues in this field lay at the convergence of several areas: low-power radio communication including medium access control and routing, data aggregation and sensor fusion for in-network or centralized processing, energy efficient localization methods and navigation/task planning and collaboration for mobile robots, in the particular the case of airborne robotic platforms. Modern approaches leverage intensively multi-agent system theory for task assignment and efficient operation of individual entities or entities groups: sensor nodes, UAVs and mixed clusters.

The rest of the paper is structured as follows. Section 2 highlights related work concerning effective task planning for data collection and communication in hybrid large scale monitoring systems and underlines the contribution of our work. Section 3 describes the reference system architecture, objectives and relevant algorithms. We present simulation results based on reference temperature/humidity data sets for ground data and UAV decision support in Sect. 4. The paper is concluded in Sect. 5, including an outlook on future work.

2 Related Work

The main context of the reported work is defined by the application-specific nature of large scale pipeline and road infrastructures which can be best described by a linear wireless sensor network (LWSN) deterministic deployment [3]. This assumes that generic communication and processing protocols can be adapted to exploit this linear deployment, or new ones can be developed altogether. It concerns the directed optimization of the communication and data collection protocols according to a linear or quasi- linear topology following one of the above mentioned infrastructure systems. The authors of [4] propose localized power-aware routing for LWSN which exploits the directed nature of the sensor node links. A multi-level system architecture specifically designed for oil pipeline monitoring is described in [5]. Key issues that are covered include the selection of the channel sharing method: TDMA versus CSMA/CA, source data fusion and compression for lowering the burden on the communication channel and increased reliability and energy supply of the nodes through energy harvesting from the monitored process.

UAV-support for heterogeneous sensor network monitoring is discussed in [6]. Here the UAV is mainly used as passive relay towards the main communication backbone for remote sensing purposes. The main objective is to design effective clustering strategies for resource balancing given the constraints of the embedded sensing devices. The challenges stemming from self-organization of such a large scale monitoring system in the military domain are presented in [7]. Through simulation results, the authors highlight the impact of parameter tuning on network connectivity and global coverage for such hybrid systems. A Markov random mobility model for the UAVs offers the best results for the relay network as compared to a pure random model.

With regard to UAV trajectory control and target tracking, a large body of knowledge has been produced over the last years. Some examples include single UAV scenarios [8] for constrained trajectory planning. It involves both an aerial robotic platform motion model and target model. This is illustrated for road network following and obstacle avoidance by using intelligent optimization methods based on genetic algorithms (GA) and particle swarm optimization (PSO). The problem is usually extended and evaluated also at the collaborative or competitive

UAV-swarm level [9]. In this case, the optimization problem is defined by the number of targets and UAVs and by defining an appropriate cost function e.g. having as goal the minimization of the time to detection, which can be solved with gradient-based approaches.

While acknowledging the current state-of-the-art, our contribution lays in advancing WSN-UAV integration at system level through in-network processing of the collected data for UAV task planning. Basic aggregation of data is carried out at the cluster head level and provided to the UAV which then autonomously decides the following waypoints based on a classification of the aggregated collected data.

3 Large Scale Deterministic System Deployment and Algorithms

The system design for large scale monitoring based on ground sensor networks with UAV support is illustrated in Fig. 1. The main components of the sensor network are the cluster head nodes N1…Nn which collect and aggregate data from local sensor nodes and interact with the UAVs D1…Dn. Apart from data collection, the drone operates as a long-range network relay towards the network control centre (NCC) and has the ability to bridge the low power wireless links in the sensor

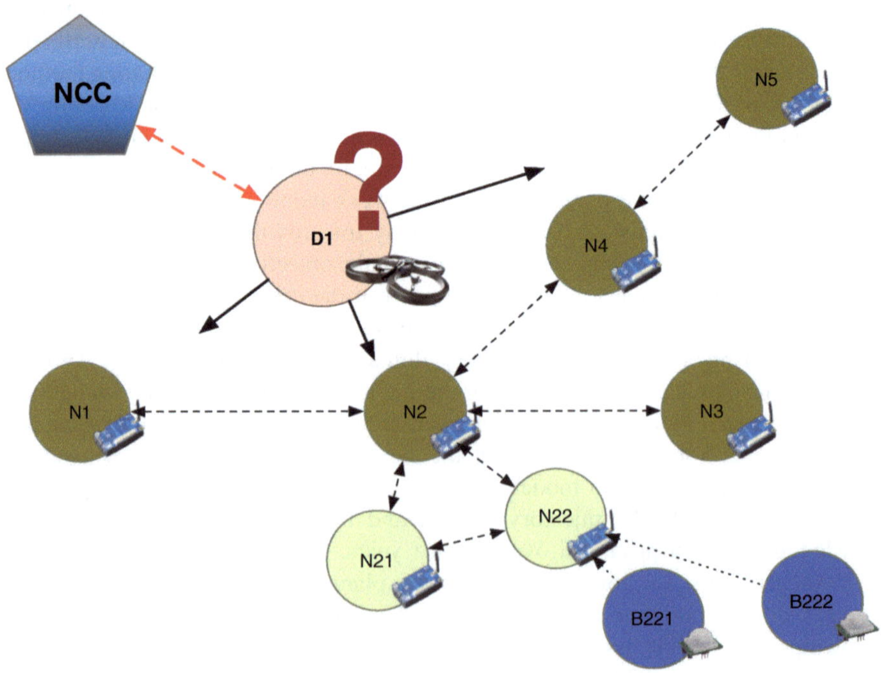

Fig. 1 System diagram for UAV-supported cluster head data collection

network. The main challenge consists of on-demand dispatching of the UAV towards the cluster head nodes which, based on a local processing algorithm, exhibit significant and persistent variations of the monitored parameters: environmental data, process variables, binary event detectors.

For large scale monitoring applications with distributed parameters like is the case of oil pipeline monitoring, this helps reducing the number of UAVs and operating costs [10]. The UAV is supposed however to have sufficient on-board storage and computing resources to keep a history of aggregated data from the nodes and update this periodically in order to run the task planning algorithm.

It is assumed that both the sensor network and the aerial robotic platform support over-the-air reprogramming in order to dynamically adjust the algorithms' parameters and mission's objectives. In the case of network congestion or faults, the task planning can revert to a static, pre-defined operation mode in which the cluster heads are periodically queried for the available data in a deterministic fashion. Introducing in-network processing of data leads to a reduction in the quantity of data transmitted over the wireless links and positively impacts battery life for both static and mobile nodes.

Another benefit of this type of multi-level monitoring system is that more complex sensors like high resolution cameras, thermal imagers and radar devices can be fitted to the UAV in order to validate information extracted from the low-cost, high density, sensor network. The human operator at the NCC is also able to adjust the WSN sampling rate and monitor network parameters relayed through the long-range wireless link.

The main contribution of the work consists of a method proposed for dynamic task planning of a UAV whose mission is to collect data from the cluster head in an opportunistic fashion. Basically the UAV has the capacity to decide by its own which cluster-head to query next depending on history, current conditions and a priori knowledge of network topology and parameter distribution.

The steps of the proposed algorithm are as follows:

Step 1 *System initialization*: The UAV receives information regarding the WSN cluster head localization, sensors, data format and the long range wireless communication channel is set-up;

Step 2 *First pass*: According to an optimal path planning based on the geographical placement of the nodes, the UAV collects initial data from each one. Aggregated information is stored locally while raw values can be streamed towards the NCC operator in real-time;

Step 3 *Cluster head ranking*: The cluster head nodes are ranked based on variations and pre-set thresholds for the process variables; a forgetting factor is used to prioritize recent events;

Step 4 *Decision*: The UAV autonomously decides which cluster head to visit next by weighting the above-mentioned ranking with both the travel distance to the next node and based on the time elapsed since the last visit at each of the nodes.

Depending on mission objectives and available resources, Step 4 loops continuously until a stop command is triggered. The scenario can be extended to multiple-UAVs by partitioning the ground sensor network into several areas which are each assigned to a single aerial robotic platform. The approach combines good response for event detection with fairness in periodically visiting each of the cluster heads. Several criteria can be added to the ranking mechanism for system tuning, e.g. for energy efficiency [11].

4 Evaluation by Means of Simulation

In order to showcase the above defined approach we carry out simulations based on real-world sensor data, stemming from the Intel Lab database [12]. This is provided by three deployed wireless sensor nodes and we focus on temperature and humidity at this point. We have chosen nodes 50, 4 and 24 respectively, with a time window of around one day. Given the sampling period of around 31 s, each data set has 1200 points.

Figure 2 illustrates the raw data stream—temperature and humidity—for the three nodes.

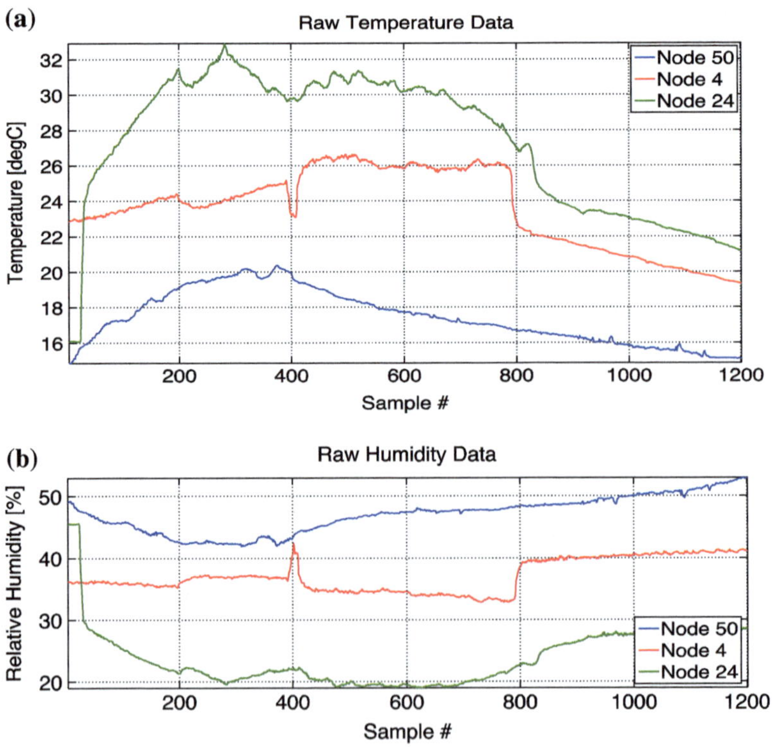

Fig. 2 Raw data collected from the sensor nodes: **a** temperature and **b** humidity

Table 1 Aggregation indicators for the raw data

Node	Parameter	AVG	MIN	MAX	σ	$σ^2$
50	Temperature	17.4279	14.863	20.3804	1.5311	2.3442
	Humidity	46.9908	41.9147	52.9368	2.8734	8.2563
4	Temperature	23.5489	19.3318	26.6132	2.2101	4.8844
	Humidity	37.0487	32.7907	42.4174	2.6548	7.0479
24	Temperature	27.3355	16.0684	32.895	3.8366	14.7192
	Humidity	23.8254	18.9947	45.5720	4.5888	21.0574

We assume that data supplied by the cluster heads has been aggregated in some form for efficient communication and storage. This includes the average value, minimum and maximum, the standard deviation and the variance of the data set. Additionally for event detection and thresholding a count field can be included in the meta-data. Cross-correlation of the data series is useful to identify phenomena that produce evolutions over a large number of nodes. The indicators for the analysed data are listed in Table 1.

Based on these assumptions, the unidirectional gradient of the data is computed. The values are smoothed using a moving average filter and squared to emphasize variations in the positive domain. Accounting for the variation of both parameters, Fig. 3 shows the final indicator as average of the squared gradients of temperature and humidity for each of the considered sensor nodes. This fact points out that node 24 at the beginning of the observation period should be periodically queried by the UAV, while later the focus should shift towards node 4. As could also be seen on the raw data, node 50 exhibits a predictable behaviour which indicates that only sporadic attention should be given to it.

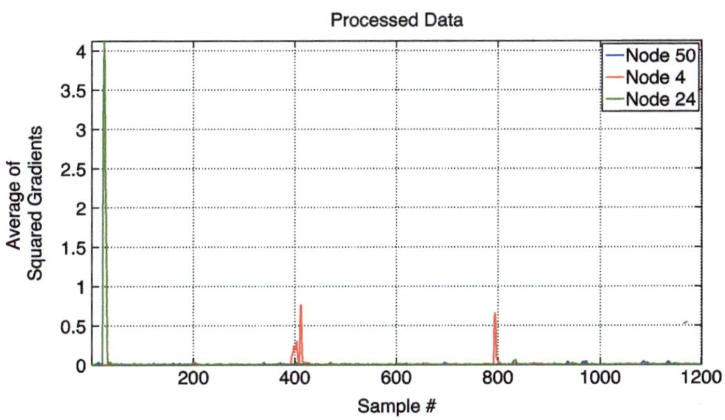

Fig. 3 Processing results for UAV task planning decision support

5 Conclusion

We have carried out initial work on multi-level large scale monitoring systems based on wireless sensor networks with aerial robotic platform support for in-network data aggregation and communication relaying. The system architecture and an algorithm were introduced to allow for effective task planning of data collection between the UAV and the WSN cluster-heads. Experimental results were provided in order to justify this approach. As future work, we aim at implementing the system based on the Ar.Drone2 platform [13] and using TelosB nodes as ground sensor network through appropriate low-power multi-hop communication protocols and data processing, which are aware of the UAV system integration.

Acknowledgments The work of Grigore Stamatescu has been funded by the Sectoral Operational Programme Human Resources Development 2007–2013 of the Ministry of European Funds through the Financial Agreement POSDRU/159/1.5/S/134398. The work of Dan Popescu and Cristian Mateescu was supported by a grant of the Romanian Space Agency, "Multisensory Robotic System for Aerial Monitoring of Critical Infrastructure Systems" (MUROS), project number 71/2013.

References

1. Athreya, A.P., Tague, P.: Network self-organization in the internet of things. In: Sensor, Mesh and Ad hoc Communications and Networks (SECON), 10th Annual IEEE Communications Society Conference, vol. 25, no. 33, pp. 24–27 (2013)
2. Stamatescu, G., Sgârciu, V.: Evaluation of wireless sensor network monitoring for indoor spaces. In: International Symposium on Instrumentation & Measurement, Sensor Network and Automation (IMSNA), 2012, vol. 1, pp. 107–111 (2012)
3. Jawhar, I., Mohamed, N., Al-Jaroodi, J., Zhang, S.: A framework for using unmanned aerial vehicles for data collection in linear wireless sensor networks. J. Intell. Robot. Syst. **74**(1), 437–453 (2014)
4. Zimmerling, M. Dargie, W., Reason, J.M.: Localized power-aware routing in linear wireless sensor networks. In: Proceedings of the 2nd ACM International Conference on Context-Awareness for Self-managing Systems (CASEMANS '08). ACM, New York, pp. 24–33 (2008)
5. Owojaiye, G., Sun, Y.: Focal design issues affecting the deployment of wireless sensor networks for pipeline monitoring. Ad Hoc Netw. **11**(3), 1237–1253 (2013). ISSN 1570-8705, http://dx.doi.org/10.1016/j.adhoc.2012.09.006
6. Antonio, P., Grimaccia, F., Mussetta, M.: Architecture and methods for innovative heterogeneous wireless sensor network applications. Remote Sens. **4**, 1146–1161 (2012)
7. Orfanus, D., Eliassen, F., de Freitas, E.P.: Self-organizing relay network supporting remotely deployed sensor nodes in military operations, ICUMT, pp. 326–333 (2014)
8. Chen, H., Chang, K.C., Agate, C.S.: A dynamic path planning algorithm for UAV tracking. In: Proceedings of SPIE 7336, Signal Processing, Sensor Fusion, and Target Recognition XVIII, p. 73360B (2009). doi:10.1117/12.819836
9. Lanillos, P., Gan, S.K., Besada-Portas, E., Pajares, G., Sukkarieh, S.: Multi-UAV target search using decentralized gradient-based negotiation with expected observation. Inf. Sci. **282**, 92–110 (2014). http://dx.doi.org/10.1016/j.ins.2014.05.054, ISSN 0020-0255

10. Stamatescu G., Popescu D., Dobrescu, R.: Cognitive radio as solution for ground-aerial surveillance through WSN and UAV infrastructure. In: Proceedings of the 6th Electronics, Computers and Artificial Intelligence Conference, IWSSS Workshop, Bucharest, 23–25 Oct 2014
11. Stamatescu, G., Chitu, C,; Vasile, C., Stamatescu, I., Popescu, D., Sgârciu, V.: Analytical and experimental sensor node energy modeling in ambient monitoring. In: Industrial Electronics and Applications (ICIEA), IEEE 9th Conference, pp. 1615–1620 (2014)
12. Intel Lab Data. Available on-line: http://db.csail.mit.edu/labdata/labdata.html
13. Bristeau, P.-J., Callou, F., Vissière, D., Petit, N.: The navigation and control technology inside the AR. Drone micro UAV. In: 18th IFAC World Congress, pp. 1477–1484. Milano (2011). doi:10.3182/20110828-6-IT-1002.02327

Improving Communication Efficiency of Hybrid UAV-WSN Systems

Maximilian Nicolae, Dan Popescu, Radu Dobrescu and Cristian Mateescu

Abstract One of the well-known topics in wireless sensor networks (WSNs) is energy consumption efficiency in regard to communication needs. An unmanned aerial vehicle (UAV) that gathers data from a ground WSN can be seen as a gateway for the WSN. The paper presents a solution for improving the communication efficiency of a hybrid UAV-WSN system by using directional antenna and synchronization mechanisms. The final goal is to increase the life of the ground nodes. The targeted architecture is described and some modelling assumptions are discussed.

Keywords UAV · Communication · Antenna pattern · WSN

1 Introduction

There is increasing interest in using relatively small unmanned aerial vehicles (UAVs) that fly at low altitudes to prove relay services for terrestrial wireless sensor networks (WSNs) whose nodes are scattered (usually by aerial launching)

M. Nicolae (✉) · D. Popescu · R. Dobrescu
Department of Automation and Industrial Informatics, Politehnica University of Bucharest, Bucharest, Romania
e-mail: max.nicolae@upb.ro

D. Popescu
e-mail: dan_popescu_2002@yahoo.com

R. Dobrescu
e-mail: rd_dobrescu@yahoo.com

C. Mateescu
Team Net International S.A, Bucharest, Romania
e-mail: cristian.mateescu@teamnet.ro

T. Borangiu (ed.), *Advances in Robot Design and Intelligent Control*,
Advances in Intelligent Systems and Computing 371,
DOI 10.1007/978-3-319-21290-6_41

in hardly accessible or inaccessible areas. The UAV acts as a decode-and-forward relay, sending the messages from the nodes on the ground to some remote base station. The work described above assumes that the ground nodes are static and the UAV is configured with only a single antenna for communication with the ground nodes.

A number of different approaches have been proposed in the literature addressing the performance of UAV-assisted communication networks. Among the first, in [1] a throughput maximization protocol was proposed for a network in which the UAV first loads data from the source node and then flies to the destination node to deliver it; but the solution is not available for real time applications. The network performance of single-antenna ground nodes communicating with a multi-antenna UAV over a multiple-access ground-to-air wireless communications link was analysed in [2]. The UAV uses beam forming to mitigate the inter-user interference and achieve spatial division multiple access (SDMA). The simulation results demonstrate the benefits of adapting the UAV heading in order to optimize the uplink communications performance. In [3], the authors propose an energy efficient communication scheme for WSNs employing UAV in data gathering, by minimizing the number of redundant sensors communicating with UAV. Special attention was paid for interpretation of WSN-UAV systems as multiagent systems, the cooperation between nodes being handled by specific techniques of Artificial Intelligence (Particle Swarm Optimization [4], Cyber-Physical Systems [5], heuristic search [6] or by using bio-inspired algorithms—evolutionary and genetic algorithms [7], and ant colony algorithms [8]).

Due to the specificity of air-to-ground and ground-to-air communications, the research effort was also focused on finding solutions that maximize the total data transmitting rate by efficient allocation of resources [9] or by adapting the antenna radiation pattern [10]. It should be noted that this work continues the research initiated by the authors and presented in [11].

The goal of this paper is to offer solutions to optimize the dual link communication performance between UAV and the terrestrial nodes, and subsequently to maximize the lifetime of the entire network. For this purpose a simulation framework was designed and several communication scenarios were tested. The simulation framework addresses the case in which the ground nodes (WSN) are deployed by the UAV which, afterwards, will collect the acquired data. In this situation it is assumed that there are no communication links between the sensor network nodes. A specific task for UAV should be to determine as accurately as possible the position of the ground nodes (localization) and through this, to compile a map of radio coverage for each ground node of the WSN. Since the WSN nodes are not communicating with each other, a method for increasing the efficiency would be making the nodes to activate their communication state only when the UAV is overflying the node's communication coverage area. The ground node having limited energy capacity, the sink node (UAV) is moving (flying) following severe constraints in regard to its trajectory.

2 Architecture of the Hybrid System

The MUROS framework [12] is first introduced with the purpose of developing the surveillance and monitoring system with unmanned aerial platforms for monitoring, preventing and mitigating incidents which have an impact on critical infrastructure such as transport routes for oil products, railways, power lines, highways etc. The main components of a system with UAV are illustrated in Fig. 1:

- GCS (Ground Control Station): data and control management;
- GDT (Ground Data Transmission): communication node between GCS and UAV;
- GSN (Ground Sensor Node): components of the deployable multisensory network;
- UAV (Unmanned Aerial Vehicle): platform gathering sensor and payload data;
- L (Launcher): device for launching the UAV.

As can be observed, the hybrid system is composed of a fixed part (ground station and multisensory wireless network) and a mobile part (the UAV).

The originality and innovative character of the UAV implementation within this framework is given mainly by the various solutions to correlate information collected by the aerial vehicle with the ground sensors network and by the flexible architecture of the whole system which allows coordinating land and aerial sensors systems.

The main idea is to design a flexible multisensory robotic system capable of monitoring critical infrastructure, evolution of phenomena due to extreme events and/or targets situated in hardly accessible areas in a semi-autonomous manner.

3 Models for Simulations

3.1 Localization of Ground Nodes

The ground nodes localization after deployment can be seen as a classical problem of radio location. Initially, nodes are in active radio listening state in which they are

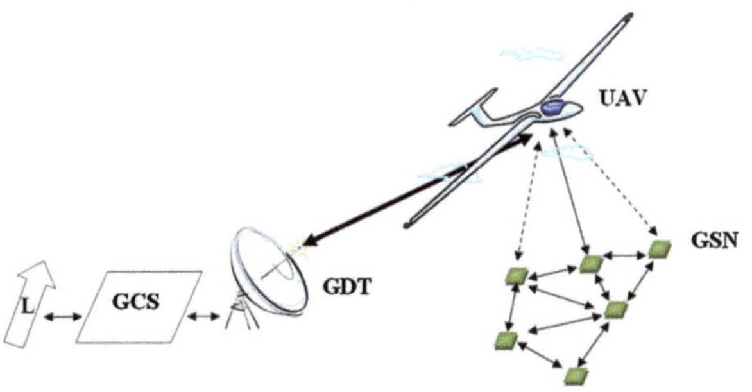

Fig. 1 Structure of the Hybrid UAV-WSN system

waiting for a beacon signal from the UAV. Immediately after the nodes are deployed, the UAV can start to go on "patrol" and send beacon signals. When a node receives the beacon, based on its ID and a protocol for collision avoidance, it will issue several answers. The UAV is able to estimate the node's location through several samples of the received signal strength or phase shift (Receiver Signal Strength Indicator—RSSI or Angle of Arrival—AoA are often used). Additionally, the node can also provide some useful information in its answer by the data it has collected from its inertial sensors during deployment (ex: accelerometers and gyros). After the handshake is established, further details can be exchanged, like the time schedule for node's state machine and transmission power level.

3.2 Radiation Pattern Model

This section refers to the antennas that will equip the ground nodes. As described in [13], an antenna is a transducer between a guided wave and a radiated wave (or vice versa). The spatial distribution of radiated energy (or received energy) is described by the antenna (radiation) pattern. This distribution is three dimensional but most often is characterized by two planes as planar section through the 3D space representation (Fig. 2). One plane is for $\theta = 90^{o}$ (θ is also known as *elevation angle*)—the xOy plane, the other one is for $\varphi = 90^{o}$ (φ referred as *azimuth angle*)—the yOz plane. These planar sections are often referred as principal plane cuts.

Isotropic antenna refers to the ideal case when the energy distribution is isotropic in all direction. Its 3D representation is a sphere, while both of the principal plane cuts will be perfect circles. Directional antennas are those that have a preferred direction where they concentrate the energy radiation. These directions are graphically described on the principal plane cuts as large lobes. Omnidirectional antennas have one of the principal plane cuts characterized by a circle. In the case of principal plane cuts which exhibit directional lobes, 3 dB-beam width is introduced (also known as half-power beam width) as the angle between the two segments which are connecting the origin with the two points on the main lobe that has 3 dB loss in power (the power is halved). By contrast, directional antennas exhibit main

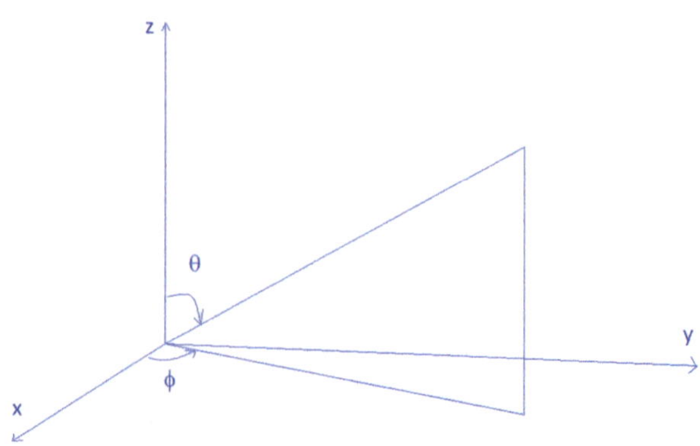

Fig. 2 Antenna measurement coordinate system (polar coordinate system)

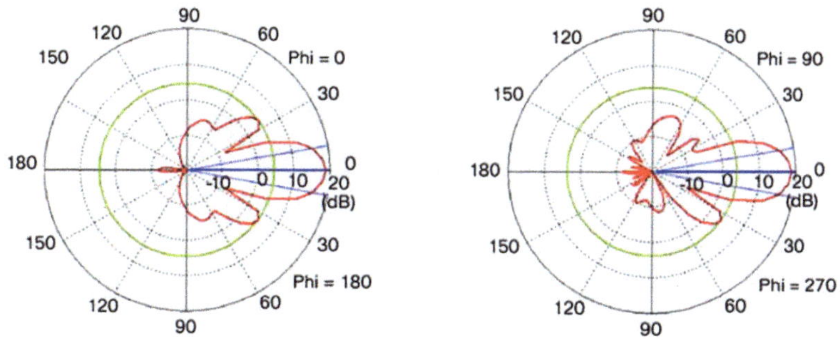

Fig. 3 Directional antenna—principal plane cuts. *Source* [13, p. 11]

lobes in both principal plane cuts (Fig. 3). Due to their possibility of concentrating the radiated energy in a particular direction of interest, with the same energy as an omnidirectional antenna, they can reach further coverage in that direction. Another benefit comes from diminishing interference zones between nearby antennas.

In our simulation framework we approximated both principal plane cuts as having one main lobe characterized by the same 3 dB-beam width. This will reduce the antenna pattern to a cone which has several parameters that can be acted upon. The cone's axis of symmetry will be considered as steering direction. We consider the position of the steering direction as the position vector in Fig. 2. Therefore we can adjust α angle describing the 3 dB-beam width and θ and φ angles as the orientation of the antenna after deployment. A study scenario with 15 deployed nodes was used for simulation.

3.3 Radio Coverage Area Model

We considered that the UAV's antenna pattern is wider than those corresponding to nodes, therefore whenever the UAV is within the communications range of the node, it can communicate with it. It is assumed that the UAV flies at a certain constant altitude H and with a constant speed v (tangential speed—v_{UAV}). The nodes are scattered at various 3D coordinates, having an arbitrary orientation. The *radio coverage area* for each node (A_{RCi}) will be determined as the intersection between its antenna pattern and the "parallel" plane situated at altitude H (flying plane—Fig. 4). Based on the volume of data needed to be transmitted by the node and the data link quality, the UAV should maintain its position in the radio coverage area of the node for a specific time. Knowing the time and velocity we can compute the optimum trajectory.

The cone cutting with a plane yields the known complex problem of conic section. There are three types of conic sections: ellipse (circle is a special case), parabola and hyperbola. For the considered simulation, conic sections are all ellipses. But there are cases corresponding to various elevation angles of steering

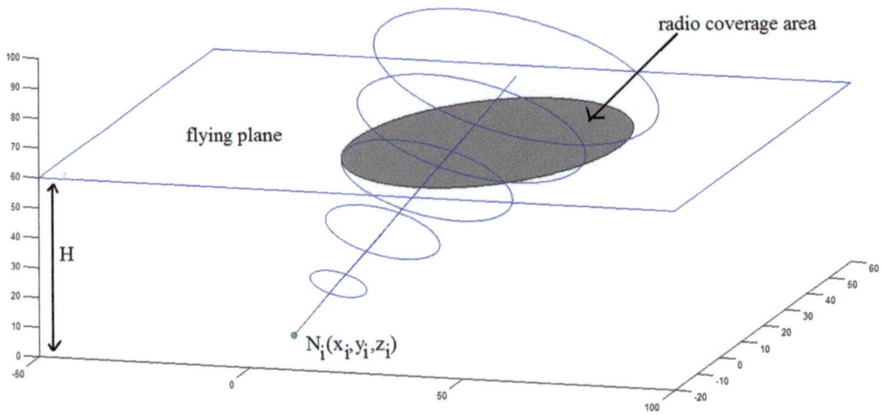

Fig. 4 The radio coverage area (conical cut—metric units scale)

direction, power of radio transmission of nodes and altitude of flying that could generates other types of sections. The equations used for computing this model allow the GCS to make the same map that our computed simulation, thus providing the map for the optimum trajectory.

3.4 Establishing the Communication Timing

The ground nodes are generally composed of a CPU, a transceiver (with antenna), sensors (with signal adaptation chain) and battery. These components can be physically distinct or partially integrated in a system on a chip (SoC). Either way, the entire node system can have a finite sate machine (FSM) which describes also the energy consumption model. The latest SoC CC2538 from Texas Instruments has among its functioning states, the following (with current consumption associated) [14]: Active-Mode RX (CPU Idle): 20 mA, Active-Mode TX at 0 dBm (CPU Idle): 24 mA, Power Mode 1: 0.6 mA, Power Mode 2 (Sleep Timer Running): 1.3 μA, Power Mode 3 (External Interrupts): 0.4 μA. Comparing the energy consumption between Active-Mode RX and Power Mode 3 (50.000:1) we can see the impact of a good management of the node's FSM. Therefore, establishing the communication timing is of extreme importance in increasing the WSN lifespan.

3.5 Communication Channel Models

There are many channel models in the literature. All of them can be plugged in our simulation. At this point we model the channel influences as corruption that appears at the data packets level. Therefore, we will introduce the following parameters:

- *Maximum data rate* (MDR): the temporal channel capacity (contains beacons and other synchronizations frames, headers and control data),
- *Effective data rate* (EDR): the ratio of payload in the entire data exchanged,
- *Actual data rate* (ADR): the ratio of payload after corruptions and retransmissions.

The local problem for a node is to determine the time the UAV will have to overfly its radio coverage without disconnections, based on the data it needs to transmit and the actual data rate.

4 Solutions for Improving Communication Performance

After deployment, the UAV will scan the area containing the ground nodes, sampling the radio signal and providing the necessary data for localization (see Sect. 3.1). When a node is localized, it will receive the new timing of its FSM. At this point, the UAV knows when each node will be available for communication. Considering that the location of the nodes was determined, our proposed algorithm for communication between UAV and ground nodes will start in two phases:

1. Preliminary flight (PF), and
2. Data gathering flight (DGF).

For the PF, it is considered that each node will communicate the same amount of data (Q_P) which includes the time interval t_p in which the UAV should overfly the node's radio coverage area (A_{RCi}). The t_p is given by the ratio between Q_P and ADR (see Sect. 3.5). Knowing the v_{UAV}, the length of the trajectory inside A_{RCi} is computed ($L_i = v_{UAV} \cdot t_p \cdot k$, where $k > 1$ is a margin factor). Q_P contains bidirectional information (node to UAV and UAV to node). The node will communicate how much extra data it has to send next time (Q_D) and the UAV will compute and send the time when it will next be in node's radio coverage. With this information, the node will go in a low power state with its radio shutoff (thus

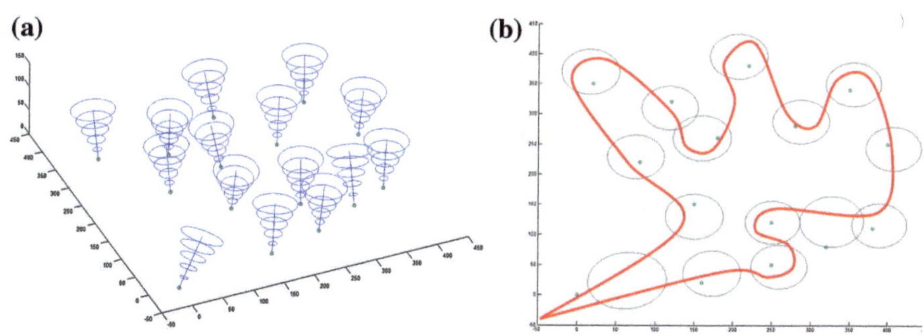

Fig. 5 Simulation images (metric units scale): **a** antenna patterns, **b** flying plane with A_{RCi} and UAV trajectory

preserving a lot of energy) and will wake-up its radio system when the UAV programmed it to do that. During the PF the DGF trajectory is computed dynamically and the timing communicated to the nodes. In DGF, the UAV will have to respect the timing it has computed during PF and dynamically compute the new timing for the following flight. The preliminary flight and data gathering can alternate. For safety reasons, each node can have an active radio interval for recovering if something went wrong and the node was desynchronized.

The algorithm presented above was tested in the simulation framework that we developed (Fig. 5). We didn't focus on determining an optimum trajectory but on validating the synchronization mechanism.

5 Conclusions

The use of directional antenna together with the synchronization of the nodes' active states improves the efficiency of the communication but also introduces additional problems. We proposed and validated a solution to overcome this new issue. For this, there was needed to introduce some modelling assumptions with the corresponding reasoning. The calculus needed for modelling and simulation with visual representation was conceived in order to be implemented on the UAV-WSN hybrid system. This will be the next step in validating our solution. Further on, we will focus on optimizing the trajectory, followed by adjusting the power transmission level of each node (changing dynamically the radio coverage area) and the speed of the UAV.

Acknowledgments The work related to nodes modelling and simulation has been funded by the Sectoral Operational Programme Human Resources Development 2007–2013 of the Ministry of European Funds through the Financial Agreement POSDRU/159/1.5/S/134398.

The work related to the hybrid system architecture has been funded by National Research Program "STAR" (Romanian Space Agency), project 71/2013: Multisensory robotic system for aerial monitoring of critical infrastructure systems-MUROS.

References

1. Cheng, C., Hsiao, P., Kung, H., Vlah, D.: Maximizing throughput of UAV-relaying networks with the load-carry-and-deliver paradigm. In: Proceedings of IEEE WCNC'07, pp. 4417–4424 (2007)
2. Feng, J., Swindlehurst, A.: Optimization of UAV heading for the ground-to-air uplink. IEEE J. Sel. Areas Commun. **30**(5), 993–1005 (2012)
3. Say, S., Kento, A., Naoto, A. et al.: Effective data gathering and energy efficient communication protocol in Wireless Sensor Networks employing UAV. In: Proceedings of the IEEE Wireless Communications and Networking Conference (WCNC), pp. 2342–2347 (2014)

4. Dac-Tu, H., Grotli, E.I., Johansen, T.A. et al.: Performance evaluation of cooperative relay and particle swarm optimization path planning for UAV and wireless sensor network. In: Proceedings of the IEEE Globecom Workshops, pp. 1403–1408 (2013)
5. Saeed, A., Neishaboori, A., Harras, K. et al.: Up and away: a visually-controlled easy-to-deploy wireless UAV cyber-physical testbed. In: Proceedings of IEEE 10th International Conference on Wireless and Mobile Computing, Networking and Communications (WiMob), pp. 578–584 (2014)
6. Lucani, D.E., Sujit, P.B., Sousa, J.B.: Joint route planning for UAV and sensor network for data retrieval. In: Proceedings of the 7th Annual IEEE Systems Conference (SysCon), pp. 688–692 (2013)
7. Zhang, L., Duan, L., Qian, Z., Huang, G.: WSN Node localization technology based on genetic algorithm. Comput. Eng. **36**(10), 85–87 (2010)
8. Freitas, E.P.: Cooperative context-aware setup and performance of surveillance missions using static and mobile wireless sensor networks. PhD Thesis, Halmstadt University (2011)
9. Fumie, O., Ryu, M., Hiroki, N. et al.: Resource allocation for data gathering in UAV-aided wireless sensor networks. In: Proceedings of the 4th IEEE International Conference on Network Infrastructure and Digital Content, pp. 11–16 (2014)
10. Mohana, S.S.: Design of Antenna pointing system (APS)—its control and communication challenges. In: Proceedings of the International Conference on Control Communication and Computing, pp. 490–495 (2013)
11. Stamatescu, G., Popescu, D., Dobrescu, R.: Cognitive radio as solution for ground-aerial surveillance through WSN and UAV infrastructure. In: Proceedings of ECAI International Conference (2014)
12. MUROS Project, Romanian Space Agency Competition C2, pp. 2013–2016. Available on-line: http://imtt.pub.ro/MUROS3/muros
13. Cisco White Paper—Antenna Patterns and Their Meaning (2007) available online
14. CC2538—A Powerful System-On-Chip for 2.4-GHz IEEE 802.15.4, 6LoWPAN and ZigBee Applications Datasheet

Top Viewing Human Tracking in Shopping Centres

Petros G. Vasileiou, Nefeli Lamprinou, Emmanouil Z. Psarakis, Giannis Tzimas and Nikos Achilleopoulos

Abstract The problem of top viewing human tracking in a closed environment, such as is the area of a shopping centre is a challenging one. In this paper we test a number of Kalman and simple particles filters-based algorithms for solving the visual tracking problem. Although results obtained from the experiments that we have conducted are promising and can be considered as adequate from all practical perspectives, more research must be done so that the filter used takes into account the existing peculiarities in the movements of human beings.

Keywords Visual tracking · Human detection and Tracking · Mapping · Kalman filter · Particles

1 Introduction

The problem of human tracking, especially in crowed environments constitutes a challenging one [1]. For an excellent experimental survey on this topic and in particular on the visual tracking, the reader is referred to [2]; here we present a system tailored for top viewing human tracking inside the area of a shopping centre. The surveillance cameras are mounted on the ceiling, more than three meters above the ground. This can be advantageous in discriminating separate humans within a

---

This research has been done on behalf of the Development Funding Program: Smartcams Integrated system for anonymous monitoring via smarcams of customer behavior, movement and responses in shopping centers (38SEMs2010).

P.G. Vasileiou · N. Lamprinou · E.Z. Psarakis (✉) · G. Tzimas
Department of Computer Engineering and Informatics, University of Patras, Patras, Greece
e-mail: psarakis@ceid.upatras.gr

N. Achilleopoulos
Dynacomp, NEO Patras, Athens 208 26442 Patras, Greece

T. Borangiu (ed.), *Advances in Robot Design and Intelligent Control*,
Advances in Intelligent Systems and Computing 371,
DOI 10.1007/978-3-319-21290-6_42

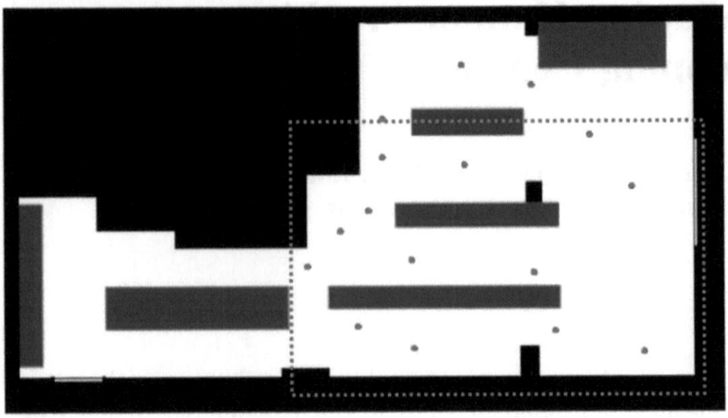

Fig. 1 Floor plan including static obstacles (*gray* and *black blocks*) and hot spots (*gray circles*) in the store hosting the system. The total field of view of the cameras is marked with *dotted line*

crowd. The head of a human will have the lowest chance of being occluded, therefore we pursued the goal of finding head candidates points representing the tops of the heads in a blob. Figure 1 shows the floor plan and visually covered space containing information about existing static obstacles, and hot spots in the store hosting the system.

1.1 The Motion Model

Let us consider the following simple motion model:

$$\boldsymbol{x}_k = \boldsymbol{x}_{k-1} + \alpha \boldsymbol{d}_k \tag{1}$$

$$\boldsymbol{d}_k = R(\theta_k)\boldsymbol{d}_{k-1} \tag{2a}$$

$$\alpha_k = (1 - \lambda)\alpha_{k-1} + \lambda\zeta_k \tag{2b}$$

$\|\boldsymbol{d}_0\| = 1$, $R(\theta_k)$ is a rotation matrix and α_k a process controlling the person's step length and the smoothness of the motion through the parameter $\lambda \in [0, 1]$ and the random variable $\zeta_k \sim \mathrm{U}[-4, 4]$. As it is clear, using the aforementioned motion model several composite movements can be synthesised, and the path followed by an object inside the area of a room to visit locations of interest, can be created. From the experiments we have conducted, paths created by this simple motion model were found to be very similar to ones obtained from real captured data.

The remainder of the paper is organized as follows. Section 2 outlines the mathematical details of the techniques used to design a tracking system appropriate for the solution of problem at hand. Section 3 contains experimental results and finally Sect. 4 contains the conclusions.

2 The Tracking System

In this section we analyse the basic components implemented to achieve our goal. In the next subsection the details of the background removal algorithm are presented.

2.1 *Background Modelling and Subtraction*

The ability to extract moving regions from video data is crucial in visual tracking of humans. The process of the background subtraction is a group of methods separating the moving pixels in the image from the static ones [3, 4]. The basic assumption underlying these approaches is that the visual appearance of the static parts in the scene remains constant, while the position of moving objects changes from frame to frame. Thus, if observed for some time, one can accumulate the image of the stationary parts of the scene and further subtract it from the future video data. Changes in lighting and dynamic changes in the scene layout, such as new fixtures being introduced, add to the complexity of background removal task. We have incorporated a background removal algorithm based on an improvement of running Gaussian average in order to achieve stable and sustainable foreground segmentation. Pfinder [3] is the most characteristic and well known algorithm of this category. According to this technique, each image pixel is considered as a random variable whose mean value and standard deviation are estimated using the intensities along the video signal (frames). Specifically, let I_k be the intensities of the k-th frame and μ_{k-1} the prior estimations of the mean values of its pixels. Then, the following convex relations are used for the adaptation of the aforementioned quantities:

$$\mu_k = \alpha I_k + (1 - \alpha)\mu_{k-1} \tag{3}$$

$$\sigma_k^2 = \alpha(I_k - \mu_k)^2 + (1 - \alpha)\sigma_{k-1}^2 \tag{4}$$

where α is a scalar weighting parameter. Having computed the aforementioned moments, the (i,j) pixel can be classified as background, that is $(i,j) \in B_k$, if the following condition holds:

$$|I_k(i,j) - \mu_k(i,j)| < \lambda\sigma_k(i,j) \tag{5}$$

where λ is a parameter that specifies the confidence interval of the sample mean.

Despite its simplicity, the aforementioned technique suffers from several problems, the most serious one being the creation of artefacts in cases where the state of a pixel is changed. In order to avoid this problem, Eqs. (3), (4) must be redefined as follows:

$$\mu_k = \beta\mu_{k-1} + (1-\beta)(\alpha I_k + (1-\alpha)\mu_{k-1}) \tag{6}$$

$$\sigma_k^2 = \beta\sigma_{k-1}^2 + (1-\beta)\left(\alpha(I_k - \mu_k)^2 + (1-\alpha)\sigma_{k-1}^2\right) \tag{7}$$

where β is a binary parameter taking the value 0 when the pixel at the previous time instant was classified as background.

The outline of the implemented background removal algorithm follows.

Algorithm 1. Background Removal Algorithm

Input: Frame $I_k(i,j)$, the prior mean μ_{k-1} and variance σ_{k-1}^2 of its pixels, the set B_{k-1} and values of parameters α, λ

For each pixel of the frame ***do***
If pixel $(i,j) \in B_{k-1}$ set $\beta = 0$
Else set $\boldsymbol{\beta} = 1$
End of if
Use (6) and (7) to update the statistics
Use test in (5) for the classification of the pixel
End of for

Output: Sets B_k, F_k, the posterior mean μ_k and variance σ_k^2.

2.2 Dynamical Systems

Let us consider a dynamical system described by the following state-space model:

$$\boldsymbol{x}_k = f_k(\boldsymbol{x}_{k-1}) + \boldsymbol{u}_{k-1} \tag{8}$$

$$\boldsymbol{y}_k = h_k(\boldsymbol{x}_k) + \boldsymbol{v}_k \tag{9}$$

where $\boldsymbol{u}_k$ and $\boldsymbol{v}_k$ are independent zero-mean white Gaussian noise processes with covariance matrices R_k and Q_k respectively. In addition, the functionals $f_k(.), h_k(.)$ denote in the general case a nonlinear transition matrix function and nonlinear measurement matrix respectively. In addition, both functionals may be time-variant.

2.3 The Linear Case

Let us consider that the dynamical system, described by Eqs. (8) and (9) is linear, and the prediction (a priori estimation) of the state vector $\boldsymbol{x}_{k|k-1}$ is known.

The requirement is to use the information contained in the new measurement y_k to update the estimate of the unknown state $\boldsymbol{x}_{k|k}$. Then, the desired a posteriori estimation is:

$$\boldsymbol{x}_{k|k} = \boldsymbol{x}_{k|k-1} + K_k\left(\boldsymbol{y}_k - H_k\boldsymbol{x}_{k|k-1}\right) \tag{10}$$

where the vector $\boldsymbol{y}_k - H_k\boldsymbol{x}_{k|k-1}$ is known as the innovation vector of the process and the matrix K_k is called the Kalman gain and is given by:

$$K_k = P_{k|k-1}H_k^T\left(H_kP_{k|k-1}H_k^T + R_k\right)^{-1} \tag{11}$$

where $P_{k|k-1}$ constitutes the prediction of the state covariance matrix. Finally, the update rule of the a posteriori state covariance matrix $P_{k|k}$ is given by:

$$P_{k|k} = (I - K_kH_k)P_{k|k-1} \tag{12}$$

The Kalman filter uses Gaussian probability density in the propagation process, the diffusion is purely linear and the density function evolves as a Gaussian pulse that translates, spreads and is reinforced, remaining Gaussian throughout. The random component of the dynamical model $\boldsymbol{u}_k$ leads to increasing uncertainty while the deterministic component $F_{k+1}\boldsymbol{x}_k$ causes the density function to drift bodily. Finally, the effect of an external observation $\boldsymbol{y}_k$ is to superimpose a reactive effect on the diffusion in which the density tends to peak in the vicinity of observations [2, 5].

2.4 *Visual Tracking Using the Kalman Filter*

The main application of the Kalman filter in robot vision is visual tracking [2]. In its general setting, visual tracking demands at each instant computing the object's position and speed. A sequence of images captured by a camera containing the object is taken as algorithm's input. After background removal it is fair to take as system state $\boldsymbol{x}_k$ the position x, y of the object at time k. Thus, one can use Kalman filter to locate more efficiently the object. Specifically, instead of looking for the object in the whole image plane we define a search window centred in the predicted location of the filter.

The outline of the proposed Kalman filter for visual tracking follows:

Algorithm 2. Kalman Filter for Visual Object Tracking

Initialization-Phase $(k = 0)$: The initial position $\boldsymbol{x}_k$ as well as initial covariance matrix of states P_k, are given.
Prediction-Phase $(k > 0)$: In this phase using the Kalman state Eq. (8) we predict the next location $\boldsymbol{x}_{k|k-1}$ of the

object. This predicted location is the centre of the circle where the object will be located.

Correction-Phase $(k > 0)$: Having defined the centre of the circle where the object will be searched, in this phase we use the measurement (real position) and the Kalman gain to compute state $\boldsymbol{x}_{k|k}$.

2.5 Visual Tracking of Multiple Moving Objects

Let us consider now the problem of tracking multiple moving objects. We exploit the aforementioned efficiency of the Kalman filter in the localization process of the object and use this efficiency in a sequential way to track one object after the other by assigning the current measurement to the nearest predicted state.

The outline of the multiple moving objects tracking algorithm is as follows:

Algorithm 3. Kalman Filter for Visual Multiple Moving Object Tracking

Input: Frame $I_k(i,j)$, the prior mean μ_{k-1} and variance σ_{k-1}^2 of each pixel, the set B_{k-1} and values of parameters α, λ

Call Algorithm 1, for background removal.
While list F_k is not empty ***do***
Call Algorithm 2, for tracking of the specific object.
Delete the object from the list F_k.
End of while

2.6 Visual Tracking Using Particles

Although Kalman filters, even in the non-linear case, are the optimal in least squares sense linear systems, they did not work so well because they cannot model accurately the underlying dynamics of the physical system. Thus, using a particle filtering based solution should be preferred to Kalman filters. However, usually particle filters need a fairly large amount of particles to be effective, so they are generally more demanding in terms of computational power than Kalman filters even from the most sophisticated ones [6]. This point can be considered as the major drawback for their use in real world applications. In addition, as it was reported in [7], often the errors are also relatively large, so one must be careful when using the estimates. In practice this means that one has to monitor the filters covariance estimate, and trust the state estimates and predictions only when the covariance estimates are low enough; but even then it is possible that the filters estimate is completely wrong.

Despite the above problems, in order to evaluate the performance in solving the problem at hand we have used the simplest form of a particle filter with its most critical part, the resampling process implemented as Resampling Algorithm 2 given in [7]. Using a number of 600 particles, the performance of this particular form of the particle filter, as stated in the next section is comparable in terms of the achieved root mean squared error with the simplest linear Kalman filter, but with an increasingly large computational cost which makes its use prohibitive in real time applications.

3 Experiments

In this section we will compare the tracking performance of the rivals in single and multiple moving objects in the area of a shopping centre. In order to be able to safely validate the performance of algorithms under comparison, we run the algorithms on simulated data produced by the motion model described in the introduction. We apply in paths created by the aforementioned motion model, three linear Kalman based systems and a simple particle filter based one. In particular, the following linear systems:

$$\begin{aligned} f_1(\boldsymbol{x}) &= I_{2\times2}\boldsymbol{x} = h_1(\boldsymbol{x}) \\ f_2(\boldsymbol{x}) &= toeplitz(\boldsymbol{e}_{4\times1}, [1 \quad 0 \quad T \quad 0])\boldsymbol{x} \end{aligned} \tag{13}$$

$$h_2(\boldsymbol{x}) = [\, I_{2\times2} \quad 0_{2\times2}\,]\boldsymbol{x} \tag{14}$$

$$\begin{aligned} f_3(\boldsymbol{x}) &= toeplitz\left(\boldsymbol{e}_{6\times1}, \left[1 \quad 0 \quad T \quad \tfrac{T^2}{2} 0\right]\right)\boldsymbol{x} \\ h_3(\boldsymbol{x}) &= [\, I_{2\times2} \quad 0_{2\times2} \quad 0_{2\times2}\,]\boldsymbol{x} \end{aligned} \tag{15}$$

where T denotes the sampling period,[1] $\boldsymbol{x}$ the state vector of size varying from 2 elements in the first system to 6 elements in the last one, $toeplitz(\boldsymbol{c}, \boldsymbol{r})$ is a toeplitz matrix having as first column $\boldsymbol{c}$ and first row $\boldsymbol{r}$, and $\boldsymbol{e}_{k\times1}$ a column vector of length k with first element 1, are used. At this point, we must stress that the state and the measurement equations used for the particle filter coincide with that of the system defined in (13).

Finally, in order to compare the performances of the rivals, we are going to use as figures of merit the *Root Mean Squared Error* that expresses the mean deviation of the estimated path from its ground truth counterpart, and its standard deviation.

[1]Assuming that the speed of a human being who is moving in a shopping centre is upper bounded by 5 km/*h*, in all the experiments we have conducted T was set to 0.1 s.

3.1 Experiment I

In this experiment we consider that in the floor map shown in Fig. 1, there is a human being who moves around the area of the shop. In particular we apply the above mentioned trackers in order to track the single path shown in Fig. 2a.

We have run the rivals for different measurement noise variances, and the obtained results are contained in Table 1. As it is clear, the tracker of (14) seems to be a good choice for the problem at hand. This is more evident in Fig. 3, where the distributions of the obtained errors for the case of noise power 10 are shown. Finally, the computational cost in terms of the mean time (in msec) needed by each one of the above mentioned filters was 7.0, 0.15, 0.55 and 0.7 respectively.

3.2 Experiment II

In this experiment we consider that in the floor map shown in Fig. 1, there are 10 humans moving around the area of the shop. The computational cost in terms of the mean time (msec) needed by each one of the above mentioned filters was 23.5, 0.7, 0.8 and 2.85 respectively. All other comments that could be made for this experiment coincide with those of Experiment I. The results obtained are contained in Table 2.

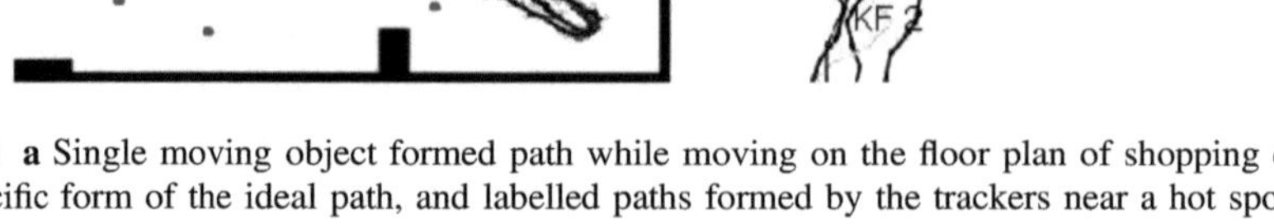

Fig. 2 **a** Single moving object formed path while moving on the floor plan of shopping center; **b** specific form of the ideal path, and labelled paths formed by the trackers near a hot spot

Table 1 Results obtained from the application of the trackers to the single path shown in Fig. 2a for different values of the power σ_V^2 of observation noise

σ_V^2	RMSE				Std			
	PF	KF1	KF2	KF3	PF	KF1	KF2	KF3
0	7.8	4.9	1.2	0.7	22.5	3.8	0.9	0.2
1	7.7	4.9	1.5	1.2	20.4	3.8	1.1	0.5
5	8.3	5.1	2.2	2.3	23.2	4.1	1.6	1.5
10	8.4	5.2	2.6	3.1	23.8	4.6	2.0	2.5
20	8.7	5.4	3.6	4.3	32.5	5.1	3.7	5.5

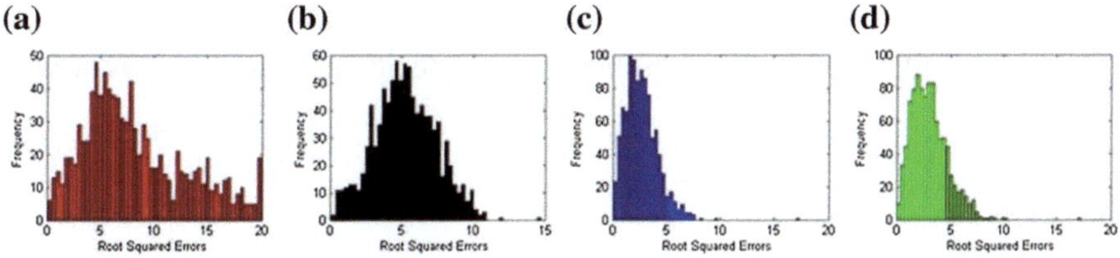

Fig. 3 Distributions of tracking errors resulting from the application of **a** Particle filter (PF), **b** Kalman filter, Eq. (13) (KF 1), **c** Kalman filter, (14) (KF 2) and **d** Kalman filter, (15) (KF 3) on simulated data contaminated by the observation noise ($\sigma_V^2 = 10$)

Table 2 Results obtained applying trackers to multiple paths (Fig. 4), for different values of the power σ_V^2 of observation noise

σ_V^2	RMSE				Std			
	PF	KF1	KF2	KF3	PF	KF1	KF2	KF3
0	4.6	1.5	0.7	0.5	330.6	5.3	2.3	1.2
1	4.5	1.5	0.8	0.7	348.3	5.4	2.5	1.4
5	4.8	1.6	1.1	1.2	350.3	5.8	3.6	2.9
10	5.4	1.7	1.4	1.7	420.9	7.3	4.3	4.5
20	5.3	1.9	1.8	2.2	415.0	7.9	7.0	7.8

Fig. 4 Formed paths of multiple objects moving on the floor plan compared with tracker paths

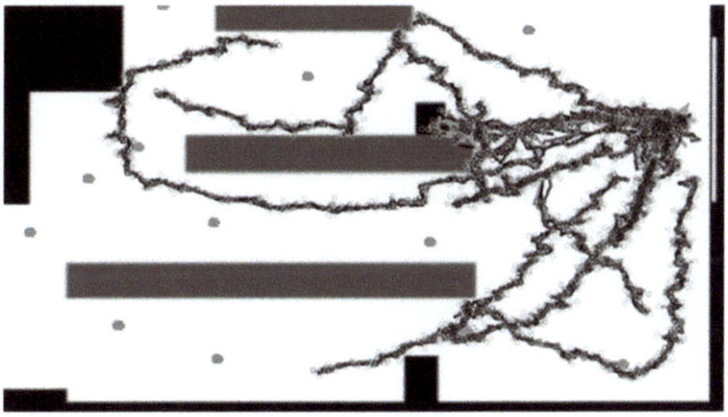

Finally, since moving objects are not tagged, a one to one correspondence between an object and a path cannot be ensured. Thus, the paths of two moving objects cannot be separated in case they meet and continue to move on from the meeting point. To overcome such difficulty, we assign to each moving object a new path. If we set aside the path mixing problem, this can be corrected using a posteriori a proper reassigning process.

4 Conclusions

In this paper the problem of top viewing human tracking in a closed environment was considered. A number of Kalman and particle based filters were tested and their performances compared. The results we obtained from the experiments performed are adequate for the solution of the problem. Future research concerns the properties of the tracking filter if more accurate and low complexity human tracking is needed.

References

1. Weina, G., Collins, R.T., Ruback, R.B.: IEEE Transactions on Vision-Based Analysis of Small Groups in Pedestrian Crowds, vol. 34, no. 5, pp. 1003–1016 (2012)
2. Smeulders, A.W., Chu, D., Cucchiara, R., Calderara, S., Dehghan, A., Shah, M.: IEEE Transactions on Visual Tracking: An Experimental Survey, vol. 36, no. 7, pp. 1442–1468 (2014)
3. Benezeth Y., Jodoin P.-M., Bruno E., Laurent H., Rosenberger C.: Comparative study of background subtraction algorithms. J. Electron. Imaging **19**(3), 033003 (2010)
4. Bowmans, T.: Recent Advanced in Statistical Background Modeling for Foreground Detection: A Systematic Survey, Recent Patents on Computer science, pp. 147–176 (2011)
5. Chen, S.Y.: Kalman filter for robot vision: a survey. IEEE Trans. Industr. Electron. **36**(7), 4409–4420 (2012)
6. Julier, S.J., Uhlmann, J.K.: Unscented filtering and nonlinear estimation. Proc. IEEE **92**(3), 401–422 (2004)
7. Arulampalam, M.S., Maskell, S., Gordon, N., Clapp, T.: NonLinear/non-gaussian bayesian tracking. IEEE Trans. Signal Process. **50**(2), 174–188 (2002)

Part IX
Autonomous Task Learning, Motion Planning and Scheduling

Motion Planning and Scheduling with Stochastic Demands

Elias K. Xidias and Philip N. Azariadis

Abstract This paper considers the combinatorial problem of motion planning and scheduling with stochastic demands. Here, an autonomous vehicle with limited capacity is requested to serve workstations in an industrial environment. Its workstation has a stochastic demand which is revealed upon the arrival of the vehicle. This combined problem is solved by optimizing the vehicle's schedule and route (minimum travel distance) under collision-free and vehicle-capacity constraints. An optimization strategy based on the combination of a genetic and micro-genetic algorithm is developed in order to determine the optimum solution. Experimental results demonstrate the effectiveness of the proposed approach.

Keywords Motion planning · Scheduling · Stochastic demands · Genetic algorithms

1 Introduction

Traditional combinatorial optimization problems require finding solutions for a single instance. However, many real life applications which are motivated from these problems involve systems that change dynamically over time. In this paper we present an extension of the single vehicle problem: simultaneously motion planning and scheduling with stochastic demands (MPSSD). Here, an autonomous vehicle which is operating in a planar indoor industrial environment is requested to serve several workstations with stochastic demands. The objective is to determine

E.K. Xidias (✉) · P.N. Azariadis
Department of Product and Systems Design Engineering,
University of the Aegean, Syros, Greece
e-mail: xidias@aegean.gr

P.N. Azariadis
e-mail: azar@aegean.gr

T. Borangiu (ed.), *Advances in Robot Design and Intelligent Control*,
Advances in Intelligent Systems and Computing 371,
DOI 10.1007/978-3-319-21290-6_43

simultaneously the shortest possible collision-free tour for the vehicle subject to its capacity.

It is emphasized that the actual demand of each workstation is revealed only when the vehicle arrives at it. Due to the uncertainty of demands at the workstations, at some point along a tour the capacity of the vehicle may be depleted before all demands of the workstations have been satisfied (tour failure) [1]. Thus, corrective actions are necessary to ensure feasibility of solutions in case of tour failure, e.g. making a return trip to the depot to restock.

The developed method is accomplished in two successive stages: in the first stage, the environment of the vehicle is mapped using the Bump-Surface concept [2] into a 3D B-spline surface [3] in $\Re^3$ that captures both the free space and the prohibited areas of the environment. In the second phase, a combination of a genetic algorithm (GA) and Micro-GA [4] is applied to determine a minimum length valid vehicle tour.

The attainment of this objective passes through the solution of two known combinatorial optimization problems: motion planning (MP) and vehicle routing and scheduling with stochastic demands (VRSSD). Both are known to be intractable. To our knowledge, no previous work in the literature has studied the two problems simultaneously, probably due to the fact that integrating motion planning and scheduling with stochastic demands forms a very challenging NP-hard combinatorial optimization problem. Vehicle motion planning and scheduling play a critical role in improving the flexibility and productivity of overall production systems [5], and therefore it is expected that this integration will initiate academic and industrial attention in the future.

There are three basic categories of VRSSD [6]: (a) stochastic customers (workstations), (b) stochastic demands and (c) stochastic travel and service times. This paper considers the case where only the workstation demand is stochastic and all other parameters are known beforehand. This problem appears in industries where a mixed-model assembly line comprises a serial set of workstations and an autonomous vehicle moving at a constant speed, which can assemble a variety of products in different models during a working shift or a working day. Traditional approaches are realized in two stages [7]. In the first stage, a solution is determined before knowing the realizations of the random variables. In the second stage, a corrective action can be taken when the values of the random variables are known. The advantage of our approach is that, (i) we can reuse the same model and solution approach to compute an optimal re-optimization policy and (ii) the vehicle's tour is generated by taking into account the environment's geometry, the depot location, the number and the location of the workstations and the scheduling algorithm. Furthermore, in contrast with traditional methods which are based on line followers, we developed a new approach that faces the combined problem of motion planning and scheduling in a "reconfigurable" shop floor environment. By the term "reconfigurable" we mean that the facilities layout in the shop floor can change rapidly according to the demands of production.

2 Problem Statement

We consider an autonomous vehicle which is operating in a 2D industrial environment (Fig. 1). The environment is cluttered with a depot, a set of workstations and with static obstacles. The vehicle has limited carrying capacity Q of the items that must be delivered to the workstations. Workstations' demands are stochastic variables ξ_i, $i = 1, \ldots, n$ independently distributed with known distributions. The actual demand of each workstation is only known when the vehicle arrives at the workstation location. It is also assumed that ξ_i does not exceed the vehicle's capacity Q. A feasible solution to the combined problem of MPSSD is a permutation of the workstations $WS = (WS_1, WS_2, \ldots, WS_n)$ which is starting from the depot (that is, $WS_1 = 0$), and it is called a "priori tour".

The vehicle visits the workstation in the order given by the "priori tour", and it has to choose, according to the actual workstations' demands, whether to proceed to the next workstation or to go to depot for restocking. The goal of this action is to avoid the risk of having a vehicle without enough loads to serve a customer and thus having to perform a back- and- forth trip to the depot to complete delivery at the workstation.

Furthermore, the following requirements and assumptions are taken into account: (i) The autonomous vehicle is a car-like robot with rectangular shape which is moving without violating the non-holonomic constraints (for detailed information on the kinematic model the reader is referred to [8]). (ii) The vehicle must serve all the workstations and each workstation should be served only once. (iii) The vehicle's motion starts from the depot and terminates its motion at the depot. (iv) The vehicle is moving with constant velocity which can be equal to zero only at the depot and the workstations. (v) The workstation locations are fixed and known. (vi) The environment is cluttered with static obstacles. The obstacles have fixed and known geometry.

Under these assumptions, the MPSSD seeks to determine a vehicle's tour in the shop floor satisfying the following the MPSSD criteria and constraints: (A) The tour should not intersect with the obstacles and should result to a safe motion. (B) The

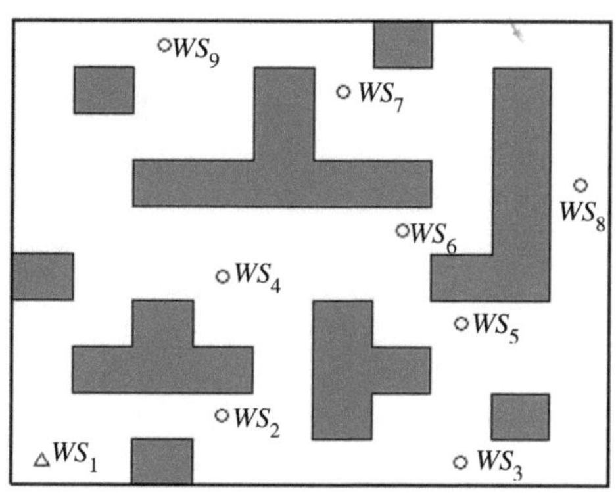

Fig. 1 The 2D industrial environment

vehicle's tour should have a minimum length starting and ending at the depot and serving all of the workstations exactly once. (C) The tour should have a lower-bounded turning radius (upper bounded curvature).

2.1 Bump-Surface Concept

Given a 2D shop floor, a normalized workspace $\boldsymbol{W}$ is constructed by linearly mapping the initial environment to $[0,1]^2$ producing what we call as normalized environment $\boldsymbol{W}$ [2]. The construction of the corresponding Bump-Surface is obtained by a straightforward extension of the Z-value algorithm. Briefly, this algorithm considers that $\boldsymbol{W}$ is discretized into uniform subintervals along its x and y orthogonal directions, respectively, forming a grid of points $\mathbf{p}_{i,j} \in [0,1]^3, i,j \in [0, N_g - 1]$ where N_g is the (user-defined) grid size. The third coordinate of each $\mathbf{p}_{i,j}$ takes a value in the interval (0,1], if the corresponding grid point lies inside an obstacle and the value 0 otherwise. In this paper, we use a (2.2)-degree B-Spline surface with optimized knot vectors [8] to represent the Bump-Surface $S : [0,1]^2 \rightarrow [0,1]^3$, which is given by,

$$\mathbf{S}(x,y) = \sum_{i=0}^{N_g-1} \sum_{j=0}^{N_g-1} N_i^2(x) N_j^2(y) \mathbf{p}_{i,j} \tag{1}$$

where $N_i^2(x)$ and $N_j^2(y)$ are the B-Spline basis functions [3]. The 3D surface S consists of 2D flat areas and 3D bumpy areas which correspond to the environment's obstacles; a visual example is shown in Fig. 2.

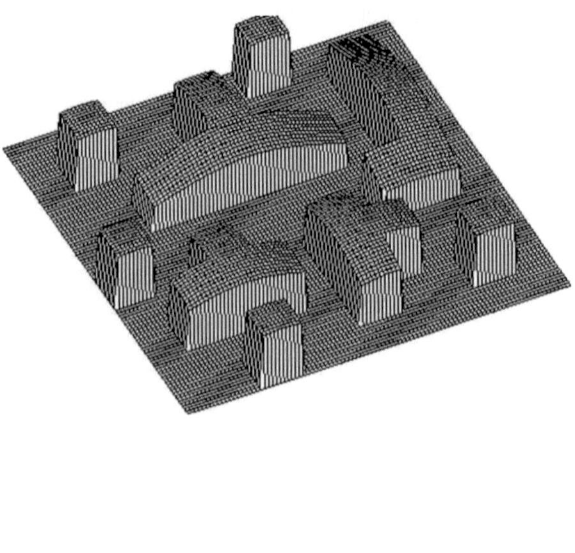

Fig. 2 The corresponding bump-surface

3 Motion Planning and Scheduling with Stochastic Demands

The integrated problem described earlier is formulated as a global optimization problem. According to [8], the midpoint of the rear wheels of the vehicle traces a route $\mathbf{R}(s) = (x(s), y(s))$ in $\boldsymbol{W}$ which is represented as a second degree NURBS curve [3],

$$\mathbf{R}(s) = \frac{\sum_{i=0}^{K-1} N_i^2(s)\omega_i \mathbf{p}_i}{\sum_{i=0}^{K-1} N_i^2(s)\omega_i}, \quad s \in [0, 1] \tag{2}$$

where $N_i^2(s)$ is the B-Spline basis function, ω_i is the weight factor and $\mathbf{p}_i$ are the K control points defined as: (i) $\mathbf{p}_0 = \mathbf{p}_{K-1}$, denotes the depot location; (ii) $\{\mathbf{p}_1, \ldots, \mathbf{p}_{K-2}\} = \begin{pmatrix} \{n \ \mathit{workstations}\, WS_i, i = 1, \ldots, n\} \cup \\ \{\mathit{intermediate\, points}\, \mathbf{g}_j, \ j = 1, \ldots, b\} \end{pmatrix}$ defined in the parametric space of S and $b + \text{cardinal}(WS_i) = K - 2$; (iii) b is the total number of the intermediate points used to facilitate collision avoidance in environments cluttered with obstacles.

The goal of the proposed approach is to determine the $K - 2$ control points $\mathbf{p}_i$ which define the requested tour $\mathbf{R}(s)$. It must be noticed that the depot and the workstations are represented with points which lie on the flat areas of the Bump-Surface.

3.1 A Valid and Safe Vehicle Tour

A valid and safe vehicle tour is a tour passing through all the workstations exactly once, not crossing obstacles and not violating the kinematic constraints.

A collision-free tour that avoids the obstacles and passes through all of the workstations should be searched in the flat area of the Bump-surface. A tour that "climbs" the bumps of the Bump-surface results in an invalid tour in the initial 2D environment because it penetrates the obstacles. By construction, the arc length of $\mathbf{R}(s)$ approximates the length of its image $S(\mathbf{R}(s))$ on S, as long as $\mathbf{R}(s)$ does not penetrate the obstacles [2]. Therefore, it is reasonable to search for a flat tour on S. The collision avoidance functional which is expressing this requirement can be written as,

$$E = e^{\left(\sum_{k=1}^{4} H_k\right)} * L \tag{3}$$

where L is the arc length of the image of $\mathbf{R}(s)$ onto S and H_k, $k = 1, \ldots, 4$ is the flatness of the image of each k-vertex ($k = 1, \ldots, 4$) of the vehicle on the

Bump-Surface [9]. E takes a value in the interval $(L, +\infty)$ if the robot collides with obstacles and a value equal to L otherwise.

In order to ensure that the planned tour does not violate the kinematic constraints, we compute at each point of $\mathbf{R}(s)$ the corresponding discrete curvature. The discrete curvature $\mathbf{C}(s)$ at any point on the motion $\mathbf{R}(s)$ is expressed by,

$$\mathbf{C}(s) = \left\|\mathbf{R}^{i-1} - 2\mathbf{R}^{i} + \mathbf{R}^{i+1}\right\|, \quad i = 1, \ldots, N_p - 1 \tag{4}$$

where $N_p - 1$ are equal sequential chords which approximate the motion $\mathbf{R}(s)$. The following condition is enforced,

$$C(s_i) \leq C_{\max}, \quad i = 1, \ldots, N_p - 1 \tag{5}$$

where $C(s_i)$ is the discrete curvature at the point $\mathbf{R}_i = \mathbf{R}(s_i)$.

Following the above analysis, the overall optimization problem with respect to the control points $\mathbf{p}_i$ is written as:

$$\begin{aligned} &\min_{\mathbf{p}_i} E \\ &\text{subject to: } C(s_i) \leq C_{\max}, \quad i = 1, \ldots, N_p - 1 \end{aligned} \tag{6}$$

4 Searching for an Optimum Solution

Traditional combinatorial problems require finding solutions for a single instance. In our case, the system changes dynamically over time. Thus, it is required to compute solutions for new problem instances derived from previous instances. Moreover, since there is a cost associated with the transition from one solution to another, the solution for the new instance must be close to original solution. Solving this re-optimization problem involves two challenges: (i) computing an optimal or near-optimal solution for a new instance, and (ii) efficiently converting the current solution to the new one.

To address the above problem, an optimization strategy is developed using a combination of GA and Micro-GA. Initially, a GA with a mixed integer and floating point chromosome is used to generate an "optimum" tour before the exact workstations demands are known. Then, upon the arrival of the vehicle at a workstation a Micro-GA is commenced to re-optimize its route by taking into account the actual demands of the current and the remaining workstations. If necessary the vehicle will return to the depot to restock. In order to accelerate the convergence of the micro-GA, the previous solution is seeded in the initial population. The micro-GA starts with an initial random population and the result of this run is used to 'seed' the initial population of the next run in the hope of starting the evolution in a more useful region of the search space. The seeding percentage is set to be 10 % of the initial population.

The main characteristics of the developed GA are analysed below. **The chromosome syntax**: A mixed integer and floating-point representation scheme was selected for use. That is, a chromosome is a string consisting of two parts: the first (left) part contains only integer numbers, while the second part contains floating-point numbers. Each chromosome corresponds to a unique vehicle's route. More specifically, each chromosome consists of the sum of the number of workstations (integer part) and the overall number of the intermediate points (real-valued part). **The fitness function**: The fitness function evaluates the quality of a chromosome, in other words, the quality of the corresponding tour. We use the following fitness function:

$$\mathcal{F} = \begin{cases} \frac{1}{E}, & \text{if } C(s_i) \leq C_{\max}, \quad i = 1, \ldots, N_p - 1 \\ 0, & \text{otherwise} \end{cases} \tag{7}$$

Genetic operators: The following three genetic operators were selected for use with the proposed GA. (a) ***Reproduction***: Reproduction is a simple copy of an individual from the population of the current generation to the one of the next generation. In this work, the proportional selection strategy is adopted, where chromosomes are selected to reproduce their structures in the next generation with a rate proportional to their fitness. (b) ***Crossover***: Crossover joins together parts of several individuals in order to produce new ones for the next generation. The individuals are randomly selected according to a user-defined probability (crossover rate). For the first part of the chromosome (with integers) the Order Crossover (OX) followed by a suitable repairing mechanism was selected for use, while for the second part (real-valued part) of the chromosome, the one-point crossover was adopted. (c) ***Mutation***: For the first part (with integers) the inversion operator is used, while for the second part a boundary mutation was used. For the Micro-GA the mutation operator is excluded.

5 Experiments

Due to space limitations, we are able to thoroughly present one simulation example implemented in Matlab. The grid size of the Bump-Surface is set to $N_g = 100$. GA run uses the settings for control parameters: population size = 250, maximum number of generations = 500, crossover rate = 0.75, inversion rate = 0.095, boundary mutation rate = 0.004. For the micro-GA the population size is set to 50 and the number of generations is set to 75. It is worth noting that the selection of the appropriate control settings was the result of extensive experimental efforts with various control schemes adopted following the indications of the literature.

The present experiment corresponds to the environment shown in Fig. 1. An autonomous vehicle with capacity $Q = 20$ is requested to deliver supplies to n = 8 workstations ws_i, $n = 1,\ldots,9$ (where ws_1 is the depot). The stochastic demand ξ_i follows a discrete probability distribution $\Pr ob(\xi_i = k)$, $k = 1, \ldots, K \leq Q$ (where

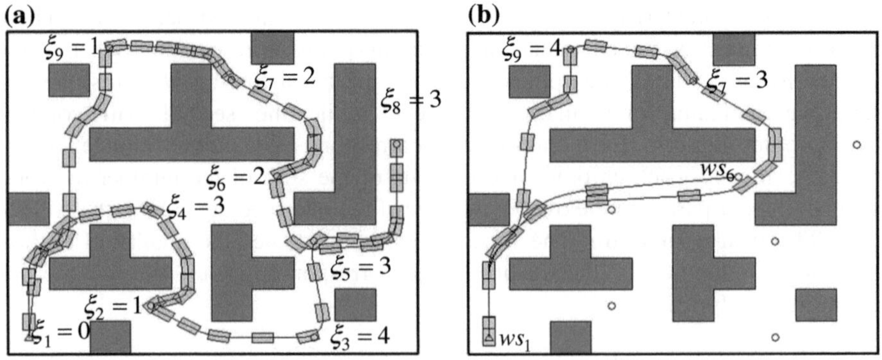

Fig. 3 **a** The resulted "priori tour" and corresponding requests for items; **b** the proposed re-optimized tour

$\xi_1 = 0$). We set $K = 4$. Furthermore, we assume that vehicles at each workstation have the ability to make a turn without violating the curvature constraint. By applying the proposed strategy, the "priori tour" was derived: $WS_1 \to WS_4 \to WS_2 \to WS_3 \to WS_5 \to WS_8 \to WS_6 \to WS_7 \to WS_9 \to WS_1$ (see Fig. 3a). When the vehicle arrives at WS_4 the Micro-GA is used to re-optimize the vehicle's route by taking into account the actual demands of the other workstations. This procedure is repeated whenever the vehicle arrives to a new workstation until its mission is completed. Due to space limitations, in Fig. 3b we present an extreme case where the vehicle is requested to return back to depot to restock. This happened when the vehicle arrived at workstation WS_6. The computed re-optimized tour is $WS_6 \to WS_1 \to WS_9 \to WS_7 \to WS_1$.

The construction of the Bump-Surface takes approximately 1.001 s, while the vehicle's tour and schedule was computed in about 4.15 min. The "priori tour" is computed in about 2.75 min while the Micro-GA needed about 0.35 min to re-optimize a tour. In the above test case the Micro-GA had to run four times.

Despite the fact that the considered problem is off-line, computational time results versus the number of workstations are of great interest. Besides, the variation of CPUs' time is indicative of the problem complexity. An interesting indication of the computational time performance is the variation of this computational time versus the number of workstations, which was investigated experimentally. In these experiments, we assume the environment of the above test case was constant while the number of workstations is changed from 6 to 18. We note that it seems that CPU time increases almost linearly with the increase in the number of workstations.

6 Conclusions

This paper presents a novel method for motion planning and scheduling with stochastic demands for an autonomous vehicle in indoor factory environment. The objective is to determine the optimum tour for the vehicle so that all the workstations are served. The proposed method is based on the Bump-Surface concept in order to express the integrated problem of MPSSD as a global constrained optimization problem. The derived optimization problem is resolved using a combination of GA and Micro GA. The method's efficiency is demonstrated with the help of a simulated experiment. Future work will be concentrated on transferring the proposed method from the simulation level to the heart of a real industrial environment.

References

1. Teodorovic′, D., Lucic, P.: Intelligent vehicle routing system. In: Proceedings of the IEEE International Conference on Intelligent Transportation Systems, pp. 482–487. Dearborn, MI, USA, 1–3 Oct 2000
2. Azariadis, P.N., Aspragathos, N.A.: Obstacle representation by bump-surfaces for optimal motion-planning. Robot. Auton. Syst. **51**(2–3), 129–150 (2005)
3. Piegl, L., Tiller, W.: The Nurbs Book. Springer, Berlin (1997)
4. Goldberg, D.E.: Genetic Algorithms in Search, Optimization and Machine Learning. Addison Wesley, Reading MA (1989)
5. Qiu, L., Hsu, W.J., Huamg, S.Y., Wang, H.: Scheduling and Routing Algorithms for AGVs: A Survey. Int. J. Prod. Res. **40**, 745–760 (2002)
6. Yang, W.H., Mathur, K., Ballou, R.H.: Stochastic vehicle routing problem with restocking. Transp. Sci. **34**, 99–112 (2000)
7. Ak, A., Erera, A.L.: A paired-vehicle recourse strategy for the vehicle routing problem with stochastic demands. Transp. Sci. **41**, 222–237 (2007)
8. LaValle, M.S.: Planning Algorithms, University of Illinois (2004)
9. Xidias, E.K., Azariadis, P.N.: Mission design for a group of autonomous guided vehicles. Robot. Auton. Syst. **59**(1), 34–43 (2011)

Autonomous Learning of Internal Dynamic Models for Reaching Tasks

Tadej Petrič, Aleš Ude and Auke J. Ijspeert

Abstract The paper addresses the problem of learning internal task-specific dynamic models for a reaching task. Using task-specific dynamic models is crucial for achieving both high tracking accuracy and compliant behaviour, which improves safety concerns while working in unstructured environment or with humans. The proposed approach uses programming by demonstration to learn new task-related movements encoded as Compliant Movement Primitives (CMPs). CMPs are a combination of position trajectories encoded in a form of Dynamic Movement Primitives (DMPs) and corresponding task-specific Torque Primitives (TPs) encoded as a linear combination of kernel functions. Unlike the DMPs, TPs cannot be directly acquired from user demonstrations. Inspired by the human sensorimotor learning ability we propose a novel method which autonomously learns task-specific TPs, based on a given kinematic trajectory in DMPs.

Keywords Compliant movement primitives · Task-specific dynamics · Learning · Dynamic movement primitives

1 Introduction

Learning of a new motor behaviour is one of the key skills that a humanoid robot should have. Programming by demonstration is a popular way to acquire new motor behaviour [1], which can be done by using different sensory systems, i.e. visual [2] or kinaesthetic guidance [3]. The main advantage is that the robot kinematics is

T. Petrič (✉) · A.J. Ijspeert
Biorobotics Laboratory, EPFL, École Polytechnique Fédérale de Lausanne, Lausanne, Switzerland
e-mail: tadej.petri@epfl.ch

T. Petrič · A. Ude
JSI, Jozef Stefan Institute, Ljubljana, Slovenia
e-mail: ales.ude@ijs.si

T. Borangiu (ed.), *Advances in Robot Design and Intelligent Control*,
Advances in Intelligent Systems and Computing 371,
DOI 10.1007/978-3-319-21290-6_44

adapted and the posture is preserved even for redundant robots. A common way of learning kinematic trajectories is by using the Dynamic Movement Primitives (DMPs) [4]. To execute the desired DMP trajectory on a robot the underlying controller, usually based on high gain feedback loop, is employed to guarantee that the trajectory is accurately executed. However, using high gains in a feedback loop makes robots inherently unsafe for interaction with the environment or humans, due to the high interaction forces that may occur during unforeseen contacts.

Different approaches can be used to minimize the interaction forces during contacts: e.g., by actively detecting them using proximity sensors or artificial skin [5]; by using passively compliant mechanical structures like artificial pneumatic muscles [6] or by using control algorithms for active compliance [7]. The latter is achieved using the control approaches based on the inverse dynamic models. Obtaining a dynamical model is a challenging and time consuming task even for experts. But it only needs to be done once. On the other hand, it is impossible to obtain the generic dynamic model, even for simple task like table wiping, due to the unknown friction between the sponge and the table. Therefore, instead of modelling the task, we propose a novel method for learning yet unknown and undefined task-specific torque primitives. The proposed method would replace the need for modelling the task dynamics and if the dynamic model is already in use, it will compensate for additional uncertainties. By learning the exact dynamic model of the task, one can achieve both the compliant robot behaviour and the accurate tracking of the desired motion. While accurate tracking is required for proper task execution, the compliant behaviour is an essential feature for safe interaction with the environment or with a human [8].

While the DMPs and their possible modulations during contact with the environment have been thoroughly analysed [9–12], their extension towards torque controlled robots was not sufficiently explored yet. Inspired by the human sensory motor ability [13–16] we propose a novel method based on Compliant Movement Primitives (CMPs), which represent a new model-free approach, while keeping simple and smooth modulation properties of the position based DMPs. Similar to what can be observed in humans [14], the proposed approach independently learns the kinematic trajectory in Cartesian space (DMPs) and the task dynamics in the joint space (TPs). The proposed approach exploits the motion pattern encoded by the DMPs to acquire the task-specific torques and encodes them as TPs.

The first step in gaining the CMPs is learning the desired motion trajectory in DMP. Next, the DMP trajectory is executed on a robot using low-feedback gains with the proposed method for learning the task-specific TPs. In each subsequent step the TPs are then used as a feedforward term, which essentially represents new learned task-specific dynamics. The stability of the proposed controller is assured by keeping the low-gain feedback loop, which ensures that the desired task is performed safely even in unstructured environment or with humans. While the proposed approach eliminates the need for dynamical modelling, the CMPs still have to learn TPs for each task variation. Once several CMPs for different task variations are learned, they can be added to a database and statistical generalization can be used as in [17]. This allows learning and performing different variations of the same task in a compliant manner without the need of any analytical models of the task or programming experts.

2 Compliant Movement Primitives

Compliant Movement Primitives (CMPs) $h(t)$ are defined as a combination of kinematic trajectories encoded in Dynamic Movement Primitives (DMPs) and corresponding task-specific dynamics encoded in Torque Primitives (TPs)

$$\boldsymbol{h}(t) = [\ddot{\boldsymbol{y}}_d(t), \dot{\boldsymbol{y}}_d(t), \boldsymbol{y}_d(t), \boldsymbol{\tau}_f(t)] \tag{1}$$

Here $\ddot{\boldsymbol{y}}_d(t), \dot{\boldsymbol{y}}_d(t), \boldsymbol{y}_d(t)$ are vectors of the desired task-space accelerations, velocities and positions respectively encoded in the DMPs, and $\boldsymbol{\tau}_f(t)$ is the vector of corresponding task-specific feedforward joint torques encoded in TPs.

First the kinematic part is obtained using the learning by demonstration approach. A short overview of the periodic version of the DMPs [4, 12] with the recursive learning algorithm is given first. The following equations are valid for one DOF; for multiple DOF the equations are used in parallel. A nonlinear system of differential equations defines DMP for periodic movements

$$\dot{z} = \Omega(\alpha_z(\beta_z(g - y) - z) + f(\phi)), \quad \dot{y} = \Omega z, \quad \dot{\phi} = \Omega, \tag{2}$$

where Ω is the frequency of motion, ϕ is the phase, α_z and β_z are the positive constants set to 8 and 2 respectively. They ensure critical damping so that the system monotonically converges to the trajectory oscillating around anchor point g. The nonlinear part $f(\phi)$ is given by

$$f(\phi) = \frac{\sum_{i=1}^{N} \psi_i w_i}{\sum_{i=1}^{N} \psi_i}, \quad \psi_i = \exp(h(\cos(\phi - c_i) - 1)), \tag{3}$$

where w_i are the weights that define the shape of the trajectory, ψ_i are the Gaussian like kernel functions where parameter h defines their width, and c_i is equally spaced between 0 and 2π. To acquire the target signal for recursive learning the Eq. (2) is rewritten in the form

$$f_t(t) = \frac{\ddot{y}_d}{\Omega^2} - \alpha_z(\beta\_z(g - y_d) - \frac{\dot{y}_d}{\Omega}), \tag{4}$$

where $\ddot{y}_d$, $\dot{y}_d$, y_d are respectively the desired acceleration, velocity and position of the desired trajectory. To update the weights w_i, an incremental regression is used:

$$w_i(t+1) = w_i(t) + \psi_i P_i(t+1) e(t), \tag{5}$$

$$P_i(t+1) = \frac{1}{\lambda}\left(P_i(t) - \frac{P_i^2(t)}{\frac{\lambda}{\psi_i} P_i(t)}\right), \tag{6}$$

$$e(t) = f_t(t) - w_i(t), \tag{7}$$

where P_i is the covariance. The initial parameters are set to $P_i = 1$, $w_i = 0$ and $\lambda = 0.995$ (see [10] for details).

To obtain the dynamical part of the CMP, i.e. the corresponding TP, the motion is executed on a robot using a low gain feedback loop and a learning algorithm for acquiring the task specific torque primitive (TP). The desired motion $\ddot{y}_d$, $\dot{y}_d$, y_d encoded in DMPs is executed using the following control law

$$\boldsymbol{\tau}_u = \mathbf{J}^{\#}\left(\mathbf{K}_p(\mathbf{y}_d - \mathbf{y}) + \mathbf{K}_d(\dot{\mathbf{y}}_d - \dot{\mathbf{y}}) + \mathbf{K}_i(\ddot{\mathbf{y}}_d - \ddot{\mathbf{y}})\right) + \mathbf{N}\mathbf{K}_n\dot{\boldsymbol{q}} + \boldsymbol{\tau}_f(\phi), \tag{8}$$

where, $\mathbf{J}^{\#}$ is the pseudo inverse, $\mathbf{N}$ is the null-space matrix and $\mathbf{K}_p$, $\mathbf{K}_d$, $\mathbf{K}_i$ and $\mathbf{K}_n$ are the constant gain matrices selected such that the robot behaves compliantly, i.e. set to match the low feedback gain requirements.

Here $\boldsymbol{\tau}_f(\phi)$ is the feedforward term encoded in TPs. For one DOF it is given by

$$\tau_f(\phi) = \frac{\sum_{i=1}^{N} \psi_i w_i}{\sum_{i=1}^{N} \psi_i}, \tag{9}$$

where the kernel functions ψ_i are defined as in Eq. (3) and the weights are updated by inserting the error given by

$$\boldsymbol{e}(t) = \mathbf{J}^{\#}(\alpha_t(y_d(t) - y(t)) + \beta_t(\dot{y}_d(t) - \dot{y}(t))) \tag{10}$$

into Eqs. (5)–(6). The rate of learning is defined by setting parameters α_t and β_t.

3 Experimental Evaluation

The proposed method was evaluated on a 3-DOF planar robot simulated in Matlab using a Planar Manipulator Toolbox [18]. The simulation step was set to 0.01 s and the link lengths of the robot were set to 1 m. The initial task space position of the robot was $y_0 = [1, 1.5]$ m. The initial robot configuration (green line) is shown in Fig. 1.

The task was to reach towards certain points in space and return back to the initial position. A proposed controlled given by Eq. (8) with low feedback gains was used. Note that inverse dynamical model for compensating the robot dynamics was not used. This task is also similar to the task used in a study of human sensory

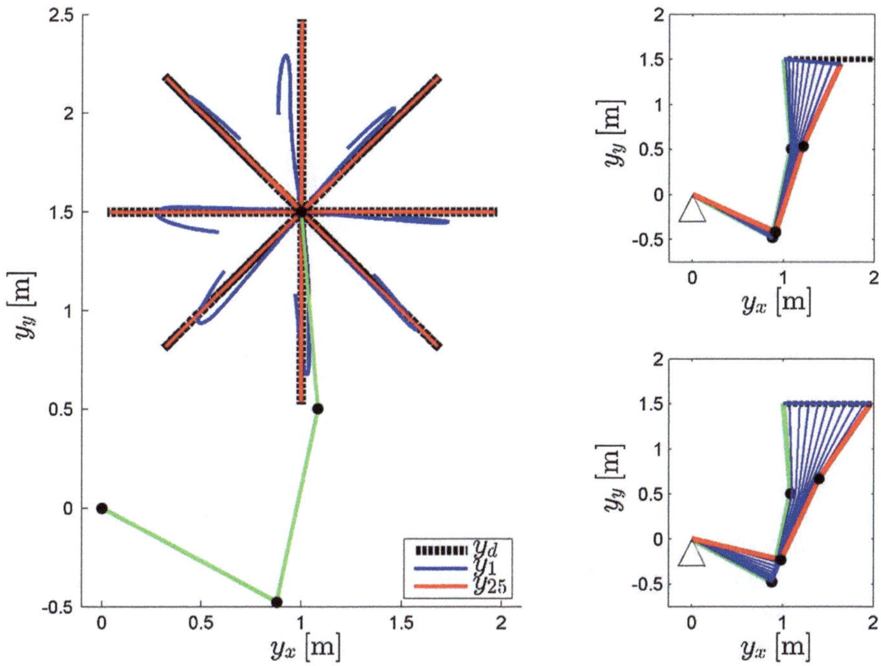

Fig. 1 Left hand side plot shows eight examples of the initial (*blue*) and the learned (*red*) robot behaviour in Cartesian space. Right hand side plots show one detailed example of the initial (*top*) and the learned (*bottom*) robot behaviour in Cartesian space

motor learning ability reported in [14]. To compare the results we chose eight different positions equally spaced on a circle with a radius of 1 m from the initial position. The desired Cartesian kinematic trajectories $\boldsymbol{y}_d$ were encoded in DMPs as a part of the CMPs. They are shown with black dotted lines on the left plot in Fig. 1.

Note that the desired kinematic trajectory is encoded in the task-space, i.e. Cartesian space, and the corresponding task-specific torques are encoded in the joint space, see Eq. (8). The mapping of the task-space error $\boldsymbol{e}(t)$ to the joint space is done by using a pseudo-inverse of the Jacobean. The learning of the TPs was performed after each iteration of the executed motion. Note that each executed motion started from the same initial position. The left hand side plot in Fig. 1 shows comparatively eight different examples between initial movement execution (blue lines) and the last iteration of learning (red line). One can see in all eight examples that the proposed approach was able to learn proper TPs in order to execute the task accurately. Note that in this example the proposed system was learning the unknown robot dynamics.

The successful learning of the torque primitives can be seen on the right hand side plots in Fig. 1 where we compare the behaviour of one representative example, i.e. moving to the right. For clarity we only show the first half of the movement. The top right plot shows the robot's behaviour for initial movement execution, i.e.

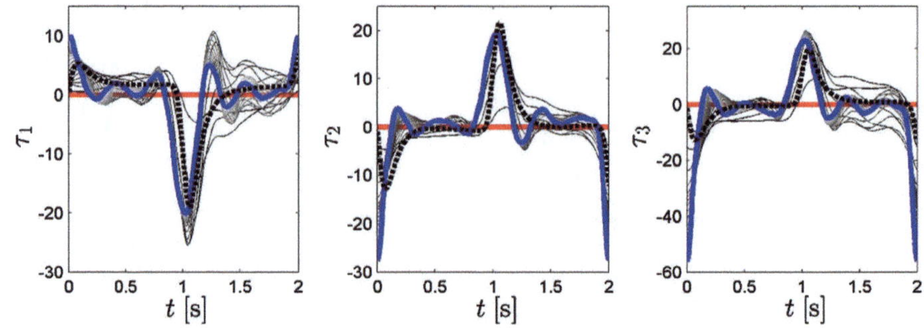

Fig. 2 Comparison with an ideal inverse dynamic model and evolution of the TPs during 25 learning iterations for all three joints. The initial TPs are in *red*, the final TPs are in *blue* and the ideal joint torques using exact dynamic models are shown with *black dotted line*

feedforward torques were zero. Here we can clearly see that the low gain feedback loop is not able to track the desired trajectory accurately. However, by applying the proposed learning algorithm to obtain the TPs, the tracking performance was improved significantly. The robot's behaviour after learning, i.e. after 25th iterations, is shown in the bottom plot, where perfect matching between the desired (black dotted line) and actual (blue line) Cartesian position can be observed.

The evolution of the TPs for all 25 iterations and for all three joints is shown in Fig. 2, were the red line shows the initial TPs, the thick blue line shows the final TPs and the black dotted line shows the ideal task-specific torques computed by using the exact inverse dynamic model.

Note that the initial torque primitives are zero. By comparing the TPs with the ideal dynamical model, we can see that they are similar in shape and amplitude. This shows that the proposed algorithm can successfully learn the required task-specific dynamics autonomously.

The proposed learning algorithm essentially minimises the error function given by Eq. (10). Thus, when $\boldsymbol{e}(t)$ in Eq. (10) is converging towards 0, $\boldsymbol{\tau}_f(\phi)$ in Eq. (9) is converging toward the exact inverse dynamic model. Ideally, if $\boldsymbol{e}(t)$ in Eq. (10) is zero then $\boldsymbol{\tau}_f(\phi)$ in Eq. (9) is the perfect inverse dynamic model.

The convergence of the proposed algorithm and the ability to re-learn the task-specific torques is also illustrated on the left hand side plot in Fig. 3, where the mean value and the standard deviation of the tracking error for all eight examples are shown. In studies about humans it has been demonstrated that they can quickly adapt to the changes in the task dynamics [14]. To show that the proposed approach can also cope with changes of task dynamics, the parameters of robot dynamics were changed after the 25th iteration. In the 26th iteration the inertia parameter of the robot was increased by a factor five and remained afterwards at that new value.

In general the learning rate of the proposed algorithm is defined with the gains α_t and β_t. If the gains are set too high, the system may potentially be destabilized, or vice versa, if the gains are too low, the learning rate would be slow and impractical. We set them empirically to $\alpha_t = 20$ and $\beta_t = 10$, which results in similar learning rate

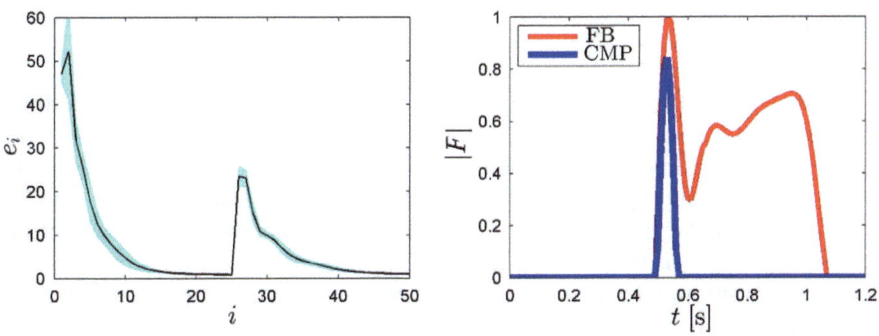

Fig. 3 Left plot shows the average error and the standard deviation (*shaded area*) for all eight examples. Right plot shows the interaction force in case of collision using the proposed control approach (*CMP*) and standard feedback (*FB*) control approach with high gains

as observed in human sensory motor learning ability reported in [14]. The results are shown on the right hand side plot in Fig. 3, where we can see that only 10 iterations were needed in order to obtain perfect tracking performance even after the dynamics was significantly altered, i.e. inertia parameters were increased by factor five. This also shows that the proposed approach can cope with sudden changes in the task dynamics and re-adapt online if necessary.

Since low feedback gains are used in combination with the CMPs we also assume that the interaction force in case of a collision with an unforeseen object will be significantly smaller compared to the general feedback approach with similar tracking performance. To investigate and compare the performance in case of unforeseen collisions, a solid wall was placed on the desired end-effector path. The results are shown on the right hand side plot in Fig. 3, where the normalized interaction force at the end effector is given. The blue line shows the results of the proposed approach and the red line shows the results of the high gain feedback control.

On the right hand side plot in Fig. 3 we can see that, at the beginning of the impact, resulting forces are similar in both cases. This was expected since the initial force is mainly a result of the robot inertia, i.e. the robot is suddenly forced to stop in both cases. This also explains why the forces are almost identical at the beginning of the impact. However, one can see that by using the proposed approach (CMPs), the impact force only appears initially and then goes towards zero. In this particulate example the normalized force impulse was only 0.035 Ns. On the contrary, in the case of the high gain feedback control, the force after impact remains rather large, i.e. it keeps almost the same value as it was at the beginning of the impact, resulting in a large force impulse of 0.33 Ns. In the latter case the force impulse was almost 10 times larger compared to the CMPs approach.

4 Conclusion

It was shown that the proposed approach (CMPs) can successfully learn both the kinematic trajectories in Cartesian space encoded with the DMPs and the corresponding torque primitives in the joint space encoded in TPs. The main contribution of the proposed approach is learning of the previously unknown task specific dynamics. With the proposed learning method, the control system was able to significantly improve the tracking accuracy in each of the subsequent iterations. Once the TPs are fully learned, the CMPs ensure accurate task execution and, at the same time, compliant robot behaviour. As such, the proposed learning framework enables simple and computationally inexpensive control of dynamically challenging tasks. Moreover, the results were also similar to human studies of sensorimotor learning abilities. Since learning of the task-specific dynamic is done autonomously by using a low gain feedback loop, the compliant behaviour is maintained during learning of task-specific torque primitives. This makes the robot safe for working in unstructured environments or with humans even during learning.

Acknowledgments The research activities leading to the results presented in this paper were supported by the Sciex-NMSCH project no. 14.069.

References

1. Nakanishi, J., Morimoto, J., Endo, G., Cheng, G., Schaal, S., Kawato, M.: Learning from demonstration and adaptation of biped locomotion. Robot. Auton. Syst. **47**(2–3), 79–91 (2004)
2. Ude, A., Atkeson, C.G., Riley, M.: Programming full-body movements for humanoid robots by observation. Robot. Auton. Syst. **47**, 93–108 (2004)
3. Hersch, M., Guenter, F., Calinon, S., Billard, A.: Dynamical system modulation for robot learning via kinaesthetic demonstrations. IEEE Trans. Rob. **24**(6), 1463–1467 (2008)
4. Schaal, S., Mohajerian, P., Ijspeert, A.: Dynamics systems vs. optimal control—a unifying view. Prog. Brain Res. **165**(1), 425–445 (2007)
5. Ulmen, J., Cutkosky, M.: A robust, low-cost and low-noise artificial skin for human-friendly robots. In: Proceedings of IEEE International Conference on Robotics and Automation, pp. 4836–4841 (2010)
6. Shin, D., Sardellitti, I., Park, Y.L., Khatib, O., Cutkosky, M.: Design and control of a bio-inspired human-friendly robot. Int. J. Robot. Res. **29**(5), 571–584 (2009)
7. Buchli, J., Stulp, F., Theodorou, E., Schaal, S.: Learning variable impedance control. Int. J. Robot. Res. **30**(7), 820–833 (2011)
8. Haddadin, S., Albu-Schaffer, A., Hirzinger, G.: Requirements for safe robots: measurements, analysis and new insights. Int. J. Robot. Res. **28**(11–12), 1507–1527 (2009)
9. Basa, D., Schneider, A.: Movement primitives learning point-to-point movements on an elastic limb using dynamic movement primitives. Robot. Auton. Syst. **66**, 55 (2015)
10. Gams, A., Petrič, T.: Adapting periodic motion primitives to external feedback: modulating and changing the motion. In: 23rd International Conference on Robotics in Alpe-Adria-Danube Region (RAAD), pp. 1–6 (2014)

11. Gopalan, N., Deisenroth, M.P., Peters, J.: Feedback error learning for rhythmic motor primitives. In: Proceedings of the IEEE International Conference on Robotics and Automation, pp. 1317–1322 (May 2013)
12. Ijspeert, A.J., Nakanishi, J., Hoffmann, H., Pastor, P., Schaal, S.: Dynamical movement primitives: learning attractor models for motor behaviors. Neural Comput. **25**, 328–373 (2013)
13. Gomi, H., Kawato, M.: Neural network control for a closed-loop system using feedback-error-learning. Neural Networks **6**, 933 (1993)
14. Krakauer, J.W., Ghilardi, M.F., Ghez, C.: Independent learning of internal models for kinematic and dynamic control of reaching. Nat. Neurosci. **2**(11), 1026–1031 (1999)
15. Oztop, E., Kawato, M., Arbib, M.: Mirror neurons and imitation: a computationally guided review. Neural Networks **19**, 254–271 (2006)
16. Wolpert, D.M., Kawato, M.: Multiple paired forward and inverse models for motor control. Neural Networks J. Int. Neural Network Soc. **11**(7–8), 1317–1329 (1998)
17. Ude, A., Gams, A., Asfour, T., Morimoto, J.: Task-specific generalization of discrete and periodic dynamic movement primitives. IEEE Trans. Rob. **26**(5), 800–815 (2010)
18. Žlajpah, L.: Simulation in robotics. Math. Comput. Simul. **79**, 879–897 (2008)

Adaptation of Motor Primitives to the Environment Through Learning and Statistical Generalization

Miha Deniša, Aleš Ude and Andrej Gams

Abstract In this paper we propose a method of adapting motion to the environment based on force feedback. Our method combines two approaches of motor primitive adaptation. Starting from a single demonstration of motion, we use iterative learning control to adapt the motion to different conditions of the environment, for example, the height of the table. The adaptation is realized through coupling terms at the velocity level of a dynamic movement primitive, and acts as a feedforward component, predetermined for the given external condition. As adaptation to each condition takes several iterations, we combine this method with statistical generalization, employing Gaussian process regression. By generating a small database of coupling terms through iterative learning, we adapt to the environment by generalizing between the coupling terms in the database, thus either already achieving an appropriate coupling term for our demonstration trajectory or providing an initial estimate for the adaptation. Consequently, the learning doesn't need to be executed for every condition of the environment, but only for a small set. In the paper we provide the details of the method and evaluate it in a simulated setting for the use case of placing a glass on a table.

Keywords Dynamic movement primitives · Force feedback · Iterative learning control · Statistical generalization · Gaussian process regression

M. Deniša (✉) · A. Ude · A. Gams
Humanoid and Cognitive Robotics Lab, Department of Automatics,
Biocybernetics and Robotics, Jožef Stefan Institute, Jamova 39, Ljubljana, Slovenia
e-mail: miha.denisa@ijs.si

A. Ude
e-mail: ales.ude@ijs.si

A. Gams
e-mail: andrej.gams@ijs.si

T. Borangiu (ed.), *Advances in Robot Design and Intelligent Control*,
Advances in Intelligent Systems and Computing 371,
DOI 10.1007/978-3-319-21290-6_45

1 Introduction

Generating robotic trajectories of motion can be implemented in different ways, for example through imitation of demonstrated trajectories [1]. Learning by demonstration (LbD) approaches cannot cover the complete space of possible trajectories, because the amount of variations with respect to an unstructured environment is too large even for a single task.

Autonomous acquisition of trajectories through un-supervised or supervised exploration has been employed in robotics [2, 3], allowing the robot to adapt to the conditions of either the task or the environment. Different properties of different approaches can be utilized based on the means of encoding the trajectories. Dynamic movement primitives (DMPs) have emerged as a method that has its background in biological systems [4], but also allows easy modulation and adaptation of periodic and discrete trajectories, including with the use of reinforcement learning [5].

Reinforcement learning method often takes a large number of repetitions to converge to the desired behaviour as a consequence of their unsupervised exploration. Alternatives, such as iterative learning control (ILC) [6], have been employed in combination with DMPs [7] in order to reduce the number of needed iterations. On the other hand, unlike in reinforcement learning, ILC needs a reference to converge to.

Even though the dimensionality of the space to explore for a certain task is infinite, many tasks can be described with a single trajectory and a parameter. For example, reaching to a target is described by the target, i.e., the query and the trajectory of reaching. As reaching to two nearby targets is similar, reaching to a target between them, if the trajectory transitions continuously, is also similar. This has been exploited for statistical generalization methods [8, 9]. Generalization methods allow generation of trajectories within the area of the demonstrations, while preserving the properties of the demonstrated trajectories. Furthermore, combining parts of trajectories from the database with the use of hierarchical action graphs [10] can be employed to synthesize new trajectories.

In this paper we combine the means of LbD, adaptation through ILC and statistical generalization to achieve fast adaptation to the environment. We propose that a single demonstration trajectory encoded as a DMP is learned. The learned trajectory is adapted through coupling terms, learned with the use of ILC, for a small number of external conditions. Gaussian process regression (GPR) is then employed to calculate the coupling term for an arbitrary query point within the area of the learned trajectories.

The novelty in the paper is in showing that statistical generalization can be applied to the coupling terms, as opposed to the demonstrated trajectories, which vastly reduces the amount of needed demonstrations to cover a large workspace of a task. Simulated robotic experiments validate the method.

This paper is organized as follows. In Sect. 2 we provide the layout of the proposed method with details on the implementation of ILC. Motion primitives are explained in the Appendix. Section 3 explains statistical generalization based on GPR. Numerical results are presented in Sect. 4, followed by a discussion and conclusions in Sect. 5.

2 The Method

The proposed method is based on three building blocks. They are: (1) dynamic movement primitives which allow different modes of modulation and adaptation, including through force feedback [7, 11]; (2) iterative learning control, which allows adaptation to the environment in a relatively small number of iterations [6]; and (3) Gaussian process regression [12], a statistical generalization method which can on-line generate trajectories with similar features from a small database.

Combining dynamic movement primitives and iterative learning, as explained in [7], allows adaptation of trajectories with minimal errors. Nevertheless, adaptation to a specific condition takes a few iterations of the exact same motion in order to achieve the desired behaviour. The number of iterations depends on the initial condition of learning. If the error of the initial, demonstrated motion is small, the adaptation will be faster. To avoid having to learn from scratch for every specific condition of the environment, we propose the following algorithm:

- Encode a desired demonstrated motion in a DMP.
- Use Iterative Learning Control (ILC) to learn the coupling terms that complete the task at several different, roughly equally spaced conditions along the task, and encode them in a set of Gaussian-like kernel functions. These coupling terms provide the database for statistical generalization.
- Estimate the Gaussian process regression (GPR) parameters from the database.
- Use GPR to generate the initial guess for ILC when encountering a new external condition (height of the object).

The initial guess for ILC will considerably reduce the number of iterations needed to achieve the same level of adaptation as in one of the tasks learned from scratch.

In the following we provide the details for the combination of DMPs with ILC. DMPs themselves have been thoroughly described in the literature [4]. For the completeness of the paper we provide a summary in the Appendix.

2.1 Modulating DMPs and Iterative Learning Control

As described in [7], one can modulate a DMP with force feedback by plugging the measured force, or the error of the desired force, at the velocity level of the DMP and the derivative of this term at the acceleration level. The equation of the DMP is thus

$$\tau \dot{z} = \alpha_z \big(\beta_z (g - y) - z \big) + f(x) + c_2 \dot{C} \tag{1}$$

$$\tau \dot{y} = z + C. \tag{2}$$

The coupling term $\boldsymbol{C}$ is composed of the feedback term and the feedforward term, i.e., $C = C_{fb} + C_{ff}$. While the feedback term is straight-forward, given by

$$C_{fb} = c(F_{des} - F_{meas}) = ce, \tag{3}$$

the feedforward coupling term is learned in several iterations using ILC, which uses the information-rich error signal from the previous iteration for improving the behavior in the next one. For details on ILC check [8]. In our case we use what is known as current-iteration ILC. The feedforward term is calculated in *i*–th iteration by

$$C_i = C_{fb} + C_{ff} = ce_i + C_{ff,i} \tag{4}$$

$$C_{ff,i} = Q(C_{ff,i-1} + Lce_{i-1,t-dt}) \tag{5}$$

where index i denotes the i-th epoch (i.e. repetition), c is the force gain, e_i is the force error, and $t - dt$ denotes a previous time-step. The tunable parameters are Q, L, c. We use $Q = 0.99$, $L = 1$, $c = 0.5$, determined empirically. For the stability of the ILC algorithm with these parameters, one can check [6, 7]. Learned coupling terms are encoded as a linear combination of basis functions

$$C_{ff} = \frac{\sum_{i=1}^{N} {}_c w_i \Psi_i(x)}{\Psi_i(x)}, \tag{6}$$

where Ψ denotes radial basis functions and ${}_c w$ weights that define the shape of the coupling term. A similar approach is used to encode trajectories as DMPs (see Appendix). Note that the phase variable x is common across DMPs and the corresponding coupling term.

3 Statistical Generalization

ILC is used to adapt a demonstrated motion to new conditions, but several iterations are needed in order to learn an appropriate coupling term. Statistical generalization can be used to gain new coupling terms adapted to new conditions. Let's assume a set of example coupling terms,

$$S = \{{}_c\boldsymbol{w}_j, q_j\}, \tag{7}$$

which transitions smoothly between each other as a function of the query point q (i.e. task descriptor), and are encoded as a linear combination of basis functions, defined by weights ${}_c\boldsymbol{w}$. For statistical generalization we use Gaussian process regression (GPR), which can be used to learn a function,

$$F_S : q \mapsto \{{}_c\boldsymbol{w}\}. \tag{8}$$

The above function computes the appropriate coupling, defined by weights, for any given query point q inside the training space of the example set.

A Gaussian process describes a distribution over functions [12]. For each component function f of the vector valued function F_S, GPR can be used to estimate f from data such as the one given in (7). In general, Gaussian process is defined as

$$f(q) \sim GP(m(q), cov(q, q')), \tag{9}$$

where $m(q)$ is the mean function and $cov\left(q, q'\right)$ is the covariance function of the process. Let's assume example outputs ${}_c w_k = f(q_k) + \in$, where ϵ is normally distributed white noise. If we are given a set of query points $\boldsymbol{w}(q^*)$ and the mean of the process is assumed zero, the joint distribution of example outputs ${}_c\boldsymbol{w}$ and new outputs ${}_c\boldsymbol{w}^*$ is given as

$$\begin{bmatrix} {}_c\boldsymbol{w} \\ {}_c\boldsymbol{w}^* \end{bmatrix} \sim N\left(0, \begin{bmatrix} \boldsymbol{\Sigma}(\boldsymbol{C}, \boldsymbol{C}) + \sigma_n^2 \boldsymbol{I} & \boldsymbol{\Sigma}(\boldsymbol{C}, \boldsymbol{C}^*) \\ \boldsymbol{\Sigma}(\boldsymbol{C}^*, \boldsymbol{C}) & \boldsymbol{\Sigma}(\boldsymbol{C}^*, \boldsymbol{C}^*) \end{bmatrix}\right), \tag{10}$$

where $\boldsymbol{C}$ denotes the matrix of example inputs, $\boldsymbol{C}^*$ the matrix of new inputs and $\boldsymbol{\Sigma}$ is obtained by a pairwise evaluation of covariance function cov. In our experiments we used the squared exponential covariance function

$$cov(q, q') = \sigma_f^2 \exp\left(\frac{\|q - q'\|^2}{2l^2}\right), \tag{11}$$

where parameter σ_f^2 defines the variance of the signal $\{{}_c w_k\}$ and l is the characteristic length-scale, i.e. roughly the distance that one has to move in the input space before the value of the function changes significantly [12]. By using (10) we can estimate the mean of the output ${}_c\boldsymbol{w}^*$ as

$$\bar{\boldsymbol{w}}^* = \boldsymbol{\Sigma}(\boldsymbol{C}^*, \boldsymbol{C})\left[\boldsymbol{\Sigma}(\boldsymbol{C}, \boldsymbol{C}) + \sigma_n^2 \boldsymbol{I}\right]_c^{-1} \boldsymbol{w} \tag{12}$$

The covariance function still needs to be defined. Equation (11) introduces free parameters that affect the prediction of the Gaussian process. These parameters, called hyper parameters $\boldsymbol{\theta} = \{\sigma_f, l, \sigma_n\}$ are ideally calculated automatically, e.g., by using standard nonlinear optimizing routines.

4 Results

This section provides the results of a simulated experiment, where the robot's task was to place a glass on a table of an arbitrary height. A KUKA LWR robot with 7 DOF was simulated in Roboworks. The experimental setup is depicted in Fig. 3.

This approach assumes that the environment will not change rapidly, i.e., a table will not change height from one execution to the other. LbD was used to teach the robot one single trajectory of motion. The trajectory shows the motion from a starting position high in the robot's workspace to a low position in its workspace. The trajectory thus describes the motion of approaching an object, for example a table, from above. It should be noted that while we use a discrete trajectory, the presented method could also be applied to periodic movement.

As the second step of the process, we let the robot build its initial database of motion, by modifying the original trajectory with coupling terms so that a predefined force of contact is achieved at a set of heights. A small force of contact is needed to place a glass on a table, so we choose the desired force of $F_d = 0$ N. Seven heights of the table were chosen and ILC was used for the learning of the coupling terms as described in Sect. 2.1. Note that the initial learning attempts should not be executed with a glass, but with a solid object which would ensure the same end-effector point of contact with the table, as high forces may occur. We also modified the algorithm to stop the trajectory at an achieved force of 100 N. Therefore, ILC learning requires a larger number of iterations. ILC ran ten iterations, where all 10 had to be below the threshold of 100 N. If the threshold was breached, more iterations took place.

Figure 1 shows the initial trajectory in green, the trajectories of iterations in red-dashed lines and the final, learned trajectories in bold dashed black. The final locations for the robot to learn to touch, i.e., the table height, were set at [1.15, 1.05, 0.95, 0.85, 0.75, 0.65, 0.65] m. Figure 1 also shows the achieved trajectories for target queries of heights [1.1, 1.0, 0.9, 0.8, 0.7, 0.6] m in the right plot. Figure 2

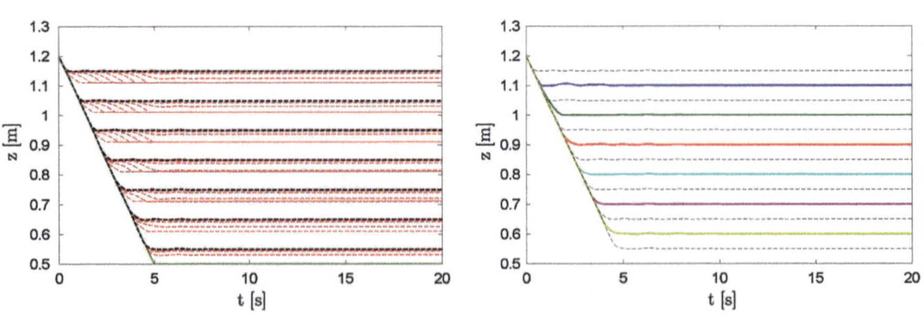

Fig. 1 *Left* database of trajectories. The *green line* shows the learned demonstrated trajectory. The *red dotted lines* show the trajectories during the ILC and the *bold black dotted lines* show the final trajectories achieved with coupling terms which make up the database. *Right* trajectories of placing a glass on a table at heights that were not learned, but generalized from the database of coupling terms which were learned for trajectories given with *dashed lines*

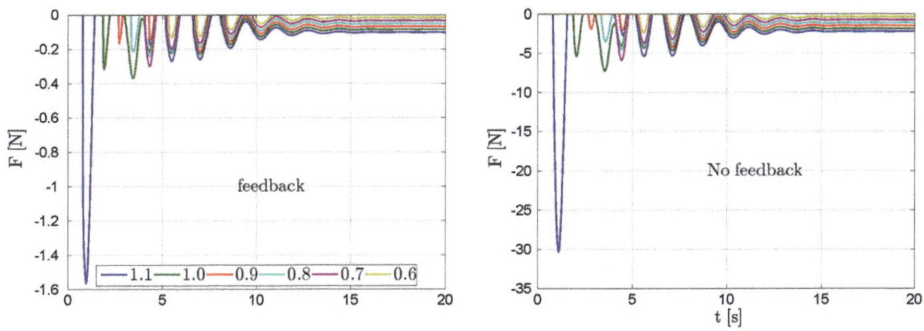

Fig. 2 The forces that occur when placing a glass on the table at different heights, specified in the text and the label. The *left plot* shows the forces when both the feedback and the feedforward coupling terms are used ($C = C_{fb} + C_{ff}$). The *right plot* shows the results when only the feedforward term is used ($C = C_{ff}$). A high force is only present for the highest table, as there the original demonstration trajectory is still very far from the target of generalization

shows in the left plot the forces that occur during the approach to these queries, when the generalized feedforward term is used in combination with a feedback term. The right plot shows the forces with only the generalized feedforward term. We can note some oscillations, which are partially the result of the ILC, but mostly of the radial basis functions which approximate the coupling term. The forces in the right plot are considerably higher, but for most cases below 5 N. The only real outlier is from the highest query, where the original demonstration and the adapted trajectory are extremely far apart, also shown in Fig. 1. A simulation of placing a glass on the table of various heights can be seen in Fig. 3.

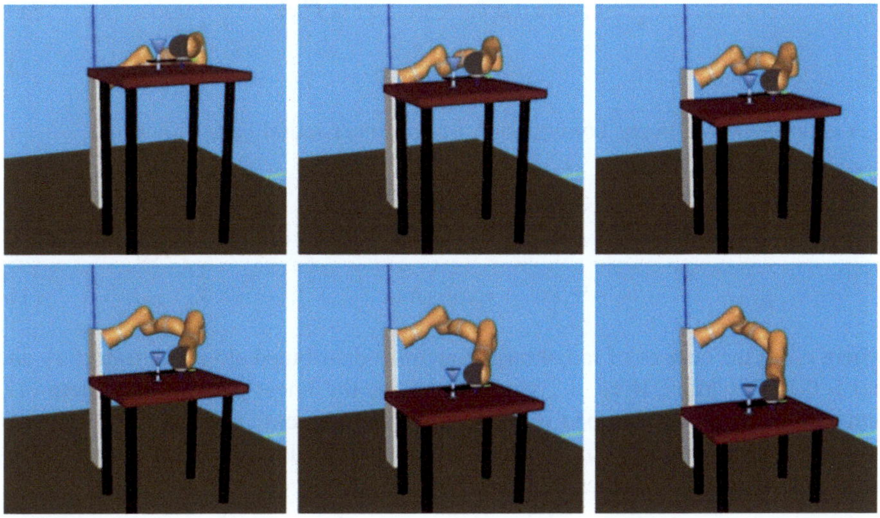

Fig. 3 Simulation of placing a glass on tables having different heights with the KUKA LWR robot. The heights are at the positions we did not explore, but generalize to them

5 Discussion and Conclusions

In this paper we have shown how the coupling terms, which modify the trajectory of the task, can be generalized to a new position. With this we can on-line produce the feedforward component of the DMP coupling term in order to adapt a demonstrated trajectory to the conditions of the environment.

Through generalization we can calculate the coupling term for any query point within the area of initial exploration. Thus, we can achieve a very good feedforward term approximation, which can either be used as it is or it can be the initial value in further exploration. The latter specifically becomes useful when the query point is not exactly accurate, but only an estimation. Thus, the need for exploration through ILC is reduced, accelerating the whole process. The results show that the generalization can achieve feedforward terms which can already be used for the final execution, specifically because the control always includes the feedback component as well.

In the future we will validate the approach on the real robot and test it for achieving a desired force of contact.

Appendix

Dynamic movement primitives have been extensively studied in robotics. For the completeness of the paper we provide a short summary based on [8]. A nonlinear system of differential equations defines DMP for discrete movements

$$\tau\dot{z} = \alpha_z(\beta_z(g - y) - z) + f(x) \tag{13}$$

$$\tau\dot{y} = z. \tag{14}$$

$f(x)$ is defined as a linear combination of nonlinear radial basis functions

$$f(x) = \frac{\sum_{i=1}^{N} w_i \Psi_i(x)}{\Psi_i(x)} x, \tag{15}$$

$$\Psi_i(x) = \exp(-h_i(x - c_i)^2), \tag{16}$$

where c_i are the centres of radial basis functions distributed along the trajectory and h_i > 0 their widths. If α_z, β_z, $\tau > 0, \alpha_z = 4\beta_z$ the linear part of the system (13) and (14) is critically damped and has a unique attractor point at $y = g$, $z = 0$. A phase variable x is used in (15) and (16). It is utilized to avoid direct dependency of $f(x)$ on time. Its dynamics is defined by

$$\tau\dot{x} = -\alpha_x x, \tag{17}$$

with initial value at x(0) = 1. α_x is a positive constant. The weight vector **w** composed of w_i defines the shape of the encoded trajectory. The learning of the weight vector is described in [8]. Multiple DOFs are realized by maintaining separate sets of (13)–(16), while a single canonical system given by (17) is used for synchronization.

References

1. Dillmann, R.: Teaching and learning of robot tasks via observation of human performance. Robot. Auton. Syst. **47**(2–3), 109–116 (2004)
2. Kober, J., Bagnell, D., Peters, J.: Reinforcement learning in robotics: a survey. Int. J. Robot. Res. **32**(11), 1238–1274 (2013)
3. Stulp, F., Sigaud, O.: Robot skill learning: from reinforcement learning to evolution strategies. Paladyn. J. Behav. Robot. **4**(1), 49–61 (2013)
4. Ijspeert, A., Nakanishi, J., Pastor, P., Hoffmann, H., Schaal, S.: Dynamical movement primitives: learning attractor models for motor behaviors. Neural Comput. **25**(2), 328–373 (2013)
5. Tamosiunaite, M., Nemec, B., Ude, A., Woergoetter, F.: Learning to pour with a robot arm combining goal and shape learning for dynamic movement primitives. Robot. Auton. Syst. **59** (11), 910–922 (2011)
6. Bristow, D., Tharayil, M., Alleyne, A.: A survey of iterative learning control. IEEE Control Syst. Mag. **26**(3), 96–114 (2006)
7. Gams, A., Nemec, B., Ijspeert, A., Ude, A.: Coupling movement primitives: interaction with the environment and bimanual tasks. IEEE Trans. Rob. **30**(4), 816–830 (2014)
8. Ude, A., Gams, A., Asfour, T., Morimoto, J.: Task-specific generalization of discrete and periodic dynamic movement primitives. IEEE Trans. Rob. **26**(5), 800–815 (2010)
9. Forte, D., Gams, A., Morimoto, J., Ude, A.: On-line motion synthesis and adaptation using a trajectory database. Robot. Auton. Syst. **60**(10), 1327–1339 (2012)
10. Deniša, M., Ude, A.: Discovering new motor primitives in transition graphs. In: Intelligent Autonomous Systems, vol. 12, pp. 219–230. Springer, Heidelberg (2013)
11. Pastor, P., Righetti, L., Kalakrishnan, M., Schaal, S.: Online movement adaptation based on previous sensor experiences. In: 2011 IEEE/RSJ International Conference on Intelligent Robots and Systems (IROS), pp. 365–371. IEEE Press, San Francisco (2011)
12. Rasmussen, C., Williams, C.: Gaussian processes for machine learning. MIT Press, Cambridge (2006)

Extended Kalman Filter (EKF)-Based Local SLAM in Dynamic Environments: A Framework

Horaţiu George Todoran and Markus Bader

Abstract In the domain of mobile robots local maps of environments are used as knowledge base for decisions allowing reactive control in order to prevent collisions when following a global trajectory. These maps are normally discrete and updated at relatively high frequency, but with no dynamic information. The proposed framework uses a sparse description of clustered scan points from a laser range scanner. These features and the system odometry are used to predict the agent's ego motion as well as feature motion using an Extended Kalman Filter. This approach is similar to the Simultaneous Localization and Mapping (SLAM) algorithm but with low-constraint features. The presented local Simultaneous Localization and Mapping (LSLAM) approach creates a decision base, holding a dynamic description which relaxes the requirement of high update rates. Simulated results demonstrate environment classification and tracking as well as self-pose correction in static and in dynamic environments.

Keywords EKF · SLAM · Adaptive filtering · Dynamic descriptors · Grouped data

1 Introduction

Despite the fact that a lot of research has been conducted on environment mapping, most of the approaches still assume it to be static. As a consequence, tasks assigned to mobile agents have been generally solved constraining the underlying algorithms

H.G. Todoran (✉)
Automation and Control Institute, Vienna University of Technology, Vienna, Austria
e-mail: george.todoran@tuwien.ac.at

M. Bader
Institute of Computer Aided Automation, Vienna University of Technology, Vienna, Austria
e-mail: markus.bader@tuwien.ac.at

T. Borangiu (ed.), *Advances in Robot Design and Intelligent Control*,
Advances in Intelligent Systems and Computing 371,
DOI 10.1007/978-3-319-21290-6_46

and approaches to such an assumption. Thus, classical approaches to Simultaneous Localization and Mapping (SLAM) such as FastSLAM [1] or ICP-based [2] encounter difficulties and often lead to divergence in dynamic unstructured environments. Regarding map-based localization tasks, considerable modification of the layout imposes offline re-mapping. Even in local path-planning, common approaches treat sensor data at each iteration as static obstacles, relying on relatively high control frequencies to deal with dynamic environments.

Local maps typically consist of close-vicinity representations of the environment. As they represent the closest layer of perception in relation with the agent's dynamic tasks (path-following, grasping etc.) [3], they are required to be accurate, online and descriptor-rich. However, traditional approaches generate local-maps without any dynamic descriptors of the local environment.

The presented framework proposes an alternative to local map creation and environment description for mobile agents in an online, fast-computing manner. Thus, the environment is modelled and grouped as rigid dynamic objects, treating object discovery, time-out, merging, splitting and symmetries. Using the acquired information regarding the objects' dynamic states, agent self-pose correction is performed enchanting local map-building to local SLAM (*LSLAM*). In addition, the framework outputs the classified objects and their dynamic descriptors for further usage in self-localization, mapping, path-planning, sensor-fusion and other robotics tasks.

Grouping of data in higher level feature objects has been widely studied in computer vision and robotics communities and recently proposed in SLAM approaches [4]. However, this work aims to include high-level features in a more complex SLAM problem, where dynamic entities are present. Dynamic object tracking has been addressed by Montessano [5] in his PhD. thesis, analysing various filtering techniques. Bar-Shalom et al. [6] analyses dynamic object tracking as well and presents a process noise for velocity-bound and acceleration-bound models assuming a Wiener-process.

This paper is organized as follows: Section 2 presents in detail the approach of the LSLAM EKF estimator. Simulated results for self-pose correction and data association are presented in Sect. 3. Conclusions are drawn in Sect. 4.

2 Framework Approach

Figure 1 presents the overview of the framework, including its modules and their input-output data. The laser sensor data undergoes a three-layered abstraction, from points (p) to segments (S) and finally objects ($\mathcal{O}$).

The input point-cloud ${}_{raw}\hat{\mathbf{p}}$ is segmented under homography constraints and its constituting segments are associated. Pose displacements corresponding to each segment pair along with uncertainty is computed by the *Covariant ICP* module. The loop is being closed an Extended Kalman Filter (EKF) type estimator implemented in the *LSLAM* module. Based on the agent predicted state $\overline{\boldsymbol{\mu}}_{\mathcal{R}}$, the correlated

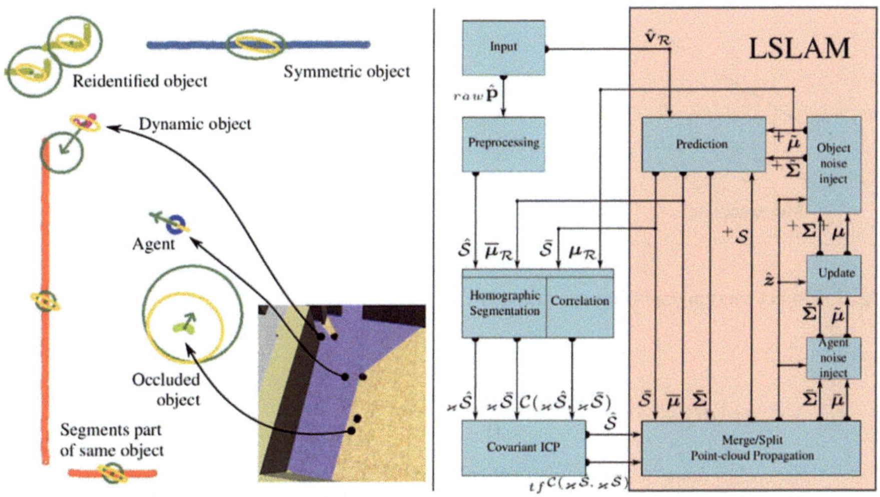

Fig. 1 Local map showing various non-trivial scenarios (*left*); framework overview (*right*)

segments information and objects observations $\hat{z}$, the estimator updates the state of the world. In the following, the underlying approach of the *LSLAM* module is presented.

2.1 Prediction

Agent prediction. Given the mapped state of the agent including its velocities and accelerations, assuming a constant acceleration model its state is predicted. The noise covariance matrix is modelled as parameterized *dt*-proportional, using two parameters β. The Jacobian of the agent's motion model with respect to the noise parameters V (1) maps the noise covariance matrix in the agent's state space (2).

$$\overline{\boldsymbol{\mu}}_{\mathcal{R}} = \mathbf{g}_{\mathcal{R}}(\boldsymbol{\mu}_{\mathcal{R}}, dt) + \overline{\boldsymbol{\alpha}}_{\mathcal{R}}(dt), \quad \overline{\boldsymbol{\alpha}}_{\mathcal{R}}(dt) = \mathbf{V}\mathbf{M}(dt)\mathbf{V}^T \tag{1}$$

$$\mathbf{V} = \frac{\partial \mathbf{g}_{\mathcal{R}}}{\partial \boldsymbol{\mu}_{\mathcal{R}_a}}, \quad \boldsymbol{\mu}_{\mathcal{R}_a} = \begin{bmatrix} a_{lin} \\ a_{ang} \end{bmatrix}, \quad \mathbf{M}(dt) = dt \begin{bmatrix} \beta_{\overline{\mathcal{R}},0} & 0 \\ 0 & \beta_{\overline{\mathcal{R}},1} \end{bmatrix} \tag{2}$$

Agent pseudo-observation. In many situations, odometry information is subject to inconsistent noise (time delay, drifts, encoder errors, wheel slips, deviation from linearization point) and furthermore imposes a refinement of the agent prediction method. Thus, at the end of the prediction step, the mapped system is being pseudo-updated with the agent's control input. The noise of the observation is modelled as having a base value and a state-velocity-proportional value (3).

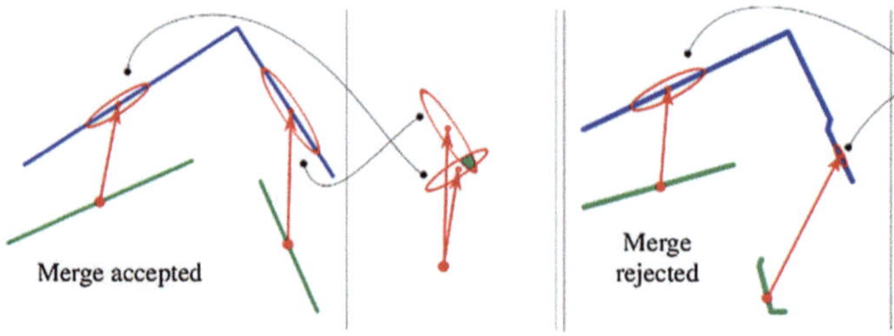

Fig. 2 Object merging/splitting scenarios

$$\hat{\boldsymbol{\alpha}}_{\mathcal{R}}(dt) = dt^2 \begin{bmatrix} (\beta_{\mathcal{R},0} + \beta_{\mathcal{R},1}|\overline{v}_{lin}|)^2 & 0 \\ 0 & (\beta_{\mathcal{R},2} + \beta_{\mathcal{R},3}|\overline{v}_{ang}|)^2 \end{bmatrix} \quad (3)$$

Object prediction. Assuming a constant-acceleration model, the objects are predicted. The object's pose prediction has no real use but pose covariance is being predicted and propagated over time. However, point-clouds that belong to an object are predicted based on its dynamic state and propagated accordingly to ensure robust segment correspondences. The object process noise $\overline{\propto}_{O}dt$ is modelled as described in [6] under the assumption of continuous-time white noise (Wiener process).

$$\overline{\boldsymbol{\mu}}_{\mathcal{O}_i} = \mathbf{g}_{\mathcal{O}_i}(\boldsymbol{\mu}_{\mathcal{O}_i}, dt) + \overline{\boldsymbol{\alpha}}_{\mathcal{O}}(dt) \quad (4)$$

2.2 Merging/Splitting Analysis

In situations when object merging or splitting is proposed (object merge/split is flagged when observing multiple segments corresponding to one or vice versa) the strength of computing transforms covariance is exploited. The mean and covariance of the proposed-to-merge segments/objects are being pair-wise combined forming a new mean $\boldsymbol{\mu}_{ij}$.[1] The cost function of 2 transforms "neighbouring level" is modelled as product of the 2 initial distributions, evaluated in $\boldsymbol{\mu}_{ij}$. The pair-wise merging occurs if the cost function is bigger than a threshold. When a flagged merge or split is valid, the state vector is resized and segments are associated to the new object configurations (Fig. 2).

[1]Merging uncertainty computed through Gaussian distributions multiplication.

2.3 Update

The general object observation function uses the computed translational and rotational deviation of the corresponding segments along with their uncertainty, as computed by the *Covariant ICP* module. As the centre of mass of the matched point-clouds changes over time, the predicted pose state of each object has to be initialized before update (5), according to its new centre of mass (computed by the *Covariant ICP* module).

$$\bar{\mu}_{\mathcal{O}_i:x,y,\theta} = \mathbf{h}^{-1}(\bar{\mathbf{z}}_{\mathcal{O}_i}), \quad \bar{\mathbf{z}}_{\mathcal{O}_i} = \begin{bmatrix} \bar{x}_x \\ \bar{x}_y \\ \bar{x}_\theta \end{bmatrix}_{\mathcal{O}_i}^{\overline{\mathcal{R}}}, \quad \hat{\mathbf{z}}_{\mathcal{O}_i} = \begin{bmatrix} \hat{x}_x \\ \hat{x}_y \\ \hat{x}_\theta \end{bmatrix}_{\mathcal{O}_i}^{\overline{\mathcal{R}}} + \hat{\boldsymbol{\alpha}}_{\mathcal{O}_i} \tag{5}$$

2.4 Adaptive Filtering

As the observed environment assumes a general low-constraint model with no correlation between objects, agent pose-correction should happen only when sufficient observed objects are in steady-state. Thus, additional uncertainty of the tracked objects is being injected as a scaled entry of their base prediction noise (6). The scale factor s_{O_i} is proportional to the Mahalanobis distance of the object residual $\mathbf{q}$ in the velocity probability distribution[2] $\sum_{O_{i_v} O_{i_v}}$. The object adaptive noise provides better object state estimation in situations of under-scaled base prediction noise in highly-dynamic tracked motions (e.g. circular trajectory with small radius at high velocity).

$$\tilde{\boldsymbol{\Sigma}}_{\mathcal{O}_i\mathcal{O}_i} = \overline{\boldsymbol{\Sigma}}_{\mathcal{O}_i\mathcal{O}_i} + s_{\tilde{\mathcal{O}}_i}\bar{\boldsymbol{\alpha}}_{\mathcal{O}}(dt) \tag{6}$$

$$s_{\tilde{\mathcal{O}}_i} = {}^t\mathbf{q}^T \boldsymbol{\Sigma}^{-1}_{\tilde{\mathcal{O}}_{i_v}\mathcal{O}_{i_v}} {}^t\mathbf{q}, \mathbf{q} = \hat{\mathbf{z}}_{\mathcal{O}_i} - \bar{\mathbf{z}}_{\mathcal{O}_i} \tag{7}$$

The agent state convergence is evaluated in a similar fashion to [7]. However, instead of using only the predicted residuals, the total Mahalanobis distances of all the objects residuals at time t and the same metric for the last N time steps is evaluated (9). Thus, large additional agent noise will not be triggered when the observed system is closer to unsteady state. In practice, the additive noise $\tilde{\alpha}_R(dt)$ injected in phase (8) is triggered if $s_{\overline{R}} > \beta_R$, having various forms depending on the expected agent's drifts from the model (time skew, slips, short distance kidnappings, frame-rate drop etc.).

[2] Evaluating the Mahalanobis distance ensures scalability and behaves equally for various pose uncertainties of the system.

$$\tilde{\boldsymbol{\Sigma}}_{\mathcal{O}_{\mathcal{R}}\mathcal{O}_{\mathcal{R}}} = \overline{\boldsymbol{\Sigma}}_{\mathcal{O}_{\mathcal{R}}\mathcal{O}_{\mathcal{R}}} + s_{\tilde{\mathcal{R}}}\tilde{\boldsymbol{\alpha}}_{\mathcal{R}}(dt) \tag{8}$$

$$s_{\tilde{\mathcal{R}}} = \beta_{\tilde{\mathcal{R}}} \frac{{}^{t}\mathbf{Q}^{T\,t}\boldsymbol{\Sigma}^{-1}_{\mathcal{O}_v\mathcal{O}_v}\,{}^{t}\mathbf{Q}}{\frac{1}{N}\sum_{i=t-N}^{t-1} {}^{i}\mathbf{Q}^{T\,i}\boldsymbol{\Sigma}^{-1}_{\mathcal{O}_v\mathcal{O}_v}\,{}^{i}\mathbf{Q}} - 1, \quad {}^{i}\mathbf{Q} = {}^{i}\hat{z}_{\mathcal{O}} - {}^{i}\overline{z}_{\mathcal{O}} \tag{9}$$

During a filter cycle, the agent adaptive noise injection process is evaluated first. After agent noise injection and all object updates, the object residuals are recomputed (using updated agent information) for object noise injection. Thus, erroneous high values of object injected noise due to pre-update agent state divergence are avoided.

2.5 Initialization and Point-Cloud Propagation

As more objects get discovered, they are appended and initialized according to the agent's uncertainty and their initial observation. Especially for moving objects, keeping their state when they are out of view is not required, their uncertainty becoming so big that correct data association is not feasible. Thus, each object is discarded when its velocity uncertainty ellipse surface exceeds certain thresholds (static out-of-view objects are remembered and potentially used for loop closures).

In the current stage, object reconstruction is not implemented; thus for short-term mapping the "longest" point-cloud from a match is propagated, given that it provides enough up-to-date information about the current observed segment.

3 Simulated Results

All experiments were conducted using Gazebo simulation environment under Robotics Operating System (ROS). The agent is Pioneer-P3DX mobile robot equipped with Hoyuko laser range sensor, its angular and linear velocities being constrained to measured acceleration limits, $\propto_{\max} \sim 0.4\,\mathrm{m/s^2}$. As ROS publishes laser and odometry data unsynchronized at 10 Hz frequency, a message-filter is used for synchronization. Thus, the framework's frequency is set to 8 Hz, ensuring input data synchronization without loop-skips. The input data from the simulator is noised with fixed Gaussian noise for laser sensor range readings $\sigma_{\emptyset} \sim 0.01\,\mathrm{m}$ and proportional Gaussian noise for observed agent velocities $\sigma_v \sim |v| \cdot 0.02\,\mathrm{m/s}$. The framework provides 3 visualization tools, one for ICP results, transforms and error metric and other 2 for agent local- and world view. Within the viewers, 3σ pose uncertainty is shown in yellow and the velocity mean and uncertainty in green. For all experiments, the agent is tele-operated.

3.1 Object Correspondences

Even though the framework provides short-term memory differential mapping, higher level features such as object correspondences are extracted. The set of experiments below shows the framework ability to track identified objects in non-trivial scenarios.

Merging/splitting. Given a finite and relatively small sensor range with respect to the environment size as well as occluded areas of environments, the agent is supposed to map segment correspondences to their true state. Thus, the experiment presents object discovery and merging with segment splitting. At time t_1 the agent observes 3 segments, assumed to be different objects. As more information is acquired, the objects get merged so that at t_2 all observed segments are evaluated as the same object (Fig. 3-left).

Swipe-occlusion detection. Most of the scenarios involving dynamic objects and static ones in the background will undergo "swiped-occlusion" situations, in which the dynamic object partially occludes from one end to the other the static background. In such cases, propagating the last observed point-cloud fails to re-identify background object after the "swipe". However, this approach achieves such semantic memory (Fig. 3-right).

Occluded tracking. Short-term object occlusion frequently occurs in dynamic environments. Especially for path-planning tasks, trajectory optimization and even collision avoidance can be achieved in case of short-term occluded memory, the object's trajectory being remembered. In this experiment, the agent is undergoing a circular motion and observes a moving object which becomes fully occluded. However, when the object reappears in the field-of-view, it is correctly matched (Fig. 4).

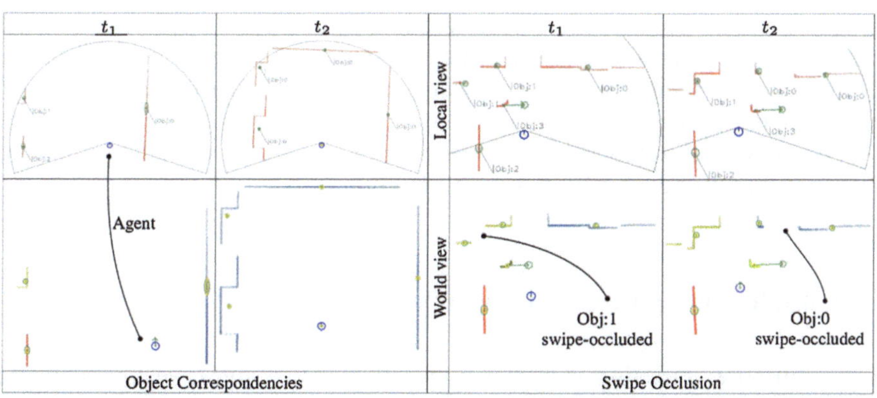

Fig. 3 Semantic memory in various scenarios

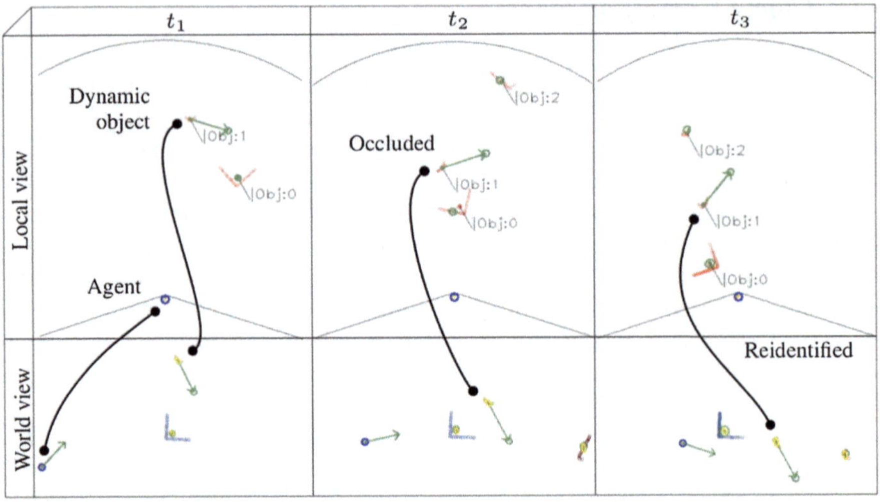

Fig. 4 Occluded tracking

3.2 *Self-Pose Correction*

The following experiments present agent self-pose correction, assuming dynamic as well as static environment. The encoder drifts have been achieved by stopping the simulation at certain time periods and manually modifying the agent pose.

As described in Sect. 2, the agent's pose is expected to be accurately corrected as long as parts of the environment are in steady state. Figure 5 presents the filtered trajectory of the robot when it undergoes short-term deviations from the motion model in various scenarios (m:m). Even though the framework was designed for agents with relatively accurate odometry values and dynamic word assumption, it can as well be operated assuming a static environment. Tests have been conducted feeding into the framework a fix measured angular and linear velocity set to *0* (odometry impairment).

Figure 6 presents estimated trajectories of the agent (left, m:m) and the estimated linear and angular agent velocities (right, m/s:s).

4 Conclusions and Future Work

Agent self-pose correction in dynamic environments is still a weakly addressed problem within the robotics community. The presented framework has proven to extract sufficient information from partially steady-state dynamic environments, even though low constraint models of the environment and agent are assumed. By limiting environment perception to 2D, the complexity of the problem is low

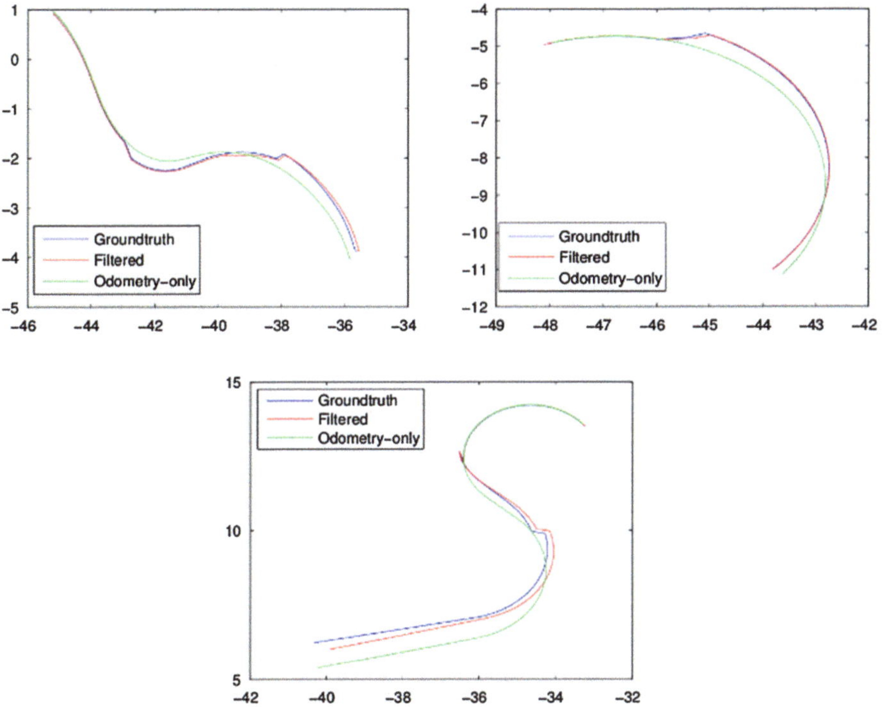

Fig. 5 Self-pose correction in dynamic environments

enough for fast, online computation. However, especially in structured environments, the reduced dimensionality makes data association difficult given high probability of symmetries. Thus, image processing techniques such as *Homographic Segmentation* and *Covariant ICP* have been developed and used to increase robustness of data association.

As laser range sensors represent a norm in mobile agents nowadays, their accuracy is assumed by the framework to extract higher-level features from the environment: Objects. Such an approach proves to simplify EKF SLAM, reducing drastically the state vector size and thus computation time, even with detailed assumed models.

The focus of the presented work is on short-term memory reasoning of state point-clouds. Long-term mapping and symmetry-robust loop closures were not addressed. However, simulated results prove that occluded tracking and gradual total occlusion are solved even with the mentioned limitations. Shifting the focus towards long-term memory tasks including offline maps as constrained static objects will improve results.

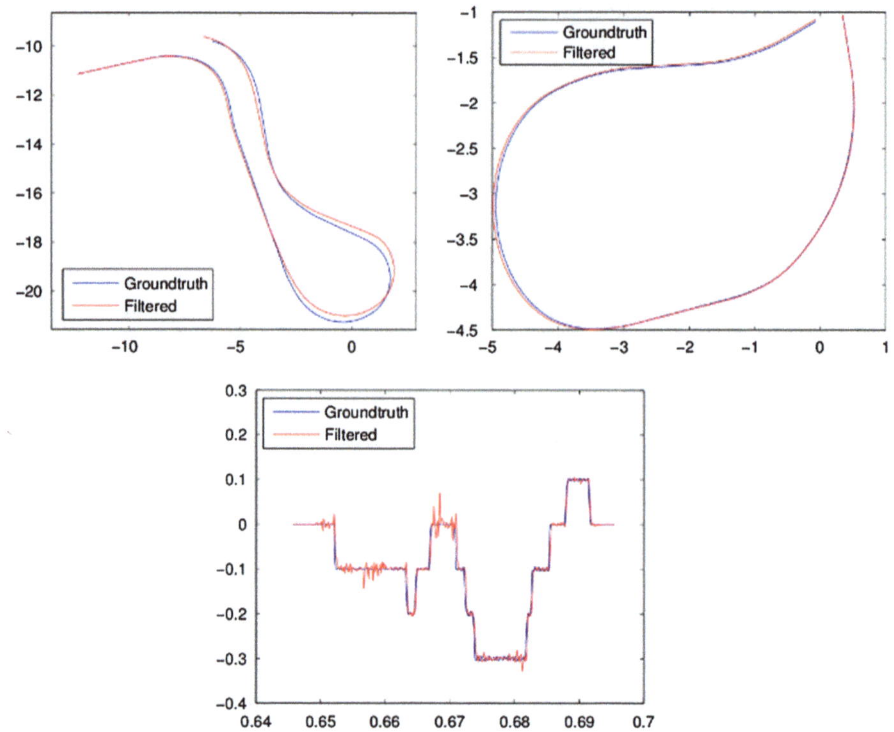

Fig. 6 Self-pose correction (*left*) and agent angular velocity (*right*) in static environments

Research can be focused on higher-level tasks such as local path-planning using approaches such as Model Predictive Control (MPC) or Dynamic Window Approach (DWA), generalized for dynamic obstacles' states. Such approaches should provide complex behaviours such as avoidance manoeuvres due to external dynamic objects.

Last but not least, when high robustness and maturity of the framework is achieved, open-source publication of the code as ROS node is desired, inviting the robotics community to take advantage of a package that provides pose-correction and dynamic descriptors of the environment for future research in various related fields.

References

1. Montemerlo, M., Thrun, S., Koller, D., Wegbreit, B.: FastSLAM: a factored solution to the simultaneous localization and mapping problem. In: Proceedings of the AAAI national conference on artificial intelligence (2002)
2. Nüchter, A., Lingemann, K., Hertzberg, J., Surmann, H.: 6D SLAM—3D mapping outdoor environments. J. Field Robot. 24, 699–722 (2007)

3. Rusu, R.B., Sucan, I.A., Gerkey, B., Chitta, S., Beetz, M., Kavraki, L.E.: Real-time perception-guided motion planning for a personal robot. In: IEEE/RSJ International Conference on Intelligent Robots and Systems, pp. 4245–4252 (2009)
4. Choudhary, S., Trevor, A.J.B., Christensen, H.I., Dellaert, F.: SLAM with object discovery, modeling and mapping. In: IEEE/RSJ International Conference on Intelligent Robots and Systems, pp. 1018–1025 (2014)
5. Montesano, L.: Detection and Tracking of Moving Objects from a Mobile Platform. Application to Navigation and Multi-robot Localization. University of Zaragoza, Zaragoza (2006)
6. Bar-Shalom, Y., Rong Li, X., Kirubarajan, T.: Estimation with applications to tracking and navigation. Wiley, New York (2001)
7. Hu, C., Chen, W., Chen, Y., Liu, D.: Adaptive Kalman filtering for vehicle navigation. J. Glob. Positioning Syst. 2(1), 42–47 (2003)

Motion Analysis of a Robotic Wheelchair

Ionuţ Geonea and Nicolae Dumitru

Abstract This paper presents the authors' contribution in designing and evaluating a transmission system for a robotic wheelchair. The kinematics of the proposed transmission is analysed in order to realize a proper gear synthesis. A 3D model of the transmission and wheelchair are designed in Solid Works, being then used for the dynamic simulation of the system in Adams software. The wheelchair motion simulation is done in Adams in two modes: traction motion and steering motion active. In case of traction motion simulation the resulting trajectory is a straight line displacement, whereas in the case of active steering motion the resulting displacement trajectory is circular. Simulation results demonstrate the efficiency of the developed transmission model and assure successful implementation of this design in a robotic wheelchair.

Keywords Robotic wheelchair · Dynamic model · Differential transmission

1 Introduction

There are many examples of assistive devices for people with manipulative and locomotive disabilities. Such devices enable disabled people to perform diverse activities of daily living thus improving their quality of life. Disabled people are increasingly able to lead an independent life and play a more productive role in society [1].

The wheelchair is still the best transportation means for disabled people, since its invention in 1595 (called *invalids chair*) for Phillip II of Spain by an unknown inventor. The device has evolved since then to complex multi-d.o.f.

I. Geonea (✉) · N. Dumitru
University of Craiova, Craiova, Romania
e-mail: igeonea@yahoo.com

N. Dumitru
e-mail: nicolae_dtru@yahoo.com

T. Borangiu (ed.), *Advances in Robot Design and Intelligent Control*,
Advances in Intelligent Systems and Computing 371,
DOI 10.1007/978-3-319-21290-6_47

electro-mechanical devices and robotic systems [1–7]. Despite the rapid scientific and technological progress, there has been very little innovation in wheelchair design over the last 200–300 years. The folding wheelchair came in 1933, and powered wheelchairs were developed in the early 1970s [4]. New materials such as plastics, fibre-reinforced composites and aluminium alloys have found their application in the design and manufacture of lighter, stronger and more reliable wheelchairs [1]. The wheelchair industry has also benefitted from the development of lighter, efficient, durable and reliable motors, better amplifiers and controllers and most important of all superior batteries. A number of computer controlled wheelchairs have been developed in recent years, including: the CALL Smart Chair, NavChair, TinMan and WALKY [1, 8–16]. On a smooth level surface, relatively little torque is required to propel the wheelchair and occupant at constants speeds of up to 5 m/s. If this were the only requirement, small motors of a few hundred watts capacity would serve. But wheelchairs must overcome obstacles usually at low speed, climb substantial grades and accelerate at reasonable rate. These conditions demand high torque, most often at low speed. Thus, the most powered wheelchairs are equipped with motors and electronic controls that are much larger than necessary for constant level propulsion and with poor efficiency most of the time since high torque is obtained only at low speed [6].

The propulsion system of a powered wheelchair typically consists of a pair of motors, one for each drive wheel, and a drive train consisting of gears, belts and other mechanical elements that couple the motor's shaft to the drive wheel shaft. Speed and torque generated by each motor is controlled by pulse width modulation (PWM) [1].

This research brings arguments for a mechanical transmission that achieves the differential movement, on which the traction and the steering components are controlled by separately motors with suitable synthesis to achieve the proper angular speed difference of wheels. The model will be equipped with a motion controller, using a pulse width modulated DC to DC converter.

2 Kinematics of Wheelchair Transmission

The kinematic scheme of the wheelchair transmission is presented in Fig. 1. Shaft I is actuated by the straight line displacement provided by the electric motor M_1; steering is performed by the motor M_2 placed on shaft V.

Bevel gears, (10, 12) and (10', 12') are planetary gears and are mounted on shafts (IV, III) respectively (IV', III') with pins assembly. Bevel gears (11, 13), and (11', 13') are satellites gears; they are mounted on needle bearings to axes fixed on differential casing, and achieve a planetary motion. To calculate the transmission ratio, the principle of motion reversing is applied (called Willis principle). For straight line displacement the traction wheels of the wheelchair spin at the same angular velocity. The motion is transmitted from worm gears 6-5 to the shaft II and by means of the final transmission with gears 4-2 to the wheelchair wheels.

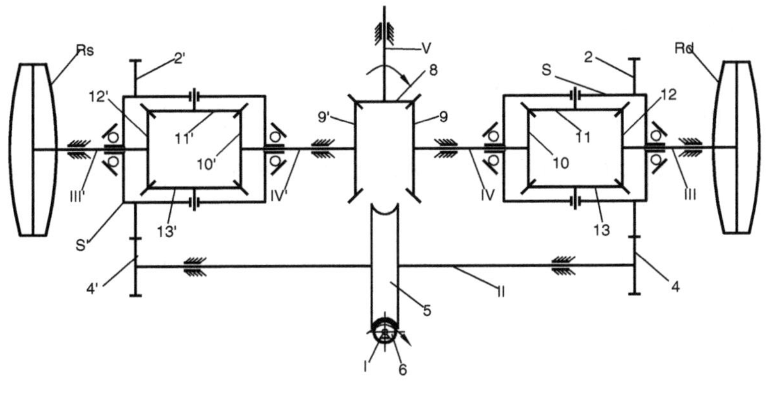

Fig. 1 Kinematics scheme of wheelchair transmission

For steering, motion is transmitted from motor M_2 to shaft V trough bevel gears (8, 9), (10, 11), (11, 12) to semi-axis III, respectively to motor wheel Rd. For the left wheel Rs the transmission flow is through bevel gears (8, 9'), (10', 11'), (11', 12') to semi-axis III'. The two wheels Rd and Rs spin with the same angular speed and opposite rotation sense. Casing S, S' and spur gear (2, 2') is not moving (M_1 is turned off).

The motion transmission chain for straight line displacement and the one for steering of the wheelchair are expressed by the Eqs. (1).

$$\begin{array}{ll} M_1 - I - i^w_{65} - II - i^c_{42} - III - Rd & M_2 - V - i^k_{89} - IV - i^k_{10\ 11} - i^k_{11\ 12} - III - Rd \\ M_1 - I - i^w_{65} - II - i^c_{4'2'} - III - Rs\text{'} & M_2 - V - i^k_{89'} - IV' - i^k_{10'\ 11'} - i^k_{11'\ 12'} - III' - Rs \end{array} \quad (1)$$

where: M_1—traction motor; i^w_{65}—worm gear ratio; i^c_{42}—final transmission ratio; Rd and Rs, right and left wheel, I, II and III—shafts; M_2—steering motor; i^k_{89} $i^k_{89'}$—bevel gear ratio; $i^k_{10\ 11}$—bevel gears 10, 11 ratio; $i^k_{11\ 12}$—bevel gears 11, 12 ratio; Rd and Rs, right and left wheel, III, IV and V—shafts.

Straight line displacement of the wheelchair when both drive wheels encounter the same resistance to ground is validated by Eq. (1). For steering, the equations of the wheels' angular velocity are deducted below. The following notations are used:

ω_{10}, ω_{12}—absolute angular velocity of bevel planetary gears 10 and 11, considered in relation to the differential casing;

$\omega^S_{10} = \omega_{10} - \omega_S$—relative angular velocity of gear 10, towards differential casing S;

$\omega^S_{12} = \omega_{12} - \omega_S$—relative angular velocity of gear 12, towards differential casing S;

ω_{11}, ω_{13}—relative angular velocity of satellite gears 11 and 13 towards differential casing S;

ω_S—absolute angular velocity of differential housing upon the wheelchair's frame;
R_W—rolling circle radius for planetary gears 10 and 12;
r_W—rolling circle radius for satellites gears 11 and 13.

According to the principle of gearing, the tangential velocities are equal in the gearing point:

$$-(\omega_{10} - \omega_s)R_W = \omega_{11}r_W = (\omega_{12} - \omega_S)R_W = \omega_{13}r_W \tag{2}$$

One obtains:

$$\omega_s = \frac{\omega_{10} + \omega_{12}}{2} \tag{3}$$

The angular velocity of the central gear (planetary) is twice the angular velocity of the differential box. If the steering motor is turned off, then $\omega_{10} = 0$, from Eq. (3) it results:

$$\omega_{12} = 2\omega_s = 2\omega_2 \tag{4}$$

For left wheel are available the equations: $\omega'_{10} = 0$, and $\omega_{12'} = 2\omega_{S'} = 2\omega_{2'}$ Considering the gears' ratio Eq. (5) is derived, where ω_S is the motion received from the traction motor.

$$\omega_S = \frac{\omega_{M_1}}{i^w_{65} \cdot i^c_{42}} [\text{rad/s}], \tag{5}$$

ω_{M_1}—angular velocity of traction motor; i^w_{65}—worm gear ratio; i^c_{42}—spur gears 4-2 transmission ratio. Tor the case of wheelchair's displacement along a nonlinear path, Eq. (6) describe the motion received from steering motor M_2:

$$\omega_{10} = \frac{n_{M_2}}{i^k_{89}}; \quad \omega'_{10} = \frac{n_{M_2}}{i^k_{89'}} \tag{6}$$

Let us assume that the right wheel Rd is outside the curve; in this case the pairs of satellite gears (11, 13) respectively (11', 13') will spin. It must be distinguished between the absolute motion of planetary gears, the transportation movement of the differential box and the relative movement of satellites gears towards differential box S or S'. Thus, the distribution of velocities in absolute motion of a planetary gear is:

$$\omega_{10} \cdot R_W = \omega_S \cdot R_W + \omega_{11} \cdot r_W \tag{7}$$

From Eq. (3), the angular velocities for satellite gear (11) and planetary gears (10), (12) are:

$$\omega_{11} = (\omega_{10} - \omega_s)\frac{R_W}{r_W}, \quad \omega_{10} = \omega_s + \omega_{11}\frac{r_W}{R_W}, \quad \omega_{12} = \omega_s - \omega_{11}\frac{r_W}{R_W} \tag{8}$$

The equations above derived are useful for the kinematic synthesis of the wheelchair transmission; they are used to establish gears ratio for the 3D model design.

3 Design of the Wheelchair's 3D Model in Solid Works

Once the gears ratio established, their geometric parameters are calculated. Gear modelling is done with Gear Trax and Solid Works software; the assembly models obtained for the transmissions are presented in Fig. 2. The transmissions are mounted on the wheelchair frame; the 3D model of the assembly is presented in Fig. 3. The weight of the entire assembly is 45 kg.

The 3D virtual prototype obtained is very important because it will be used to simulate in Adams the wheelchair motion trajectory and to study the motion dynamics. The virtual prototype is useful in further design stages to optimize the construction (weight minimization, ergonomics studies).

4 Dynamic Model of the Wheelchair

The virtual prototype of the wheelchair, presented in Fig. 3, is transferred to the Adams multi-body model using the transfer interface embedded in Solid Works. Through a suitable modelling the dynamic analysis of wheelchair model was done. The steps taken in order to define the dynamic model are: (1) specify the mass properties of the kinematic elements, (2) define kinematics joints, (3) specify

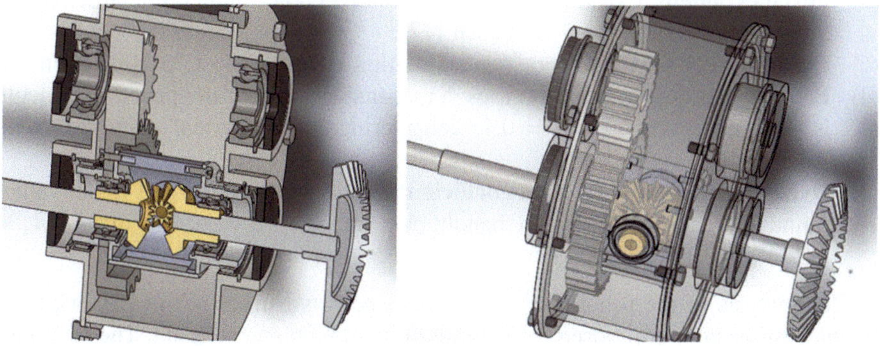

Fig. 2 CAD model of differential transmission

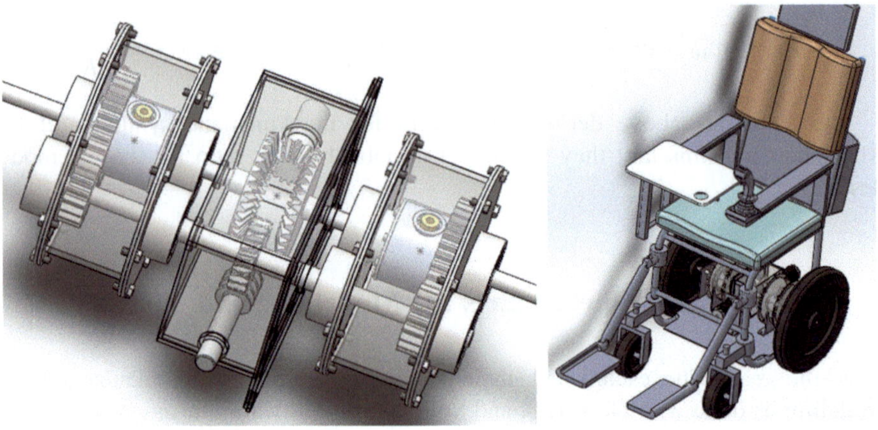

Fig. 3 Wheelchair transmission assembly

friction models (friction in joints and wheels-ground contact), (4) specify the human load of 70 kg, and (5) most important -define gear type connections.

The contact forces in the gear set are described by a mechanical contact model for which the parameters: stiffness, force exponent, damping and friction coefficients and penetration depth are specified. These parameters were defined by studying [11].

Contact parameters are calculated as follows: considering computational efficiency and accuracy, the impact method was adopted to define gears contact. The parameters associated to for this method are further explained.

The contact between ground and wheels is specified according to the technical literature: $\Psi_a = 0.5\ldots0.6$ for old asphalt road, old concrete. It should be noted that the front wheels are self-directed, being modelled by rotation joints with friction.

(a) **Stiffness *K***: The materials for the transmission's pinions and gears are alloy steel; the values for the Poisson ratio and the Young's modulus are: $\upsilon = 0.3$ and $E = 2.1 \times 10^5\ \mathrm{N/mm^2}$.
(b) **Force exponent**: Considering numerical convergence and computation speed, a force exponent e = 1.5 is chosen [17].
(c) **Damping coefficient C**: For this simulation it was set to C = 1000 Ns/mm.
(d) **Penetration depth**: In most cases, a reasonable value of the penetration depth is 0.01 mm. We used d = 0.1, considering the numerical convergence in Adams.
(e) **Dynamic and static friction coefficient and viscous velocity**: Typical values found in mechanical design handbooks are: $\mu_s = 0.1$; $\mathrm{v}_s = 1\ \mathrm{mm/s}$, $\mu_d = 0.08$ $\mathrm{v}_d = 10\ \mathrm{mm/s}$. [11]

The analysis of the robotic system workspace was carried out in Adams, for a combined trajectory composed by a straight line motion and steering. The functions used in Adams to define straight line motion and steering motion are given by Eq. 9.

$$\begin{aligned}&\text{IF(time}-2:32\{\text{rad/s}\},0,\ \text{IF(time}-4:0,32\{\text{rad/s}\},32\{\text{rad/s}\}))-\text{traction motion}\\&\text{IF(time}-2:0,03\{\text{rad/s}\})-\text{steering motion}\end{aligned} \tag{9}$$

The second case of simulation considers active only the steering motion, with value $\omega_2 = 12$ rad/ sec (applied to shaft V). The simulation uses WSTIFF solver with SI2 integration. Figure 4 shows the wheelchair motion trajectories obtained in both cases.

For the combined traction and steering motion simulation, the right and left wheel angular velocity (respectively, planetary gear 12 and 12') were obtained, see Fig. 5.

Analyzing Fig. 5 one can observe that in the first 2 s of simulation, when only the traction motion is active, the wheels spin with the same angular velocity and the wheelchair motion trajectory is a straight line. In the simulation time interval from 2 s to 4 s, only the steering motion is active and the wheels spin in opposite directions, the motion trajectory being a circle. In the last 3 s of simulation both motions are active, steering along any type of path (motion curve).

Figure 6 shows that for steering motion the wheelchair wheels spins with the same angular velocity but in opposite directions; hence the wheelchair is moving in circle.

Figure 7 depicts the variation of the traction motor torque. It results that the traction torque is zero when this motion is inactive, corresponding to the simulation time 2–4 s. Also, the steering torque is near zero when the steering motion is inactive.

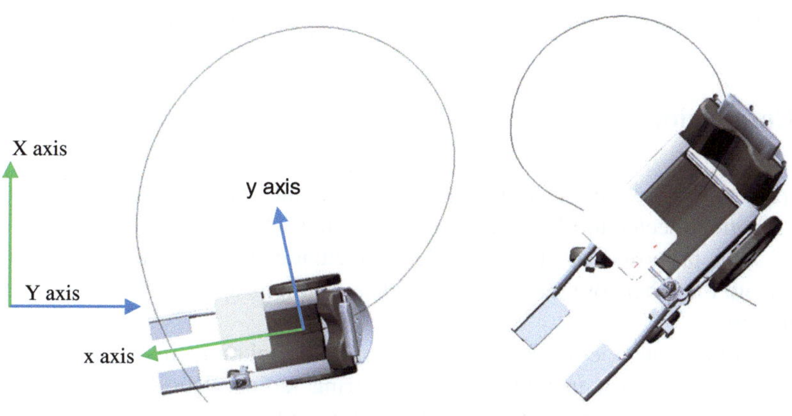

Fig. 4 Wheelchair trajectory: combined motion and steering motion

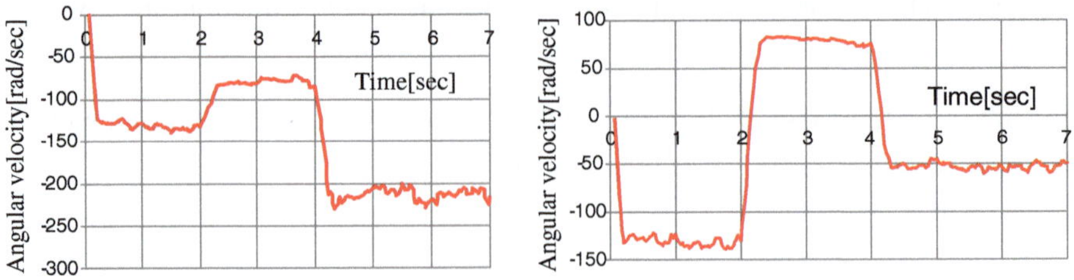

Fig. 5 *Right* and *left* wheel angular velocity in case of combined motion

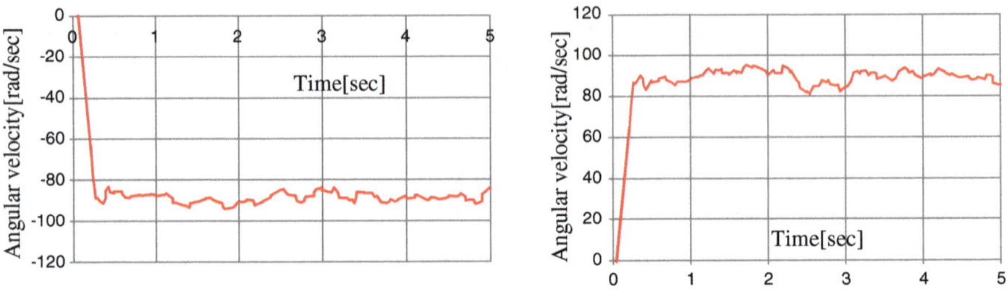

Fig. 6 *Right* and *left* wheel angular velocity in case of steering simulation

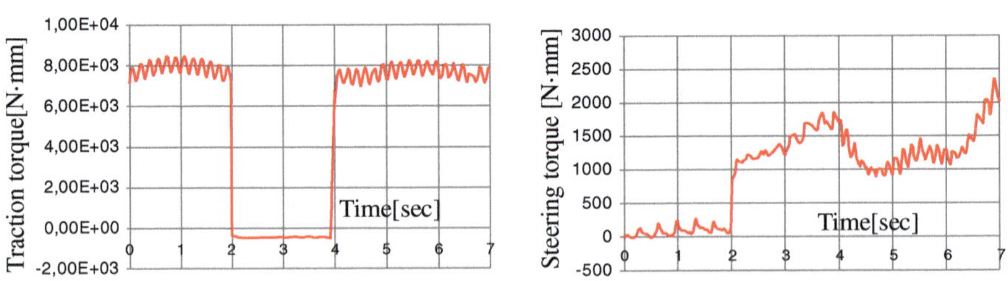

Fig. 7 Traction torque variation in case of combined motion, and steering torque

5 Conclusions

This paper describes the development of a kinematic and dynamic analysis model for a robotic wheelchair. The virtual model of the wheelchair is designed in Solid Works. The proposed solution uses two kinematic chains, one for straight line motion and the other for steering motion; these motions are realized respectively by two motors.

The motors can operate independently or in correlation; in the latter case the wheelchair can be driven with different steering radius. The dynamic simulation reveals the motion trajectories of wheelchair and the kinematic and dynamic parameters of the system: running speed, propulsion and steering torque. From Adams simulation it was determined that the necessary motor torque for propulsion when the wheelchair carries a 70 kg human is 8 Nm and the steering torque is 2.5 Nm.

Acknowledgments This work was supported by the strategic grant POSDRU/159/1.5/S/133255, Project ID 133255 (2014), co-financed by the European Social Fund within the Sectorial Operational Program Human Resources Development 2007–2013.

References

1. Modak, G.S.: Evolution of a stair-climbing power wheelchair (Review Article). IOSR J. Mech. Civil Eng. (IOSR-JMCE), 36–41 (2009)
2. Gonzalez-Rodriguez, A.: Mechanical synthesis for easy and fast operation in climbing and walking robots. Climbing and Walking Robots, Intech Vienna (2010)
3. Gurpude, R.: Design, synthesis and simulation of four bar mechanism for guiding wheel for climbing, Int. J. Eng Res Appl. (2012)
4. Kumar, V.: Assistive devices for people with motor disabilities. Wiley Encyclopaedia Electr. Electron. Eng. (1997)
5. Razak, S.: Design and implementation electrical wheel chair for disable able to stairs climbing by using hydraulic jack. IOSR J. **7**(3), 82–92 (2013)
6. Reswick, J.B.: Automatic transmission for electric wheelchairs, J. Rehabil. Res. Dev. (1985)
7. Salmin, H.: Design and implementation of an electric wheel-chair, Bangladesh. Int. J. Multi. Sci. Eng. (2014)
8. Sheng, L.J.: Wireless brain computer interface for electric wheelchairs with EEG and eye-blinking signals. In: International Journal of Innovative Computing, Information and Control (2012)
9. Wellman, P.: Design of a wheelchair with legs for people with motor disabilities. IEEE Trans. Rehabil. Eng. **3**, 343–353 (1995)
10. Wolm, P.: Dynamic stability control of front wheel drive wheelchairs using solid state accelerometers and gyroscopes. Ph.D. thesis, University of Canterbury (2009)
11. Kong, D.: Nonlinear contact analysis of gear teeth for malfunction diagnostics. In: Conference and Exposition on Structural Dynamics (2008)
12. Dumitru, N., Malciu, R., Geonea, I.: Differential transmission for robotizing a powered wheelchair. In: Proceedings of OPTIROB'10, pp. 47–51, 28–30 May 2010
13. Abdulkadir, H.: Design and fabrication of a motorized prototype tricycle for disabled persons. IOSR J. Eng. **2**(5), 1071–1074 (2012)
14. Pires, G.: Autonomous wheelchair for disabled people. In: Proceedings of the IEEE International Symposium on Industrial Electronics (ISIE97), Guimarães, pp. 797–801 (1997)
15. Rajasekar, R.: Design and fabrication of staircase climbing wheelchair. Int. J. Mech. Eng. Robot. Res. (2013)
16. Shanu, S.: User centric designed mechanism for stairs-climbing wheelchair. In: 15th National Conference on Machines and Mechanisms NaCoMM'11 (2011)
17. MSC Inc., MSC Adams 2013 reference manual

On Building Remotely Operated Underwater Robot-Explorer with Bi-manual Poly-articular System

Aleksandar Rodić, Ilija Stevanović, Miloš D. Jovanović and Đorđe Urukalo

Abstract The paper addresses the mechanical design of a river underwater robot (remotely operated vehicle—ROV) suitable for implementation in delicate and risky underwater tasks. The main factors that determine the mechanical design of the robot are hydrodynamic drags and low underwater visibility. The robot body is inspired by biological models of fishes as well as by the "golden ratio"-a natural geometry proportion that commonly appears in nature. The underwater ROV presented in the paper has two redundant, poly-articular, tendon-driven robot arms, suitable for use in submarine tasks due to their flexibility and light mechanical structure. The paper explains how robot propulsion is determined and which kind of thruster motors are chosen for this purpose. The designed mechanical structure is evaluated by corresponding simulation tests and the results of which are analysed. The paper finally presents concluding remarks and objectives of future work.

Keywords Remotely operated vehicle · Underwater robot · Hydrodynamic drags · Poly-articular robot arm

A. Rodić (✉) · I. Stevanović · M.D. Jovanović · Đ. Urukalo
University of Belgrade, Mihailo Pupin Institute, Robotics Laboratory, Belgrade, Serbia
e-mail: aleksandar.rodic@pupin.rs

I. Stevanović
e-mail: ilija.stevanovic@pupin.rs

M.D. Jovanović
e-mail: milos.jovanovic@pupin.rs

Đ. Urukalo
e-mail: djordje.urukalo@pupin.rs

T. Borangiu (ed.), *Advances in Robot Design and Intelligent Control*,
Advances in Intelligent Systems and Computing 371,
DOI 10.1007/978-3-319-21290-6_48

1 Introduction

The subject of research and development presented in this paper concerns prototyping the development of a remotely operated vehicle (ROV) for underwater tasks. This work is focused on the development of river underwater robot-grebe intended for remotely operated underwater search, camera shooting, monitoring and inspection of submerged objects and infrastructure in the water, rescue missions, ecologic cleaning tasks, waste disposal removing and sampling of river bottom material as well as assistance in extracting of submerged objects from the river. The prototype of the underwater robot-explorer considered in this paper will be used in the big, smooth (not trouble and highland) rivers. These are mainly lowland rivers with large water potential (flow), whose streaming neither overcomes 5.5 km/h speed and 20 m depth.

These watercourses are characterized by high quantities of sediments (small particles of various river materials) carried by the water current, that cause relatively low underwater visibility of only 1–2 m. River watercourses frequently push (roll down) inside the water the bigger pieces of biological, industrial or constructional waste (fallen trees, stumps, dead animals, industrial material, plastic packaging, etc.), too. On the river bottom, there can be frequently found submerged and impaired sailing crafts and vehicles (e.g. ships, boats, cars, etc.), dangerous objects (e.g. unexploded bombs from the former wars), toxic and harmful industrial material (thrown in the water or lost from barges), residuals of former infrastructure (buildings, bridges, water fronts, etc.), immersed archaeological locations, canalization pipes, etc. Rivers, by their huge flow power, often produce water vortexes that additionally complicate any kind of underwater searching and rescue tasks.

The river underwater robot-explorer "Danube grebe" is estimated to be a viable response and prevention means for early detection of problems in the water and accomplishing underwater tasks instead of using specialized divers whenever needed.

2 State-of-the-Art

Recent times witnessed ever expanding applications of robots. Under-water vehicles were initially taken into operation for military applications. Dated back to 1950s, the Royal Navy of Great Britain employed a submersible for recovering torpedoes and removing undersea mines. In 1960s, a Cable-Controlled Underwater Recovery Vehicle (CURV) intended for rescue and recovery operations in a deep-sea was realized by US Navy, being succeeded by CURV II and CURV III.

ROV has substantial use in services for humans for exploring the oceanic world and dealing with underwater deployments. The scientific literature does not pin point the first ROV developed historically. The CUTLET ROV has been categorically highlighted as a pioneer remotely operated oceanic fellow that was developed and introduced by the Royal Navy in 1950s to recover practice torpedoes [1]. This vehicle remained operational until the 1980s. The first time ROV's gained significant recognition [2] was due to role of CURV in recovery operations in the Mediterranean Sea [1]. The most prominent operation conducted by CURV III was retrieval of a manned underwater vehicle from Irish Sea near Cork in 1973 saving the pilots of Pisces [2].

For visual analysis of underwater structures, an observation class ROV design was proposed in 2010. The system was devised for Federal Mexican Oil Company to inspect pipelines, oil production units and other structures in deep waters [3].

The Human-Sized ROV with Dual-Arm is capable of working like a diver by performing its tasks with two hands, as the name suggests. It has been tested at the depth of 10–20 m in Biwa Lake, Japan [4]. The Hybrid ROV Marine Environmental Sciences MARUM (University of Bremen, Germany) has been designed as ROV/Autonomous Underwater Vehicle (AUV) to work in high-risk zones like under-ice operations, hydrothermal vents and harsh topography near the seabed [5].

3 Technical Requests and Concept Outline

The ROV for river underwater tasks described in this paper is designed to operate in slow streaming (0–3.0 km/h) or moderately fast (3.0–5.5 km/h) lowland rivers. The robot is designed to operate at depths that not overcome 20 m. The robot should be built of durable, impact-resistive, water-proof composite materials—carbon fibre and epoxy resin. The operation range (workspace) is limited to maximally 50 m from the basis station (boat on the river surface) due to the strong drags caused by freshwater streaming.

The robot is permanently in physical contact with (connected to) the base station (Fig. 1), due to safety reasons (prevent loss), power supply and permanent cable-communication. This operating range of 50 m represents a trade-off between system reachability, safety and power supply feasibility. Full system autonomy (battery operated robot) is disregarded in this case, because robot operation in the water current is extremely demanding and risky tasks require steering by wire using power-cord supply. Because of that, the underwater ROV is always connected to the base station. An AC/DC electric generator installed on the base station provides enough energy to the robot by use of the buoyant-neutral power cord (Fig. 1). Optical cables are used for signal and video transmission from the robot to the base station.

The robot must have fine mobility and manoeuvrability in the water to behave as a dexterous underwater device capable to accomplish various tasks. This assumes that the robot can move in all directions: up and down, forward and backward, right and left, and twisting about the vertical axis of symmetry. Visibility in the river water is commonly less than 1–2 m. Due to these conditions the robot has to be

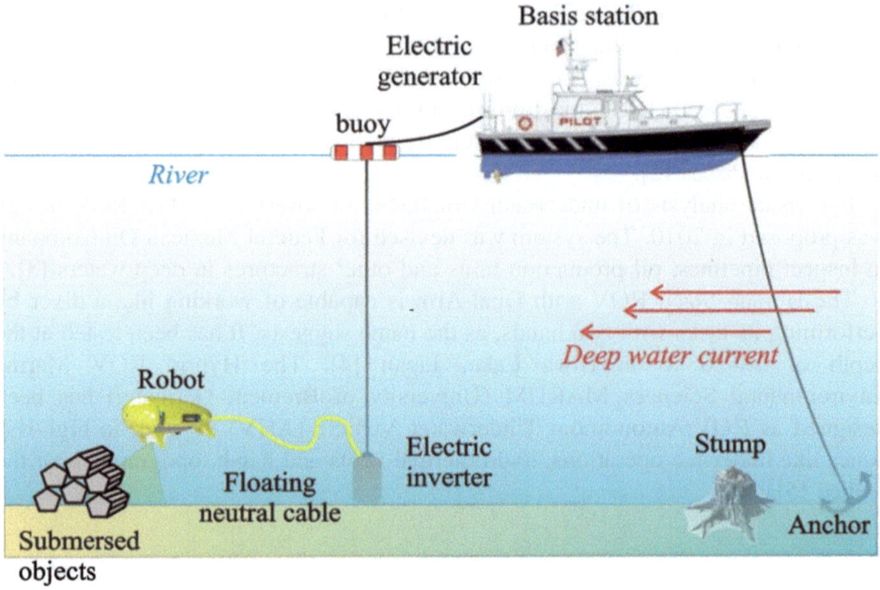

Fig. 1 Principle of operation of the ROV—underwater robot

equipped with underwater low-light sensitive cameras and forward sonar sensor that can enable reliable navigation and motion in low visibility conditions.

The operating principle of the underwater robotic system developed is presented in Fig. 1.

4 Mechanical Design

The system presented in this paper is designed according to the technical requirements previously defined and finding inspiration in the nature by imitating biological systems and principles of underwater motion.

4.1 Robot Structure

The mechanical design of the underwater robot needs to satisfy strict requirements of hydrodynamics, buoyancy stability in conditions of strong water current and drags, balance of masses with aim to enable reliable hovering and laminar motion (dive and ascent accomplishing), fine manoeuver capabilities in restricted task space, etc. (Fig. 2a, b). The robot has a supporting system (flappers, Fig. 2a) such as to enable reliable landing to the river bottom but also to improve horizontal stability.

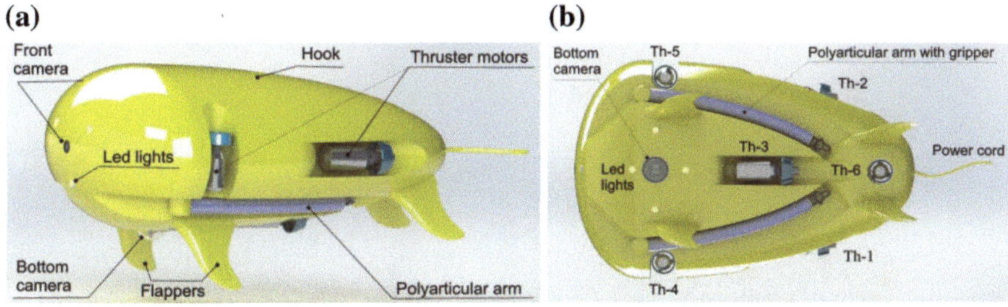

Fig. 2 Underwater robot with its components, **a** side view, **b** bottom view

The shape of the underwater robot is optimized in the sense to have minimal hydrodynamic resistance (Fig. 3a). The robot is similar to a fish having a flattened body [6]. Two flexible poly-articulated robotic arms are integrated into the mechanical structure (Fig. 4). The arms have length of 1.2 m each, which provides a relatively large workspace for underwater tasks. Six robust thruster motors (Figs. 2a, b and 3b) move the robot in 6 coordinate directions. The first group of motors (Th-1, Th-2 and Th-3) is mounted horizontally and is parallel with the longitudinal axis of robot body. This group of motors is responsible for forward-backward robot motion and for robot rotation around the vertical axis passing through the centre of mass of the robot body, and partially for the rotation about the system pitch axis. The second group of motors (Th-4, Th-5 and Th-6) is mounted vertically. They are distributed: two on the sides and one in the tail of the ROV. These motors are responsible for robot up and down motion, but they also provide rotation about the roll-axis. The tail motor (that ensures horizontal stability) partially affects the rotational movement about the system pitch-axis. The robot ventral fins (flappers, Fig. 2) have a dual role:

(i) To improve horizontal stability of the system and
(ii) To serve as backbone—the landing gear when the robot is descending to the river bottom.

The robot body is made of composite materials (fibre glass textile and epoxy resin as bonding material) which have sufficient strength, durability and possibility of fine modelling (shaping). The robot geometry proportion was computed according to the "golden ratio", which represents a specific relationship between two quantities as follows: two quantities, a and b are in the golden ratio if their ratio is the same as the ratio of their sum to the larger of the two quantities a and b.

$$\frac{a+b}{a} = \frac{a}{b} = \varphi = 1.6180 \tag{1}$$

The hydro dynamically optimized underwater robot (Fig. 2a, b) was tested upon the drag forces by modelling in Solid Works. The results of this simulation

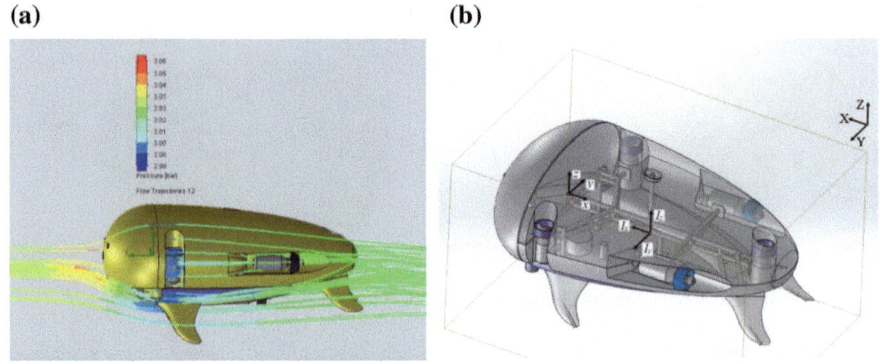

Fig. 3 **a** Hydrodynamic analysis of the submarine body; **b** Transparent 3D rigid-body model of the underwater ROV with thruster motors, ballasts and metal frame

experiment and corresponding streamlines are presented in Fig. 3a (the case when the streaming speed is 10 km/h and the robot keeps position). The simulation results are systematized in Table 1 presenting hydrodynamic resistances in different directions.

The propulsion of the robot is determined by taking into account the maximal hydrodynamic drags displayed in Table 1 as well as by simulating the robot dynamics. The robot is requested to float against the water current of up to 5.5 km/h with floating speed that does not overcome 1.5 m/s. Motor thrusters are chosen to ensure the calculated propulsion but to keep minimal margin of energy costs.

4.2 *Bi-manual Manipulation System*

The remotely operated underwater robot has two flexible, poly-articular robot arms of 1.2 m length each. One arm is the master arm and the second one is slave. The master arm is responsible for gripping objects and rending them available to the slave arm. The slave robot-arm has integrated a small led lamp and a small camera to enable better visibility and precision. Both arms can easy change end-effectors: grippers, hooks, needles, scissors, etc. Robot arms consist of 10 segments, 9 bi-directional flexion joints, 1 yaw and 1 twist joint and a gripper as presented in Fig. 4a, b. Flexion joints are powered by non-tensile tendons (Fig. 4b) and 6 electric motors located in the robot basis. The third, sixth and ninth joint are connected to the particular DC-motors. Particular motors simultaneously articulate 3 successive robot links in 2 possible orthogonal directions (Fig. 4a). The robot's work space is a hollow ellipsoid.

Table 1 Robot hydrodynamic drags obtained by simulation in solid works

Speed of streaming (km/h)		Resistance force during the robot movement (N)					
		1	2	3	5	7.5	10
Robot movement direction	Forward	8.355808128	–	75.90494183	210.4932276	461.4344195	804.9717383
	Backward	9.297515187	–	87.22105725	241.1581502	544.6749591	970.8956312
	Laterally	23.5458368	–	213.0559673	–	–	–
	Up	49.29869231	197.0768946	–	–	–	–
	Down	39.22048346	212.0448867	–	–	–	–

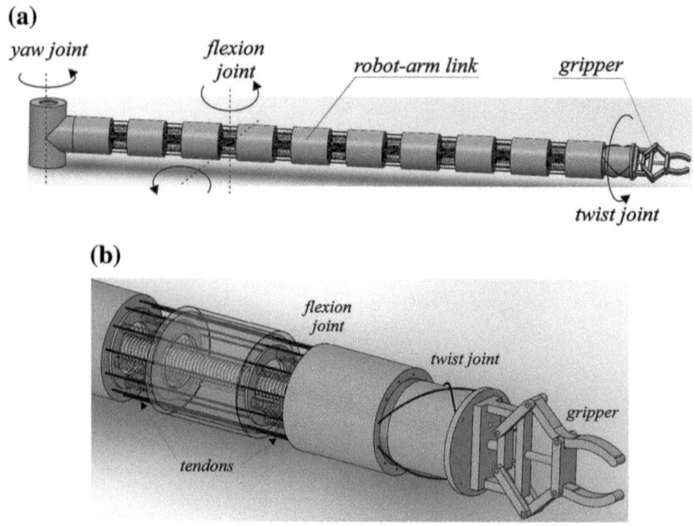

Fig. 4 Poly-articular flexible robot-arm and robot-gripper

5 Simulation Experiments

In order to prove feasibility to robot design and to identify power requirements, corresponding model simulation was accomplished. The following coordinate systems are assumed (Fig. 3b): (i) global coordinate system OXYZ attached to the river surface; (ii) local coordinate system Oxyz attached to the cross section of the longitudinal and transversal axes of symmetry; (iii) coordinate system attached to the mass centre of the robot body. Robot rigid body dynamics in Fig. 3b can be defined in the form:

$$\tau = H(q)\ddot{q} + h_{ccf}(q, \dot{q}) + h_g - \sigma - J^T(q)W(\dot{q}) - J^T(q)\emptyset \tag{2}$$

$$q = [\mathrm{x} \quad \mathrm{y} \quad \mathrm{z} \quad \varphi \quad \theta \quad \varepsilon]^T \tag{3}$$

$$F = A(\varphi, \theta, \varepsilon)^{-1} \cdot \tau \tag{4}$$

$$F = [F_1 \quad F_2 \quad F_3 \quad F_4 \quad F_5 \quad F_6], \tag{5}$$

where τ represents 6×1 vectors of the generalized forces and torques reduced to the MC of robot body. This vector takes into account successively forces and moments of inertia $H(q)\ddot{q}$, Centrifugal and Coriolis forces/torques $h_{ccf}(q, \dot{q})$, gravity forces/torques h_g, hydraulic thrust forces σ upon the submerged mass of robot, hydrodynamic drags $J^T(q)W(\dot{q})$ due to the motion resistance in the fluid and external payload forces $J^T(q)\emptyset$ acting upon the system. The state vector q includes

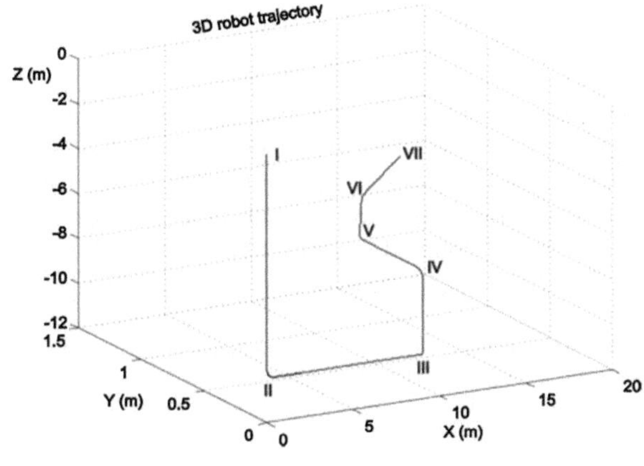

Fig. 5 Robot trajectory considered in simulation experiment

linear positions and corresponding Euler angles. The motor thrust forces (5) are calculated from (3) where A is a thruster payload distribution matrix.

To test robot dynamics in motion through water stream, a path is chosen (Fig. 5) which includes: dive (segment I–II, 10 s), movement in longitudinal direction (segment II–III, 10 s), upward movement in vertical direction (segment III–IV, 5 s), sliding sideways in horizontal plane (segment IV–V, 5 s), diagonal movement in horizontal plane (segment V–VI, 5 s) and inclined buoyant movement (segment VI–VII, 5 s). The robot diving speed along this trajectory is shown in Fig. 6a.

By analysing the results presented in Figs. 5 and 6 the appropriate thrust motors KZ-25 K and KZ-12 K were chosen from [7]. The required power of the thrust motors Th-1 to Th-3 is 1000 W, corresponding to a maximal thrust force of 25 kg. The power of motors Th-4 to Th-6 is 400 W, with maximal thrust force of 12 kg. The chosen motors satisfy the criterion of power and enable robot motion in conditions of deep water streaming of 5.0 km/h. The realization of motion in particular directions requires synchronized operation of thrusters provided the robot dynamics is coupled.

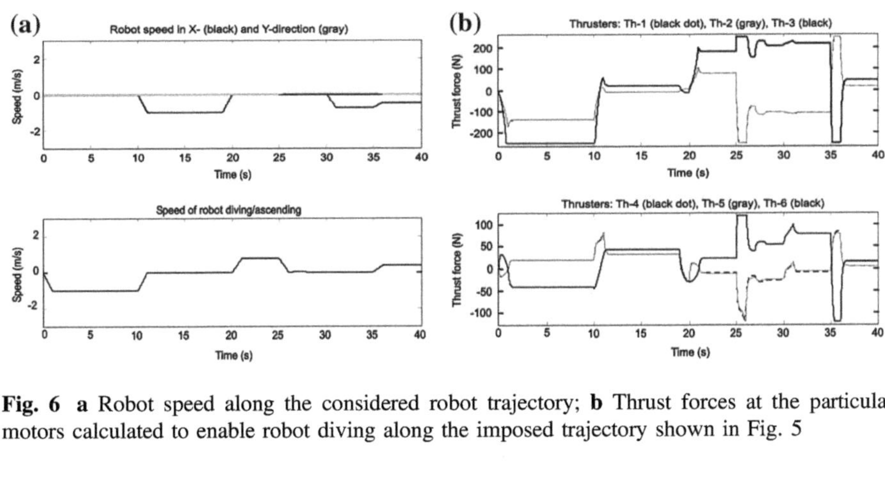

Fig. 6 **a** Robot speed along the considered robot trajectory; **b** Thrust forces at the particular motors calculated to enable robot diving along the imposed trajectory shown in Fig. 5

6 Conclusions and Future Work

The mechanical design of the robot was adapted in such a way to achieve a trade-off between hydrodynamic drags and technical requirements of the system. Future work concerns building the prototype of the robot (mechanical structure, electronics, sensorial system) as well as investigating the best control strategy taking into account coupled system dynamics and hydrodynamic drags. An optimal mechanical design will lead to achieving reliable control and energy efficiency.

Acknowledgments This project is realized within the Serbia-China bilateral project under the grant 451-03-02338/2012-14/12 funded by the Serbian Ministry of Science and partially supported by the EU research and innovation program HORIZON 2020 "Research night" under the grant agreement no. 633398, 2014–2015.

References

1. Matsumoto, G.I., Potts, T.A.: Two ways of researching the underwater world. Brooks/Cole Enrichment Module, 26 (2011)
2. Murphree, T., Michel, D., Sullivan, D., Zande, J.: Remotely operated vehicles. Knowledge and skill guidelines for marine science and technology, vol. 3. Marine Advanced Technology Center (2011)
3. Jimenez, T.S., Lopez, J.L..G., Soto, L.F.M., Lopez, E., Gonzalez, P.A.R., Sanchez, M.B.: Deep water ROV design for the Mexican oil industry. In: IEEE Oceans Conference, pp. 1–6 (2010)
4. Sakagami, N., Shibata, M., Hashizume, H., Hagiwara, Y., Ishimaru, K., Inoue, T., Onishi, H., Murakami, S.: Development of a human-sized ROV with dual-arm. In: IEEE Oceans Conference, pp. 1–6 (2010)
5. Meinecke, G., Ratmeyer, V., Renken, J.: Hybrid ROV development of a new underwater vehicle for high-risk areas. In: IEEE Oceans Conference, pp. 1–6 (2011)
6. Fish, F.E., et al.: Hydrodynamic control in marine mammals. Integr. Comp. Biol. **48**(6), 788–800 (2008)
7. http://www.aliexpress.com/item/KZ1000-Underwater-Thruster-with-ducted-propeller-for-ROV-AUV-water-leisure-equipments-aquaculture-underwater-cleaing-robot/1869857127.html

Author Index

A
Achilleopoulos, Nikos, 417
Andris, Pavel, 369
Anton, Florin, 13, 127
Anton, Silvia, 127
Aspragathos, Nikos A., 235
Azariadis, Philip N., 429

B
Bader, Markus, 459
Balaz, Vladimir, 277
Bertetto, Andrea Manuello, 115
Besalduch, Luigi Antonio, 115
Borangiu, Theodor, 13, 127, 245
Brad, Stelian, 255
Burlacu, Adrian, 185
Butera, Luca Girolamo, 105
Butnariu, Silviu, 339

C
Cadeddu, Andrea, 115
Cafolla, Daniele, 307
Cameron, Stephen, 347
Carbone, Giuseppe, 307
Ceccarelli, Marco, 307
Chatzilygeroudis, Konstantinos I., 235
Chevalier, Pauline, 225
Cojocaru, Dorian, 317
Conduraru, Ionel, 157

D
Deniša, Miha, 449
Dobrescu, Radu, 407
Dobrovodský, Karol, 369
Dolga, Valer, 213
Doroftei, Ioan, 157
Drăgoicea, Monica, 245
Dumitru, Nicolae, 471

F
Felde, Imre, 137
Ferraresi, Carlo, 115, 357
Florescu, Mihaela, 297
Fulea, Mircea, 255

G
Gîrbacia, Florin, 339
Gîrbacia, Teodora, 339
Galati, Rocco, 387
Gams, Andrej, 449
Geonea, Ionuţ, 471
Gherman, Bogdan, 329, 339
Giannoccaro, Ivan, 387
Gottschalk, Michael, 27

H
Hajduk, Mikuláš, 89
Hajimirzaalian, Hamidreza, 357
Horga, Vasile, 157
Hricko, Jaroslav, 37

I
Ijspeert, Auke J., 439
Isableu, Brice, 225
Ivanescu, Mircea, 297

J
Jatsun, Sergey, 165, 173
Jovanović, Miloš D., 481

K
Kacso, Gabriel, 329
Kallweit, Stephan, 27
Kasanický, Tomáš, 147
Koustoumpardis, Panagiotis N., 235
Kozlowski, Piotr, 347

T. Borangiu (ed.), *Advances in Robot Design and Intelligent Control*,
Advances in Intelligent Systems and Computing 371,
DOI 10.1007/978-3-319-21290-6

L
Lamprinou, Nefeli, 417
Lazar, Corneliu, 185
Loktionova, Oksana, 173

M
Maffiodo, Daniela, 97, 357
Malchikov, Andrei, 165
Marghitu, Dan B., 317
Martin, Jean-Claude, 225
Mateescu, Cristian, 397, 407
Messina, Arcangelo, 387
Mocan, Bogdan, 255
Mogan, Gheorghe-Leonte, 47
Moldovan, Constantin Cătălin, 195

N
Nádai, László, 137
Nicolae, Maximilian, 407
Niola, Vincenzo, 79
Niţulescu, Mircea, 267

O
Olaru, Sorin Mănoiu, 267

P
Panescu, Doru, 3
Penta, Francesco, 71
Petrič, Tadej, 439
Pîslă, Doina, 329, 339
Plitea, Nicolae, 329, 339
Popescu, Dan, 379, 397, 407
Potapov, Pavel, 79
Psarakis, Emmanouil Z., 417

Q
Quaglia, Giuseppe, 105

R
Raileanu, Silviu, 13, 127
Raparelli, Terenziano, 97
Reina, Giulio, 387
Ricciu, Roberto, 115
Rodić, Aleksandar, 481
Rossi, Cesare, 59, 71, 79
Rudas, Imre J., 137
Ružič, Anton, 203

S
Savin, Sergei, 165
Savino, Sergio, 71, 79
Savu, Diana, 213
Sbanca, Mădălin-Petru, 47
Semjon, Jan, 277
Stamatescu, Grigore, 397
Stareţu, Ionel, 195
Stănescu, Tony, 213
Stevanović, Ilija, 481
Stoian, Viorel, 287
Stoican, Florin, 379
Synodinos, Aris I., 235

T
Tar, József K., 137
Todoran, Horaţiu George, 459
Tzimas, Giannis, 417
Ţăpuş, Adriana, 225

U
Ude, Aleš, 439, 449
Urukalo, Đorđe, 481

V
Vagas, Marek, 277
Vaida, Călin, 339
Varga, Jozef, 89
Varvara, Gabriela, 3
Vasileiou, Petros G., 417
Vladu, Ileana, 287
Vladu, Ionel Cristian, 287
Voiculescu, Irina, 347
Vorochaev, Alexander, 173
Vorochaeva, Lyudmila, 173

W
Walenta, Robert, 27
Wang, Mingfeng, 307

X
Xidias, Elias K., 429

Y
Yatsun, Andrey, 165

Z
Zabarauskas, Manfred, 347
Zelenka, Ján, 147

MIX
Papier aus verantwortungsvollen Quellen
Paper from responsible sources
FSC® C105338

If you have any concerns about our products,
you can contact us on
ProductSafety@springernature.com

In case Publisher is established outside the EU,
the EU authorized representative is:
Springer Nature Customer Service Center GmbH
Europaplatz 3, 69115 Heidelberg, Germany

Printed by Libri Plureos GmbH
in Hamburg, Germany